Thin-film optical filters

Second edition

H A Macleod

Professor of Optical Sciences
University of Arizona

Adam Hilger Ltd, Bristol

British Library Cataloguing in Publication Data

Macleod, H. A.
Thin-film optical filters—2nd ed.
1. Light filters 2. Thin filters—Optical properties
I. Title
621.36 QC373.L5

ISBN 0-85274-784-5

Consultant Editor: **Professor W T Welford**
Imperial College, London

Published by Adam Hilger Ltd
Techno House, Redcliffe Way, Bristol BS1 6NX

Phototypeset by Macmillan India Ltd, Bangalore
Printed in Great Britain by J W Arrowsmith Ltd, Bristol

Preparing a plant for the manufacture of narrowband filters. (Courtesy of Walter Nurnberg FIEP FRPS, the editors of *Engineering*, and Sir Howard Grubb, Parsons & Co Ltd.)

To
my Mother and Father
Agnes Donaldson Macleod

Contents

Foreword

A great deal has happened in the subject of optical coatings since the first edition of this book. This is especially true of facilities for thin-film calculations. In 1969 my thin-film computing was performed on an IBM 1130 computer that had a random access memory of 10 kbytes. Time had to be booked in advance, sometimes days in advance. Calculations remote from this computer were performed either by slide rule, log tables or electromechanical calculator. Nowadays my students scarcely know what a slide rule is, my pocket calculator accommodates programs that can calculate the properties of thin-film multilayers and I have on my desk a microcomputer with a random access memory of 0.5 Mbytes, which I can use as and when I like. The earlier parts of this revision were written on a mechanical typewriter. The final parts were completed on my own word processor. These advances in data processing and computing are without precedent and, of course, have had a profound and irreversible effect on many aspects of everyday life as well as on the whole field of science and technology.

There have been major developments, too, in the deposition of thin-film coatings, and although these lack the spectacular, almost explosive, character of computing progress, nevertheless important and significant advances have been made. Electron-beam sources have become the norm rather than the exception, with performance and reliability beyond anything available in 1969. Pumping systems are enormously improved, and the box-coater is now standard rather than unusual. Microprocessors control the entire operation of the pumping system and, frequently, even the deposition process. We have come to understand that many of our problems are inherent in the properties of our thin films rather than in the complexity of our designs. Microstructure and its influence on material properties is especially important. Ultimate coating performance is determined by the losses and instabilities of our films rather than the accuracy and precision of our monitoring systems.

My own circumstances have changed too. I wrote the first edition in industry. I finish the second as a university professor in a different country.

All this change has presented me with difficult problems in the revision of this book. I want to bring it up to date but do not want to lose what was useful in the first edition. I believe that in spite of the great advances in computers, there is still an important place for the appreciation of the fundamentals of thin-film coating design. Powerful synthesis and refinement techniques are available and are enormously useful, but an understanding of thin-film coating performance and the important design parameters is still an essential ingredient of success. The computer frees us from much of the previous drudgery and puts in our hands more powerful tools for improving our understanding. The availability of programmable calculators and of micro-computers implies easy handling of more complex expressions and formulae in design and performance calculations. The book, therefore, contains many more of these than did the first edition. I hope they are found useful. I have included a great deal of detail on the admittance diagram and admittance loci. I use them in my teaching and research and have taken this opportunity to write them up. SI units, rather than Gaussian, have been adopted, and I think chapter 2 is much the better for the change. There is more on coatings for oblique incidence including the admittance diagram beyond the critical angle, which explains and predicts many of the resonant effects that are observed in connection with surface plasmons, effects used by Greenland and Billington (chapter 8, reference 12) in the late 1940s and early 1950s for monitoring thin-film deposition.

Inevitably, the first edition contained a number of mistakes and misprints and I apologise for them. Many were picked up by friends and colleagues who kindly pointed them out to me. Perhaps the worse mistake was in figure 9.4 on uniformity. The results were quoted as for a flat plate but, in fact, referred to a spherical work holder. These errors have been corrected in this edition and I hope that I have avoided making too many fresh ones. I am immensely grateful to all the people who helped in this correction process. I hope they will forgive me for not including the huge list of their names here. My thanks are also due to J H Apfel, G DeBell, E Pelletier and W T Welford who read and commented on various parts of the manuscript.

To the list in the foreword of the first edition of organisations kindly providing material should be added the names Leybold-Heraeus GmbH, and Optical Coating Laboratory Inc. Airco-Temescal is now known as Temescal, a Division of the BOC Group Inc, and the British Scientific Instrument Research Association as Sira Institute.

My publisher is still the same Adam Hilger, but now part of the Institute of Physics. I owe a very great debt to Neville Goodman who was responsible for the first edition and who also persuaded and encouraged me into the second. He retired while it was still in preparation, and the task of extracting the final manuscript from me became Jim Revill's. Ian Kingston and Brian McMahon did a tremendous job on the manuscript at a distance of 3000 miles. Their patience with me in the delays I have caused them has been amazing.

My wife and family have once again been a great source of support and encouragement.

Newcastle-upon-Tyne, England
and
Tucson, Arizona, USA
1985

Angus Macleod

F	A function used in the theory of the Fabry–Perot interferometer (p 159).
$\mathscr{F}$	Finesse—the ratio of the separation of adjacent fringes to the fringe halfwidth in the Fabry–Perot interferometer (p 160).
g	$g = \lambda_0/\lambda = v/v_0$ sometimes called the relative wavelength or the relative wavenumber or the wavelength ratio. λ_0 and v_0 are usually chosen to be the wavelength or wavenumber, respectively, at which the optical thicknesses of the more important layers in the assembly are quarter-waves. The phase thickness, δ, of quarter-wave layers is given by $\delta = (\pi/2)g$. (p 91).
$\mathscr{H}$	The magnetic amplitude (p 22).
$\boldsymbol{H}$	The magnetic vector in the electromagnetic field (p 11).
H	The amplitude of the tangential component of magnetic field, that is the field parallel to a boundary (p 23).
H	Represents a quarter-wave of high index (p 46).
I	The intensity of the wave. A measure of the energy per unit area per unit time carried by the wave (p 16).
k	The extinction coefficient. The complex refractive index is given by $N = n - ik$. A finite value of k for a medium denotes the presence of absorption. See also the absorption coefficient α.
L	Represents a quarter-wave of low index (p 46).
M	Represents a quarter-wave of intermediate index (p 46).
M_a	A symbol denoting the elements of the characteristic matrix of layer a (p 44).
N	The complex refractive index. $N = n - ik$. (p 13).
n	The real part of the refractive index (p 13).
n^*	The effective index, that is the index of an equivalent layer that shifts in wavelength by the same amount as a narrowband filter when tilted with respect to the incident light (p 260).
p	Packing density of a film (p 399).
p	Indicates the plane of polarisation in which the electric vector is parallel to the plane of incidence. Equivalent to TM. (p 22).
R	The reflectance. The ratio at a boundary of the reflected intensity to the incident intensity. At oblique incidence the components normal to the boundary are used. (p 21).

s	Indicates the plane of polarisation in which the electric vector is normal to the plane of incidence. (From the German *senkrecht*). Equivalent to TE. (p 22).
T	The transmittance. The ratio at a boundary of the transmitted intensity to the incident intensity. At oblique incidence the components normal to the boundary are used. (p 21).
TE	Transverse electric. The plane of polarisation in which the electric vector is normal to the plane of incidence. Equivalent to s-polarisation. (p 22).
TM	Transverse magnetic. The plane of polarisation in which the magnetic vector is normal to the plane of incidence. Equivalent to p-polarisation. (p 22).
x, y, z	The three axes defining the orientation of a thin-film assembly. z is normally taken normal to the interfaces and with positive direction in the sense of the propagation of the incident wave, x and y in the plane of the interfaces with x also in the plane of incidence. x, y and z form a right-handed set. (p 17).
$X+\mathrm{i}Z$	The optimum exit admittance for a metal layer in order to achieve the maximum potential transmittance (p 296).
$\mathscr{Y}$	The admittance of free space (p 15).
y	The admittance of a medium. In SI units y is measured in siemens. $y=N\mathscr{Y}$ and so is numerically equal to the refractive index if measured in free space units (p 15).
Y	The admittance of a surface or multilayer. It is given by C/B. (p 35).
y_0	The admittance of the incident medium (p 19).
y_m (y_{sub} or y_s)	The admittance of the substrate upon which the film system is deposited (p 36).
α	The absorption coefficient. The inverse of the distance along the direction of propagation in which the intensity of a wave falls to 1/e times its original value. $\alpha=4\pi k/\lambda$ where k is the extinction coefficient. (p 16).
α	A symbol used to represent $2\pi nd/\lambda$ (p 184).
α, β, γ	The three direction cosines (p 14).
$(\alpha-\mathrm{i}\beta)$	Symbols used to represent the admittance of a metal. Similar to $n-\mathrm{i}k$. (p 140).
β	A symbol used to represent $2\pi kd/\lambda$ (p 184).

γ	The equivalent phase thickness of a symmetrical assembly (p 192).
Δ_q	(η_p/η_s) where η_p and η_s are modified admittances (p 316). This is a quantity used in the design of polarisation-free coatings. (p 339).
ε	Indicates a small error or a departure from a reference value of a number (p 74).
ε	The permittivity of a medium (p 12).
η	The tilted optical admittance (p 25).
η_m	The tilted admittance of the substrate. See y_m. (p 36).
η_E	The equivalent admittance of a symmetrical assembly. See also E. (p 120).
θ	The angle of incidence in a medium (p 17).
θ_0	The angle of incidence in the incident medium (p 17).
λ	The wavelength of the light, usually the wavelength in free space (p 14).
λ_0	The reference wavelength. See g. (p 76).
ν_0	The reference wavenumber. $\nu_0 = 1/\lambda_0$. See g.
ρ	The amplitude reflection coefficient (p 20).
ρ	The electric charge density (p 14).
γ	The amplitude transmission coefficient (p 20).
ϕ	The phase shift on reflection (p 37).
ψ	Potential transmittance. $\psi = T/(1-R)$. (p 39).
ψ	Used in some limited calculations on pages 199 and 200 to represent $2\delta_p/\delta_q$ (p 199).

1 Introduction

This book is intended to form an introduction to thin-film optical filters for both the manufacturer and the user. It does not pretend to present a detailed account of the entire field of thin-film optics, but it is hoped that it will form a supplement to those works already available in the field and which only briefly touch on the principles of filters. For the sake of a degree of completeness, it has been thought desirable to repeat again some of the information that will be found elsewhere in textbooks, referring the reader to more complete sources for greater detail. The topics covered are a mixture of design, manufacture, performance and application, including enough of the basic mathematics of optical thin films for the reader to carry out thin-film calculations. The aim has been to present, as far as possible, a unified treatment, and there are some alternative methods of analysis which are not discussed. For a much more complete study of thin-film calculations, the reader should consult the book by Knittl[1]. Similarly, some of the manufacturing methods are not dealt with in depth because there is an excellent textbook on the vacuum deposition of thin films by Holland[2] which, although it was written more than 20 years ago, is still relevant and topical. For further information on coatings involving a few layers, such as beam splitters, high reflectance coatings and metal–dielectric filters, together with information on alternative deposition techniques, the book by Anders[3] will be found useful. There is the well known book by Heavens[4] which deals principally with the basic optical properties, mainly of a single film, and includes much information on thin-film calculations. More recently, there has been a survey paper by Lissberger[5], the reports of two meetings devoted entirely to thin-film optical coatings[6,7], and an excellent review of the entire field of optical filters by Dobrowolski[8]. There is also the recent and useful book on filter design by Liddel[9].

In a work of this size, it is not possible to cover the entire field of thin-film optical devices in the detail that some of them may deserve. The selection of topics is due, at least in part, to the author's own preferences and knowledge. In this book, optical filters have been interpreted fairly broadly to include such items as antireflection and high-reflectance coatings.

EARLY HISTORY

The earliest of what might be called modern thin-film optics was the discovery, independently, by Robert Boyle and Robert Hooke of the phenomenon known as 'Newton's rings'. The explanation of this is nowadays thought to be a very simple matter, being due to interference in a single thin film of varying thickness. However, at that time, the theory of the nature of light was not sufficiently far advanced, and the explanation of this and a number of similar observations made in the same period by Sir Isaac Newton on thin films eluded scientists for almost a further 150 years. Then, on 12 November 1801, in a Bakerian Lecture to the Royal Society, Thomas Young enunciated the principle of the interference of light and produced the first satisfactory explanation of the effect. As Henry Crew[10] has put it, 'This simple but tremendously important fact that two rays of light incident upon a single point can be added together to produce darkness at that point is, as I see it, the one outstanding discovery which the world owes to Thomas Young.'

Young's theory was far from achieving universal acceptance. Indeed Young became the victim of a bitter personal attack, against which he had the greatest difficulty defending himself. Recognition came slowly and depended much on the work of Augustin Jean Fresnel[11] who, quite independently, also arrived at a wave theory of light. Fresnel's discovery, in 1816, that two beams of light which are polarised at right angles could never interfere, established the transverse nature of light waves. Then Fresnel combined Young's interference principle and Huygens' ideas of light propagation into an elegant theory of diffraction. It was Fresnel who put the wave theory of light on such a firm foundation that it has never been shaken. For the thin-film worker, Fresnel's laws, governing the amplitude and phase of light reflected and transmitted at a single boundary, are of major importance. It has been pointed out recently by Knittl[12], that it was Fresnel who first summed an infinite series of rays to determine the transmittance of a thick sheet of glass and that it was Simeon Denis Poisson, in correspondence with Fresnel, who included interference effects in the summation to arrive at the important results that a half-wave thick film does not change the reflectance of a surface, and that a quarter-wave thick film of index $(n_0 n_s)^{1/2}$ will reduce to zero the reflectance of a surface between two media of indices n_s and n_0. Fresnel died in 1827, at the early age of 39.

In 1873, the great work of James Clerk Maxwell, *A Treatise on Electricity and Magnetism*[13], was published, and in his system of equations we have all the basic theory for the analysis of thin-film optical problems.

Meanwhile, in 1817, Joseph Fraunhofer had made what are probably the first ever antireflection coatings. It is worth quoting his observations at some length because they show the considerable insight which he had, even at that early date, into the physical causes of the effects which were produced. The following is a translation of part of the paper as it appears in the collected works[14].

Before I quote the experiments which I have made on this I will give the method which I have made use of to tell in a short time whether the glass will withstand the influence of the atmosphere. If one grinds and then polishes as finely as possible, one surface of glass which has become etched through long exposure to the atmosphere, then wets one part of the surface, for example half, with concentrated sulphuric or nitric acid and lets it work on the surface for twenty-four hours, one finds after cleaning away the acid that that part of the surface on which the acid was, reflects much less light than the other half, that is it shines less although it is not in the least etched and still transmits as much light as the other half, so that one can detect no difference on looking through. The difference in the amount of reflected light will be most easily detected if one lets the light strike approximately vertically. It is the greater the more the glass is liable to tarnish and become etched. If the polish on the glass is not very good this difference will be less noticeable. On glass which is not liable to tarnish, the sulphuric and nitric acid does not work. . . . Through this treatment with sulphuric or nitric acid some types of glasses get on their surfaces beautiful vivid colours which alter like soap bubbles if one lets the light strike at different angles.

Then, in an appendix to the paper added in 1819:

Colours on reflection always occur with all transparent media if they are very thin. If for example, one spreads polished glass thinly with alcohol and lets it gradually evaporate towards the end of the evaporation, colours appear as with tarnished glass. If one spreads a solution of gum-lac in a comparatively large quantity of alcohol very thinly over polished warmed metal the alcohol will very quickly evaporate, and the gum-lac remains behind as a transparent hard varnish which shows colours if it is thinly enough laid on. Since the colours, in glasses which have been coloured through tarnishing, alter themselves if the inclination of the incident light becomes greater or smaller, there is no doubt that these colours are quite of the same nature as those of soap bubbles, and those which occur through the contact of two polished flat glass surfaces, or generally as thin transparent films of material. Thus there must be on the surface of tarnished glass which shows colours, a thin layer of glass which is different in refractive power from the underlying. Such a situation must occur if a component is partly removed from the surface of the glass or if a component of the glass combines at the surface with a related material into a new transparent product.

It seems that Fraunhofer did not follow up this particular line into the development of an antireflection coating for glass, perhaps because optical components were not, at that time, sufficiently complicated for the need for antireflection coatings to be obvious. Possibly the important point that not only was the reflectance less but the transmittance also greater had escaped him.

In 1886, Lord Rayleigh reported to the Royal Society an experimental verification of Fresnel's reflection law at near-normal incidence[15]. In order to attain a sufficiently satisfactory agreement between measurement and prediction, he had found it necessary to use freshly polished glass because the reflectance of older material, even without any visible signs of tarnish, was too low. One possible explanation which he suggested was the formation, on the surface, of a thin layer of different refractive index from the underlying material. He was apparently unaware of the earlier work of Fraunhofer.

Then, in 1891, Dennis Taylor published the first edition of his famous book *On the Adjustment and Testing of Telescopic Objectives* and mentioned that[16]

> As regards the tarnish which we have above alluded to as being noticeable upon the flint lens of an ordinary objective after a few years of use, we are very glad to be able to reassure the owner of such a flint that this film of tarnish, generally looked upon with suspicion, is really a very good friend to the observer, inasmuch as it increases the transparency of his objective.

In fact, Taylor went on to develop a method of artificially producing the tarnish by chemical etching[17]. This work was followed up by Kollmorgen, who developed the chemical process still further for different types of glasses[18].

At the same time, in the nineteenth century, a great deal of progress was being made in the field of interferometry. The most significant development, from the thin-film point of view, was the Fabry–Perot interferometer[19], described in 1899, which has become one of the basic types of structure for thin-film filters.

Developments became much more rapid in the 1930s, and indeed it is in this period that we can recognise the beginnings of modern thin-film optical coating. In 1932, Rouard[20] observed that a very thin metallic film reduced the internal reflectance of a glass plate, although the external reflectance was increased. In 1934, Bauer[21], in the course of fundamental investigations of the optical properties of halides, produced reflection-reducing coatings, and Pfund[22] evaporated zinc sulphide layers to make low-loss beam splitters for Michelson interferometers, noting, incidentally, that titanium dioxide could be a better material. In 1936, John Strong[23] produced antireflection coatings by evaporation of fluorite to give inhomogeneous films which reduced the reflectance of glass to visible light by as much as 89 %, a most impressive figure. Then, in 1939, Geffken[24] constructed the first thin-film metal–dielectric interference filters.

The most important factor in this sudden expansion of thin-film optical coatings was the manufacturing process. Although sputtering was discovered about the middle of the nineteenth century, and vacuum evaporation around the beginning of the twentieth, they were not considered as useful manufacturing processes. The main difficulty was the lack of really suitable pumps, and it was not until the early 1930s that the work of C R Burch on diffusion pump oils made it possible for this process to be used satisfactorily.

Since then, tremendous strides have been made, particularly in the last few years. Filters with perhaps one hundred layers are not uncommon and uses have been found for them in almost every branch of science and technology.

THIN-FILM FILTERS

To understand in a qualitative way the performance of thin-film optical devices, it is necessary to accept several simple statements. The first is that the amplitude reflectance of light at any boundary between two media is given by $(1-\rho)/(1+\rho)$, where ρ is the ratio of the refractive indices at the boundary (the intensity reflectance is the square of this quantity). The second is that there is a phase shift of 180° when the reflectance takes place in a medium of lower refractive index than the adjoining medium and zero if the medium has a higher index than the one adjoining it. The third is that if light is split into two components by reflection at the top and bottom surfaces of a thin film, then the beams will recombine in such a way that the resultant amplitude will be the difference of the amplitudes of the two components if the relative phase shift is 180°, or the sum of the amplitudes if the relative phase shift is either zero or a multiple of 360°. In the former case, we say that the beams interfere destructively and in the latter constructively. Other cases where the phase shift is different will be intermediate between these two possibilities.

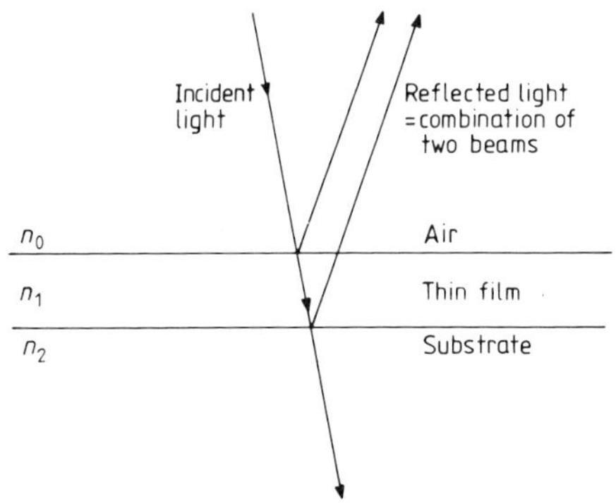

Figure 1.1 A single thin film.

The antireflection coating depends for its operation on the more or less complete cancellation of the light reflected at the upper and lower of the two surfaces of the thin film. Let the index of the substrate be n_2, that of the film n_1, and that of the incident medium, which will in almost all cases be air, n_0. For complete cancellation of the two beams of light, the intensities of the light reflected at the upper and lower boundaries of the film should be equal, which implies that the ratios of the refractive indices at each boundary should be equal, i.e. $n_0/n_1 = n_1/n_2$ or $n_1 = (n_0 n_2)^{1/2}$, which shows that the index of the thin film should be intermediate between the indices of air, which may be taken as unity, and of the substrate, which may be taken as at least 1.52.

Part of the incident light will be reflected at the top and bottom surfaces of the antireflection film, and in both cases the reflection will take place in a medium of lower refractive index than the adjoining medium. Thus, to ensure that the relative phase shift is 180°, the optical thickness of the film should be made one quarter wavelength when the total difference in phase between the two beams will correspond to twice one quarter wavelength, that is 180°.

A simple antireflection coating should, therefore, consist of a single film of refractive index equal to the square root of that of the substrate, and of optical thickness one quarter of a wavelength. As will be explained in the chapter on antireflection coatings, there are other improved coatings covering wider wavelength ranges involving greater numbers of layers.

Another basic type of thin-film structure is a stack of alternate high- and low-index films, all one quarter wavelength thick (see figure 1.2). Light which is reflected within the high-index layers will not suffer any phase shift on reflection, while those beams reflected within the low-index layers will suffer a change of 180°. It is fairly easy to see that the various components of the incident light produced by reflection at successive boundaries throughout the assembly will reappear at the front surface all in phase so that they will recombine constructively. This implies that the effective reflectance of the assembly can be made very high indeed, as high as may be desired, merely by increasing the number of layers. This is the basic form of the high-reflectance coating. When such a coating is constructed, it is found that the reflectance remains high over only a limited range of wavelengths, depending on the ratio of high and low refractive indices. Outside this zone, the reflectance changes abruptly to a low value. Because of this behaviour, the quarter-wave stack is used as a basic building block for many types of thin-film filters. It can be used as a longwave-pass filter, a shortwave-pass filter, a straightforward high-reflectance coating, for example in laser mirrors, and as a reflector in a thin-film Fabry–Perot interferometer (figure 1.3), which is another basic filter type described in some detail in chapters 5 and 7. Here, it is sufficient to say that it consists of a spacer layer which is usually half a wavelength thick, bounded by two high-reflectance coatings. Multiple-beam interference in the spacer layer causes the transmission of the filter to be extremely high over a narrow band of wavelengths around that for which the spacer is a multiple of one half wavelength thick. It is possible, as with lumped electric circuits, to couple two or more Fabry–Perot filters in series to give a more rectangular pass band.

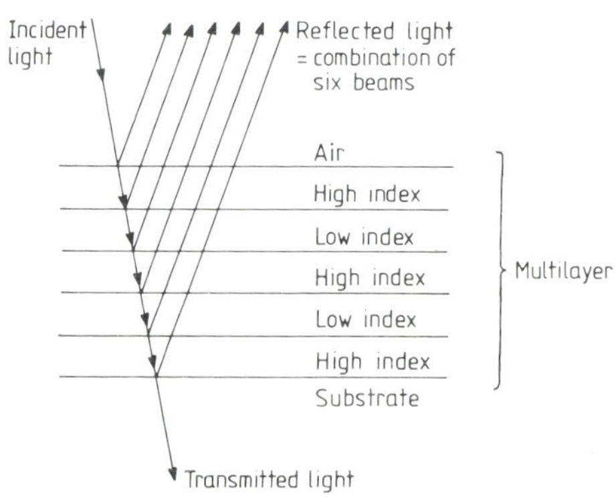

Figure 1.2 A multilayer.

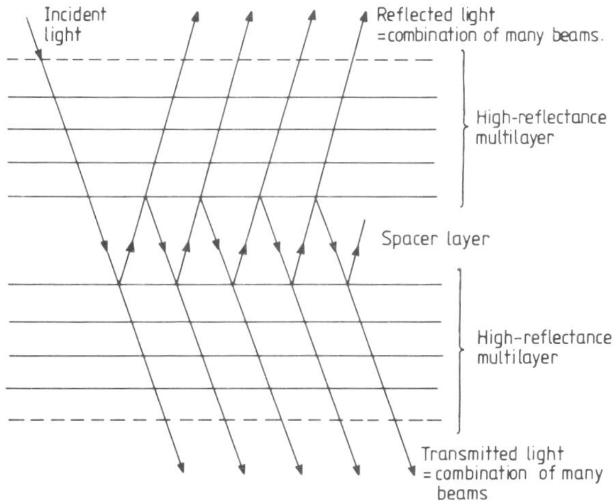

Figure 1.3 A Fabry–Perot filter showing multiple reflections in the spacer layer.

In the great majority of cases the thin films are completely transparent, so that no energy is absorbed. The filter characteristic in reflection is the complement of that in transmission. This fact is used in the construction of such devices as dichroic beam splitters for colour primary separation in, for example, colour television cameras.

This brief description has neglected the effect of multiple reflections in most of the layers and, for an accurate evaluation of the performance of a filter, these extra reflections must be taken into account. This involves extremely complex calculations and an alternative, and more effective, approach has been found in the development of entirely new forms of solution of Maxwell's equations in stratified media. This is, in fact, the principal method used in chapter 2 where basic mathematics are considered. The solution appears as a very elegant product of 2×2 matrices, each matrix representing a single film. Unfortunately, in spite of the apparent simplicity of the matrices, calculation by hand of the properties of a given multilayer, particularly if there are absorbing layers present and a wide spectral region is involved, is an extremely tedious and time-consuming task. The advent of the electronic pocket calculator has greatly reduced the necessary labour, and programmable versions can be used for calculations of multilayers, provided the number of layers and of wavelength or frequency points is limited. The preferred method, when layers are numerous and especially absorbing or when the calculation is over a wide spectral region, is to use a computer. A particularly significant advantage of the matrix method is that it has made possible the development of exceedingly powerful design techniques based on the algebraic manipulation of the matrices themselves.

In the design of a thin-film multilayer, we are required to find an arrangement of layers which will give a performance specified in advance, and this is much more difficult than straightforward calculation of the properties of a given multilayer. There is no analytical solution to the general problem. The normal method of design is to arrive at a possible structure for a filter, using techniques which will be described, and which consist of a mixture of analysis, experience and the use of well known building blocks. The evaluation is then completed by calculating the performance on a computer. Depending on the results of the computations, adjustments to the proposed design may be made, then recomputed, until a satisfactory solution is found. This adjustment process can itself be undertaken by a computer and is often known by the term refinement.

The successful application of refinement techniques depends largely on a starting solution which has a performance close to that required. Under these conditions it has been made to work well. Methods of completely automatic synthesis of designs without any starting solution are less frequently used, although great progress has been made. In common with refinement techniques, they operate to adjust the parameters of the system to minimise a merit coefficient representing the gap between the performance achieved by the design at any stage and the desired performance. A major problem is the enormous number of parameters which can potentially be involved. Refinement is kept within bounds by limiting the search to small changes in an almost acceptable starting design, but with no starting design the possibilities are virtually infinite, and so the rules governing the search procedure have to be very carefully organised. Automatic design synthesis is undoubtedly increasing in importance with developments in computers, but this branch of the subject is much more a matter of computing techniques rather than fundamental to the understanding of thin-film filters, and so it has been considered outside the scope of this book. The recent book by Liddell[9] gives a full account of the various methods. The real limitation to what is, at the present time, possible in optical thin-film filters and coatings, is the capability of the manufacturing process to produce layers of precisely the correct optical constants and thickness, rather than any deficiency in design techniques.

The basic manufacturing method for the construction of thin-film filters is that of vacuum deposition and this is the principal process described in this book. Further information on the vacuum deposition process is given in the book by Holland[2], now over twenty years old and unfortunately out of print but still very useful. Then, while this book was in the proof stage, the exceptionally detailed book on the physics and chemistry involved in thin-film coatings on glass by Pulker[25] appeared. It also contains much information on cleaning techniques, alternative deposition methods, and testing methods.

Optical filters are used in almost every conceivable field. To cover the whole range of applications would clearly be impossible, so several typical uses have been selected and described along with some of the important points to watch in designing such applications.

REFERENCES

1 Knittl Z 1976 *Optics of Thin Films* (London: Wiley)

2 Holland L 1956 *Vacuum Deposition of Thin Films* (London: Chapman and Hall)

3 Anders H 1965 *Dünne Schichten für die Optik* (Stuttgart: Wissenschaftliche Verlagsgesellschaft) (Engl. transl. 1967 *Thin films in optics* (London: Focal))

4 Heavens O S 1955 *Optical Properties of Thin Solid Films* (London: Butterworths). Reprinted 1965 (New York: Dover)

5 Lissberger P H 1970 Optical applications of dielectric thin films *Rep. Prog. Phys.* **33** 197–268

6 DeBell G W and Harrison D H eds 1974 Optical Coatings, Applications and Utilization *Proc. Soc. Photo-optical Instrum. Eng.* **50**

7 DeBell G W and Harrison D H eds 1978 Optical Coatings II (Applications and Utilization) *Proc. Soc. Photo-optical Instrum. Eng.* **140**

8 Dobrowolski J A 1978 *Coatings and Filters* in *Handbook of Optics* ed. Driscoll W G (New York: McGraw-Hill) pp. 8-1–8-124

9 Liddell H M 1981 *Computer-aided Techniques for the Design of Multilayer Filters* (Bristol: Adam Hilger)

10 Crew H 1930 Thomas Young's place in the history of the wave theory of light *J. Opt. Soc. Am.* **20** 3–10

11 de Senarmont H, Verdet E and Fresnel L 1866–70 *Oeuvres complètes d'Augustin Fresnel* (Paris: Impériale)

12 Knittl Z 1978 Fresnel historique et actuel *Opt. Acta* **25** 167–73

13 Maxwell J C 1873 *A Treatise on Electricity and Magnetism.* First edition published in 1873. The third edition, originally published by the Clarendon Press in 1891, was republished in unabridged form in 1954 (New York: Dover)

14 von Fraunhofer J 1817 *Versuche über die Ursachen des Anlaufens und Mattwerdens des Glases und die Mittel denselben zuvorzukommen.* Taken from *Joseph von Fraunhofer's Gesammelte Schriften* 1888 (Munich). The extracts appear on pages 35 and 46

15 Lord Rayleigh 1886 On the intensity of light reflected from certain surfaces at nearly perpendicular incidence *Proc. R. Soc.* **41** 275–94

16 Taylor H D 1983 *On the adjustment and testing of Telescopic Objectives.* First published in 1891. The quotation is taken from the second edition, p 62, published by T Cooke & Sons in 1896. The third edition, 1921, was later republished unchanged by Sir Howard Grubb, Parsons & Company Limited in 1946. The quotation may now be found on p 59 of the fifth edition (1983) (Bristol: Adam Hilger)

17 Taylor H D 1904 Lenses *UK Patent Specification* 29 561

18 Kollmorgen F 1916 Light transmission through telescopes *Trans. Am. Illum. Eng. Soc.* **11** 220–8

19 Fabry C and Perot A 1899 Théorie et applications d'une nouvelle méthode de spectroscopie interférentielle *Ann. Chim. Phys., Paris* 7th series **16** 115–44

20 Rouard P 1932 Sur le pouvoir réflecteur des métaux en lames très minces *C. R. Acad. Sci., Paris* **195** 868–71

21 Bauer G 1934 Absolutwerte der optischen Absorptionskonstanten von Alkalihalogenidkristallen im Gebiet ihrer ultravioletten Eigenfrequenzen *Ann. Phys. Lpz.* 5th series **19** 434–64

22 Pfund A H 1934 Highly reflecting films of zinc sulphide *J. Opt. Soc. Am.* **24** 99–102

23 Strong J 1936 On a method of decreasing the reflection from non-metallic substances *J. Opt. Soc. Am.* **26** 73–4

24 Geffken W 1939 Interferenzlichtfilter *Deutsches Reich Patentschrift* 716 153

25 Pulker H K 1984 *Coatings on Glass* (Oxford: Elsevier)

2 Basic theory

The next part of the book is a long and rather tedious account of some basic theory which is necessary in order to make calculations of the properties of multilayer thin-film coatings. It is perhaps worth reading just once, or when some deeper insight into thin-film calculations is required. In order to make it easier for those who have read it to find the basic results, or, for those who do not wish to read it at all, to proceed with the remainder of the book, the principal results are summarised, beginning on page 40.

MAXWELL'S EQUATIONS AND PLANE ELECTROMAGNETIC WAVES

For those readers who are still with us we begin our attack on thin-film problems by solving Maxwell's equations together with the appropriate material equations. In isotropic media these are

$$\operatorname{curl} \boldsymbol{H} = \boldsymbol{j} + \partial \boldsymbol{D}/\partial t \tag{2.1}$$

$$\operatorname{curl} \boldsymbol{E} = -\partial \boldsymbol{B}/\partial t \tag{2.2}$$

$$\operatorname{div} \boldsymbol{D} = \rho \tag{2.3}$$

$$\operatorname{div} \boldsymbol{B} = 0 \tag{2.4}$$

$$\boldsymbol{j} = \sigma \boldsymbol{E} \tag{2.5}$$

$$\boldsymbol{D} = \varepsilon \boldsymbol{E} \tag{2.6}$$

$$\boldsymbol{B} = \mu \boldsymbol{H}. \tag{2.7}$$

In anisotropic media, equations (2.5)–(2.7) become much more complicated with σ, ε and μ being tensor rather than scalar quantities.

The International System of Units (SI) is used as far as possible throughout this book. Table 2.1 shows the definitions of the quantities in the equations together with the appropriate SI units.

Table 2.1

Symbol	Physical quantity	SI unit	Symbol for SI unit
$\boldsymbol{E}$	Electric field strength	volts per metre	$\mathrm{V\,m^{-1}}$
$\boldsymbol{D}$	Electric displacement	coulombs per square metre	$\mathrm{C\,m^{-2}}$
$\boldsymbol{H}$	Magnetic field strength	amperes per metre	$\mathrm{A\,m^{-1}}$
$\boldsymbol{j}$	Electric current density	amperes per square metre	$\mathrm{A\,m^{-2}}$
$\boldsymbol{B}$	Magnetic flux density or magnetic induction	tesla	T
ρ	Electric charge density	coulombs per cubic metre	$\mathrm{C\,m^{-3}}$
σ	Electric conductivity	siemens per metre	$\mathrm{S\,m^{-1}}$
μ	Permeability	henries per metre	$\mathrm{H\,m^{-1}}$
ε	Permittivity	farads per metre	$\mathrm{F\,m^{-1}}$

To the equations we can add

$$\varepsilon = \varepsilon_r \varepsilon_0 \tag{2.8}$$

$$\mu = \mu_r \mu_0 \tag{2.9}$$

$$\varepsilon_0 = 1/\mu_0 c^2 \tag{2.10}$$

where ε_0 and μ_0 are the permittivity and permeability of free space respectively. ε_r and μ_r are the relative permittivity and permeability, and c is a constant which can be identified as the velocity of light in free space. ε_0, μ_0 and c are important constants, the values of which are given in table 2.2.

Table 2.2

Symbol	Physical quantity	Value
c	Speed of light in a vacuum	$2.997925 \times 10^8\ \mathrm{m\,s^{-1}}$
μ_0	Permeability of a vacuum	$4\pi \times 10^{-7}\ \mathrm{H\,m^{-1}}$
ε_0	Permittivity of a vacuum ($= \mu_0^{-1} c^{-2}$)	$8.8541853 \times 10^{-12}\ \mathrm{F\,m^{-1}}$

The following analysis is brief and incomplete. For a full, rigorous treatment of the electromagnetic field equations, the reader is referred to Born and Wolf[1]. Normally there will be no space charge in the medium, and equation (2.3) becomes

$$\mathrm{div}\,\boldsymbol{D} = 0$$

and, solving for $\boldsymbol{E}$

$$\nabla^2 \boldsymbol{E} = \varepsilon\mu \frac{\partial^2 \boldsymbol{E}}{\partial t^2} + \mu\sigma \frac{\partial \boldsymbol{E}}{\partial t}. \tag{2.11}$$

A similar expression holds for $\boldsymbol{H}$.

First of all, we look for a solution of equation (2.11) in the form of a plane-polarised plane harmonic wave, and we choose the complex form of this wave, the physical meaning being associated with the real part of the expression.

$$\boldsymbol{E} = \boldsymbol{\mathscr{E}} \exp[i\omega(t - x/v)] \tag{2.12}$$

represents such a wave propagating along the x axis with velocity v. $\boldsymbol{\mathscr{E}}$ is the vector amplitude and ω the angular frequency of this wave. The advantage of the complex form of the wave is that phase changes can be dealt with very readily by including them in a complex amplitude. If $\mathscr{E}$, the magnitude of $\boldsymbol{\mathscr{E}}$, is given by $|\boldsymbol{\mathscr{E}}| \exp(i\phi)$, then the real part of the expression (2.12) is of the form

$$|\boldsymbol{\mathscr{E}}| \cos[\omega(t - x/v) + \phi].$$

For equation (2.12) to be a solution of equation (2.11) it is necessary that

$$\omega^2/v^2 = \omega^2\varepsilon\mu - i\omega\mu\sigma. \tag{2.13}$$

In a vacuum we have $\sigma = 0$ and $v = c$, so that from equation (2.13)

$$c^2 = 1/\mu_0\varepsilon_0$$

which is identical to equation (2.10). Multiplying equation (2.13) by equation (2.10) and dividing through by ω^2 we obtain

$$\frac{c^2}{v^2} = \frac{\varepsilon\mu}{\varepsilon_0\mu_0} - i\frac{\mu\sigma}{\omega\varepsilon_0\mu_0}$$

where c/v is clearly a dimensionless parameter of the medium, which we denote by N:

$$N^2 = \varepsilon_r\mu_r - i\,\mu_r\sigma/\omega\varepsilon_0. \tag{2.14}$$

This implies that N is of the form

$$N = c/v = n - ik. \tag{2.15}$$

There are two possible values of N from equation (2.14), but for physical reasons we choose that which gives a positive value of n. N is known as the complex refractive index, n as the real part of the refractive index (or often simply as the refractive index, because N is real in an ideal dielectric material), and k is known as the extinction coefficient.

From equation (2.14)

$$n^2 - k^2 = \varepsilon_r/\mu_r \tag{2.16}$$

$$2nk = \frac{\mu_r\sigma}{\omega\varepsilon_0}. \tag{2.17}$$

Equation (2.11) can now be written

$$\boldsymbol{E} = \mathscr{E}\ \exp\{\mathrm{i}[\omega t - (2\pi N/\lambda)x]\} \tag{2.18}$$

where we have introduced the wavelength in free space, $\lambda(=2\pi c/\omega)$.

Substituting $n - \mathrm{i}k$ for N in equation (2.18) gives

$$\boldsymbol{E} = \mathscr{E}\ \exp[-(2\pi k/\lambda)x]\exp\{\mathrm{i}[\omega t - (2\pi n/\lambda)x]\} \tag{2.19}$$

and the significance of k emerges as being a measure of the absorption in the medium. The distance $\lambda/2\pi k$ is that in which the amplitude of the wave falls to $1/\mathrm{e}$ of its initial value. The way in which the intensity falls off will be considered shortly.

The change in phase produced by a traversal of distance x in the medium is the same as that produced by a distance nx in a vacuum. Because of this, nx is known as the optical distance, as distinct from the physical or geometrical distance. Generally, in thin-film optics one is more interested in optical distances and optical thicknesses than in geometrical ones.

Equation (2.18) represents a plane-polarised plane wave propagating along the x axis. For a similar wave propagating in a direction given by direction coefficient (α, β, γ) the expression becomes

$$\boldsymbol{E} = \mathscr{E}\ \exp\{\mathrm{i}[\omega t - (2\pi N/\lambda)(\alpha x + \beta y + \gamma z)]\}. \tag{2.20}$$

This is the simplest type of wave in an absorbing medium. In an assembly of absorbing thin films, we shall see that we are occasionally forced to adopt a slightly more complicated expression for the wave.

There are some important relationships for this type of wave which can be derived from Maxwell's equations. Let the direction of propagation of the wave be given by unit vector $\hat{\boldsymbol{s}}$ where

$$\hat{\boldsymbol{s}} = \alpha\boldsymbol{i} + \beta\boldsymbol{j} + \gamma\boldsymbol{k}$$

and where $\boldsymbol{i}, \boldsymbol{j}$ and $\boldsymbol{k}$ are unit vectors along the x, y and z axes, respectively. From equation (2.20) we have

$$\partial\boldsymbol{E}/\partial t = \mathrm{i}\omega\boldsymbol{E}$$

and from equations (2.1), (2.5) and (2.6)

$$\begin{aligned}\operatorname{curl}\boldsymbol{H} &= \sigma\boldsymbol{E} + \varepsilon\,\partial\boldsymbol{E}/\partial t\\ &= (\sigma + \mathrm{i}\omega\varepsilon)\boldsymbol{E}\\ &= \mathrm{i}\frac{\omega N^2}{c^2\mu}\boldsymbol{E}.\end{aligned}$$

Now

$$\operatorname{curl} \equiv \left(\frac{\partial}{\partial x}\boldsymbol{i} + \frac{\partial}{\partial y}\boldsymbol{j} + \frac{\partial}{\partial z}\boldsymbol{k}\right)\times$$

where × denotes the vector product. But

$$\frac{\partial}{\partial x} = -\mathrm{i}\frac{2\pi N}{\lambda}\alpha = -\mathrm{i}\frac{\omega N}{c}\alpha$$

$$\frac{\partial}{\partial y} = -\mathrm{i}\frac{\omega N}{c}\beta \qquad \frac{\partial}{\partial z} = -\mathrm{i}\frac{\omega N}{c}\gamma$$

so that

$$\mathrm{curl}\,\boldsymbol{H} = -\mathrm{i}\frac{\omega N}{c}(\hat{\boldsymbol{s}} \times \boldsymbol{H}).$$

Then

$$-\mathrm{i}\frac{\omega N}{c}(\hat{\boldsymbol{s}} \times \boldsymbol{H}) = \mathrm{i}\frac{\omega N^2}{c^2\mu}\boldsymbol{E}$$

i.e.

$$(\hat{\boldsymbol{s}} \times \boldsymbol{H}) = -\frac{N}{c\mu}\boldsymbol{E} \tag{2.21}$$

and similarly

$$\frac{N}{c\mu}(\hat{\boldsymbol{s}} \times \boldsymbol{E}) = \boldsymbol{H}. \tag{2.22}$$

For this type of wave, therefore, $\boldsymbol{E}$, $\boldsymbol{H}$ and $\hat{\boldsymbol{s}}$ are mutually perpendicular and form a right-handed set. The quantity $N/c\mu$ has the dimensions of an admittance and is known as the characteristic optical admittance of the medium, written y. In free space it can be readily shown that the optical admittance is given by

$$\mathscr{Y} = (\varepsilon_0/\mu_0)^{1/2} = 2.6544 \times 10^{-3}\,\mathrm{S}.$$

Now

$$\mu = \mu_r\mu_0$$

and at optical frequencies μ_r is unity so that we can write

$$y = N\mathscr{Y} \tag{2.23}$$

and

$$\boldsymbol{H} = y(\hat{\boldsymbol{s}} \times \boldsymbol{E}) = N\mathscr{Y}(\hat{\boldsymbol{s}} \times \boldsymbol{E}). \tag{2.24}$$

THE POYNTING VECTOR

An important feature of electromagnetic radiation is that it is a form of energy transport, and it is the energy associated with the wave which is observed. The instantaneous rate of flow of energy across unit area is given by the Poynting vector

$$\boldsymbol{P} = \boldsymbol{E} \times \boldsymbol{H}.$$

The direction of the vector is the direction of energy flow.

This expression is non-linear and so we cannot use directly the complex form of the wave. Either the real or the imaginary part of the wave expression should be inserted. The instantaneous value of the Poynting vector oscillates at twice the frequency of the wave and it is its mean value which is significant. This is defined as the intensity. For a harmonic wave we can derive a similar expression for the intensity which also uses the complex form of the wave. This is

$$\boldsymbol{I} = \tfrac{1}{2}\,\mathrm{Re}\,(\boldsymbol{E} \times \boldsymbol{H}^*) \qquad (2.25)$$

where * denotes complex conjugate. It should be emphasised that the complex form *must* be used in equation (2.25). The intensity $\boldsymbol{I}$ is written in equation (2.25) as a vector quantity when it has the same direction as the flow of energy of the wave. The more usual scalar intensity I is simply the magnitude of $\boldsymbol{I}$. Since $\boldsymbol{E}$ and $\boldsymbol{H}$ are perpendicular, equation (2.25) can be written

$$I = \tfrac{1}{2}\,\mathrm{Re}\,(EH^*)$$

where E and H are the scalar magnitudes.

It is important to note that the electric and magnetic vectors in equation (2.25) should be the total resultant fields due to all the waves which are involved. This is implicit in the derivation of the Poynting vector expression. We will return to this point when calculating reflectance and transmittance.

For a single, homogeneous, harmonic wave of the form (2.20):

$$\boldsymbol{H} = y(\hat{\boldsymbol{s}} \times \boldsymbol{E})$$

so that

$$\begin{aligned} \boldsymbol{I} &= \mathrm{Re}\,(\tfrac{1}{2}y\,EE^*\hat{\boldsymbol{s}}) \\ &= \tfrac{1}{2}n\mathscr{Y}\,EE^*\,\hat{\boldsymbol{s}}. \end{aligned}$$

Now, from equation (2.20), the magnitude of E is given by

$$\begin{aligned} E &= \mathscr{E}\exp\left[\mathrm{i}\left(\omega t - \frac{2\pi(n-\mathrm{i}k)}{\lambda}(\alpha x + \beta y + \gamma z)\right)\right] \\ &= \mathscr{E}\exp\left(-\frac{2\pi k}{\lambda}(\alpha x + \beta y + \gamma z)\right)\exp\left[\mathrm{i}\left(\omega t - \frac{2\pi n}{\lambda}(\alpha x + \beta y + \gamma z)\right)\right] \end{aligned}$$

implying

$$EE^* = \mathscr{E}\mathscr{E}^*\exp[-(4\pi k/\lambda)(\alpha x + \beta y + \gamma z)]$$

and

$$I = \tfrac{1}{2}n\mathscr{Y}\,|\mathscr{E}|^2\exp[-(4\pi k/\lambda)(\alpha x + \beta y + \gamma z)].$$

The expression $(\alpha x + \beta y + \gamma z)$ is simply the distance along the direction of propagation, and thus the intensity drops to $1/\mathrm{e}$ of its initial value in a distance given by $\lambda/4\pi k$. The inverse of this distance is defined as the absorption coefficient α, that is

$$\alpha = 4\pi k/\lambda. \qquad (2.26)$$

The absorption coefficient α should not be confused with the direction cosine.
However,

$$|\mathscr{E}| \exp[-(2\pi k/\lambda)(\alpha x + \beta y + \gamma z)]$$

is really the amplitude of the wave at the point (x, y, z) so that a much simpler way of writing the expression for intensity is

$$I = \tfrac{1}{2} n \mathscr{Y} \text{(amplitude)}^2$$

or

$$I \propto n \times \text{(amplitude)}^2. \tag{2.27}$$

This expression is a better form than the more usual

$$I \propto \text{(amplitude)}^2.$$

The expression will frequently be used for comparing intensities, in calculating reflectance or transmittance, for example, and if the media in which the two waves are propagating are of different index, then errors will occur unless n is included as above.

THE SIMPLE BOUNDARY

Thin-film filters usually consist of a number of boundaries between various homogeneous media and it is the effect which these boundaries will have on an incident wave which we will wish to calculate. A single boundary is the simplest case. First of all we consider absorption-free media, i.e. $k = 0$. The arrangement is sketched in figure 2.1. At a boundary, the tangential components of $\boldsymbol{E}$ and $\boldsymbol{H}$, that is, the components along the boundary, are continuous across it. In this case, the boundary is defined by $z = 0$, and the tangential components must be continuous for all values of x, y and t.

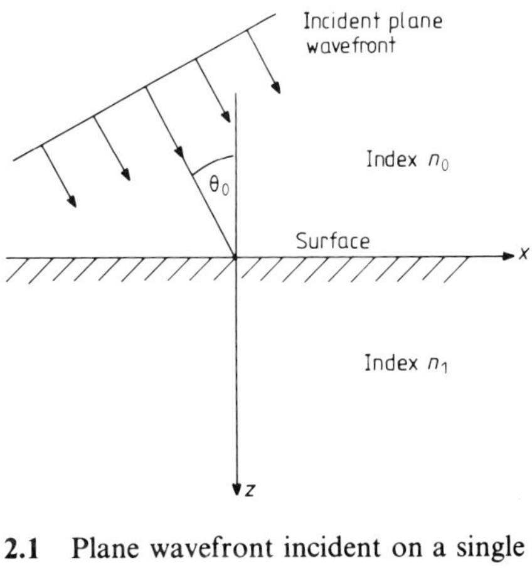

Figure 2.1 Plane wavefront incident on a single surface.

Let us retain our plane-polarised plane harmonic form for the incident wave; we can be safe in assuming that this wave will be split into a reflected wave and a transmitted wave at the boundary and our objective is the calculation of the parameters of these waves. Without specifying their exact form for the moment, we can, however, be certain that they will consist of an amplitude term and a phase factor. The amplitude terms will not be functions of x, y or t, any variations due to these being included in the phase factors.

Let the direction cosines of the $\hat{s}$ vectors of the transmitted and reflected waves be $(\alpha_t, \beta_t, \gamma_t)$ and $(\alpha_r, \beta_r, \gamma_r)$ respectively. We can then write the phase factors in the form:

Incident wave	$\exp\{i[\omega_i t - (2\pi n_0/\lambda)(x \sin\theta_0 + z \cos\theta_0)]\}$
Reflected wave	$\exp\{i[\omega_r t - (2\pi n_0/\lambda)(\alpha_r x + \beta_r y + \gamma_r z)]\}$
Transmitted wave	$\exp\{i[\omega_t t - (2\pi n_1/\lambda)(\alpha_t x + \beta_t y + \gamma_t z)]\}$.

The relative phases of these waves are included in the complex amplitudes. For waves with these phase factors to satisfy the boundary conditions for all x, y, t and for $z = 0$, implies that

$$\omega_i \equiv \omega_r \equiv \omega_t$$

that is, there is no change of frequency in reflection or refraction and hence

$$\lambda = \lambda_r = \lambda_t \qquad n_0\beta_r \equiv n_1\beta_t \equiv 0.$$

That is, the directions of the reflected and transmitted or refracted beams are confined to the plane of incidence. It implies also that

$$n_0 \sin\theta_0 = n_0\alpha_r = n_1\alpha_t$$

so that if the angles of reflection and refraction are θ_r and θ_t respectively, then

$$\theta_0 = \theta_r$$

that is, the angle of reflection equals the angle of incidence, and

$$n_0 \sin\theta_0 = n_1 \sin\theta_t.$$

The result appears more symmetrical if we replace θ_t by θ_1, giving

$$n_0 \sin\theta_0 = n_1 \sin\theta_1$$

which is the familiar relationship known as *Snell's law*. γ_r and γ_t are then given by

$$\alpha_r^2 + \gamma_r^2 = 1 \qquad \text{and} \qquad \alpha_t^2 + \gamma_t^2 = 1.$$

Normal incidence

Let us limit our initial discussion to normal incidence and let the incident wave be a plane-polarised plane harmonic wave. The coordinate axes are shown in

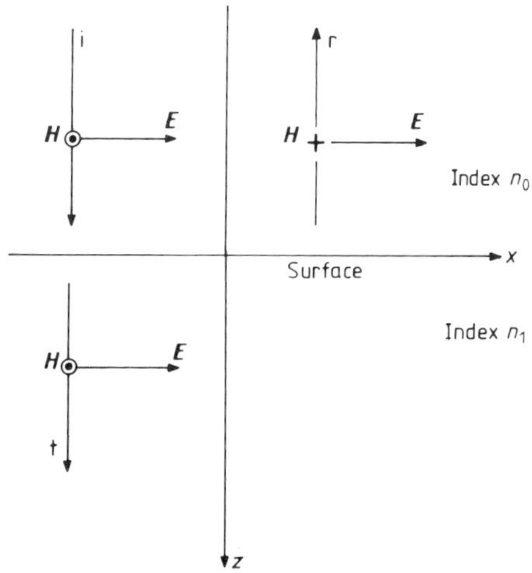

Figure 2.2 Convention defining positive directions of the electric and magnetic vectors for reflection and transmission at an interface at normal incidence.

figure 2.2. The xy plane is the plane of the boundary. The incident wave we can take as propagating along the z axis with the positive direction of the $\boldsymbol{E}$ vector along the x axis. Then the positive direction of the $\boldsymbol{H}$ vector will be the y axis. It is clear that the only waves which satisfy the boundary conditions are plane-polarised in the same plane as the incident wave. Before proceeding further, we need to define a sign convention for the electric and magnetic vectors in order that we can have a reference for any phase changes that may occur. This is merely a convention, and we have complete freedom of choice. We must simply ensure that once we have made our choice we adhere to it. Since we will be emphasising the electric vector, the most convenient sign convention is to choose the positive direction of $\boldsymbol{E}$ along the x axis for all the beams which are involved. Because of this choice, the positive direction of the magnetic vector will be along the y axis for the incident and transmitted waves, but along the negative direction of the y axis for the reflected wave.

We are now in a position to apply the boundary conditions. Since we have already made sure that the phase factors are satisfactory, we have only to consider the amplitudes, and we will be including any phase changes in these.
(*a*) Electric vector continuous across the boundary.

$$\mathscr{E}_{\mathrm{i}} + \mathscr{E}_{\mathrm{r}} = \mathscr{E}_{\mathrm{t}}. \tag{2.28}$$

(*b*) Magnetic vector continuous across the boundary. Here we use equation (2.24) to give

$$y_0\mathscr{E}_{\mathrm{i}} - y_0\mathscr{E}_{\mathrm{r}} = y_1\mathscr{E}_{\mathrm{t}} \tag{2.29}$$

where

$$y_0 = n_0 \mathscr{Y} \qquad \text{and} \qquad y_1 = n_1 \mathscr{Y}.$$

We can eliminate $\mathscr{E}_t$ to give

$$y_1(\mathscr{E}_i + \mathscr{E}_r) = y_0(\mathscr{E}_i - \mathscr{E}_r)$$

i.e.

$$\frac{\mathscr{E}_r}{\mathscr{E}_i} = \frac{y_0 - y_1}{y_0 + y_1} = \frac{n_0 - n_1}{n_0 + n_1}.$$

Similarly, eliminating $\mathscr{E}_r$

$$\frac{\mathscr{E}_t}{\mathscr{E}_i} = \frac{2y_0}{y_0 + y_1} = \frac{2n_0}{n_0 + n_1}.$$

These quantities are called the Fresnel amplitude reflection and transmission coefficients and are denoted by ρ and τ respectively. Thus

$$\rho = \frac{y_0 - y_1}{y_0 + y_1} = \frac{n_0 - n_1}{n_0 + n_1}$$
$$\tau = \frac{2y_0}{y_0 + y_1} = \frac{2n_0}{n_0 + n_1}. \tag{2.30}$$

In this particular case, these two quantities are real. τ is always a positive real number, indicating that according to our phase convention there is no phase shift between the incident and transmitted beams at the interface. The behaviour of ρ indicates that there will be no phase shift between the incident and reflected beams at the interface provided $n_0 > n_1$, but that if $n_0 < n_1$ then there will be a phase change of π because the value of ρ becomes negative.

We now examine the energy balance at the boundary. Since the boundary is of zero thickness, it can neither supply energy to nor extract energy from the various waves. The Poynting vector will therefore be continuous across the boundary, so that we can write

$$\begin{aligned}\text{net intensity} &= \mathrm{Re}[\tfrac{1}{2}(\mathscr{E}_i + \mathscr{E}_r)(y_0\mathscr{E}_i - y_0\mathscr{E}_r)^*] \\ &= \mathrm{Re}[\tfrac{1}{2}\mathscr{E}_t(y_1\mathscr{E}_t)^*]\end{aligned}$$

(using $\mathrm{Re}\,(\frac{1}{2}\,\boldsymbol{E}\times\boldsymbol{H}^*)$ and equations (2.28) and (2.29)). Now

$$\mathscr{E}_r = \rho\mathscr{E}_i \qquad \text{and} \qquad \mathscr{E}_t = \tau\mathscr{E}_i$$

i.e.

$$\text{net intensity} = \tfrac{1}{2}y_0\mathscr{E}_i\mathscr{E}_i^*(1-\rho^2) = \tfrac{1}{2}y_0\mathscr{E}_i\mathscr{E}_i^*(y_1/y_0)\tau^2. \tag{2.31}$$

Now, $\frac{1}{2}y_0\mathscr{E}_i\mathscr{E}_i^*$ is the intensity of the incident beam I_i. We can identify $\rho^2\frac{1}{2}y_0\mathscr{E}_i\mathscr{E}_i^* = \rho^2 I_i$ as the intensity of the reflected beam I_r and $(y_1/y_0)\times\tau^2\frac{1}{2}y_0\mathscr{E}_i\mathscr{E}_i^* = (y_1/y_0)\tau^2 I_i$ as the intensity of the transmitted beam I_t. We define the reflectance R as the ratio of the reflected and incident intensities and the transmittance T as the ratio of the transmitted and incident intensities. Then

$$T = \frac{I_t}{I_i} = \frac{y_1}{y_0}\tau^2 = \frac{4y_0 y_1}{(y_0+y_1)^2} = \frac{4n_0 n_1}{(n_0+n_1)^2}$$

$$R = \frac{I_r}{I_i} = \rho^2 = \left(\frac{y_0-y_1}{y_0+y_1}\right)^2 = \left(\frac{n_0-n_1}{n_0+n_1}\right)^2 . \qquad (2.32)$$

From equation (2.31) we have, using equations (2.32)

$$(1-R) = T. \qquad (2.33)$$

Equations (2.31), (2.32) and (2.33) are therefore consistent with our ideas of splitting the intensities into incident, reflected and transmitted intensities which can be treated as separate waves, the energy flow into the second medium being simply the difference of the incident and reflected intensities.

Oblique incidence

Now let us consider oblique incidence. For any general direction of the vector amplitude of the incident wave we quickly find that the application of the boundary conditions leads us into complicated and difficult expressions for the vector amplitudes of the reflected and transmitted waves. It turns out that there are two orientations of the incident wave which lead to reasonably straightforward calculations, the vector electrical amplitudes aligned in the plane of incidence (i.e. the xy plane of figure 2.1) and the vector electrical amplitudes aligned normal to the plane of incidence (i.e. parallel to the y axis in figure 2.1). In each of these cases, the orientations of the transmitted and reflected vector amplitudes are the same as for the incident wave. Any incident wave of arbitrary polarisation can therefore be split into two components having these simple orientations. The transmitted and reflected components can be calculated for each orientation separately and then combined to yield the resultant. Since, therefore, it is necessary to consider two orientations only, they have been given special names. A wave with the electric vector in the plane of incidence is known as p-polarised or as TM (for transverse magnetic) and a wave with the electric vector normal to the plane of incidence as s-polarised or TE (for transverse electric). p and s are derived from the German *parallel* and *senkrecht* (perpendicular). Before we can actually proceed to the calculation of the reflected and transmitted amplitudes, we must choose the various reference directions of the vectors from which any phase differences will be calculated. We have, once again, complete freedom of choice, but once we have established the convention we must adhere to it, just as in the normal incidence case. The conventions which we will use in this book are illustrated in figure 2.3. They have been chosen to be compatible with those for normal incidence already established. In some works, an opposite convention for the p-polarised

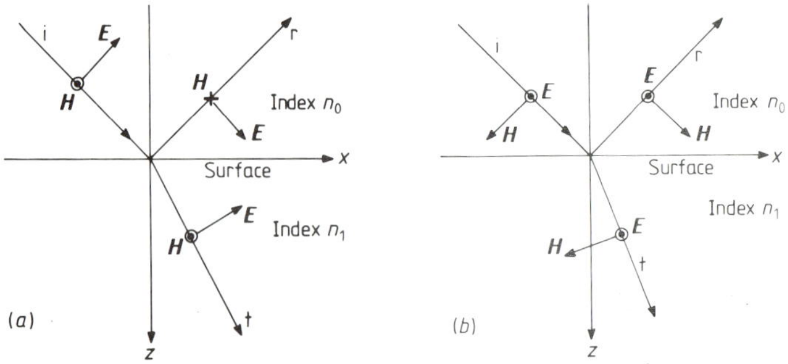

Figure 2.3 (*a*) Convention defining the positive directions of the electric and magnetic vectors for p-polarised light (TM waves). (*b*) Convention defining the positive directions of the electric and magnetic vectors for s-polarised light (TE waves).

reflected beam has been adopted, but this leads to an incompatibility with results derived for normal incidence, and we prefer to avoid this situation.

We can now apply the boundary conditions. Since we have already ensured that the phase factors will be correct, we need only consider the vector amplitudes.

p-polarised light

(*a*) Electric component parallel to the boundary continuous across it.

$$\mathscr{E}_i \cos\theta_0 + \mathscr{E}_r \cos\theta_0 = \mathscr{E}_t \cos\theta_1. \tag{2.34}$$

(*b*) Magnetic component parallel to the boundary continuous across it. Here we need to calculate the magnetic vector amplitudes, and we do this by using equation (2.24) with N replaced by n

$$\mathscr{H} = y(\hat{s} \times \mathscr{E}).$$

Since $\mathscr{H}$ is perpendicular to $\hat{s}$ and $\mathscr{E}$ it will be in the y direction for each of the waves and we can write

$$y_0\mathscr{E}_i - y_0\mathscr{E}_r = y_1\mathscr{E}_t. \tag{2.35}$$

At first sight it seems logical just to eliminate first $\mathscr{E}_t$ and then $\mathscr{E}_r$ from these two equations to obtain $\mathscr{E}_r/\mathscr{E}_i$ and $\mathscr{E}_t/\mathscr{E}_i$, i.e.

$$\left.\begin{aligned}\frac{\mathscr{E}_r}{\mathscr{E}_i} &= \frac{y_0\cos\theta_1 - y_1\cos\theta_0}{y_0\cos\theta_1 + y_1\cos\theta_0}\\[2ex] \frac{\mathscr{E}_t}{\mathscr{E}_i} &= \frac{2y_0\cos\theta_0}{y_0\cos\theta_1 + y_1\cos\theta_0}\end{aligned}\right\} \tag{2.36}$$

and then simply to set

$$R = \left(\frac{\mathscr{E}_r}{\mathscr{E}_i}\right)^2 \qquad \text{and} \qquad T = \frac{y_1}{y_0}\left(\frac{\mathscr{E}_t}{\mathscr{E}_i}\right)^2$$

but when we calculate the expressions which result, we find that $R + T \neq 1$. In fact, there is no mistake in the calculations. We have computed the intensities measured along the direction of propagation of the waves and the transmitted wave is inclined at an angle which differs from that of the incident wave. This leaves us with the problem that to adopt these definitions will involve the rejection of the $(R + T = 1)$ rule.

We can correct this situation by modifying the definition of T to include this angular dependence, but an alternative, and preferable, approach which is generally adopted in optical thin-film calculations is to consider the components of the waves and the energy flows which are normal to the boundary. The vectors are then parallel to the boundary and enter quite naturally into the boundary conditions which we use in deriving the relationship between the various intensities.

The tangential components of $\boldsymbol{E}$ and $\boldsymbol{H}$, that is, the components parallel to the boundary, have already been calculated for use in equations (2.34) and (2.35). However, it is convenient to introduce special symbols for the tangential amplitudes E and H.

Then we can write

$$E_i = \mathscr{E}_i \cos\theta_0 \qquad H_i = \mathscr{H}_i = y_0 \mathscr{E}_i = \frac{y_0}{\cos\theta_0} E_i$$

$$E_r = \mathscr{E}_r \cos\theta_0 \qquad H_r = \frac{y_0}{\cos\theta_0} E_r$$

$$E_t = \mathscr{E}_t \cos\theta_1 \qquad H_t = \frac{y_1}{\cos\theta_1} E_t.$$

The orientations of these vectors are exactly the same as for incident light. Equations (2.34) and (2.35) can then be written as follows.

(*a*) Electric field parallel to the boundary

$$E_i + E_r = E_t.$$

(*b*) Magnetic field parallel to the boundary

$$\frac{y_0}{\cos\theta_0} H_i - \frac{y_0}{\cos\theta_0} H_r = \frac{y_1}{\cos\theta_1} H_t$$

giving us, by a process exactly similar to that we have already used for normal incidence,

$$\rho_p = \frac{E_r}{E_i} = \left(\frac{y_0}{\cos\theta_0} - \frac{y_1}{\cos\theta_1}\right)\left(\frac{y_0}{\cos\theta_0} + \frac{y_1}{\cos\theta_1}\right)^{-1}$$

$$\tau_p = \frac{E_t}{E_i} = \left(\frac{2y_0}{\cos\theta_0}\right)\left(\frac{y_0}{\cos\theta_0} + \frac{y_1}{\cos\theta_1}\right)^{-1}$$

$$R_p = \left[\left(\frac{y_0}{\cos\theta_0} - \frac{y_1}{\cos\theta_1}\right)\left(\frac{y_0}{\cos\theta_0} + \frac{y_1}{\cos\theta_1}\right)^{-1}\right]^2 \tag{2.37}$$

$$T_p = \frac{y_1/\cos\theta_1}{y_0/\cos\theta_0}\tau_p^2 = \left(\frac{4y_0y_1}{\cos\theta_0\cos\theta_1}\right)\left(\frac{y_0}{\cos\theta_0} + \frac{y_1}{\cos\theta_1}\right)^{-2}$$

where $y_0 = n_0\mathscr{Y}$ and $y_1 = n_1\mathscr{Y}$ and the $(R+T=1)$ rule is retained. The suffix p has been used in the above expressions to denote p-polarisation.

It should be noted that the expression for τ_p is now different from that in equations (2.36). (2.36) is the form often quoted for the Fresnel amplitude transmission coefficient, but the version in equations (2.37) is consistent with optical thin-film usage. Fortunately, the reflection coefficients in equations (2.36) and (2.37) are identical, and since much more use is made of reflection coefficients confusion is rare.

s-polarised light

In the case of s-polarisation the amplitudes of the components of the waves parallel to the boundary are

$$E_i = \mathscr{E}_i \qquad H_i = \mathscr{H}_i\cos\theta_0 = y_0\cos\theta_0 E_i$$
$$E_r = \mathscr{E}_r \qquad H_r = \mathscr{H}_r\cos\theta_0 = y_0\cos\theta_0 E_r$$
$$E_t = \mathscr{E}_t \qquad H_t = y_1\cos\theta_1 E_t$$

and here we have again an orientation exactly as for normally incident light, and so a similar analysis leads to

$$\rho_s = \frac{E_r}{E_i} = \left(\frac{y_0\cos\theta_0 - y_1\cos\theta_1}{y_0\cos\theta_0 + y_1\cos\theta_1}\right)$$

$$\tau_s = \frac{E_t}{E_i} = \frac{2y_0\cos\theta_0}{(y_0\cos\theta_0 + y_1\cos\theta_1)}$$

$$R_s = \left(\frac{y_0\cos\theta_0 - y_1\cos\theta_1}{y_0\cos\theta_0 + y_1\cos\theta_1}\right)^2 \tag{2.38}$$

$$T_s = \frac{4y_0\cos\theta_0\, y_1\cos\theta_1}{(y_0\cos\theta_0 + y_1\cos\theta_1)^2}$$

where

$$y_0 = n_0\,\mathscr{Y} \quad \text{and} \quad y_1 = n_1\,\mathscr{Y}.$$

Once again, at the boundary, $R+T=1$.

The optical admittance for oblique incidence

The expressions which we have derived so far have been in their traditional form, involving the refractive indices of the various media, together with the admittance of free space $\mathscr{Y}$. However, the notation is becoming increasingly cumbersome and will appear even more so when we consider the behaviour of thin films.

Equation (2.24) gives $\boldsymbol{H} = y(\hat{\boldsymbol{s}} \times \boldsymbol{E})$ where $y = N\mathscr{Y}$ is the optical admittance. We have found it convenient to deal with E and H, the components of $\mathscr{E}$ and $\mathscr{H}$ parallel to the boundary, and so we introduce a modified optical admittance η which connects E and H

$$\eta = H/E. \tag{2.39}$$

At normal incidence $\eta = y = n\mathscr{Y}$ while at oblique incidence

$$\eta_p = \frac{y}{\cos\theta} = \frac{n\mathscr{Y}}{\cos\theta}$$

$$\eta_s = y\cos\theta = n\mathscr{Y}\cos\theta.$$

Then, in all cases, we can write

$$\begin{aligned} \rho &= \frac{\eta_0-\eta_1}{\eta_0+\eta_1} & \tau &= \frac{2\eta_0}{\eta_0+\eta_1} \\ R &= \left(\frac{\eta_0-\eta_1}{\eta_0+\eta_1}\right)^2 & T &= \frac{4\eta_0\eta_1}{(\eta_0+\eta_1)^2}. \end{aligned} \tag{2.40}$$

These expressions can be used to compute the variation of reflectance of simple boundaries between extended media. Examples are shown in figure 2.4 of the variation of reflectance with angle of incidence. In this case, there is no absorption in the material and it can be seen that the reflectance for p-polarised light (TM) falls to zero at a definite angle. This particular angle is known as the Brewster angle and is of some importance, especially in the field of gas lasers, where losses in the end windows of the tube containing the active gas must be as low as possible, otherwise the device will not operate. It is usual to attach the windows to the tube so that the light is incident at the Brewster angle where p-polarised light is transmitted without loss (or in practice with only very slight residual loss due to errors in alignment and scattering from imperfections).

The expression for the Brewster angle can be derived as follows. For the

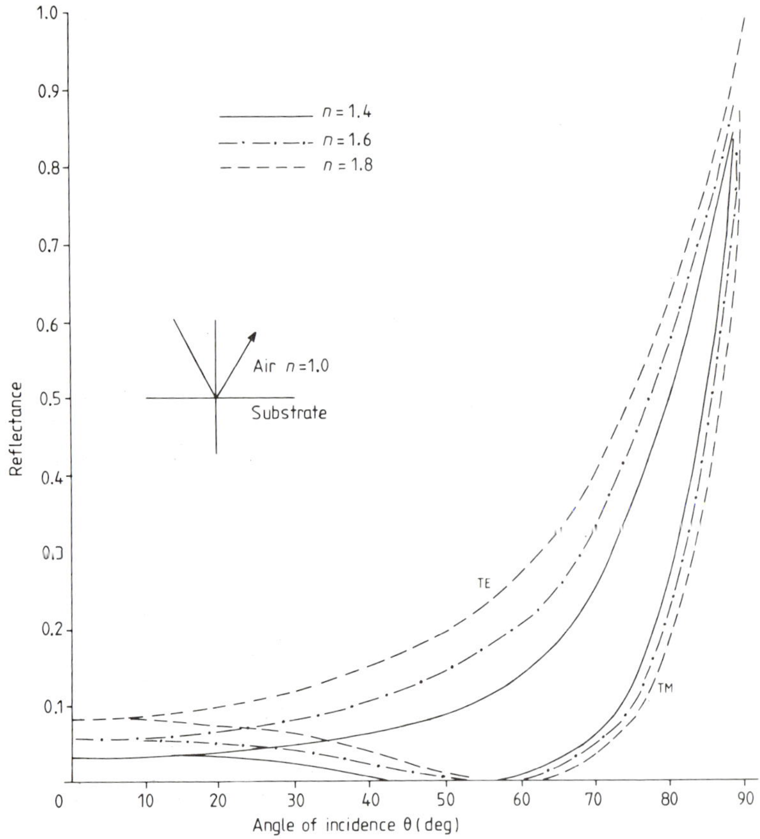

Figure 2.4 Variation of reflectance with angle of incidence for various values of refractive index.

p-reflectance (TM) to be zero, from equation (2.37)

$$\frac{n_0}{\cos\theta_0} = \frac{n_1}{\cos\theta_1}.$$

Snell's law gives another relationship between θ_0 and θ_1:

$$n_0 \sin\theta_0 = n_1 \sin\theta_1.$$

Eliminating θ_1 from these two equations gives an expression for θ_0

$$\tan\theta_0 = n_1/n_0.$$

Nomograms which connect the angle of incidence θ referred to an incident medium of unit refractive index, the refractive index of a dielectric film and the optical admittance of the film at θ are reproduced in figure 2.5.

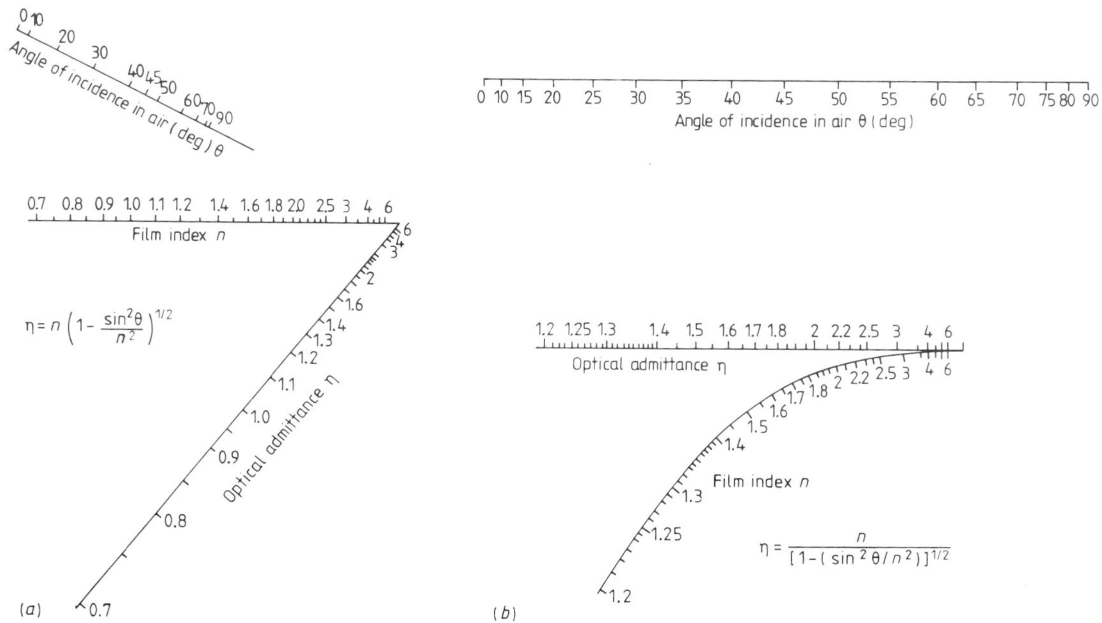

Figure 2.5 (*a*) Nomogram giving variation of optical admittance with angle of incidence for s-polarised light (TE waves). (*b*) Nomogram giving variation of optical admittance with angle of incidence for p-polarised light (TM waves).

Normal incidence in absorbing media

We must now consider how our results must be modified in the presence of absorption. First we consider the case of normal incidence and write

$$N_0 = n_0 - \mathrm{i}k_0$$
$$N_1 = n_1 - \mathrm{i}k_1$$
$$y_0 = N_0\mathscr{Y} = (n_0 - \mathrm{i}k_0)\mathscr{Y}$$
$$y_1 = N_1\mathscr{Y} = (n_1 - \mathrm{i}k_1)\mathscr{Y}.$$

The analysis follows that for absorption-free media. The boundaries are, as before,

(*a*) Electric vector continuous across the boundary:

$$\mathscr{E}_\mathrm{i} + \mathscr{E}_\mathrm{r} = \mathscr{E}_\mathrm{t}$$

(*b*) Magnetic vector continuous across the boundary:

$$y_0\mathscr{E}_\mathrm{i} - y_0\mathscr{E}_\mathrm{r} = y_1\mathscr{E}_\mathrm{t}$$

i.e.

$$N_0\mathscr{Y}\mathscr{E}_\mathrm{i} - N_0\mathscr{Y}\mathscr{E}_\mathrm{r} = N_1\mathscr{Y}\mathscr{E}_\mathrm{t}$$

and eliminating first $\mathscr{E}_\mathrm{t}$ and then $\mathscr{E}_\mathrm{r}$ we obtain the expressions for the Fresnel coefficients

$$\rho = \frac{\mathscr{E}_\mathrm{r}}{\mathscr{E}_\mathrm{i}} = \frac{N_0 - N_1}{N_0 + N_1} = \frac{(n_0 - n_1) - \mathrm{i}(k_0 - k_1)}{(n_0 + n_1) + \mathrm{i}(k_0 + k_1)}$$
$$\tau = \frac{\mathscr{E}_\mathrm{t}}{\mathscr{E}_\mathrm{i}} = \frac{2N_0}{N_0 + N_1} = \frac{2(n_0 - \mathrm{i}k_0)}{(n_0 + n_1) - \mathrm{i}(k_0 + k_1)}. \tag{2.41}$$

Our troubles begin when we try to extend this to reflectance and transmittance. Following the method for the absorption-free case, we compute the Poynting vector at the boundary in each medium, using the components of $\boldsymbol{E}$ and $\boldsymbol{H}$ parallel to the boundary, and equate the two values obtained. In the incident medium the resultant electric and magnetic fields are

$$E_\mathrm{i} + E_\mathrm{r} = E_\mathrm{i}(1+\rho)$$

and

$$H_\mathrm{i} + H_\mathrm{r} = N_0\mathscr{Y}(1-\rho)E_\mathrm{i}$$

respectively, where we have used the notation for tangential components, and in the second medium the fields are

$$\tau E_\mathrm{i} \qquad \text{and} \qquad N_1\mathscr{Y}\tau E_\mathrm{i}$$

respectively. Then the net intensity on either side of the boundary is

Medium 0: $I = \mathrm{Re}\{\tfrac{1}{2}[E_\mathrm{i}(1+\rho)][N_0^*\mathscr{Y}(1-\rho^*)E_\mathrm{i}^*]\}$

Medium 1: $I = \mathrm{Re}\{\tfrac{1}{2}[\tau E_\mathrm{i}][N_1^*\mathscr{Y}\tau^* E_\mathrm{i}^*]\}.$

We then equate these two values which gives, at the boundary,

$$\mathrm{Re}[\tfrac{1}{2}N_0^* \mathscr{Y} E_i E_i^* (1+\rho-\rho^*-\rho\rho^*)] = \tfrac{1}{2} n_1 \mathscr{Y} \tau\tau^* E_i E_i^*$$

i.e.

$$\tfrac{1}{2} n_0 \mathscr{Y} E_i E_i^* [1-\rho\rho^* + \mathrm{i}(k_0/n_0)(\rho-\rho^*)] = \tfrac{1}{2} n_0 \mathscr{Y} E_i E_i^* [(n_1/n_0)\tau\tau^*].$$

(Note: since $\rho-\rho^*$ is imaginary, $\mathrm{i}(k_0/n_0)(\rho-\rho^*)$ is real.) Now, it seems reasonable to associate $\tfrac{1}{2} n_0 \mathscr{Y} E_i E_i^*$ with the incident intensity, R with $\rho\rho^*$ and T with $(n_1/n_0)\tau\tau^*$. Then we find

$$(1-R) + \mathrm{i}(k_0/n_0)(\rho-\rho^*) = T.$$

If ρ is complex, the second term will be non-zero and $1-R \neq T$. The intensities involved in the analysis are those actually at the boundary, which is of zero thickness, and it is impossible that it should either remove or donate energy to the waves. Our assumption that the intensities can be divided into separate incident, reflected and transmitted intensities is therefore incorrect. The source of the difficulty is a coupling between the incident and reflected fields which occurs only in an absorbing medium and which must be taken into account when computing energy transport. The expressions for the amplitude coefficients are perfectly correct. This explanation has followed that of Berning[2], who should be consulted for further information.

Although we will be dealing with absorbing media in thin-film assemblies, our incident media will never be heavily absorbing and it will not be a serious lack of generality if we assume that our incident media are absorption-free. Since our expressions for the amplitude coefficients are valid, then any calculations of amplitudes in absorbing media will be correct. We simply have to ensure that calculations of reflectances are carried out in a transparent medium. With this restriction, then, we have

$$R = \rho\rho^* \left(\frac{N_0-N_1}{N_0+N_1}\right)\left(\frac{N_0-N_1}{N_0+N_1}\right)^* = \left(\frac{y_1-y_0}{y_1+y_0}\right)\left(\frac{y_1-y_0}{y_1+y_0}\right)^*$$

$$T = \frac{n_1}{n_0}\tau\tau^* = \frac{4n_0n_1}{(N_0+N_1)(N_0+N_1)^*} = \frac{4(\mathrm{Re}\, y_0)(\mathrm{Re}\, y_1)}{(y_0+y_1)(y_0+y_1)^*} \tag{2.42}$$

where

$$N_0 = n_0 \quad \text{(i.e. real)} \qquad N_1 = n_1 - \mathrm{i}k_1$$

$$y_0 = N_0 \mathscr{Y} \qquad y_1 = N_1 \mathscr{Y}.$$

Oblique incidence in absorbing media

Remembering what we said in the previous section, we limit this to a transparent incident medium and an absorbing second medium. Our first aim must be to ensure that the phase factors are consistent. Taking advantage of some of the earlier results, we can write the phase factors as:

Incident: $\exp\{\mathrm{i}[\omega t - (2\pi n_0/\lambda)(x \sin\theta_0 + z\cos\theta_0)]\}$

Reflected: $\exp\{i[\omega t-(2\pi n_0/\lambda)(x\sin\theta_0-z\cos\theta_0)]\}$

Transmitted: $\exp\{i[\omega t-[2\pi(n_1-ik_1)/\lambda](\alpha x+\gamma z)]\}$

where α and γ in the transmitted phase factors are the only unknowns. The phase factors must be identically equal for all x and t with $z=0$. This implies

$$\alpha=\frac{n_0\sin\theta_0}{(n_1-ik_1)} \tag{2.43}$$

and, since $\alpha^2+\gamma^2=1$

$$\gamma=(1-\alpha^2)^{1/2} \tag{2.44}$$

There are two solutions to this equation and we must decide which is to be adopted.

$$\begin{aligned}(n_1-ik_1)\gamma &= [(n_1-ik_1)^2-n_0^2\sin^2\theta_0]^{1/2}\\ &= (n_1^2-k_1^2-n_0^2\sin^2\theta_0-i2n_1k_1)^{1/2}.\end{aligned}$$

The quantity within the square root is in either the third or fourth quadrant and so the square roots are in the second quadrant (of the form $-a+ib$) and in the fourth quadrant (of the form $a-ib$). If we consider what happens when these values are substituted into the phase factors, we see that the fourth quadrant solution must be correct because this leads to an exponential fall-off with z of amplitude. The second quadrant solution would lead to an increase with z which would not be physically correct. The fourth quadrant solution is consistent with the solution for the absorption-free case. The transmitted phase factor is therefore of the form

$$\begin{aligned}&\exp\{i[\omega t-(2\pi n_0\sin\theta_0 x/\lambda)-(2\pi/\lambda)(a-ib)z]\}\\ &\qquad=\exp(-2\pi bz/\lambda)\exp\{i[\omega t-(2\pi n_0\sin\theta_0 x)/\lambda-(2\pi az/\lambda)]\}\end{aligned}$$

where

$$(a-ib)=(n_1^2-k_1^2-n_0\sin^2\theta_0-i2n_1k_1)^{1/2}. \tag{2.45}$$

A wave which possesses such a phase factor is known as inhomogeneous. The exponential fall-off in amplitude is along the z axis, while the propagation direction in terms of phase is determined by the direction cosines, which can be extracted from

$$(2\pi n_0\sin\theta_0 x)/\lambda+2\pi az/\lambda.$$

The existence of such waves is another good reason for our choosing to consider the components of the fields parallel to the boundary and the flow of energy normal to the boundary.

We should note at this stage that provided we include the possibility of complex angles, the formulation of the absorption-free case applies equally well to absorbing media and we can write

$$\begin{aligned}(n_1-ik_1)\sin\theta_1 &= n_0\sin\theta_0\\ \alpha &= \sin\theta_1\\ \gamma &= \cos\theta_1\end{aligned}$$

and

$$(a - ib) = (n_1 - ik_1)\cos\theta_1.$$

The calculation of amplitudes follows the same pattern as before. However, we have not previously examined the implications of an inhomogeneous wave. Our main concern is the calculation of the tilted admittance connected with such a wave. Since the x, y and t variations of the wave are contained in the phase factor, we can write

$$\begin{aligned}\text{curl} &\equiv \left(\frac{\partial}{\partial x}\boldsymbol{i} + \frac{\partial}{\partial y}\boldsymbol{j} + \frac{\partial}{\partial z}\boldsymbol{k}\right)\times \\ &\equiv \left(-\mathrm{i}\frac{2\pi N}{\lambda}\alpha\boldsymbol{i} - \mathrm{i}\frac{2\pi N}{\lambda}\gamma\boldsymbol{k}\right)\times\end{aligned}$$

and

$$\frac{\partial}{\partial t} \equiv \mathrm{i}\omega.$$

For p-waves the $\boldsymbol{H}$ vector is parallel to the boundary in the y direction and so $\boldsymbol{H} = H_y\boldsymbol{j}$. The component of $\boldsymbol{E}$ parallel to the boundary will then be in the x direction, $E_x\boldsymbol{i}$. We follow the analysis leading up to equation (2.21) and as before

$$\begin{aligned}\text{curl}\,\boldsymbol{H} &= \sigma\boldsymbol{E} + \varepsilon\frac{\partial \boldsymbol{E}}{\partial t} \\ &= (\sigma + \mathrm{i}\omega\varepsilon)\boldsymbol{E} \\ &= \frac{\mathrm{i}\omega N^2}{c^2\mu}\boldsymbol{E}.\end{aligned}$$

Now the tangential component of curl $\boldsymbol{H}$ is in the x direction so that

$$-\mathrm{i}\frac{2\pi N}{\lambda}\gamma(\boldsymbol{k}\times\boldsymbol{j})H_y = \mathrm{i}\frac{\omega N^2}{c^2\mu}E_x\boldsymbol{i}.$$

But

$$-(\boldsymbol{k}\times\boldsymbol{j}) = \boldsymbol{i}$$

so that

$$\begin{aligned}\eta_{\mathrm{p}} &= \frac{H_y}{E_x} = \frac{\omega N\lambda}{2\pi c^2\mu\gamma} = \frac{N}{c\mu\gamma} \\ &= \frac{N\mathscr{Y}}{\gamma} = \frac{y}{\gamma}.\end{aligned}$$

For the s-waves we use

$$\text{curl}\,\boldsymbol{E} = -\frac{\partial \boldsymbol{B}}{\partial t} = -\mu\frac{\partial \boldsymbol{H}}{\partial t}.$$

E is now along the y axis and a similar analysis to that for p-waves yields

$$\eta_s = \frac{H_x}{E_y} = N\mathscr{Y}\gamma = y\gamma.$$

Now γ can be identified as $\cos\theta$, provided that θ is permitted to be complex, and so

$$\begin{aligned} \eta_p &= y/\cos\theta \\ \eta_s &= y\cos\theta. \end{aligned} \tag{2.46}$$

Thus the amplitude and intensity coefficients become as before

$$\begin{aligned} \rho &= \frac{\eta_0 - \eta_1}{\eta_0 + \eta_1} \\ \tau &= \frac{2\eta_0}{\eta_0 + \eta_1} \\ R &= \left(\frac{\eta_0 - \eta_1}{\eta_0 + \eta_1}\right)\left(\frac{\eta_0 - \eta_1}{\eta_0 + \eta_1}\right)^* \\ T &= \frac{4\,\mathrm{Re}(\eta_0)\,\mathrm{Re}(\eta_1)}{(\eta_0 - \eta_1)(\eta_0 + \eta_1)^*}. \end{aligned} \tag{2.47}$$

These expressions are valid for absorption-free media as well.

THE REFLECTANCE OF A THIN FILM

A simple extension of the above analysis occurs in the case of a thin, plane parallel film of material covering the surface of a substrate. The presence of two (or more) interfaces means that a number of beams will be produced by successive reflections and the properties of the film will be determined by the summation of these beams. We say that the film is thin when interference effects can be detected in the reflected or transmitted light, that is, when the path difference between the beams is less than the coherence length of the light, and thick when the path difference is greater than the coherence length. The same film can appear thin or thick depending entirely on the illumination conditions. The thick case can be shown to be identical with the thin case integrated over a sufficiently wide wavelength range or a sufficiently large range of angles of incidence. Normally, we will find that the films on the substrates can be treated as thin while the substrates supporting the films can be considered thick. Thick films and substrates will be considered towards the end of this chapter. Here we concentrate on the thin case.

The arrangement is illustrated in figure 2.6. At this stage it is convenient to introduce a new notation. We denote waves in the direction of incidence by the

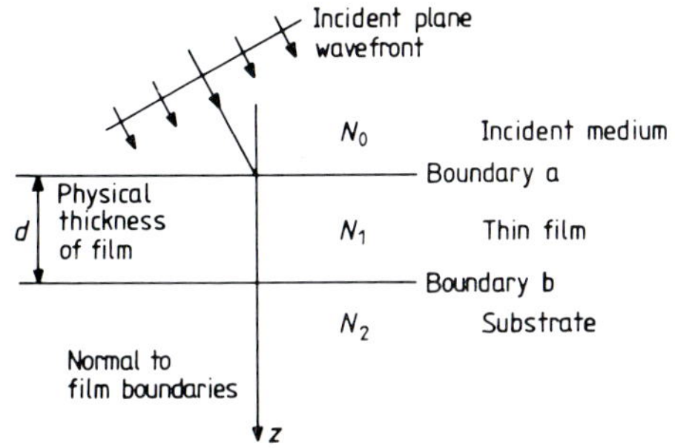

Figure 2.6 Plane wave incident on a thin film.

symbol + (that is, positive-going) and waves in the opposite direction by − (that is, negative-going).

The interface between the film and the substrate, denoted by the symbol b, can be treated in exactly the same way as the simple boundary already discussed. We consider the tangential components of the fields. There is no negative-going wave in the substrate and the waves in the film can be summed into one resultant positive-going wave and one resultant negative-going wave. At this interface, then, the tangential components of **E** and **H** are

$$E_b = E_{1b}^+ + E_{1b}^- \tag{2.48}$$

$$H_b = \eta_1 E_{1b}^+ - \eta_1 E_{1b}^- \tag{2.49}$$

where we are neglecting the common phase factors and where E_b and H_b represent the resultants. Hence

$$E_{1b}^+ = \tfrac{1}{2}(H_b/\eta_1 + E_b) \tag{2.50}$$

$$E_{1b}^- = \tfrac{1}{2}(-H_b/\eta_1 + E_b) \tag{2.51}$$

$$H_{1b}^+ = \eta_1 E_{1b}^+ = \tfrac{1}{2}(H_b + \eta_1 E_b) \tag{2.52}$$

$$H_{1b}^- = -\eta_1 E_{1b}^- = \tfrac{1}{2}(H_b - \eta_1 E_b). \tag{2.53}$$

The fields at the other interface a at the same instant and at a point with identical x and y coordinates can be determined by altering the phase factors of the waves to allow for a shift in the z coordinate from 0 to $-d$. The phase factor of the positive-going wave will be multiplied by exp (iδ) where

$$\delta = 2\pi N_1 d \cos\theta_1/\lambda \tag{2.54}$$

and θ_1 may be complex, while the negative-going phase factor will be multiplied by exp (−iδ). We imply that this is a valid procedure when we say that the film is thin. The values of E and H at the interface are now, using equations (2.50)–(2.53),

$$
\begin{aligned}
E_{1a}^{+} &= E_{1b}^{+}e^{i\delta} = \tfrac{1}{2}(H_b/\eta_1 + E_b)e^{i\delta} \\
E_{1a}^{-} &= E_{1b}^{-}e^{-i\delta} = \tfrac{1}{2}(-H_b/\eta_1 + E_b)e^{-i\delta} \\
H_{1a}^{+} &= H_{1b}^{+}e^{i\delta} = \tfrac{1}{2}(H_b + \eta_1 E_b)e^{i\delta} \\
H_{1a}^{-} &= H_{1b}^{-}e^{-i\delta} = \tfrac{1}{2}(H_b - \eta_1 E_b)e^{-i\delta}
\end{aligned}
$$

so that

$$
\begin{aligned}
E_a &= E_{1a}^{+} + E_{1a}^{-} \\
&= E_b\left(\frac{e^{i\delta}+e^{-i\delta}}{2}\right) + H_b\left(\frac{e^{i\delta}-e^{-i\delta}}{2\eta_1}\right) \\
&= E_b\cos\delta + H_b\frac{i\sin\delta}{\eta_1} \\
H_a &= H_{1a}^{+} + H_{1a}^{-} \\
&= E_b\eta_1\left(\frac{e^{i\delta}-e^{-i\delta}}{2}\right) + H_b\left(\frac{e^{i\delta}+e^{-i\delta}}{2}\right) \\
&= E_b\, i\eta_1\sin\delta + H_b\cos\delta.
\end{aligned}
$$

This can be written in matrix notation as

$$
\begin{bmatrix} E_a \\ H_a \end{bmatrix} = \begin{bmatrix} \cos\delta & (i\sin\delta)/\eta_1 \\ i\eta_1\sin\delta & \cos\delta \end{bmatrix} \begin{bmatrix} E_b \\ H_b \end{bmatrix}. \tag{2.55}
$$

Since the tangential components of ***E*** and ***H*** are continuous across a boundary, and since there is only a positive-going wave in the substrate, this relationship connects the tangential components of ***E*** and ***H*** at the incident interface with the tangential components of ***E*** and ***H*** which are transmitted through the final interface. The 2 × 2 matrix on the right-hand side of equation (2.55) is known as the characteristic matrix of the thin film.

We define the input optical admittance of the assembly by analogy with equation (2.39) as

$$
Y = H_a/E_a \tag{2.56}
$$

when the problem becomes merely that of finding the reflectance of a simple interface between an incident medium of admittance η_0 and a medium of admittance Y, i.e.

$$
\begin{aligned}
\rho &= \frac{\eta_0 - Y}{\eta_0 + Y} \\
R &= \left(\frac{\eta_0 - Y}{\eta_0 + Y}\right)\left(\frac{\eta_0 - Y}{\eta_0 + Y}\right)^{*}.
\end{aligned} \tag{2.57}
$$

Equation (2.55) can be written

$$
E_a\begin{bmatrix} 1 \\ Y \end{bmatrix} = \begin{bmatrix} \cos\delta & (i\sin\delta)/\eta_1 \\ i\eta_1\sin\delta & \cos\delta \end{bmatrix} \begin{bmatrix} 1 \\ \eta_2 \end{bmatrix} E_b \tag{2.58}
$$

which gives

$$Y = \frac{\eta_2 \cos\delta + i\eta_1 \sin\delta}{\cos\delta + i(\eta_2/\eta_1)\sin\delta}.$$

Normally, Y is the parameter which is of interest and the matrix product on the right-hand side of equation (2.58) gives sufficient information for calculating it:

$$\begin{bmatrix} B \\ C \end{bmatrix} = \begin{bmatrix} \cos\delta & (i\sin\delta)/\eta_1 \\ i\eta_1 \sin\delta & \cos\delta \end{bmatrix} \begin{bmatrix} 1 \\ \eta_2 \end{bmatrix} \tag{2.59}$$

where

$$\begin{bmatrix} B \\ C \end{bmatrix}$$

is defined as the characteristic matrix of the assembly. Clearly, $Y = C/B$.

THE REFLECTANCE OF AN ASSEMBLY OF THIN FILMS

Let another film be added to the single film of the previous section so that the final interface is now denoted by c, as shown in figure 2.7. The characteristic matrix of the film nearest the substrate is

$$\begin{bmatrix} \cos\delta_2 & (i\sin\delta_2)/\eta_2 \\ i\eta_2 \sin\delta_2 & \cos\delta_2 \end{bmatrix}$$

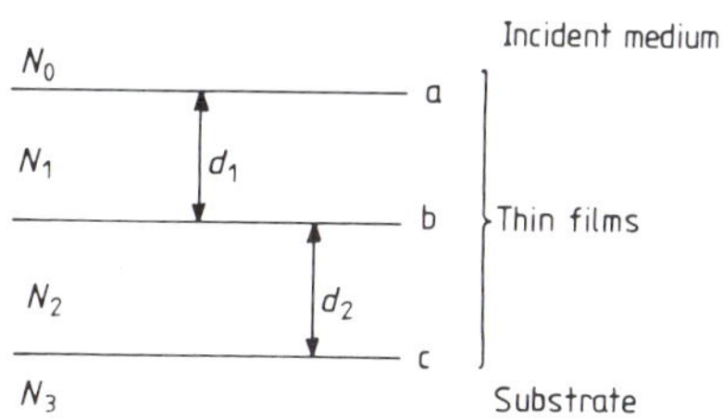

Figure 2.7 Notation for two films on a surface.

and from equation (2.51)

$$\begin{bmatrix} E_b \\ H_b \end{bmatrix} = \begin{bmatrix} \cos\delta_2 & (i\sin\delta_2)/\eta_2 \\ i\eta_2 \sin\delta_2 & \cos\delta_2 \end{bmatrix} \begin{bmatrix} E_c \\ H_c \end{bmatrix}.$$

We can apply equation (2.51) again to give the parameters at interface a, i.e.

$$\begin{bmatrix} E_a \\ H_a \end{bmatrix} = \begin{bmatrix} \cos\delta_1 & (i\sin\delta_1)/\eta_1 \\ i\eta_1 \sin\delta_1 & \cos\delta_1 \end{bmatrix} \begin{bmatrix} \cos\delta_2 & i\sin\delta_2/\eta_2 \\ i\eta_2 \sin\delta_2 & \cos\delta_2 \end{bmatrix} \begin{bmatrix} E_c \\ H_c \end{bmatrix}$$

and the characteristic matrix of the assembly, by analogy with equation (2.59), is

$$\begin{bmatrix} B \\ C \end{bmatrix} = \begin{bmatrix} \cos\delta_1 & (\mathrm{i}\sin\delta_1)/\eta_1 \\ \mathrm{i}\eta_1\sin\delta_1 & \cos\delta_1 \end{bmatrix} \begin{bmatrix} \cos\delta_2 & (\mathrm{i}\sin\delta_2)/\eta_2 \\ \mathrm{i}\eta_2\sin\delta_2 & \cos\delta_2 \end{bmatrix} \begin{bmatrix} 1 \\ \eta_3 \end{bmatrix}.$$

Y is, as before, C/B, and the amplitude reflection coefficient and the reflectance are, from equations (2.57),

$$\rho = \frac{\eta_0 - Y}{\eta_0 + Y}$$

$$R = \left(\frac{\eta_0 - Y}{\eta_0 + Y}\right)\left(\frac{\eta_0 - Y}{\eta_0 + Y}\right)^*.$$

This result can be immediately extended to the general case of an assembly of q layers, when the characteristic matrix is simply the product of the individual matrices taken in the correct order, i.e.

$$\begin{bmatrix} B \\ C \end{bmatrix} = \left(\prod_{r=1}^{q} \begin{bmatrix} \cos\delta_r & (\mathrm{i}\sin\delta_r)/\eta_r \\ \mathrm{i}\eta_r\sin\delta_r & \cos\delta_r \end{bmatrix}\right) \begin{bmatrix} 1 \\ \eta_\mathrm{m} \end{bmatrix} \qquad (2.60)$$

where

$$\delta_r = \frac{2\pi \mathrm{N}_r \mathrm{d}_r \cos\theta_r}{\lambda}$$

$$\eta_r = \mathcal{Y} N_r \cos\theta_r \qquad \text{for s-polarisation (TE)}$$

$$\eta_r = \mathcal{Y} N_r/\cos\theta_r \qquad \text{for p-polarisation (TM)}$$

and where we have now used the suffix m to denote the substrate or exit medium.

$$\eta_\mathrm{m} = \mathcal{Y} N_\mathrm{m} \cos\theta_\mathrm{m} \qquad \text{for s-polarisation (TE)}$$

$$\eta_\mathrm{m} = \frac{\mathcal{Y} N_\mathrm{m}}{\cos\theta_\mathrm{m}} \qquad \text{for p-polarisation (TM)}.$$

If θ_0, the angle of incidence, is given, the values of θ_r can be found from Snell's law, i.e.

$$N_0 \sin\theta_0 = N_r \sin\theta_r = N_\mathrm{m} \sin\theta_\mathrm{m}.$$

The expression (2.60) is of prime importance in optical thin-film work and forms the basis of almost all calculations.

A useful property of the characteristic matrix of a thin film is that the determinant is unity. This means that the determinant of the product of any number of these matrices is also unity.

It avoids difficulties over signs and quadrants if, in the case of absorbing media, the scheme used for computing phase thicknesses and admittances is:

$$\delta_r = (2\pi/\lambda) d_r (n_r^2 - k_r^2 - n_0^2 \sin^2\theta_0 - 2\mathrm{i} n_r k_r)^{1/2}$$

the correct solution being in the fourth quadrant. Then

$$\eta_{rs} = \mathscr{Y}\,(n_r^2 - k_r^2 - n_0^2 \sin^2\theta_0 - 2\mathrm{i}n_r k_r)^{1/2}$$

and

$$\eta_{rp} = \frac{y_r^2}{\eta_{rs}} = \frac{\mathscr{Y}^2 (n_r - \mathrm{i}k_r)^2}{\eta_{rs}}.$$

It is useful to examine the phase shift associated with the reflected beam. Let $Y = a + \mathrm{i}b$. Then with η_0 real

$$\rho = \frac{\eta_0 - a - \mathrm{i}b}{\eta_0 + a + \mathrm{i}b}$$

$$= \frac{(\eta_0^2 - a^2 - b^2) - \mathrm{i}(2b\eta_0)}{(\eta_0 + a)^2 + b^2}$$

i.e.

$$\tan\phi = \frac{(-2b\eta_0)}{(\eta_0^2 - a^2 - b^2)} \tag{2.61}$$

where ϕ is the phase shift. This must be interpreted, of course, on the basis of the sign convention we have already established in figure 2.3. It is important to preserve the signs of the numerator and denominator separately as shown, otherwise the quadrant cannot be uniquely specified. The rule is simple. It is the quadrant in which the vector associated with ρ lies and the following scheme can be derived by treating the denominator as the x coordinate and the numerator as the y coordinate.

Numerator	+	+	−	−
Denominator	+	−	+	−
Quadrant	1st	2nd	4th	3rd

REFLECTANCE, TRANSMITTANCE AND ABSORPTANCE

Sufficient information is included in equation (2.60) to allow the transmittance and absorptance of a thin-film assembly to be calculated. For this to have a physical meaning, as we have already seen, the incident medium should be transparent, that is, η_0 must be real. The substrate need not be transparent, but the transmittance calculated will be the transmittance into, rather than through, the substrate.

First of all, we calculate the net intensity at the exit side of the assembly, which we take as the kth interface. This is given by

$$I_k = \tfrac{1}{2}\,\mathrm{Re}\,(E_k H_k^*)$$

where, once again, we are dealing with the component of intensity normal to the interfaces.

$$\begin{aligned} I_k &= \tfrac{1}{2}\,\mathrm{Re}(E_k \eta_m^* E_k^*) \\ &= \tfrac{1}{2}\,\mathrm{Re}(\eta_m) E_k E_k^*. \end{aligned} \tag{2.62}$$

If the characteristic matrix of the assembly is

$$\begin{bmatrix} B \\ C \end{bmatrix}$$

then the net intensity at the entrance to the assembly is

$$I_a = \tfrac{1}{2}\,\mathrm{Re}\,(BC^*) E_k E_k^*. \tag{2.63}$$

Let the incident intensity be denoted by I_i; then equation (2.63) represents the intensity actually entering the assembly, which is $(1-R)I_i$:

$$(1-R)I_i = \tfrac{1}{2}\,\mathrm{Re}\,(BC^*) E_k E_k^*$$

i.e.

$$I_i = \frac{\mathrm{Re}\,(BC^*) E_k E_k^*}{2(1-R)}.$$

Equation (2.62) represents the intensity leaving the assembly and entering the substrate and so the transmittance T is

$$T = \frac{I_k}{I_i} = \frac{\mathrm{Re}\,(\eta_m)(1-R)}{\mathrm{Re}\,(BC^*)}. \tag{2.64}$$

The absorptance A in the multilayer is connected with R and T by the relationship

$$1 = R + T + A$$

so that

$$A = 1 - R - T = (1-R)\left(1 - \frac{\mathrm{Re}\,(\eta_m)}{\mathrm{Re}\,(BC^*)}\right). \tag{2.65}$$

In the absence of absorption in any of the layers it can readily be shown that the above expressions are consistent with $A = 0$ and $T + R = 1$, for the individual film matrices will have determinants of unity and the product of any number of these matrices will also have a determinant of unity. The product of the matrices can be expressed as

$$\begin{bmatrix} \alpha & \mathrm{i}\beta \\ \mathrm{i}\gamma & \delta \end{bmatrix}$$

where $\alpha\delta + \gamma\beta = 1$ and, because there is no absorption, α, β, γ and δ are all real.

$$\begin{bmatrix} B \\ C \end{bmatrix} = \begin{bmatrix} \alpha & \mathrm{i}\beta \\ \mathrm{i}\gamma & \delta \end{bmatrix} \begin{bmatrix} 1 \\ \eta_m \end{bmatrix} = \begin{bmatrix} \alpha + \mathrm{i}\beta\eta_m \\ \delta\eta_m + \mathrm{i}\gamma \end{bmatrix}$$

$$\begin{aligned} \mathrm{Re}\,(BC^*) &= \mathrm{Re}\,[(\alpha + \mathrm{i}\beta\eta_m)(\delta\eta_m^* - \mathrm{i}\gamma)] = (\alpha\delta + \gamma\beta)\,\mathrm{Re}\,(\eta_m) \\ &= \mathrm{Re}\,(\eta_m) \end{aligned}$$

and the result follows.

We can manipulate equations (2.64) and (2.65) into slightly better forms. From equation (2.57)

$$R = \left(\frac{\eta_0 B - C}{\eta_0 B + C}\right)\left(\frac{\eta_0 B - C}{\eta_0 B + C}\right)^* \tag{2.66}$$

so that

$$(1 - R) = \frac{2\eta_0\,(BC^* + B^*C)}{(\eta_0 B + C)(\eta_0 B + C)^*}.$$

Inserting this result in equation (2.64) we obtain

$$T = \frac{4\eta_0\,\mathrm{Re}\,(\eta_m)}{(\eta_0 B + C)(\eta_0 B + C)^*} \tag{2.67}$$

and in equation (2.65)

$$A = \frac{4\eta_0\,\mathrm{Re}\,(BC^* - \eta_m)}{(\eta_0 B + C)(\eta_0 B + C)^*}. \tag{2.68}$$

Equations (2.66), (2.67) and (2.68) are the most useful forms of the expressions for R, T and A.

An important quantity which we shall discuss in a later section of this chapter is $T/(1-R)$, known as the potential transmittance ψ. From equation (2.64)

$$\psi = \frac{T}{(1-R)} = \frac{\mathrm{Re}\,(\eta_m)}{\mathrm{Re}\,(BC^*)}. \tag{2.69}$$

The phase change on reflection (equation (2.61)) can also be put in a form compatible with equations (2.66)–(2.69).

$$\phi = \tan^{-1}\left(\frac{\mathrm{i}\eta_m\,(CB^* - BC^*)}{(\eta_m^2 BB^* - CC^*)}\right).$$

The quadrant of ϕ is given by the same scheme of signs of numerator and denominator as equation (2.61).

UNITS

We have been using the International System of Units (SI) in the work so far. In this system η and Y are measured in siemens. The majority of thin-film literature to date has been written in gaussian units. In gaussian units, $\mathscr{Y}$ is unity and so, since $\eta = N\mathscr{Y}$, η (the optical admittance) and N (the refractive index) are numerically equal at normal incidence, although N is a number without units. The position is different in SI units, where $\mathscr{Y}$ is 2.6544×10^{-3} S. We could, if we choose, measure η in units of $\mathscr{Y}$ siemen, which we can call free space units, and in this case η at normal incidence becomes numerically equal to N, just as in the gaussian system. This is a perfectly valid procedure, and all the expressions for ratioed quantities, notably reflectance, transmittance, absorptance and potential transmittance, are unchanged. We must simply take due care when calculating absolute rather than relative intensity and also when deriving the magnetic field. In particular, equation (2.55) becomes

$$\begin{bmatrix} E_a \\ H_a/\mathscr{Y} \end{bmatrix} = \begin{bmatrix} \cos\delta & (\mathrm{i}\sin\delta)/\eta \\ \mathrm{i}\eta\sin\delta & \cos\delta \end{bmatrix} \begin{bmatrix} E_b \\ H_b/\mathscr{Y} \end{bmatrix}$$

where η is now measured in free space units. In most cases in this book, either scheme can be used. In some cases, particularly where we are using graphical techniques, we shall use free space units, because otherwise the scales become quite cumbersome.

SUMMARY OF IMPORTANT RESULTS

We have now covered all the basic theory necessary for the understanding of the remainder of the book. It has been a somewhat long and involved discussion and so we now summarise the principal results. The statement numbers refer to those in the text where the particular quantities were originally introduced.

Refractive index is defined as the ratio of the velocity of light in free space c to the velocity of light in the medium v. It is denoted by N and is often complex.

$$N = c/v = n - \mathrm{i}k. \tag{2.15}$$

N is often called the complex refractive index, n the real refractive index (or often simply refractive index) and k the extinction coefficient. N is always a function of λ.

k is related to the absorption coefficient α by

$$\alpha = 4\pi k/\lambda. \tag{2.27}$$

Light waves are electromagnetic and a homogeneous, plane, plane-polarised harmonic (or monochromatic) wave may be represented by expressions of the form

$$\boldsymbol{E} = \boldsymbol{\mathscr{E}} \exp\{\mathrm{i}[\omega t - (2\pi N/\lambda)x + \phi]\} \tag{2.18}$$

where x is the distance along the direction of propagation, $\boldsymbol{E}$ is the electric field,

$\mathscr{E}$ the electric amplitude and ϕ an arbitrary phase. A similar expression holds for $\boldsymbol{H}$, the magnetic field:

$$\boldsymbol{H} = \mathscr{H} \exp\{\mathrm{i}[\omega t - (2\pi N/\lambda)x + \phi']\}$$

where ϕ, ϕ' and N are not independent. The physical significance must be attached to the real parts of the above expressions.

The phase change suffered by the wave on traversing a distance d of the medium is, therefore,

$$-\frac{2\pi N d}{\lambda} = -\frac{2\pi n d}{\lambda} + \mathrm{i}\frac{2\pi k d}{\lambda}$$

and the imaginary part can be interpreted as a reduction in amplitude (by substituting in equation (2.18)).

The optical admittance is defined as the ratio of the magnetic and electric fields

$$y = H/E \tag{2.21)–(2.24}$$

and y is usually complex. In free space, y is real and is denoted by $\mathscr{Y}$

$$\mathscr{Y} = 2.6544 \times 10^{-3}\ \mathrm{S}.$$

The optical admittance of a medium is connected with the refractive index by

$$y = N\mathscr{Y}. \tag{2.23}$$

(In gaussian units $\mathscr{Y}$ is unity and y and N are numerically the same. In SI units we can make y and N numerically equal by expressing y in units of $\mathscr{Y}$, i.e. free space units. All expressions for reflectance, transmittance etc involving ratios will remain valid, but care must be taken when computing absolute intensities, although these are not often needed in thin-film optics, except where damage studies are involved.)

The intensity of the light, defined as the mean rate of flow of energy per unit area carried by the wave, is given by

$$I = \tfrac{1}{2}\,\mathrm{Re}\,(EH^*). \tag{2.25}$$

This can also be written

$$I = \tfrac{1}{2} n \mathscr{Y} EE^*$$

where * denotes complex conjugate.

At a boundary between two media, denoted by suffix 0 for the incident medium and by suffix 1 for the exit medium, the incident beam is split into a reflected beam and a transmitted beam. For *normal incidence* we have

$$\rho = \text{amplitude reflection coefficient} = \frac{y_0 - y_1}{y_0 + y_1} = \frac{N_0 - N_1}{N_0 + N_1} \tag{2.41}$$

$$\tau = \text{amplitude transmission coefficient} = \frac{2y_0}{y_0 + y_1} = \frac{2N_0}{N_0 + N_1}.$$

There are fundamental difficulties associated with the definitions of reflectance and transmittance unless the incident medium is absorption-free, i.e. N_0 and y_0 are real. For that case:

$$R=\rho\rho^*=\left(\frac{y_0-y_1}{y_0+y_1}\right)\left(\frac{y_0-y_1}{y_0+y_1}\right)^*=\left(\frac{N_0-N_1}{N_0+N_1}\right)\left(\frac{N_0-N_1}{N_0+N_1}\right)^*.$$

$$T=\frac{n_1}{n_0}\tau\tau^*=\frac{4\,\mathrm{Re}\,(y_0)\,\mathrm{Re}\,(y_1)}{(y_0+y_1)\,(y_0+y_1)^*}=\frac{4n_0n_1}{(N_0+N_1)\,(N_0+N_1)^*}. \tag{2.42}$$

Oblique incidence calculations are simpler if the wave is split into two plane-polarised components, one with the electric vector in the plane of incidence, known as p-polarised (or TM, for transverse magnetic field) and one with the electric vector normal to the plane of incidence, known as s-polarised (or TE, for transverse electric field). The propagation of each of these two waves can be treated quite independently of the other. Calculations are further simplified if only energy flows normal to the boundaries and electric and magnetic fields parallel to the boundaries are considered, because then we have a formulation which is equivalent to a homogeneous wave.

We must introduce the idea of a tilted optical admittance η, which is given by

$$\eta_p=\frac{N\mathscr{Y}}{\cos\theta} \qquad \text{(for p-waves)}$$

$$\eta_s=N\mathscr{Y}\cos\theta \qquad \text{(for s-waves)} \tag{2.46}$$

where N and θ denote either N_0, θ_0 or N_1, θ_1 as appropriate. θ_1 is given by Snell's law, in which complex angles may be included

$$N_0\sin\theta_0=N_1\sin\theta_1.$$

Denoting η_p or η_s by η we have, for either plane of polarisation,

$$\rho=\frac{\eta_0-\eta_1}{\eta_0+\eta_1}$$

$$\tau=\frac{2\eta_0}{\eta_0+\eta_1} \tag{2.47}$$

and, if η_0 is real, we can write

$$R=\left(\frac{\eta_0-\eta_1}{\eta_0+\eta_1}\right)\left(\frac{\eta_0-\eta_1}{\eta_0+\eta_1}\right)^*$$

$$T=\frac{4\,\mathrm{Re}\,(\eta_0)\,\mathrm{Re}(\eta_1)}{(\eta_0+\eta_1)(\eta_0+\eta_1)^*}. \tag{2.47}$$

The phase shift experienced by the wave as it traverses a distance d normal to the boundary is then given by $-2\pi Nd\cos\theta/\lambda$.

The reflectance of an assembly of thin films is calculated through the concept of optical admittance. We replace the multilayer by a single surface which has an input optical admittance Y which is given by

$$Y = C/B$$

where

$$\begin{bmatrix} B \\ C \end{bmatrix} = \left(\prod_{r=1}^{q} \begin{bmatrix} \cos\delta_r & (\mathrm{i}\sin\delta_r)/\eta_r \\ \mathrm{i}\eta_r \sin\delta_r & \cos\delta_r \end{bmatrix} \right) \begin{bmatrix} 1 \\ \eta_m \end{bmatrix}. \tag{2.60}$$

$\delta_r = 2\pi N_r d_r \cos\theta_r/\lambda$ and η_m = substrate admittance.
The order of multiplication is important. If q is the layer next to the substrate then the order is

$$\begin{bmatrix} B \\ C \end{bmatrix} = [M_1][M_2]\ldots[M_q] \begin{bmatrix} 1 \\ \eta_m \end{bmatrix}.$$

M_1 indicates the matrix associated with layer 1, and so on. Y and η are in the same units. If η is in siemens then so is Y, or if η is in free space units (i.e. units of $\mathscr{Y}$) then Y will be in free space units also. As in the case of a single surface, η_0 must be real for reflectance and transmittance to have a valid meaning. With that proviso, then

$$R = \left(\frac{\eta_0 B - C}{\eta_0 B + C} \right) \left(\frac{\eta_0 B - C}{\eta_0 B + C} \right)^* \tag{2.66}$$

$$T = \frac{4\eta_0 \operatorname{Re}(\eta_m)}{(\eta_0 B + C)(\eta_0 B + C)^*} \tag{2.67}$$

$$A = \frac{4\eta_0 \operatorname{Re}(BC^* - \eta_m)}{(\eta_0 B + C)(\eta_0 B + C)^*} \tag{2.68}$$

$$\psi = \text{potential transmittance} = \frac{\operatorname{Re}(\eta_m)}{\operatorname{Re}(BC^*)}. \tag{2.69}$$

POTENTIAL TRANSMITTANCE

The potential transmittance of a layer or an assembly of layers is the ratio of the intensity leaving by the rear, or exit, interface to that entering by the front interface. The concept was introduced by Berning and Turner[3] and we will make considerable use of it in designing metal–dielectric filters and in calculating losses in all-dielectric multilayers. Potential transmittance is denoted by ψ and is given by

$$\psi = \frac{I_e}{I_i}$$

where the symbols are defined in figure 2.8. The potential transmittance of a

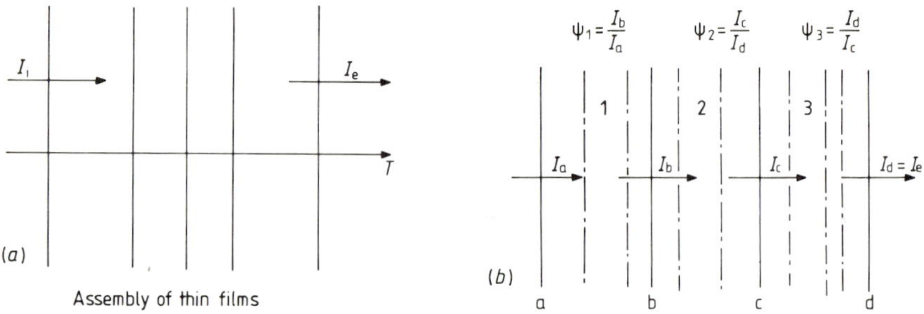

Figure 2.8 (*a*) An assembly of thin films. (*b*) The potential transmittance of an assembly of thin film consisting of a number of subunits.

series of subassemblies of layers is simply the product of the individual potential transmittances. Again, with reference to figure 2.8,

$$\psi = \frac{I_e}{I_i} = \frac{I_b}{I_i}\frac{I_c}{I_b}\frac{I_e}{I_c} = \psi_1\psi_2\psi_3.$$

The potential transmittance is fixed by the parameters of the layer, or combination of layers, involved, and by the characteristics of the structure at the exit interface, and it represents the transmittance which this particular combination would give if there were no reflection losses. Thus it is a measure of the maximum transmittance which could be expected from the arrangement. By definition, the potential transmittance is unaffected by any transparent structure deposited over the front surface—which can affect the transmittance as distinct from the potential transmittance—and to ensure that the transmittance is equal to the potential transmittance the layers added to the front surface must maximise the intensity actually entering the assembly. This implies reducing the reflectance of the complete assembly to zero or, in other words, adding an antireflection coating. The potential transmittance is, however, affected by any changes in the structure at the exit interface and it is possible to maximise the potential transmittance of a subassembly in this way.

We now show that the parameters of the layer, or subassembly of layers, together with the optical admittance at the rear surface, are sufficient to define the potential transmittance. Let the complete multilayer performance be given by

$$\begin{bmatrix} B \\ C \end{bmatrix} = [M_1][M_2]\ldots[M_a][M_b][M_c]\ldots[M_p][M_q]\begin{bmatrix} 1 \\ \eta_m \end{bmatrix}$$

where we want to calculate the potential transmittance of the subassembly $[M_a][M_b][M_c]$. Let the product of the matrices to the right of the subassembly be given by

$$\begin{bmatrix} B_e \\ C_e \end{bmatrix}.$$

Now, if

$$\begin{bmatrix} B_i \\ C_i \end{bmatrix} = [M_a][M_b][M_c]\begin{bmatrix} B_e \\ C_e \end{bmatrix} \tag{2.70}$$

then

$$\psi = \frac{\mathrm{Re}(B_e C_e^*)}{\mathrm{Re}(B_i C_i^*)}. \tag{2.71}$$

By dividing equation (2.70) by B_e we have

$$\begin{bmatrix} B_i' \\ C_i' \end{bmatrix} = [M_a][M_b][M_c]\begin{bmatrix} 1 \\ Y_e \end{bmatrix}$$

where $Y_e = C_e/B_e$, $B_i' = B_i/B_e$, $C_i' = C_i/B_e$ and the potential transmittance is

$$\begin{aligned} \psi &= \frac{\mathrm{Re}(Y_e)}{\mathrm{Re}(B_i' C_i'^*)} \\ &= \frac{\mathrm{Re}(C_e/B_e)}{\mathrm{Re}[(B_i/B_e)(C_i^*/B_e^*)]} = \frac{B_e B_e^* \,\mathrm{Re}(C_e/B_e)}{\mathrm{Re}(B_i C_i^*)} \\ &= \frac{\mathrm{Re}(B_e^* C_e)}{\mathrm{Re}(B_i C_i^*)} = \frac{\mathrm{Re}(B_e C_e^*)}{\mathrm{Re}(B_i C_i^*)} \end{aligned}$$

which is identical with equation (2.71). Thus the potential transmittance of any subassembly is determined solely by the characteristics of the layer or layers of the subassembly together with the optical admittance of the structure at the exit interface.

Further expressions involving potential transmittance will be derived as they are required.

FURTHER COMMENTS ON EXPRESSION (2.60)

In spite of the apparent simplicity of expression (2.60), numerical calculations without some automatic aid are tedious in the extreme. Even with the help of a calculator, the labour involved in determining the performance of an assembly of more than a very few transparent layers at one or two wavelengths is completely discouraging. At the very least, a programmable calculator of reasonable capacity is required. Extended calculations are usually carried out on a computer.

However, insight into the properties of thin-film assemblies cannot easily be gained simply by feeding the calculations into a computer, and insight is necessary if filters are to be designed and if their limitations in use are to be fully understood. Studies have been made of the properties of the characteristic matrices and some results which are particularly helpful in this context have been obtained. Approximate methods, especially graphical ones, have also been found useful.

Quarter- and half-wave optical thicknesses

The characteristic matrix of a dielectric thin film takes on a very simple form if the optical thickness is an integral number of quarter or half waves. That is, if

$$\delta = n(\pi/4) \qquad n = 0, 1, 2, 3 \ldots$$

For n even, $\cos\delta = \pm 1$ and $\sin\delta = 0$, so that the layer is an integral number of half wavelengths thick, and the matrix becomes

$$\pm\begin{bmatrix} 1 & 0 \\ 0 & 1 \end{bmatrix}.$$

This is the unity matrix and can have no effect on the reflectance or transmittance of an assembly. It is as if the layer were completely absent. This is a particularly useful result and, because of it, half-wave layers are sometimes referred to as absentee layers. In the computation of the properties of any assembly, layers which are an integral number of half wavelengths thick can be omitted completely without altering the result.

For n odd, $\sin\delta = \pm 1$ and $\cos\delta = 0$, so that the layer is an odd number of quarter wavelengths thick, and the matrix becomes

$$\pm\begin{bmatrix} 0 & \mathrm{i}/\eta \\ \mathrm{i}\eta & 0 \end{bmatrix}.$$

This is not quite as simple as the half-wave case, but such a matrix is still easy to handle in calculations. In particular, if a substrate or combination of thin films has an admittance of Y, then addition of an odd number of quarter-waves of admittance η alters the admittance of the assembly to η^2/Y. This makes the properties of a succession of quarter-wave layers very easy to calculate. The admittance of, say, a stack of five quarter-wave layers is

$$Y = \frac{\eta_1^2}{\eta_2^2}\frac{\eta_3^2}{\eta_4^2}\frac{\eta_5^2}{\eta_\mathrm{m}}$$

where the symbols have their usual meanings.

Because of the simplicity of assemblies involving quarter- and half-wave optical thicknesses, designs are often specified in terms of fractions of quarter-waves at a reference wavelength. Usually only two, or perhaps three, different materials are involved in designs and a convenient shorthand notation for quarter-wave optical thicknesses is H, M or L where H refers to the highest of the three indices, M the intermediate and L the lowest. Half-waves are denoted by HH, MM or LL and so on.

A theorem on the transmission of a thin-film assembly

The transmittance of a thin-film assembly is independent of the direction of propagation of the light. This applies regardless of whether or not the layers are absorbing.

A proof of this result, due to Abelès[4], who was responsible for the development of the matrix approach to the analysis of thin films, follows quickly from the properties of the matrices.

Let the matrices of the various layers in the assembly be denoted by

$$[M_1], [M_2], \ldots, [M_q]$$

and let the two products of these corresponding to the two possible directions of propagation be

$$[M] = [M_1][M_2][M_3]\ldots[M_q]$$

and

$$[M'] = [M_q][M_{q-1}]\ldots[M_2][M_1].$$

Now, because the form of the matrices is such that the diagonal terms are equal, regardless of whether there is absorption or not, we can show that if

$$[M] = [a_{ij}] \qquad \text{and} \qquad [M'] = [a'_{ij}]$$

then

$$a_{ij} = a'_{ij}(i \neq j),\ a_{11} = a'_{22} \text{ and } a_{22} = a'_{11}.$$

This can be proved simply by induction.

We denote the medium on one side of the assembly by η_0 and on the other by η_m, where η_0 is next to layer 1. In the case of the first direction the characteristic matrix is given by (equation (2.60))

$$\begin{bmatrix} B \\ C \end{bmatrix} = [M]\begin{bmatrix} 1 \\ \eta_m \end{bmatrix}$$

and

$$B = a_{11} + a_{12}\eta_m \qquad C = a_{21} + a_{22}\eta_m.$$

In the second case

$$B = a'_{11} + a'_{12}\eta_0 = a_{22} + a_{12}\eta_0$$
$$C = a'_{21} + a'_{22}\eta_0 = a_{21} + a_{11}\eta_0.$$

The two expressions for the transmittance of the assembly are then, from equation (2.67),

$$T = \frac{4\,\mathrm{Re}(\eta_0)\mathrm{Re}(\eta_m)}{|\eta_0(a_{11} + a_{12}\eta_m) + a_{21} + a_{22}\eta_m|^2}$$

$$T' = \frac{4\,\mathrm{Re}(\eta_m)\mathrm{Re}(\eta_0)}{|\eta_m(a_{22} + a_{12}\eta_0) + a_{21} + a_{11}\eta_0|^2}$$

which are identical.

This rule does not, of course, apply to the reflectance of an assembly, which will necessarily be the same on both sides of the assembly only if there is no absorption in any of the layers.

Amongst other things, this expression shows that the one-way mirror, which allows light to travel through it in one direction only, cannot be constructed from simple optical thin films. The common so-called one-way mirror has a high reflectance with some transmittance and relies for its operation on an appreciable difference in the illumination conditions existing on either side.

The Herpin index

An extremely important result for filter design is derived in chapter 6, which deals with edge filters. Briefly, this is the fact that any symmetrical product of three thin-film matrices can be replaced by a single matrix which has the same form as that of a single film and which therefore possesses an equivalent thickness and an equivalent optical admittance. Of course, this is a mathematical device rather than a case of true physical equivalence, but the result is of considerable use in giving an insight into the properties of a great number of filter designs which can be split into a series of symmetrical combinations. The method also allows the replacement, under certain conditions, of a layer of intermediate index by a symmetrical combination of high and low index material. This is especially useful in the design of antireflection coatings, which frequently require quarter-wave thicknesses of unobtainable intermediate indices. These difficult layers can be replaced by symmetrical combinations of existing materials with the additional advantage of limiting the total number of materials required for the structure.

The equivalent admittance is frequently known as the Herpin index, after the originator, and the symmetrical combination as an Epstein period, after the author of two of the most important early papers dealing with the application of the result to the design of filters.

The detailed derivation of the relevant formulae is left until chapter 6, which will make considerable use of the concept.

THE VECTOR METHOD

The vector method is a valuable technique, especially in design work associated with antireflection coatings. Two assumptions are involved: first, that there is no absorption in the layers, and second, that the behaviour of a multilayer can be determined by considering one reflection of the incident wave at each interface only. The errors involved in using this method can, in some cases, be significant, especially where high overall reflectance from the multilayer exists, but they are small in most types of antireflection coating.

Consider the assembly sketched in figure 2.9. If there is no absorption in the layers, then $N_r = n_r$ and $k_r = 0$. The amplitude reflection coefficient at each interface is given by

$$\rho = \frac{n_{r-1} - n_r}{n_{r-1} + n_r} \tag{2.72}$$

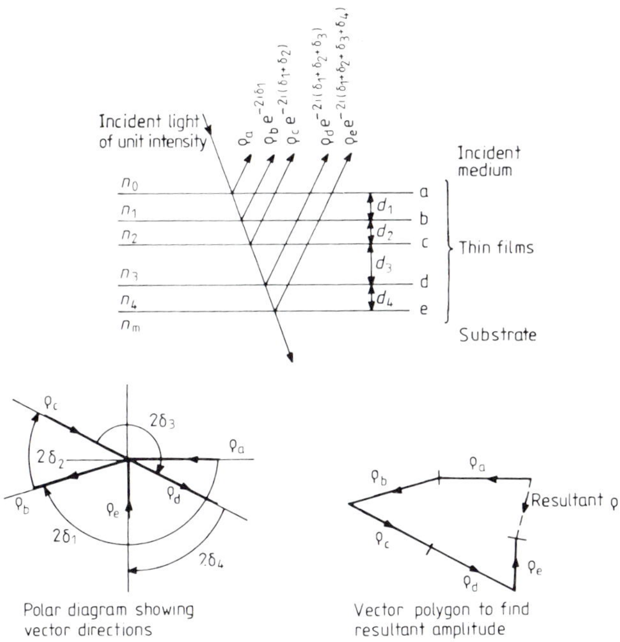

Figure 2.9 The vector method. The lengths of the vectors and the phase angles are given by:

$$\rho_a = (n_0 - n_1)/(n_0 + n_1) \qquad \delta_1 = 2\pi n_1 d_1/\lambda$$

$$\rho_b = (n_1 - n_2)/(n_1 + n_2) \qquad \delta_2 = 2\pi n_2 d_2/\lambda$$

etc. Note that the sign of the expression for the vector lengths is important and must be included. In the diagram ρ_a, ρ_c and ρ_e are shown as of negative sign. Note also that the angles between successive vectors are phase *lags*, so that the sense of all the angles in the polar diagram, $2\delta_1$, $2\delta_2$, etc, is also negative.

which may be positive or negative depending on the relative magnitudes of n_{r-1} and n_r.

The phase thicknesses of the layers are given by $\delta_1, \delta_2, \ldots$, where

$$\delta_r = 2\pi n_r \cos\theta_r d_r/\lambda.$$

A quarter-wave optical thickness is represented by 90° and a half-wave by 180°.

As the diagram shows, the resultant amplitude reflection coefficient is given by the vector sum of the coefficients for each interface, where each is associated with the appropriate phase lag corresponding to the passage of the wave from the front surface to the interface and back to the front surface again.

$$\rho = \rho_a + \rho_b \exp(-2i\delta_1) + \rho_c \exp[-2i(\delta_1+\delta_2)] + \rho_d \exp[-2i(\delta_1+\delta_2+\delta_3)] + \ldots \qquad (2.73)$$

The sum can be found analytically, or, as is more usual, graphically. The graphical case is easier because the angles between successive vectors are merely $2\delta_1$, $2\delta_2$, $2\delta_3$, and so on.

The calculation of the angles for any wavelength is simplified if, as is usual, the optical thicknesses of the layers are given in terms of quarter-wave optical thicknesses at a reference wavelength λ_0. If the optical thickness of the rth layer is t_r quarter-waves at λ_0, then the value of δ_r at λ is just $\delta_r = (90t_r\lambda_0/\lambda)$ degrees of arc.

In practice it will be found extremely easy to confuse angles and directions, particularly where negative reflectances are involved. The task of drawing the vector diagram is greatly eased by plotting first the vectors with directions on a polar diagram and then transferring the vectors to a vector polygon rather than attempting to draw the vector polygon straight away. An important point to remember is that the resultant vector represents the amplitude reflection coefficient and its length must be squared in order to give the reflectance.

A typical arrangement is shown in Figure 2.9. The vector method is used to a considerable extent in chapter 3, which deals with antireflection coatings.

ALTERNATIVE METHOD OF CALCULATION

The success of the vector method prompts one to ask whether it can be made more accurate by considering second and subsequent reflections at the various boundaries instead of just one. In fact, an alternative solution of the thin-film problem can be obtained in this way. It is simpler to consider normal incidence only. The expressions can be adapted for non-normal incidence quite simply when the materials are transparent and with some difficulty when they are absorbing. We consider first the case of a single film. Figure 2.10 defines the various parameters.

The resultant amplitude reflection coefficient is given by

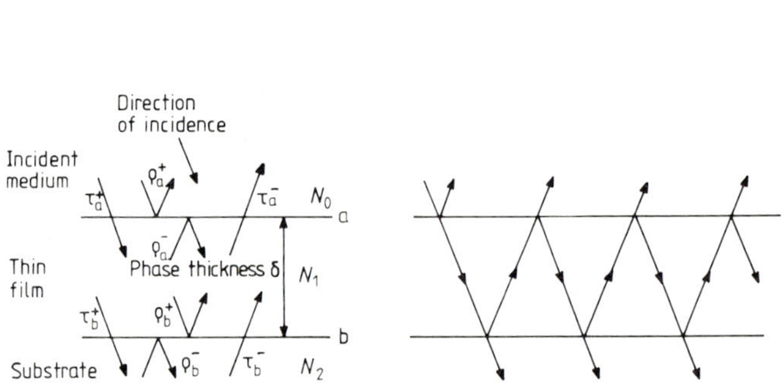

Figure 2.10

$$\rho^+ = \rho_a^+ + \tau_a^+ \rho_b^+ \tau_a^- e^{-2i\delta} + \tau_a^+ \rho_b^+ \rho_a^- \rho_b^+ \tau_a^- e^{-4i\delta}$$
$$+ \tau_a^+ \rho_b^+ \rho_a^- \rho_b^+ \rho_a^- \rho_b^+ \tau_a^- e^{-6i\delta}$$
$$= \rho_a^+ + \frac{\rho_b^+ \tau_a^+ \tau_a^- e^{-2i\delta}}{1 - \rho_b^+ \rho_a^- e^{-2i\delta}}. \tag{2.74}$$

However

$$\tau_a^+ \tau_a^- = \frac{4N_0 N_1}{(N_0 + N_1)^2} = 1 - \rho_a^{+2}$$

and $\rho_a^- = -\rho_a^+$ so that

$$\rho^+ = \frac{\rho_a^+ + \rho_b^+ e^{-2i\delta}}{1 + \rho_b^+ \rho_a^+ e^{-2i\delta}}. \tag{2.75}$$

Similarly,

$$\tau^+ = \tau_a^+ \tau_b^+ e^{-i\delta} + \tau_a^+ \rho_b^+ \rho_a^- \tau_b^+ e^{-3i\delta} + \tau_a^+ \rho_b^+ \rho_a^- \rho_b^+ \rho_a^- \tau_b^+ e^{-5i\delta} + \ldots$$

which reduces to

$$\tau^+ = \frac{\tau_a^+ \tau_b^+ e^{-i\delta}}{1 - \rho_a^- \rho_b^+ e^{-2i\delta}} \tag{2.76}$$

$$= \frac{\tau_a^+ \tau_b^+}{1 + \rho_a^+ \rho_b^+ e^{-2i\delta}}. \tag{2.77}$$

These expressions can be used in calculations of assemblies of more than one film by applying them successively, first to the final two interfaces which can then be replaced by a single interface with the resultant coefficients, and then to this equivalent interface and the third last interface, and so on.

The resultant amplitude transmission and reflection coefficients τ^+ and ρ^+ can be converted into transmittance and reflectance using the expressions

$$R = (\rho^+)(\rho^+)^*$$
$$T = \frac{n_2}{n_0}(\tau^+)(\tau^+)^*.$$

n_2 and n_0 are the refractive indices of the substrate, or exit medium, and the incident medium respectively. For these expressions to be meaningful we must, as before, restrict the incident medium to be transparent so that $N_0 = n_0$. No such restriction applies to the exit medium which can have complex $N_2 = n_2 - ik_2$, the real part being used in the above expression for T.

It is also possible to develop a matrix approach along these lines. The electric field vectors E_0^+ and E_0^- in medium 0 at interface a can be expressed in terms of E_1^+ and E_1^- in film 1 at interface b (see figure 2.11).

$$\begin{bmatrix} E_0^+ \\ E_0^- \end{bmatrix} = \frac{1}{\tau_a^+} \begin{bmatrix} e^{i\delta_1} & \rho_a^+ e^{-i\delta_1} \\ \rho_a^+ e^{i\delta_1} & e^{-i\delta_1} \end{bmatrix} \begin{bmatrix} E_1^+ \\ E_1^- \end{bmatrix}. \tag{2.78}$$

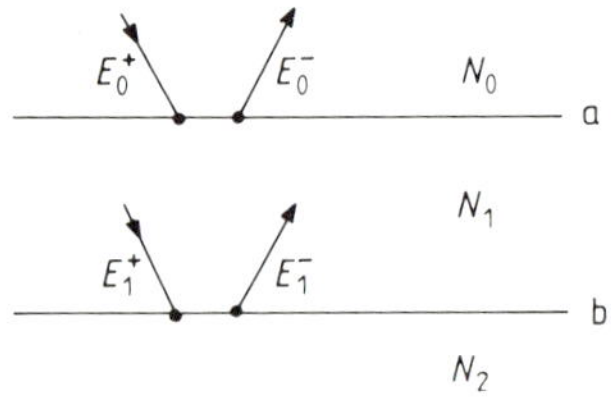

Figure 2.11

If E_2^+ is the tangential component of amplitude in medium 2, then, since there is only a positive-going wave in that medium

$$\begin{bmatrix} E_1^+ \\ E_1^- \end{bmatrix} = \frac{1}{\tau_b^+} \begin{bmatrix} 1 \\ \rho_b^+ \end{bmatrix} E_2^+ . \tag{2.79}$$

Equations (2.78) and (2.79) can be extended in the normal way to cover the case of many layers. The only point to watch is that ρ_a^+ and τ_a^+ must refer to the coefficient of the boundary in the correct medium. That is, all the reflection coefficients ρ and transmission coefficients τ must be calculated for the boundaries as they exist in the multilayer. Thus, if we take an existing multilayer and add an extra layer, not only do we add an extra interface but we alter the amplitude reflection and transmission coefficients of what now becomes the second last interface. Thus two layers must be recomputed and not just one.

If absorption is included, the formulae remain the same but the parameters ρ, τ and δ become complex.

SMITH'S METHOD OF MULTILAYER DESIGN

In 1958, Smith[5], then of the University of Reading, published a useful design method based on equation (2.76). The technique is also known as the *method of effective interfaces*. It consists of choosing any layer in the multilayer and then considering multiple reflections within it, the reflection and transmission coefficients at its boundaries being the resultant coefficients of the complete structures on either side. The method of summing multiple beams is, of course, quite old and the novel feature of the present technique is the way in which it is applied. Although the technique described by Smith was principally concerned with dielectric multilayers, it can be extended to deal with absorbing layers. As before, we limit ourselves, in the derivation, to normal incidence. When the layers are transparent, the expressions can be extended to oblique incidence without major difficulty. The notation is illustrated in figure 2.12.

From equation (2.76):

$$\tau^+ = \frac{\tau_a^+ \tau_b^+ e^{-i\delta}}{(1 - \rho_a^- \rho_b^+ e^{-2i\delta})}$$

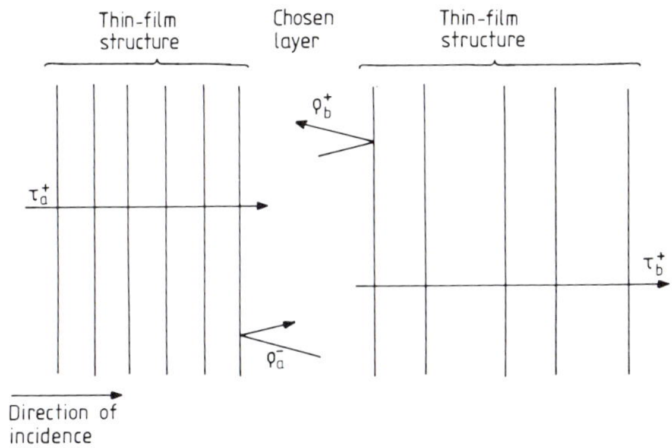

Figure 2.12

where

$$\delta = 2\pi N d/\lambda.$$

Now $N = n - ik$ and we can write δ as

$$\delta = 2\pi(n - ik)d/\lambda = \alpha + i\beta$$

and

$$e^{-i\delta} = e^{-\beta}e^{-i\alpha}$$

where $\alpha = 2\pi nd/\lambda$, the phase thickness of the layer, and $\beta = 2\pi kd/\lambda$. Now,

$$T = \frac{n_m}{n_0}(\tau^+)(\tau^+)^*$$

where n_m is the real part of the exit medium index and n_0 is the refractive index of the incident medium.

$$T = \frac{n_m}{n_0}\frac{(\tau_a^+)(\tau_a^+)^*(\tau_b^+)(\tau_b^+)^* e^{-2\beta}}{(1 - \rho_a^- \rho_b^+ e^{-2\beta} e^{-2i\alpha})(1 - \rho_a^- \rho_b^+ e^{-2\beta} e^{-2i\alpha})^*}.$$

Now, let

$$\tau_a^+ = |\tau_a^+| e^{i\phi_a'} \qquad \rho_a^- = |\rho_a^-| e^{i\phi_a}$$

$$\tau_b^+ = |\tau_b^+| e^{i\phi_b'} \qquad \rho_b^+ = |\rho_b^+| e^{i\phi_b}.$$

Then,

$$T = \frac{n_m}{n_0}\frac{|\tau_a^+|^2 |\tau_b^+|^2 e^{-2\beta}}{(1 - |\rho_a^-||\rho_b^+| e^{i(\phi_a+\phi_b)} e^{-2\beta} e^{-2i\alpha})(1 - |\rho_a^-||\rho_b^+| e^{-i(\phi_a+\phi_b)} e^{-2\beta} e^{2i\alpha})}$$

i.e.

$$T = \frac{n_m}{n_0}\frac{|\tau_a^+|^2 |\tau_b^+|^2 e^{-2\beta}}{[1 + |\rho_a^-|^2 |\rho_b^+|^2 e^{-4\beta} - 2|\rho_a^-||\rho_b^+| e^{-2\beta}\cos(\phi_a + \phi_b - 2\alpha)]}.$$

A marginally more convenient form of the expression can be obtained by substituting $1-2\sin^2[(\phi_a+\phi_b)/2-\alpha]$ for $\cos(\phi_a+\phi_b-2\alpha)$, and with some rearrangement

$$T=\frac{n_m}{n_0}\frac{|\tau_a^+|^2|\tau_b^+|^2 e^{-2\beta}}{(1-|\rho_a^-||\rho_b^+|e^{-2\beta})^2}\left[1+\frac{4|\rho_a^-||\rho_b^+|e^{-2\beta}}{(1-|\rho_a^-||\rho_b^+|e^{-2\beta})^2}\times\sin^2\left(\frac{\phi_a+\phi_b}{2}-\frac{2\pi nd}{\lambda}\right)\right]^{-1}. \tag{2.80}$$

If there is no absorption in the chosen layer, i.e. $\beta=0$, then the restrictions on reflectances in absorbing media no longer apply and we can write

$$T_a=\frac{n}{n_0}|\tau_a^+|^2 \qquad R_a^-=|\rho_a^-|^2$$

$$T_b=\frac{n_m}{n}|\tau_b^+|^2 \qquad R_b^+=|\rho_b^+|^2.$$

Then,

$$T=\frac{T_aT_b}{[1-(R_a^-R_b^+)^{1/2}]^2}\left[1+\frac{4R_a^-R_b^+}{[1-(R_a^-R_b^+)^{1/2}]^2}\sin^2\left(\frac{\phi_a+\phi_b}{2}-\frac{2\pi nd}{\lambda}\right)\right]^{-1} \tag{2.81}$$

which is the more usually quoted version.

The usefulness of this method is mainly in providing an insight into the properties of a particular type of filter, and it is of considerable value in design. It is certainly not the easiest method of determining the performance of a given multilayer—this is best tackled by a straightforward application of the matrix method. What equations (2.80) or (2.81) do is to make it possible to isolate a layer, or a combination of several layers, and to examine the influence which these layers and any changes in them have on the performance of the filter as a whole. Smith's original paper includes a large number of examples of this approach and repays close study.

THE SMITH CHART

The Smith chart is one of a number of different devices of the same broad type which were originally intended to simplify calculation. The Smith chart is the one which appears most frequently in the literature and so it is included here, although little use is made of it in the remainder of the book. The method depends on three properties of a thin-film structure.

(1) Since the tangential components of $\boldsymbol{E}$ and $\boldsymbol{H}$ are continuous across a boundary, so also is the equivalent admittance. This has been implied in the section dealing with the matrix method, but has not, perhaps, been explicitly stated there.

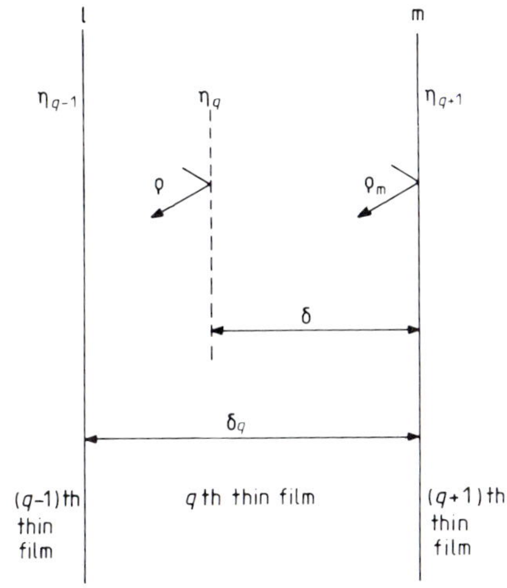

Figure 2.13

(2) In any thin film, for example layer q in figure 2.13, the amplitude reflectance ρ at any plane within the layer is related to that at the edge of the layer remote from the incident wave ρ_m by

$$\rho = \rho_m e^{-2i\delta} \tag{2.82}$$

where δ is the phase thickness of that part of the layer between the far boundary m and the plane in question.

This second point is almost self evident, but may be shown by putting $\rho_a^+ = 0$ in equation (2.75), since the boundary under consideration is an imaginary one between two media of identical admittance.

(3) The amplitude reflection coefficient of any thin-film assembly, with optical admittance at the front surface of Y, is given by equation (2.57), i.e.

$$\rho = \frac{\eta_0 - Y}{\eta_0 + Y} = \frac{1 - Y/\eta_0}{1 + Y/\eta_0} \tag{2.83}$$

where η_0 is the admittance of the incident medium. Y/η_0 is sometimes known as the reduced admittance.

The procedure for calculating the effect of any layer in a thin-film assembly by using these properties is as follows.

(1) ρ_m, the amplitude reflection coefficient at the boundary of the layer remote from the side of incidence, is given.

(2) The amplitude reflection coefficient within the layer just inside the

boundary 1 is then given by equation (2.82):

$$\rho = \rho_{\mathrm{m}} e^{-2i\delta_q}. \tag{2.84}$$

(3) The optical admittance just inside the boundary 1 is given by equation (2.83):

$$\rho = \frac{1 - Y/\eta_q}{1 + Y/\eta_q}$$

i.e.

$$\frac{Y}{\eta_q} = \frac{1-\rho}{1+\rho}. \tag{2.85}$$

(4) The optical admittance on the incident side of the boundary 1 is still Y because of condition (1) above. The reduced admittance is Y/η_{q-1} where

$$\frac{Y}{\eta_{q-1}} = \frac{\eta_q}{\eta_{q-1}} \frac{Y}{\eta_q}. \tag{2.86}$$

(5) The amplitude reflection coefficient ρ_1 on the incident side of the boundary 1 is given by

$$\rho_1 = \frac{1 - Y/\eta_{q-1}}{1 + Y/\eta_{q-1}}. \tag{2.87}$$

Calculation of the amplitude reflection coefficient of any thin-film assembly is merely the successive application of equations (2.84), (2.85), (2.86) and (2.87) to each layer in the system, starting with that at the end of the assembly remote from the incident wave.

The calculation can be carried out in any convenient way, and can be used as the basis for a computer program. The problem is similar to one found in the study of transmission lines, principally in the microwave region, and a simple graphical approach has been devised. The most awkward parts of the calculation are in equations (2.85) and (2.87). A chart connecting values of X and Z, where

$$X = \frac{1-Z}{1+Z}$$

is shown in figure 2.14 and is known as a Smith chart after the originator P H Smith (not to be confused with the S D Smith of the previous section). Z is plotted in polar coordinates on the diagram and the corresponding real and imaginary parts of X are read off from the sets of orthogonal circles. A slide rule is capable of the other part of the calculation, the multiplication by η_q/η_{q-1}.

A scale is provided around the outside of the chart to enable the calculation involved in equation (2.84) to be very simply carried out by rotating the point corresponding to ρ_{m} around the centre of the chart through the appropriate angle $2\delta_q$. The scale is calibrated in terms of optical thickness measured in fractions of a wavelength, taking into account that the angle is actually $2 \times \delta_q$.

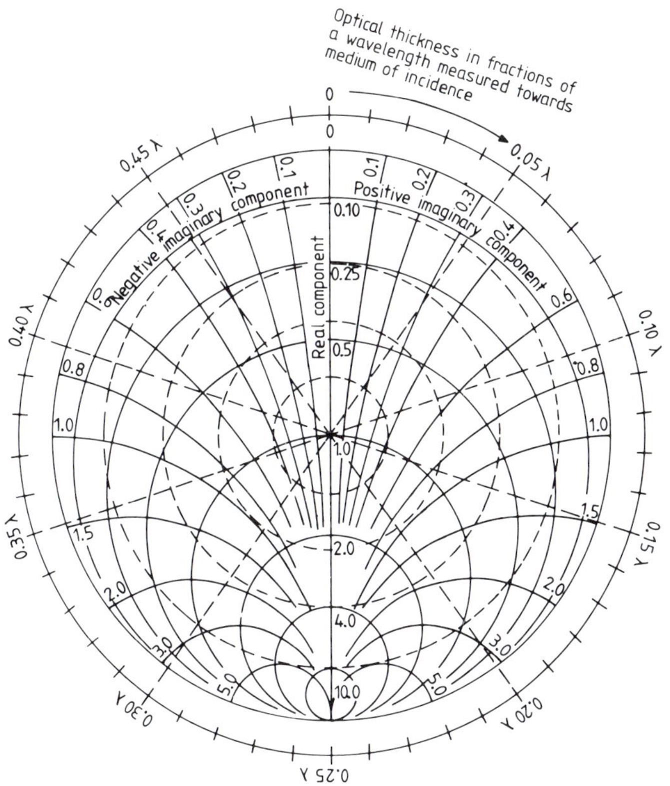

Figure 2.14 The Smith chart. Broken circles are circles of constant amplitude reflection coefficient ρ. From the smallest to the largest they correspond to $\rho = 0.2, 0.4$, 0.6, 0.8 and 1.0, the outer solid circle. Solid circles are circles of constant real part and constant imaginary part of the reduced optical admittance. Note: an optical thickness of 0.25λ corresponds to a phase thickness of 90°. (This Smith chart was constructed using the details given in *High frequency transmission lines*, Willis Jackson, pp 129 and 146 (3rd edition, London: Methuen 1951).)

CIRCLE DIAGRAMS

There are several other graphical techniques which are based on an exact solution of the appropriate equations. In these, we imagine that the multilayer is gradually built up on the substrate layer by layer, immersed all the time in the final incident medium. As each layer in turn increases from zero thickness to its final value, some parameter of the multilayer at that stage of its construction is calculated and the locus is plotted. In the techniques described here, the loci for dielectric layers take the form of a series of circular arcs or even complete circles, each corresponding to a single layer, which are connected at points corresponding to the interfaces between the different layers. Although the techniques can be used for quantitative calculation, and indeed in some cases were derived for that purpose, they cannot compete with a small programmable calculator, and their great value is in the visualisation of the characteristics of a particular multilayer.

Reflection circles

We first mention a technique, originally published by Berning[2], the use of which in coating design has been developed and described in detail by Apfel[6]. According to Apfel, Frank Rock originated this technique in the mid-1950s.

Equation (2.75) gives an expression for calculating the change in amplitude reflection coefficient resulting from the addition of a single layer:

$$\rho^+ = \frac{\rho_a^+ + \rho_b^+ e^{-2i\delta}}{1 + \rho_a^+ \rho_b^+}.$$

We can calculate the properties of a multilayer by successive applications of this formula, as has already been indicated. Let us imagine that we have arrived at the pth layer in the calculation. The quantities involved are indicated in figure 2.15. ρ_f^+ is the amplitude reflection coefficient of the $(p-1)$th layer at the outer interface, which we have labelled f.

$$\rho_f^+ = \frac{\eta_{p-1} - \eta_p}{\eta_{p-1} + \eta_p}.$$

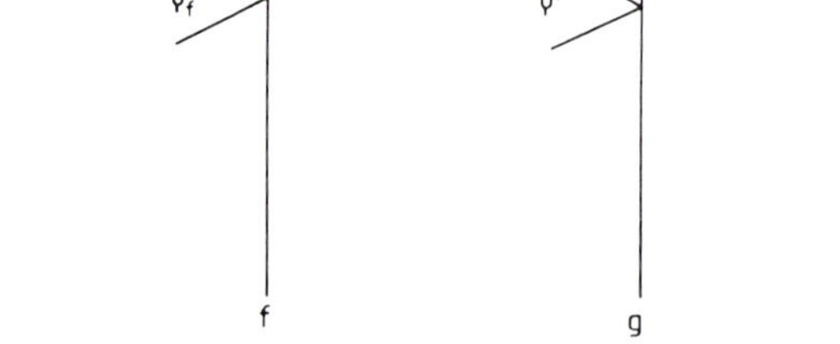

Figure 2.15

ρ' is the resultant amplitude reflection coefficient at the inner interface of the pth layer due to the entire structure on that side and is not to be confused with ρ_g, the amplitude reflection coefficient of the gth interface. Then the resultant amplitude reflection coefficient ρ at the fth interface is given by

$$\rho = \frac{\rho_f^+ + \rho' \mathrm{e}^{-2\mathrm{i}\delta}}{1 + \rho_f^+ \rho' \mathrm{e}^{-2\mathrm{i}\delta}}. \tag{2.88}$$

Provided we are dealing with dielectric materials, ρ_f^+ will be real. ρ' may be complex but we may include any phase angle due to ρ' in the factor $\mathrm{e}^{-2\mathrm{i}\delta}$. Let us plot the locus of ρ in the complex plane as δ varies. To simplify the analysis, we can replace ρ by $x + \mathrm{i}y$ and $\rho' \mathrm{e}^{-2\mathrm{i}\delta}$ by $\alpha + \mathrm{i}\beta$, where

$$(\alpha^2 + \beta^2)^{1/2} = |\rho'|.$$

Then

$$x + \mathrm{i}y = \frac{\rho_f^+ + \alpha + \mathrm{i}\beta}{1 + \rho_f^+ (\alpha + \mathrm{i}\beta)}.$$

Multiplying both sides by the denominator of the right-hand side and then equating real and imaginary parts of the resulting expressions yields

$$x(1 + \rho_f^+ \alpha) - y\rho_f^+ \beta = (\rho_f^+ + \alpha)$$

$$y(1 + \rho_f^+ \alpha) + x\rho_f^+ \beta = \beta$$

i.e.

$$(x - \rho_f^+) = \alpha(1 - x\rho_f^+) + \beta y \rho_f^+$$

$$y = -\alpha y \rho_f^+ + \beta(1 - x\rho_f^+).$$

To find the locus, we square and add these equations to give

$$\begin{aligned}(x - \rho_f^+)^2 + y^2 &= (\alpha^2 + \beta^2)[(1 - x\rho_f^+)^2 + (\rho_f^+ y)^2] \\ &= |\rho'|^2[(1 - x\rho_f^+)^2 + (\rho_f^+ y)^2]\end{aligned}$$

which can be manipulated to

$$x^2(1 - |\rho'|^2 \rho_f^{+2}) + y^2(1 - |\rho'|^2 \rho_f^{+2}) - 2x\rho_f^+(1 - |\rho'|^2) + \rho_f^{+2} - |\rho'|^2 = 0.$$

This is the equation of a circle with centre

$$\left(\frac{\rho_f^+(1 - |\rho'|^2)}{1 - |\rho'|^2 \rho_f^{+2}}, 0\right)$$

i.e. on the real axis, and radius

$$\frac{|\rho'|(1 - \rho_f^{+2})}{(1 - |\rho'|^2 \rho_f^{+2})}.$$

The locus of the reflection coefficient as the layer thickness is allowed to increase steadily from zero is therefore a circle. A half-wave layer traces out a complete circle, while a quarter-wave layer, if it starts on the real axis, will trace out a semicircle; otherwise it will be slightly more or less than a semicircle, depending on the exact starting point. In all cases, the circle is traced clockwise.

The locus corresponding to a single layer is straightforward. The plotting of the locus corresponding to two or more layers is slightly more complicated. The form of the locus of each layer is an arc of a circle traced from the terminal point of the previous layer. The complication arises from the subsidiary calculation which must be performed each time to calculate the current value of ρ' from the terminal value of the previous layer. An example will serve to illustrate the point.

Let us consider a glass substrate of index 1.52, on which is deposited first a layer of zinc sulphide of index 2.35 and thickness of one quarter-wave followed by a layer of cryolite of index 1.35 and of thickness also one quarter-wave. Air, of index 1.0, is the incident medium.

Calculation of the circles is most easily performed by using equation (2.88) to calculate the terminal points. The starting point is known and that, together with the fact that the centre is on the real axis, completes the specification of the circles.

The values of ρ_f^+ and ρ' for the first layer are

$$\rho_f^+ = \frac{1.0 - 2.35}{1.0 + 2.35} = -0.4030$$

$$\rho' = \frac{2.35 - 1.52}{2.35 + 1.52} = 0.2144.$$

The starting point for the layer is

$$\rho = \frac{\rho_f^+ + \rho'}{1 + \rho_f^+ \rho'} = -0.2063$$

which corresponds to the amplitude reflection coefficient of bare glass in air.

For a quarter-wave layer $e^{-2i\phi} = -1$ and so the terminal value of ρ is given by

$$\rho = \frac{\rho_f^+ - \rho'}{1 - \rho_f^+ \rho'} = -0.5683$$

and the locus up to this point is a semicircle. This value of ρ corresponds to the amplitude reflection coefficient of a quarter-wave of zinc sulphide on glass in air. To continue the locus into the next layer, we need new values of ρ_f^+ and ρ'.

$(\rho_f^+)_{\text{new}}$ is straightforward, being the external reflection coefficient at an air–cryolite boundary:

$$(\rho_f^+)_{\text{new}} = \frac{1.0 - 1.52}{1.0 + 1.52} = -0.1489.$$

$(\rho')_{\text{new}}$ is more difficult. This is the amplitude reflection coefficient which the substrate plus a quarter-wave of zinc sulphide will have, no longer in a medium of air, but in one of cryolite. It can be calculated either using the normal matrix method or simply by inverting the equation

$$\rho = (\rho)_{\text{old}} = \frac{(\rho_f^+)_{\text{new}} + (\rho')_{\text{new}}}{1 + (\rho_f^+)_{\text{new}} (\rho')_{\text{new}}}$$

which must be satisfied if the start of the new layer is to coincide with $(\rho)_{\text{old}}$, the termination of the old.

$$(\rho')_{\text{new}} = \frac{(\rho)_{\text{old}} - (\rho_f^+)_{\text{new}}}{1 - (\rho)_{\text{old}}(\rho_f^+)_{\text{new}}}$$

and in this case $(\rho)_{\text{old}}$ is -0.5683, so that

$$(\rho')_{\text{new}} = \frac{-0.5683 - (-0.1489)}{1 - (-0.5683)(-0.1489)} = -0.4582.$$

The new locus, which is another semicircle, then starts at the point -0.5683 on the real axis and terminates at

$$\rho = \frac{(\rho_f^+)_{\text{new}} - (\rho')_{\text{new}}}{1 - (\rho_f^+)_{\text{new}}(\rho')_{\text{new}}} = 0.3319.$$

The loci are shown in figure 2.16.

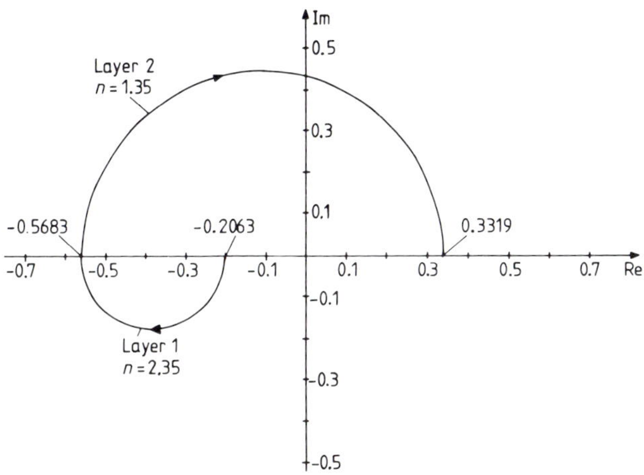

Figure 2.16 Reflection circles, or amplitude reflection locus, for the coating: Air $|LH|$ Glass, where L indicates a quarter-wave of index 1.35, H of 2.35, and the indices of air and glass are 1.00 and 1.52 respectively.

Many examples of the use of this technique in design are given by Apfel[6] who has also extended it to include absorbing layers such as metals. The loci of metals are, of course, no longer circles. The advantage of the technique over the Smith chart is especially that the locus is a continuous one, since the termination of each layer is the starting point for the next. We will, however, make more use of an allied technique, based on optical admittance rather than amplitude reflection coefficient, that is based on equation (2.60) rather than equation (2.78).

Admittance loci

Here, as the reference plane moves from the surface of the substrate to the front surface of the multilayer, we plot the variation of the input optical admittance calculated at the reference plane.

Equation (2.60) is

$$\begin{bmatrix} B \\ C \end{bmatrix} = \left(\prod_{r=1}^{q} \begin{bmatrix} \cos\delta_r & (\mathrm{i}\sin\delta_r)/\eta_r \\ \mathrm{i}\eta_r\sin\delta_r & \cos\delta_r \end{bmatrix} \right) \begin{bmatrix} 1 \\ \eta_m \end{bmatrix}$$

where $Y = C/B$ is the input optical admittance of the assembly. For the rth layer we can write

$$\begin{bmatrix} B \\ C \end{bmatrix} = \begin{bmatrix} \cos\delta_r & (\mathrm{i}\sin\delta_r)/\eta_r \\ \mathrm{i}\eta_r\sin\delta_r & \cos\delta_r \end{bmatrix} \begin{bmatrix} B' \\ C' \end{bmatrix}$$

and since it is optical admittance we are interested in we can divide throughout by B' to give

$$\begin{bmatrix} B/B' \\ C/B' \end{bmatrix} = \begin{bmatrix} \cos\delta_r & (\mathrm{i}\sin\delta_r)/\eta_r \\ \mathrm{i}\eta_r\sin\delta_r & \cos\delta_r \end{bmatrix} \begin{bmatrix} 1 \\ Y' \end{bmatrix}$$

where $Y' = C'/B'$ and represents the admittance of the structure at the exit side of the layer. We now find the locus of the input admittance

$$Y = \frac{C}{B} = \frac{C/B'}{B/B'}.$$

Let

$$Y = x + \mathrm{i}y$$

and

$$Y' = \alpha + \mathrm{i}\beta.$$

Then

$$Y = x + \mathrm{i}y = \frac{(\alpha + \mathrm{i}\beta)\cos\delta_r + \mathrm{i}\eta_r\sin\delta_r}{\cos\delta_r + (\alpha + \mathrm{i}\beta)(\mathrm{i}\sin\delta_r)/\eta_r}$$

$$= \frac{\alpha\cos\delta_r + \mathrm{i}(\beta\cos\delta_r + \eta_r\sin\delta_r)}{[\cos\delta_r - (\beta/\eta_r)\sin\delta_r] + \mathrm{i}(\alpha/\eta_r)\sin\delta_r}.$$

Equating real and imaginary parts:

$$x[\cos\delta_r - (\beta/\eta_r)\sin\delta_r] - (y\alpha/\eta_r)\sin\delta_r = \alpha\cos\delta_r \tag{2.89}$$

$$y[\cos\delta_r - (\beta/\eta_r)\sin\delta_r] + (x\alpha/\eta_r)\sin\delta_r = \beta\cos\delta_r + \eta_r\sin\delta_r. \tag{2.90}$$

Eliminating δ_r yields

$$x^2 + y^2 - x[(\alpha^2 + \beta^2 + \eta_r^2)/\alpha] + \eta_r^2 = 0$$

which is the equation of a circle with centre $((\alpha^2+\beta^2+\eta_r^2)/2\alpha, 0)$, i.e. on the real axis and with radius such that it passes through the point (α, β), i.e. its starting point. The circle is traced out in a clockwise direction, which can be shown by setting $\beta = 0$ in equation (2.90).

We can plot the locus in an Argand diagram in the same way as the locus of the amplitude reflection coefficient.

The scale of δ_r can also be plotted on the diagram. Let $\beta = 0$ and then, from equations (2.89) and (2.90),

$$x - (y\alpha/\eta_r)\tan\delta_r = \alpha$$
$$y + (x\alpha/\eta_r)\tan\delta_r = \eta_r \tan\delta_r.$$

Eliminating α, we have

$$x^2 + y^2 - y\eta_r(\tan\delta_r - 1/\tan\delta_r) - \eta_r^2 = 0.$$

This is a circle with centre

$$(0, (\eta_r/2)(\tan\delta_r - 1/\tan\delta_r))$$

i.e. on the imaginary axis and passing through the point $(\eta_r, 0)$. The simplest contours of equal δ_r are $\delta_r = 0, \pi/2, \pi, 3\pi/2, \ldots$, which coincide with the real axis, and $\delta_r = \pi/4, 3\pi/4, 5\pi/4, \ldots$, which is the circle with centre the origin and which passes through the point $(\eta_r, 0)$. For layers which start at a point not on the real axis the same set of contours of equal δ_r will still apply, with a correction to the value of δ_r which each represents.

Figure 2.17(*a*) shows the locus of a film which is deposited on a transparent substrate of admittance α. The starting point is $(\alpha, 0)$ and as the thickness is increased to a quarter-wave a semicircle is traced out clockwise which re-intersects the real axis in the point $(\eta_r^2/\alpha, 0)$. A second quarter-wave completes the circle. We could have had any point on the locus as starting point without changing its form. The only difference would have been an offset in the scale of δ_r.

We could add isoreflectance contours to the diagram if we wished. These are circles with centres on the real axis, centres and radii being given by

$$(\eta_0(1+R)/(1-R), 0) \qquad \text{and} \qquad 2\eta_0(R)^{1/2}/(1-R)$$

respectively, where η_0 is the admittance of the incident medium.

The phase of the reflectance can also be important and isophase contours are not unlike the contours of constant δ_r. We can carry through a similar procedure to determine the contours and the most important ones are $0, \pi/2, \pi$ and $3\pi/2$, that is, the boundaries between the quadrants. The boundary between the first and fourth and between the second and third is simply the real axis, while that between the first and second and the third and fourth is a circle with centre the origin which passes through the point $(\eta_0, 0)$. These contours are shown in figure 2.17(*b*) where the various quadrants are labelled.

For the purpose of drawing an admittance diagram, it is most convenient to

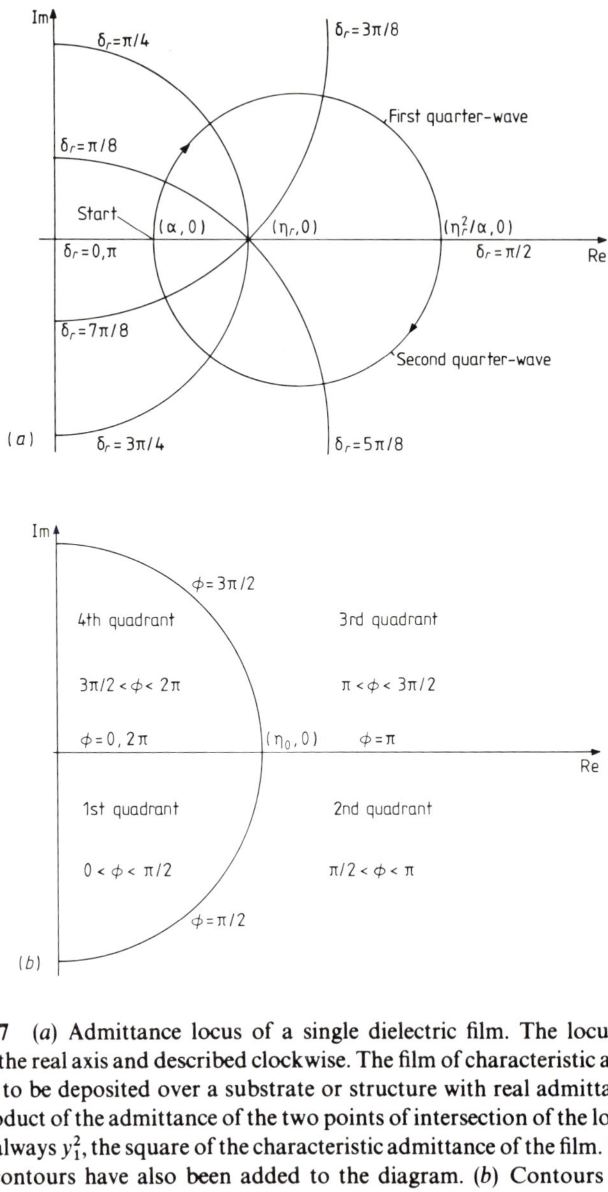

Figure 2.17 (*a*) Admittance locus of a single dielectric film. The locus is a circle centred on the real axis and described clockwise. The film of characteristic admittance y is assumed to be deposited over a substrate or structure with real admittance α. Note that the product of the admittance of the two points of intersection of the locus with the real axis is always y_1^2, the square of the characteristic admittance of the film. Equi-phase-thickness contours have also been added to the diagram. (*b*) Contours of constant phase shift on reflection ϕ can be added to the admittance diagram. These contours are all circles with centres on the imaginary axis and passing through the point on the real axis corresponding to the admittance y_0 of the incident medium. The four most important contours correspond to 0, $\pi/2$, π, $3\pi/2$, and these are represented by portions of the real axis and the circle centred on the origin and passing through the point y_0. These are indicated on the diagram and the regions corresponding to the various quadrants of ϕ are marked.

set η in units of $\mathscr{Y}$, the admittance of free space. Then the optical admittances will have the same numerical value as the refractive indices (at normal incidence only, of course).

The method can be illustrated by the same example as in the amplitude reflection coefficient loci

$$\text{Air}|HL|\text{Glass}$$

where glass has index 1.52, air 1.0, and H and L are quarter-waves of zinc sulphide ($n = 2.35$) and cryolite ($n = 1.35$) respectively.

In free space units, the starting admittance is simply 1.52, the admittance of glass. The termination of the first layer, since it is a quarter-wave, will be at an admittance of $2.35^2/1.52 = 3.633$ on the real axis, and of the second, which is also a quarter-wave, at $1.35^2/3.633 = 0.5016$ on the real axis. The circles are traced out clockwise and each is a semicircle with centre on the real axis. Figure 2.18 shows the complete locus.

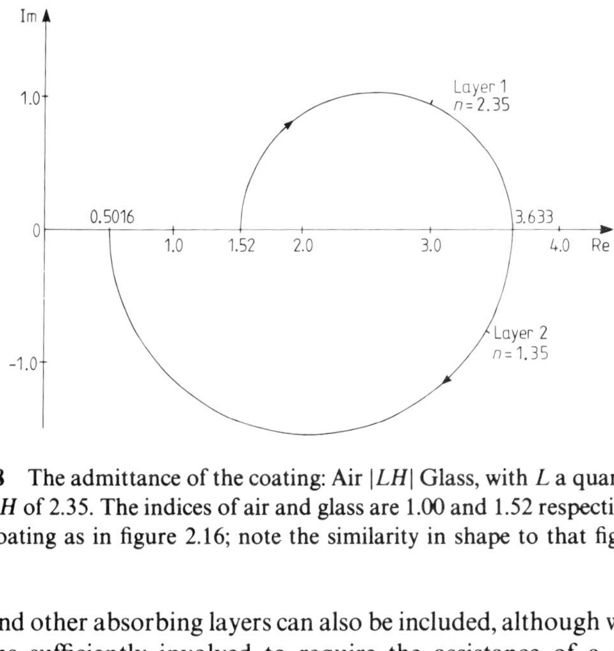

Figure 2.18 The admittance of the coating: Air $|LH|$ Glass, with L a quarter-wave of index 1.35, H of 2.35. The indices of air and glass are 1.00 and 1.52 respectively. This is the same coating as in figure 2.16; note the similarity in shape to that figure.

Metal and other absorbing layers can also be included, although we find the calculations sufficiently involved to require the assistance of a computer. Figure 2.19 shows two loci applying to metal layers, one starting from an admittance of 1.0 and the other from 1.52 (free space units). The higher the ratio k/n for the metal, the nearer the locus is to a circle with centre the origin. In the case of figure 2.19 the locus is somewhat distorted from the ideal case, with a loop bowing out along the direction of the real axis. If we were to add isoreflectance contours to the diagram, corresponding to admittances of 1.52 for the starting admittance of 1.0, and of 1.0 for the starting admittance of 1.52,

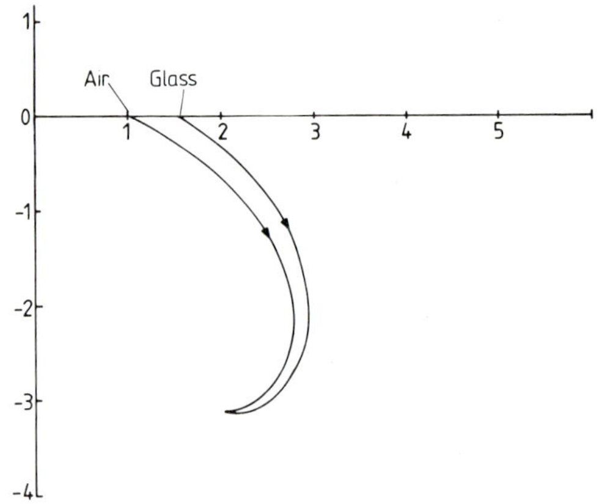

Figure 2.19 Admittance loci corresponding to a metal such as chromium with $n-ik = 2-i3$. Loci are shown for starting points 1.00 and 1.52, corresponding to air and glass respectively. Note that the initial direction towards the lower right of the diagram implies that in the case of the internal reflectance of the film deposited on glass (i.e. air as substrate and glass as incident medium and the left of the two loci) the reflectance initially falls and then rises, whereas the external reflectance (glass as substrate and air as incident medium and the right of the two admittance loci) always increases, even for very thin layers. When the layers are very thick, they terminate at the point $2-i3$, so that the film is optically indistinguishable from the bulk material.

so that the loci correspond to internal and external reflection from such a metal layer on glass in air, we would see that the observed reduction in internal reflectance when the metal is very thin is predicted by the diagram as well as the constantly increasing external reflectance for the same range of thicknesses (we can see such an effect in figure 4.7). Metals with still lower ratios of k/n depart still further from the ideal circle and in fact those starting at 1.0 can initially loop into the first quadrant so that they actually cut the real axis again, even sometimes at the point 1.52 to give zero internal reflectance.

We have gained much in simplicity by choosing to deal in terms of optical admittance throughout the assembly. It has not affected in any way our ability to calculate either the amplitude reflection coefficient or reflectance. Transmittance is another matter. We need to preserve the values of B and C in the matrix calculation. The optical admittance is not sufficient. For dielectric assemblies, we know that the transmittance is given by $(1-R)$, but for assemblies containing absorbing layers, subsidiary calculations are necessary. For many purposes, reflectance is sufficient and, furthermore, the graphical technique is used for visualisation rather than calculation and the lack of transmission information is not a serious defect.

INCOHERENT REFLECTION AT TWO OR MORE SURFACES

So far, we have treated substrates as being one-sided slabs of material of infinite depth. In almost all practical cases, the substrate will have finite depth with rear surfaces which reflect some of the energy and affect the performance of the assembly.

The depth of the substrate will usually be much greater than the wavelength of the light and variations in the flatness and parallelism of the two surfaces will be appreciable fractions of a wavelength. Generally the incident light will not be particularly well collimated. Under these conditions it will not be possible with a finite aperture to observe interference effects between light reflected at the front and rear surfaces of the substrate, and because of this the substrate is known as thick. The waves reflected successively at the front and back surfaces add incoherently instead of coherently. The resultant is the sum of the various intensities instead of the vector sum of the amplitudes. It can be shown that incoherent addition yields the same result as the averaging of the coherent result over any moderate geometrical area or wavelength interval or range of angles of incidence, such that an appreciable number of fringes is included in the interval.

The symbols used are illustrated in figure 2.20. Waves are reflected successively at the front and rear surfaces. The sums of the intensities are given by

$$\begin{aligned} R &= R_a^+ + T_a^+ R_b^+ T_a^- [1 + R_a^- R_b^+ + (R_a^- R_b^+)^2 + \ldots] \\ &= R_a^+ + [T_a^+ T_a^- R_b^+ / (1 - R_a^- R_b^+)] \end{aligned}$$

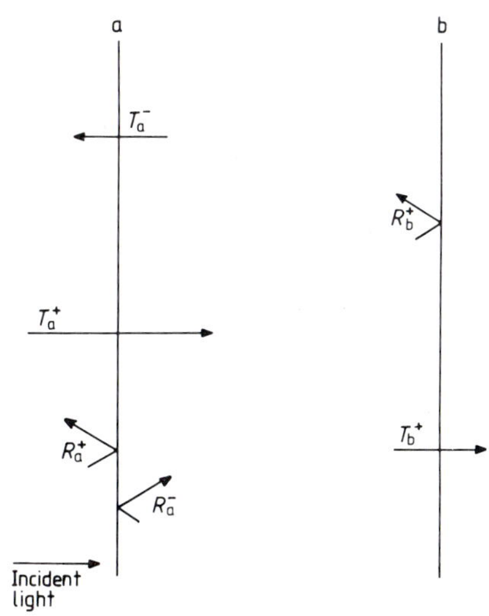

Figure 2.20

i.e., since T^+ and T^- are always identical,

$$R = \frac{R_a^+ + R_b^+(T_a^2 - R_a^+ R_a^-)}{1 - R_a^- R_b^+}. \tag{2.91}$$

If there is no absorption in the layers,

$$R_a^+ = R_a^- = R_a, \qquad T_a^+ = T_a^- = T_a \qquad \text{and} \qquad 1 = R_a + T_a$$

so that

$$R = \frac{R_a + R_b - 2R_a R_b}{1 - R_a R_b}. \tag{2.92}$$

Similarly

$$T = T_a^+ T_b^+ [1 + R_a^- R_b^+ + (R_a^- R_b^+)^2 + \ldots]$$

$$= \frac{T_a T_b}{1 - R_a^- R_b^+} \tag{2.93}$$

and again, if there is no absorption,

$$T = \frac{T_a T_b}{1 - R_a R_b} \tag{2.94}$$

or

$$T = \left(\frac{1}{T_a} + \frac{1}{T_b} - 1\right)^{-1} \tag{2.95}$$

since

$$R_a = 1 - T_a \qquad R_b = 1 - T_b.$$

A nomogram for solving equation (2.95) can easily be constructed. Two axes at right angles are laid out on a sheet of graph paper and, taking the point of intersection as the zero, two linear equal scales of transmittance are marked out on the axes. One of these is labelled T_a and the other T_b. The angle between T_a and T_b is bisected by a third axis which is to have the T scale marked out on it. To do this, a straight edge is placed so that it passes through the 100% transmission point on, say, the T_a-axis and any chosen point on the T_b-axis. The value of T to be associated with the point where the straight edge crosses the T-axis is then that of the intercept with the T_b-axis. The entire scale can be marked out in this way. A completed nomogram of this type is shown in figure 2.21.

In the absence of absorption, the analysis can be very simply extended to further surfaces. Consider the case of two substrates, i.e. four surfaces. These we can label T_a, T_b, T_c and T_d. Then, from equation (2.95), we have for the first substrate

$$T_1 = \left(\frac{1}{T_a} + \frac{1}{T_b} - 1\right)^{-1}$$

i.e.

$$\frac{1}{T_1} = \frac{1}{T_a} + \frac{1}{T_b} - 1$$

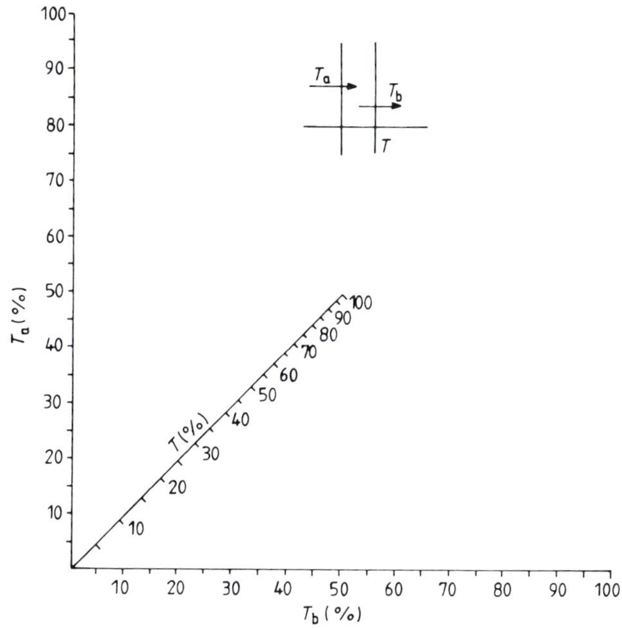

Figure 2.21 A nomogram for calculating the overall transmittance of a thick transparent plate given the transmittance of each individual surface.

and similarly for the second

$$\frac{1}{T_2}=\frac{1}{T_c}+\frac{1}{T_d}-1.$$

The transmittance through all four surfaces is then obtained by applying equation (2.95) once again:

$$\frac{1}{T}=\frac{1}{T_1}+\frac{1}{T_2}-1$$

i.e.

$$T=\left(\frac{1}{T_a}+\frac{1}{T_b}+\frac{1}{T_c}+\frac{1}{T_d}-3\right)^{-1}.$$

The analysis of the behaviour of multiple surfaces with and without absorption has been considered in detail by Baumeister *et al*[7].

FURTHER INFORMATION

Further information on the calculation of the properties of thin-film multilayers is available from many sources. In addition to those cited already in this chapter, references 8–11 will be found useful, but in no way represent anything approaching a complete list.

REFERENCES

1 Born M and Wolf E 1975 *Principles of Optics* 5th edn (Oxford: Pergamon)
2 Berning P H 1963 Theory and calculations of optical thin films *Physics of Thin Films*, vol 1, ed G Hass (New York: Academic) pp 69–121
3 Berning P H and Turner A F 1957 Induced transmission in absorbing films applied to band pass filter design *J. Opt. Soc. Am.* **47** 230–9
4 Abelès F 1950 Recherches sur la propagation des ondes électromagnetiques sinusoidales dans les milieus stratifiés *Ann. Phys., Paris* 12th series, **5** 596–640 and 706–84. The result quoted comes from p 631
5 Smith S D 1958 Design of multilayer filters by considering two effective interfaces *J. Opt. Soc. Am.* **48** 43–50
6 Apfel J H 1972 Graphics in optical coating design *Appl. Opt.* **11** 1303-12
7 Baumeister P W, Hahn R and Harrison D 1972 The radiant transmittance of tandem arrays of filters *Opt. Acta* **19** 853–64
8 Vasicek A 1960 *Optics of Thin Films* (Amsterdam: North-Holland)
9 Anders H 1965 *Dünne Schichten für die Optik* (Stuttgart: Wissenschaftliche Verlagsgesellschaft). Engl. transl. 1967: *Thin Films in Optics* (London: Focal)
10 Heavens O S 1955 *Optical Properties of Thin Solid Films* (London: Butterworth). Reprinted 1975 (New York: Dover)
11 Welford W T (writing as W Weinstein) 1954 Computations in thin film optics *Vacuum* **4** 3–19

3 Antireflection coatings

As has already been mentioned in chapter 1, antireflection coatings were the principal objective of much of the early work in thin-film optics. Of all the possible applications, antireflection coatings have had the greatest impact on technical optics, and even today, in sheer volume of production, they still exceed all other types of coating. In some applications, antireflection coatings are simply required for the reduction of surface reflection. In others, not only must surface reflection be reduced, but the transmittance must also be increased. The crown glass elements in a compound lens have a transmittance of only 96 % per untreated surface, while the flint components can have a surface transmittance as low as 90 %. The net transmittance of even a modest number of untreated elements in series can therefore be quite low. Additionally, part of the light reflected at the various surfaces eventually reaches the focal plane, where it appears as ghosts or as a veiling glare, thus reducing the contrast of the images. This is especially true of the zoom lenses used in television or photography, where twenty or more elements may be included, and which would be completely unusable without antireflection coatings.

Antireflection coatings can range from a simple single layer, having virtually zero reflectance at just one wavelength, to a multilayer system of more than a dozen layers, having virtually zero reflectance over a range of several octaves. The type used in any particular application will depend on a variety of factors, including the substrate material, the wavelength region, the required performance and the cost.

In the visible region, crown glass, which has a refractive index of around 1.52, is most commonly used. As we shall see, this presents a very different problem from infrared materials, which can have very much higher refractive indices. It is convenient, therefore, to split what follows into antireflection coatings for low-index substrates and antireflection coatings for high-index substrates, corresponding roughly to the visible and infrared. Since, from the point of view of design, antireflection coatings for high-index substrates are more straightforward, they are considered first.

There is no systematic method for the design of antireflection coatings. Trial and error, assisted by approximate techniques (frequently one or other of the graphical methods mentioned in chapter 2), backed up by accurate computer calculation, is frequently employed. Very promising designs can be further improved by computer refinement. Several different approaches are used in this chapter, partly to illustrate their use and partly because they are complementary. All the performance curves have been computed by application of the matrix method. In most cases, the materials are considered to be completely transparent.

The vast majority of antireflection coatings are required for matching an optical element into air. Air has an index of around 1.0003 at standard temperature and pressure which, for practical purposes, can be considered as unity.

The earliest antireflection coatings were on glass for use in the visible region of the spectrum. As shall become apparent later, a single-layer antireflection coating on glass, for the centre of the visible region, has a distinct magenta tinge when examined visually in reflection. This gives an appearance not unlike tarnish, and indeed in chapter 1 we mentioned the beneficial effects of the tarnish layer on aged flint objectives, and so the term 'bloom', in the sense of tarnish, has been used in this connection. The action of applying the coating is referred to as 'blooming' and the element is said to be 'bloomed'.

ANTIREFLECTION COATINGS ON HIGH-INDEX SUBSTRATES

The term high-index in this context cannot be defined precisely in the sense of a range with a definite lower bound. It simply means that the substrate has an index sufficiently higher than the available thin-film materials to enable the design of high-performance antireflection coatings consisting entirely, or almost entirely, of layers with indices lower than that of the substrate. These high-index substrates are principally of use in the infrared. Semiconductors, such as germanium, with an index of around 4.0, giving a reflection loss of around 36 % per surface, and silicon, with index around 3.5 and reflection loss of 31 %, are common, and it would be completely impossible to use them in the vast majority of applications without some form of antireflection coating. For many purposes, the reduction of a 30 % reflection loss to one of a few per cent would be considered adequate. It is only in a limited number of applications where the reflection loss must be reduced to less than one per cent.

The single-layer antireflection coating

The simplest form of antireflection coating is a single layer. Consider figure 3.1. Here we have a vector diagram which, since two interfaces are involved, contains two vectors, each representing the amplitude reflection coefficient at

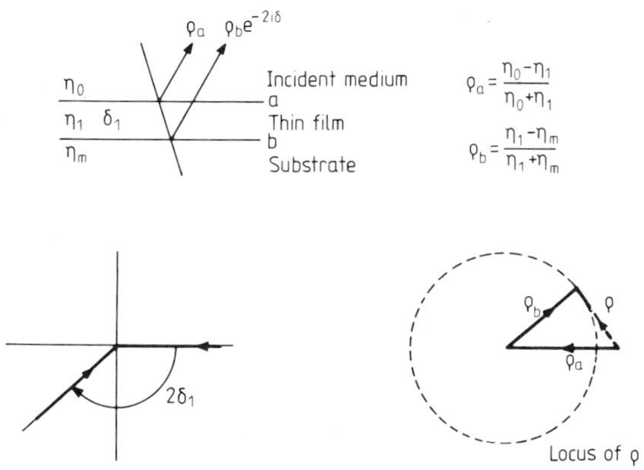

Figure 3.1 Vector diagram of a single-layer antireflection coating.

an interface. If the incident medium is air, then, provided the index of the film is lower than the index of the substrate, the reflection coefficient at each interface will be negative, denoting a phase change of 180°. The resultant locus is a circle with a minimum at the wavelength for which the phase thickness of the layer is 90°, that is, a quarter-wave optical thickness, when the two vectors are completely opposed. Complete cancellation at this wavelength, that is, zero reflectance, will occur if the vectors are of equal length. This condition, in the notation of figure 3.1, is

$$\frac{\eta_0-\eta_1}{\eta_0+\eta_1}=\frac{\eta_1-\eta_m}{\eta_1+\eta_m}$$

which requires

$$\frac{\eta_1}{\eta_0}=\frac{\eta_m}{\eta_1}$$

or

$$\eta_1=(\eta_0\eta_m)^{1/2}$$

which, for normal incidence, can also be written

$$n_1=(n_0 n_m)^{1/2}.$$

The condition for a perfect single-layer antireflection coating is, therefore, a quarter-wave optical thickness of material with optical admittance equal to the square root of the product of the admittances of substrate and medium.

It is seldom possible to find a material of exactly the optical admittance which is required. It was shown in chapter 2 that the optical admittance of a substrate coated with a quarter-wave optical thickness of material is

$$Y=\eta_f^2/\eta_m$$

where η_f is the admittance of the film material and η_m that of the substrate. The reflectance is therefore given by

$$R = \left(\frac{\eta_0 - Y}{\eta_0 + Y}\right)^2 = \left(\frac{\eta_0 - \eta_f^2/\eta_m}{\eta_0 + \eta_f^2/\eta_m}\right)^2.$$

If there is an error ε in η_f such that

$$\eta_f = (1 + \varepsilon)\,(\eta_0 \eta_m)^{1/2}$$

then

$$R = \left(\frac{2\varepsilon - \varepsilon^2}{2 + 2\varepsilon + \varepsilon^2}\right)^2 \simeq \varepsilon^2$$

provided that ε is small. A 10% error in η_f, therefore, leads to a residual reflectance of 1%.

Zinc sulphide has an index of around 2.2 at 2 μm and 2.15 at 15 μm. It has sufficient transparency for use as a quarter-wave antireflection coating over the range 0.4–25 μm. Germanium, silicon, gallium arsenide, indium arsenide and indium antimonide can all be treated satisfactorily by a single layer of zinc sulphide. The procedure to be followed for hard, rugged zinc sulphide films is described in a paper by Cox and Hass[1]. The substrate should be maintained at around 150°C during coating and cleaned by a glow discharge immediately before coating. The transmittance of a germanium plate with a single-layer zinc sulphide antireflection coating is shown in figure 3.2.

Zinc sulphide, even deposited under the best conditions, can deteriorate after prolonged exposure to humid atmospheres. Somewhat harder and more robust coatings are produced with cerium oxide or silicon monoxide. Cerium

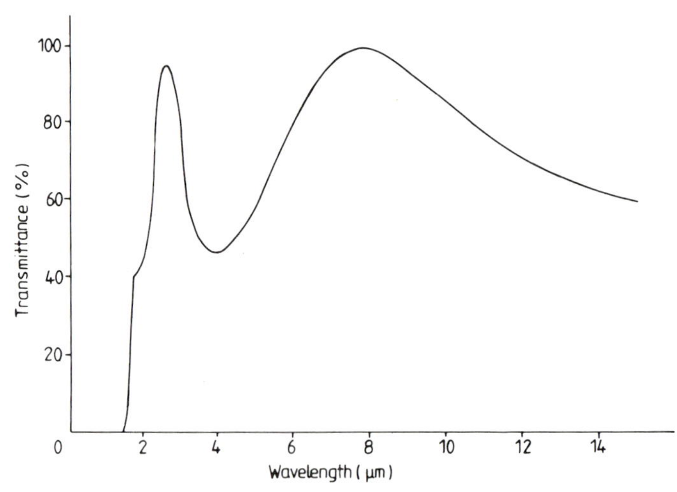

Figure 3.2 Transmittance of a germanium plate bloomed on both sides with zinc sulphide for 8 μm. (Courtesy of Sir Howard Grubb, Parsons & Co Ltd.)

oxide, when deposited at a substrate temperature of 200°C or more, forms very hard and durable films of refractive index 2.2 at 2 μm. Unfortunately, it displays a slight absorption band at 3 μm owing to adsorbed water vapour. Silicon monoxide does not show this water vapour band to the same degree, and so Cox and Hass recommend this material as the most satisfactory for coating germanium and silicon in the near infrared. The index of silicon monoxide evaporated in a good vacuum at a high rate is around 1.9. The transmittance of a silicon plate coated on both sides with silicon monoxide is shown in figure 3.3.

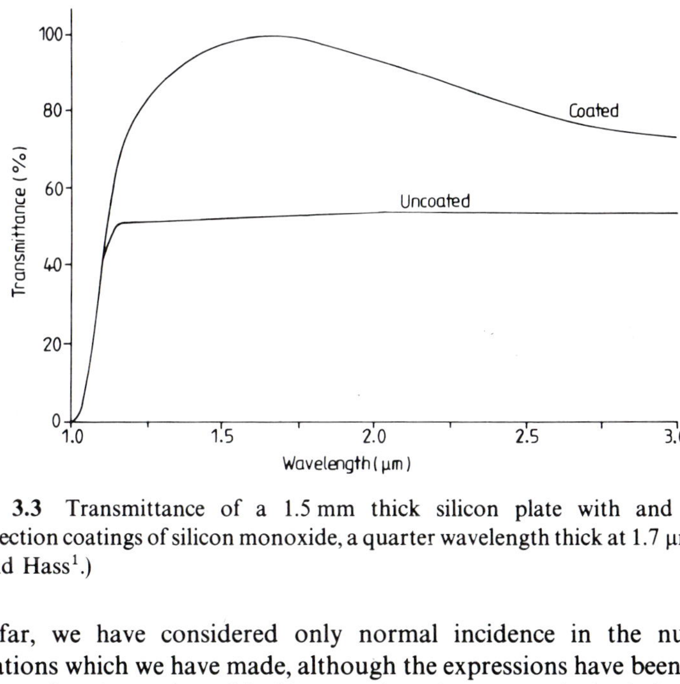

Figure 3.3 Transmittance of a 1.5 mm thick silicon plate with and without antireflection coatings of silicon monoxide, a quarter wavelength thick at 1.7 μm. (After Cox and Hass[1].)

So far, we have considered only normal incidence in the numerical calculations which we have made, although the expressions have been written in terms of η. At angles of incidence other than normal, the behaviour is similar, but the effective phase thickness of the layer is reduced and so the optimum wavelength is shorter. For the optical admittance we must use the appropriate η_p or η_s, and, as these are different, polarisation effects become evident. For high-index substrates and coatings the effects are much less than for the low-index coatings for the visible region, as we shall see later. Figure 3.4 shows the calculated variation with angle of incidence of the performance of a zinc sulphide coating ($n = 2.2$) on a germanium substrate ($n = 4.0$).

Such calculations are relatively straightforward. We use the matrix method. The characteristic matrix of a single film on a substrate is given by

$$\begin{bmatrix} B \\ C \end{bmatrix} = \begin{bmatrix} \cos\delta_1 & (\mathrm{i}\sin\delta_1)/\eta_1 \\ \mathrm{i}\eta_1\sin\delta_1 & \cos\delta_1 \end{bmatrix} \begin{bmatrix} 1 \\ \eta_m \end{bmatrix}$$

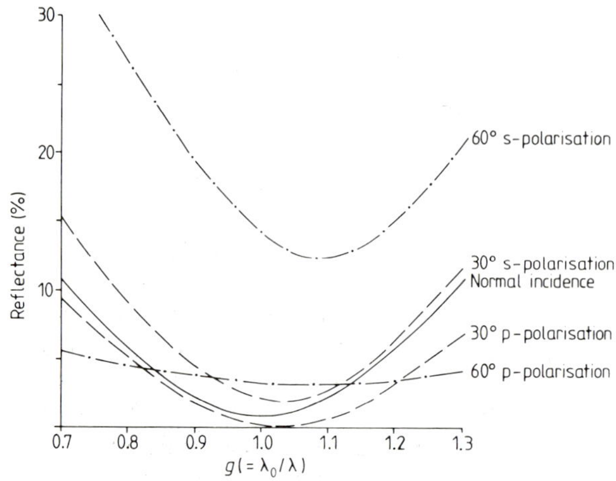

Figure 3.4 Calculated performance at various angles of incidence of a zinc sulphide coating (n = 2.2) on a germanium substrate (n = 4.0).

i.e.

$$\begin{bmatrix} B \\ C \end{bmatrix} = \begin{bmatrix} \cos\delta_1 + i(\eta_s/\eta_1)\sin\delta_1 \\ \eta_m\cos\delta_1 + i\eta_1\sin\delta_1 \end{bmatrix}$$

where the symbols have the meanings, defined in chapter 2,

$$\left.\begin{aligned} \eta_p &= \eta/\cos\theta \\ \eta_s &= \eta\cos\theta \end{aligned}\right\} \text{for each material}$$

$$\delta_1 = 2\pi\eta_1 d_1 \cos\theta_1/\lambda$$

and where

$$\eta_0 \sin\theta_0 = \eta_1 \sin\theta_1 = \eta_m \sin\theta_m.$$

If λ_0 is the wavelength for which the layer is a quarter-wave optical thickness at normal incidence, then $n_1 d_1 = \lambda_0/4$ and

$$\delta_1 = \frac{\pi}{2}\left(\frac{\lambda_0}{\lambda}\right)\cos\theta_1$$

so that the new optimum wavelength is $\lambda_0 \cos\theta_1$.

The amplitude reflection coefficient is

$$\rho = \frac{\eta_0 - Y}{\eta_0 + Y} = \frac{\eta_0 - C/B}{\eta_0 + C/B}$$

$$= \frac{(\eta_0 - \eta_m)\cos\delta_1 + i[(\eta_0\eta_m/\eta_1) - \eta_1]\sin\delta_1}{(\eta_0 + \eta_m)\cos\delta_1 + i[(\eta_0\eta_m/\eta_1) + \eta_1]\sin\delta_1} \tag{3.1}$$

and the reflectance

$$R = \frac{(\eta_0 - \eta_m)^2 \cos^2 \delta_1 + [(\eta_0 \eta_m / \eta_1) - \eta_1]^2 \sin^2 \delta_1}{(\eta_0 + \eta_m)^2 \cos^2 \delta_1 + [(\eta_0 \eta_m / \eta_1) + \eta_1]^2 \sin^2 \delta_1}. \tag{3.2}$$

It is instructive to prepare an admittance diagram (figure 3.5) for the single-layer coating. We recall that admittance loci were discussed in chapter 2 in the section beginning on page 62. We consider normal incidence only and use free space units for the admittances so that they are numerically equal to the refractive indices. The locus for the single layer is a circle beginning at the point 4.0 on the real axis, corresponding to the admittance of germanium. The centre of the circle is on the real axis and the circle cuts the real axis again at the point $2.2^2/4.0 = 1.21$, corresponding to a quarter-wave optical thickness. We can mark a scale of δ_1 along the locus. Since $\delta_1 = 2\pi n_1 d_1/\lambda$, we can either assume λ constant and replace the scale with one of optical thickness, or, provided that we assume that the refractive index remains constant with wavelength, for a given layer optical thickness we can mark the scale in terms of $g(= \lambda_0/\lambda)$. These various scales have been added. The scale of g assumes that λ_0 is the wavelength for which the layer has an optical thickness of one quarter-wave.

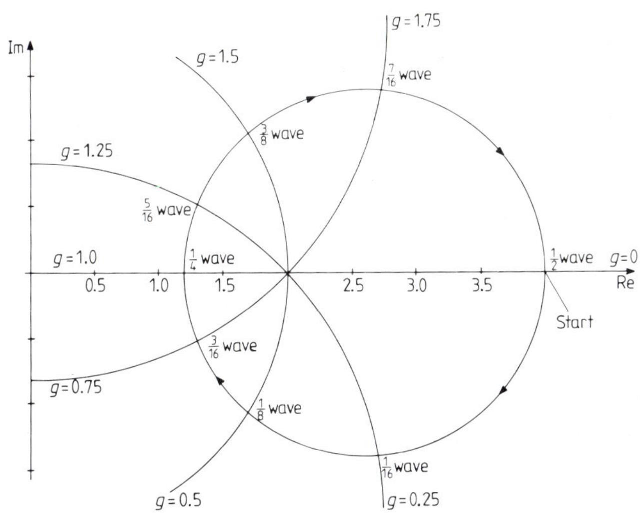

Figure 3.5 Admittance diagram for a single-layer zinc sulphide ($n = 2.2$) coating on germanium ($n = 4.0$).

This is a particularly simple admittance locus and it is included principally to illustrate the method. We will make some use of admittance diagrams in this chapter. Normally these will be drawn for one value of wavelength and for one value of optical thickness for each layer.

Double-layer antireflection coatings

The disadvantage of the single-layer coating, as far as the design is concerned, is the limited number of adjustable parameters. We can see from the admittance locus of figure 3.5 that only where the locus passes through the point (1, 0) will zero reflectance be obtained (or more generally when the locus passes through the point (η_0, 0)) and this must correspond to a semicircle or a quarter-wave optical thickness (or, strictly, an odd integral multiple thereof). The refractive index, or optical admittance, of the layer is also uniquely determined as $\eta_1 = (\eta_0\eta_m)^{1/2}$. There is thus no room for manoeuvre in the design of a single-layer coating. In practice, the refractive index is not a parameter which can be varied at will. Materials suitable for use as thin films are limited in number and the designer has to use what is available. A more rewarding approach, therefore, is to use more layers, specifying obtainable refractive indices for all layers at the start, and to achieve zero reflectance by varying the thickness. Then, too, there is the limitation that the single-layer coating can give zero reflectance at one wavelength only and low reflectance over a narrow region. A wider region of high performance demands additional layers.

We will consider first the problem of ensuring zero reflectance at one single wavelength and we shall attempt to achieve this with a two-layer coating. Since we are dealing with high-index substrates we look initially at combinations of layers having refractive indices lower than that of the substrate.

A vector diagram of one possibility is shown in figure 3.6. Provided the vectors are not such that any one is greater in length than the sum of the other two, then there are two sets of thicknesses for which zero reflectance can be obtained at one wavelength. The thinner combination, as in figure 3.6(*a*), will give the broadest characteristic and should normally be chosen.

In some ways, it is easier to visualise the design using an admittance plot. As usual, we plot admittance in free space units so that it is numerically the same as refractive index. Two possible arrangements are shown in figure 3.7 which can be obtained simply by drawing the circle corresponding to index n_1 passing through the point n_0, and the circle corresponding to index n_2 passing through the point n_m. Provided these circles intersect, then an antireflection coating using these particular indices is possible. The two sets of thicknesses correspond to the two points of intersection.

This is a very important coating with wider implications than just the blooming of a high-index substrate and so it is worth examining in greater detail. We use the matrix method and follow an analysis by Catalan[2], changing the notation to agree with the system used here. The characteristic matrix of the assembly is

$$\begin{bmatrix} B \\ C \end{bmatrix} = \begin{bmatrix} \cos\delta_1 & (\mathrm{i}\sin\delta_1)/\eta_1 \\ \mathrm{i}\eta_1\sin\delta_1 & \cos\delta_1 \end{bmatrix} \begin{bmatrix} \cos\delta_2 & (\mathrm{i}\sin\delta_2)/\eta_2 \\ \mathrm{i}\eta_2\sin\delta_2 & \cos\delta_2 \end{bmatrix} \begin{bmatrix} 1 \\ \eta_m \end{bmatrix}$$

$$= \begin{bmatrix} \cos\delta_1[\cos\delta_2 + \mathrm{i}(\eta_m/\eta_2)\sin\delta_2] + \mathrm{i}\sin\delta_1(\eta_m\cos\delta_2 + \mathrm{i}\eta_2\sin\delta_2)/\eta_1 \\ \mathrm{i}\eta_1\sin\delta_1[\cos\delta_2 + \mathrm{i}(\eta_m/\eta_2)\sin\delta_2] + \cos\delta_1(\eta_m\cos\delta_2 + \mathrm{i}\eta_2\sin\delta_2) \end{bmatrix}.$$

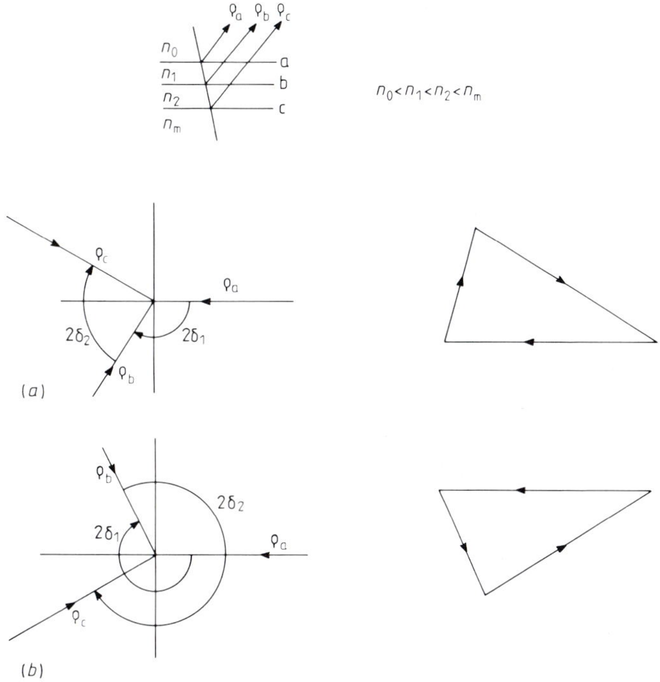

Figure 3.6 Vector diagram for double-layer antireflection coating. The thickness of the layers can be chosen to close the vector triangle and give zero reflectance in two ways, (*a*) and (*b*).

The reflectance will be zero if the optical admittance Y is equal to η_0, i.e.

$$\begin{aligned} &\mathrm{i}\eta_1 \sin\delta_1[\cos\delta_2 + \mathrm{i}(\eta_m/\eta_2)\sin\delta_2] + \cos\delta_1(\eta_m\cos\delta_2 + \mathrm{i}\eta_2\sin\delta_2) \\ &\quad = \eta_0\{\cos\delta_1[\cos\delta_2 + \mathrm{i}(\eta_m/\eta_2)\sin\delta_2] + \mathrm{i}\sin\delta_1(\eta_m\cos\delta_2 + \mathrm{i}\eta_2\sin\delta_2)/\eta_1\}. \end{aligned}$$

The real and imaginary parts of these expressions must be equated separately giving

$$\begin{aligned} -(\eta_1\eta_m/\eta_2)\sin\delta_1\sin\delta_2 + \eta_m\cos\delta_1\cos\delta_2 &= \eta_0\cos\delta_1\cos\delta_2 - (\eta_0\eta_2/\eta_1)\sin\delta_1\sin\delta_2 \\ \eta_1\sin\delta_1\cos\delta_2 + \eta_2\cos\delta_1\sin\delta_2 &= (\eta_0\eta_m/\eta_2)\cos\delta_1\sin\delta_2 \\ &\quad + (\eta_0\eta_m/\eta_1)\sin\delta_1\cos\delta_2 \end{aligned}$$

i.e.

$$\begin{aligned} \tan\delta_1\tan\delta_2 &= (\eta_m - \eta_0)/[(\eta_1\eta_m/\eta_2) - (\eta_0\eta_2/\eta_1)] \\ &= \eta_1\eta_2(\eta_m - \eta_0)/(\eta_1^2\eta_m - \eta_0\eta_2^2) \end{aligned} \tag{3.3}$$

and

$$\tan\delta_2/\tan\delta_1 = \eta_2(\eta_0\eta_m - \eta_1^2)/[\eta_1(\eta_2^2 - \eta_0\eta_m)] \tag{3.4}$$

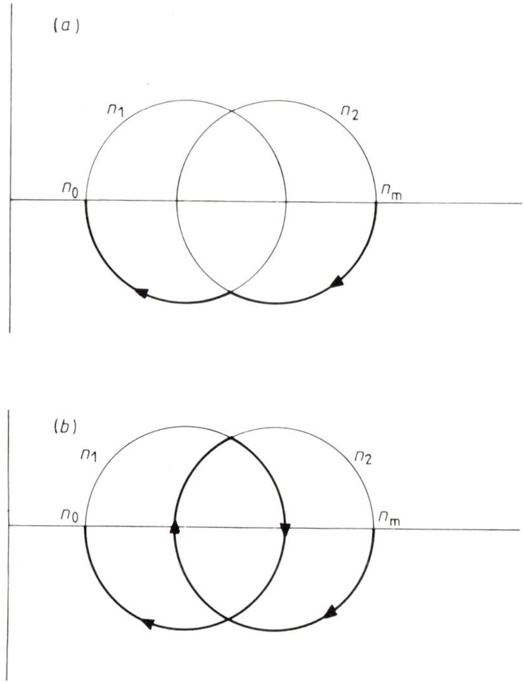

Figure 3.7 Admittance diagram for the double-layer antireflection coating. The two possible solutions are shown in (*a*) and (*b*).

giving

$$\tan^2\delta_1 = \frac{(\eta_m-\eta_0)(\eta_2^2-\eta_0\eta_m)\eta_1^2}{(\eta_1^2\eta_m-\eta_0\eta_2^2)(\eta_0\eta_m-\eta_1^2)} \tag{3.5}$$

$$\tan^2\delta_2 = \frac{(\eta_m-\eta_0)(\eta_0\eta_m-\eta_1^2)\eta_2^2}{(\eta_1^2\eta_m-\eta_0\eta_2^2)(\eta_2^2-\eta_0\eta_m)}. \tag{3.6}$$

The values of δ_1 and δ_2 found from these equations must be correctly paired and this is most easily done either by ensuring that they also satisfy the two preceding equations or by sketching a rough admittance diagram.

For solutions to exist, or, putting it in another way, for the circles in the admittance diagram to intersect, the right-hand sides of equations (3.5) and (3.6) must be positive. δ_1 and δ_2 are then real. This requires that, of the expressions

$$(\eta_2^2-\eta_0\eta_m) \tag{3.7}$$

$$(\eta_1^2\eta_m-\eta_0\eta_2^2) \tag{3.8}$$

$$(\eta_0\eta_m-\eta_1^2) \tag{3.9}$$

either all three must be positive, or, any two are negative and the third positive. This can be summarised in a useful diagram (figure 3.8) known as a Schuster diagram after one of the originators[3]. The bottom right-hand part of the diagram corresponds to the validity conditions given in figure 3.7.

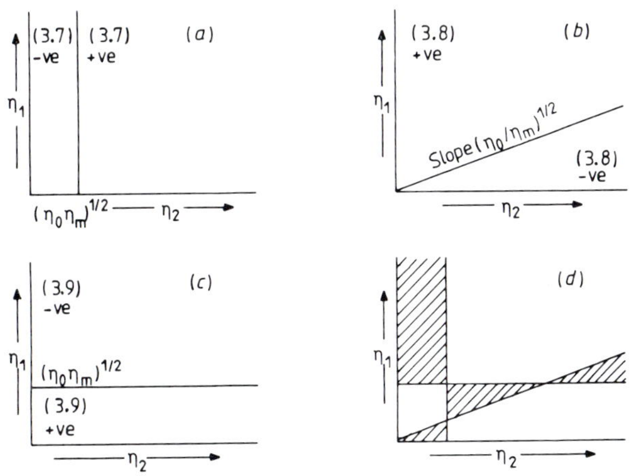

Figure 3.8 The construction of a Schuster diagram. (*a*), (*b*) and (*c*) are combined in one diagram in (*d*) and the shaded areas are those in which real solutions exist.

One useful coating is given by the area at the top left-hand edge of the diagram where $\eta_1 \geqslant (\eta_0\eta_m)^{1/2} \geqslant \eta_2$. For germanium at normal incidence in air,

$$(\eta_0\eta_m)^{1/2} = 2.0.$$

There is no upper limit to the magnitude of η_1, which can be conveniently chosen to be germanium with index 4.0, while η_2 can be magnesium fluoride with index 1.38, didymium fluoride with index 1.57, cerium fluoride with index 1.59, or any other similar material. The advantage of this arrangement is that the low-index film, which tends to be less robust, is protected by the high-index layer. Germanium layers are particularly good in this respect. Figure 3.9 gives an example of this type of coating. Generally, the total thickness, as in the example, is rather thinner than a quarter-wave, which adds to the durability. Cox[4] has published a paper discussing a number of different possibilities along these lines.

Unfortunately, this type of double-layer coating tends to have rather narrower useful ranges than the single-layer coating, which may itself not be broad enough for certain applications. It is possible to broaden the region of low reflectance by using two, or even more, layers. The approach generally adopted is to choose layer thicknesses which are whole numbers of quarter-waves, and then to determine the refractive indices which should be used to give the desired performance.

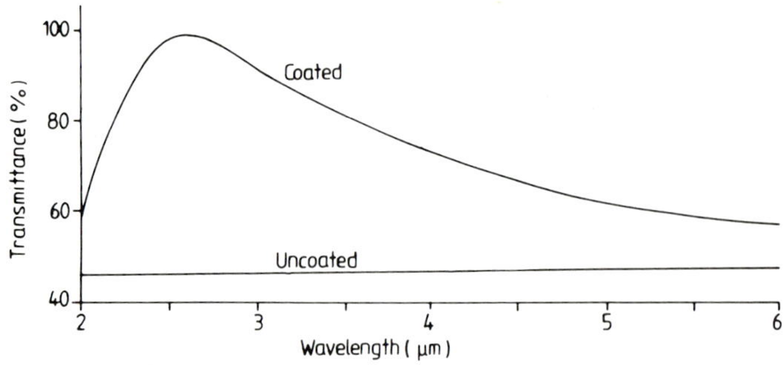

Figure 3.9 Transmittance of a germanium plate with two-layer antireflection coatings of MgF_2 ($nd = \lambda/4$ at 1.03 μm) and germanium ($nd = \lambda/4$ at 0.61 μm), the germanium being the outermost layer. (After Cox[4]).

An effective coating is one consisting of two quarter-wave layers (see figure 3.10). The appearances of the vector diagram at three different wavelengths is shown in (*a*), (*b*) and (*c*). At $\lambda = \frac{3}{4}\lambda_0$ and $\lambda = \frac{3}{2}\lambda_0$ the three vectors in the triangle are inclined at 60° to each other. Provided the vectors are all of equal length, the triangles will be closed and the reflectance will be zero at these wavelengths. This condition can be written

$$\frac{\eta_1}{\eta_0} = \frac{\eta_2}{\eta_1} = \frac{\eta_m}{\eta_2}$$

and solved for η_1 and η_2:

$$\eta_1^3 = \eta_0^2\eta_m \tag{3.10}$$

$$\eta_2^3 = \eta_0\eta_m^2. \tag{3.11}$$

The reflectance at the reference wavelength λ_0 where the layers are quarter-waves is given by

$$R = \left(\frac{\eta_0 - (\eta_1^2/\eta_2^2)\eta_m}{\eta_0 + (\eta_1^2/\eta_2^2)\eta_m}\right)^2$$

$$= \left(\frac{1 - (\eta_m/\eta_0)^{1/3}}{1 + (\eta_m/\eta_0)^{1/3}}\right)^2$$

which is a considerable improvement over the bare substrate.

For germanium of refractive index 4.0 in air, at normal incidence, the values required for the indices are

$$n_1 = 1.59$$

$$n_2 = 2.50$$

and the reflectance at λ_0 is 5.6%. The theoretical curve of this coating is shown in figure 3.11(*a*). Theoretical and measured curves of a similar coating on arsenic trisulphide and triselenide are given in figures 3.11(*b*) and (*c*).

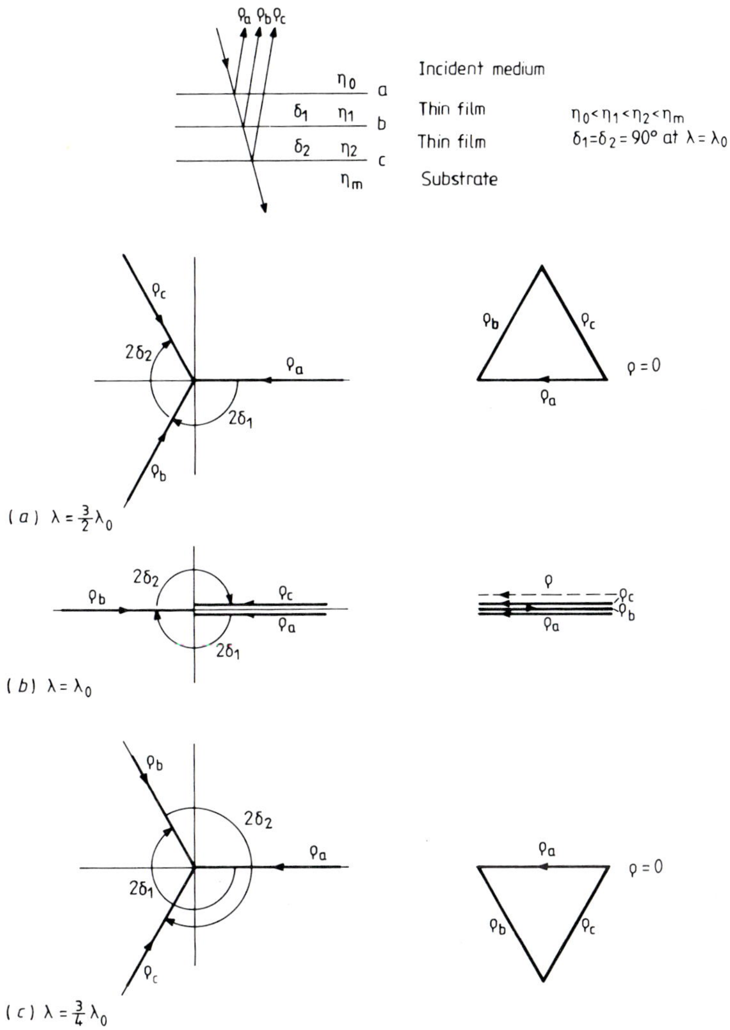

Figure 3.10 Vector diagrams for quarter–quarter antireflection coatings on a high-index substrate.

The coating just described is a special case of a general coating where the layers are of equal thickness. To compute the general conditions it is easiest to return to the analysis leading up to equations (3.5) and (3.6).

Let δ_1 be set equal to δ_2 and denoted by δ, where we recall that if λ_0 is the wavelength for which the layers are quarter-waves then

$$\delta = \frac{\pi}{2}\left(\frac{\lambda_0}{\lambda}\right).$$

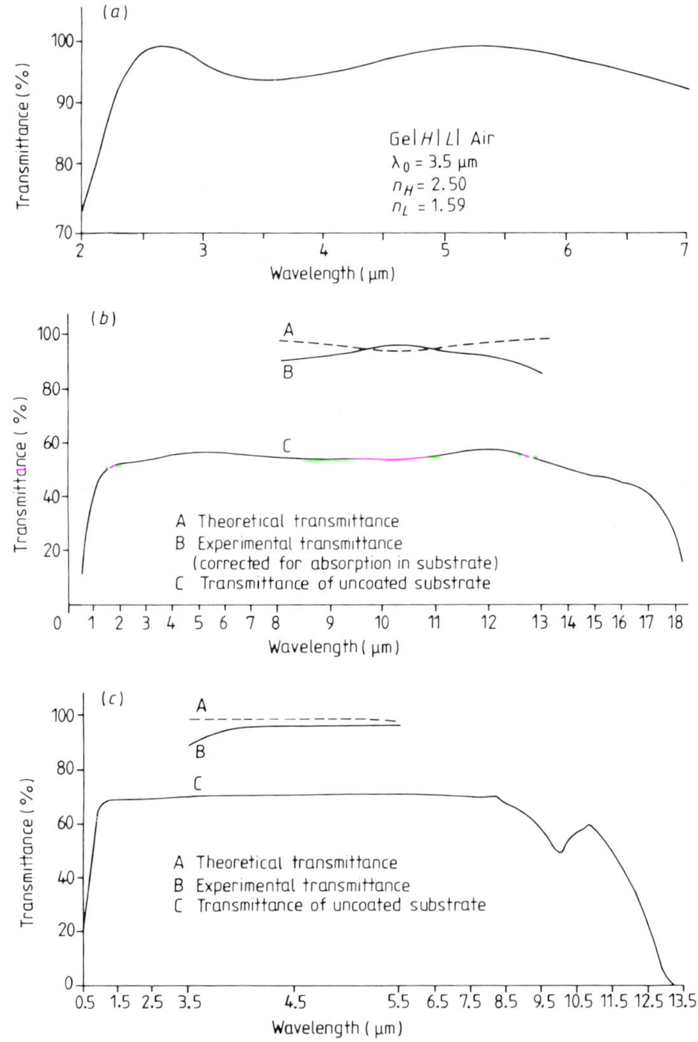

Figure 3.11 Double-layer antireflection coatings for high-index substrates. (*a*) Theoretical transmittance of a quarter–quarter coating on germanium (single surface). (*b*) Theoretical and measured transmittance of a similar coating on arsenic trisulphide glass (double surface). (*c*) Theoretical and measured transmittance of a similar coating on arsenic triselenide glass (double surface). ((*b*) and (*c*) by courtesy of Barr and Stroud Ltd.)

From equation (3.4)

$$\eta_2(\eta_0\eta_m - \eta_1^2) = \eta_1(\eta_2^2 - \eta_0\eta_m)$$

i.e.

$$\eta_0\eta_m = \eta_1\eta_2$$

which is a necessary condition for zero reflectance.

From equation (3.3) we find the wavelengths λ corresponding to zero reflectance:

$$\tan^2\delta = \frac{\eta_1\eta_2(\eta_m - \eta_0)}{\eta_1^2\eta_m - \eta_0\eta_2^2} = \frac{\eta_0\eta_m(\eta_m - \eta_0)}{\eta_1^2\eta_m - \eta_0\eta_2^2}.$$

If δ' is the solution in the first quadrant, then there are two solutions:

$$\delta = \delta' \qquad \text{or} \qquad \pi - \delta'$$

and the two values of λ are:

$$\lambda = \left(\frac{\pi/2}{\delta}\right)\lambda_0.$$

In all practical cases, η_m will be greater than η_0 and the above equation for $\tan^2\delta$ will have a real solution provided

$$\eta_1^2\eta_m - \eta_0\eta_2^2$$

is positive or zero. This expression is identical to expression (3.8).

Figure 3.12 gives the allowed values of η_1 and η_2 for germanium in air

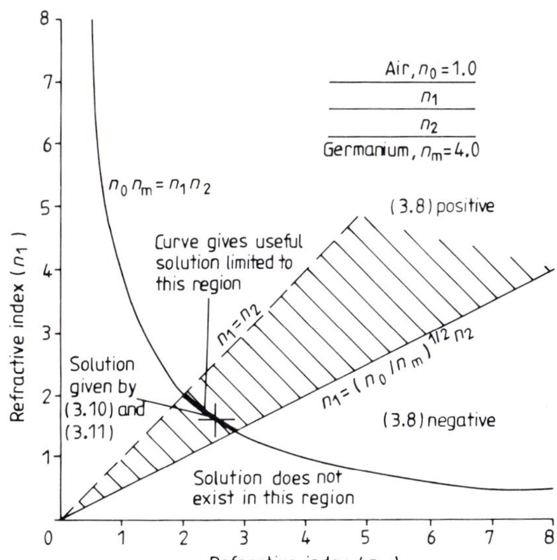

Figure 3.12 A Schuster diagram showing possible values of film indices for a quarter–quarter coating on germanium.

plotted on a Schuster diagram assuming normal incidence. The optical admittances η are then numerically equal to the refractive indices n. The form of the coating is similar to that of figure 3.11. The reflectance rises to a maximum value at the reference wavelength λ_0 which is situated between the two zeros. The reflectance at λ_0 can be found quite simply. At this wavelength, $\delta = \pi/2$ and the layers are quarter-waves. The optical admittance is given, therefore, by

$$\frac{\eta_1^2}{\eta_2^2}\eta_m$$

and the reflectance by

$$R = \left(\frac{\eta_0 - (\eta_1^2/\eta_2^2)\eta_m}{\eta_0 + (\eta_1^2/\eta_2^2)\eta_m}\right)^2.$$

We are considering cases where η_m is large. For $\eta_1 = \eta_2$ the reflectance at λ_0 is that of the bare substrate. If $\eta_1 > \eta_2$ the reflectance is even higher. Thus, for the solution to be at all useful, η_1 should be less than η_2 and the region where this condition holds is indicated on the diagram.

Multilayer coatings

Figure 3.13 shows a vector diagram for a three-layer coating on germanium. Each layer is a quarter-wave thick at λ_0. If $\eta_m > \eta_3 > \eta_2 > \eta_1 > \eta_0$, then the vectors will oppose each other, as shown, at $\frac{2}{3}\lambda_0$, λ_0, and $2\lambda_0$, and, provided the vectors are all of equal length, will completely cancel at these wavelengths, giving zero reflectance.

This coating is similar to the quarter–quarter coating of figure 3.10, but where the two zeros of the two-layer coating are situated at $\frac{3}{4}\lambda_0$ and $\frac{3}{2}\lambda_0$, those of this three-layer coating stretch from $\frac{2}{3}\lambda_0$ to $2\lambda_0$, a much broader region.

The condition for the vectors to be of equal length is

$$\frac{\eta_1}{\eta_0} = \frac{\eta_2}{\eta_1} = \frac{\eta_3}{\eta_2} = \frac{\eta_m}{\eta_3}$$

which with some manipulation becomes

$$\eta_1^4 = \eta_0^3\eta_m$$
$$\eta_2^4 = \eta_0^2\eta_m^2$$
$$\eta_3^4 = \eta_0\eta_m^3.$$

For germanium in air at normal incidence

$$n_0 = 1.00 \qquad n_m = 4.0$$

and the refractive indices required for the layers are

$$n_1 = 1.41$$
$$n_2 = 2.00$$
$$n_3 = 2.83.$$

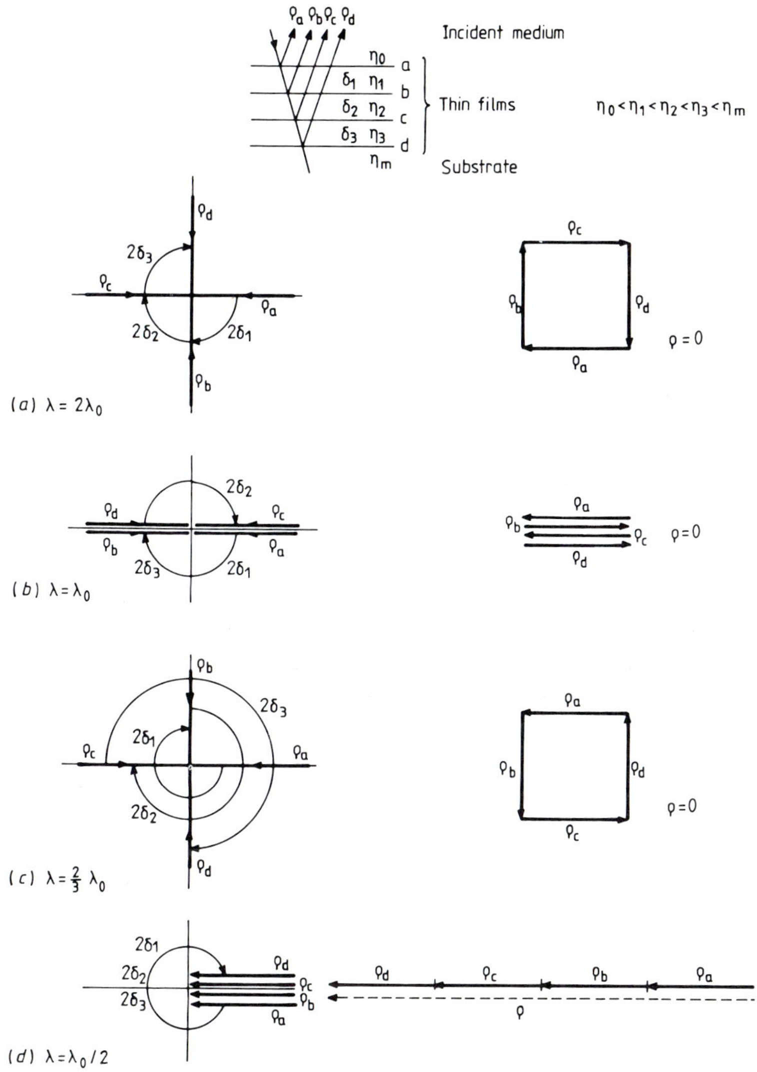

Figure 3.13 Vector diagram for a quarter–quarter–quarter coating on a high-index substrate.

A coating which is not far removed from these theoretical figures is silicon, next to the substrate, of index 3.3, followed by cerium oxide of index 2.2, followed by magnesium fluoride, index 1.35. The performance of such a coating with $\lambda_0 = 3.5$ μm is shown in figure 3.14. This coating, along with other one- and two-layer coatings for the infrared, is described by Cox *et al*[5]. The exact theory of this coating may be developed in the same way as that of the two-layer coating, but the calculations are more involved.

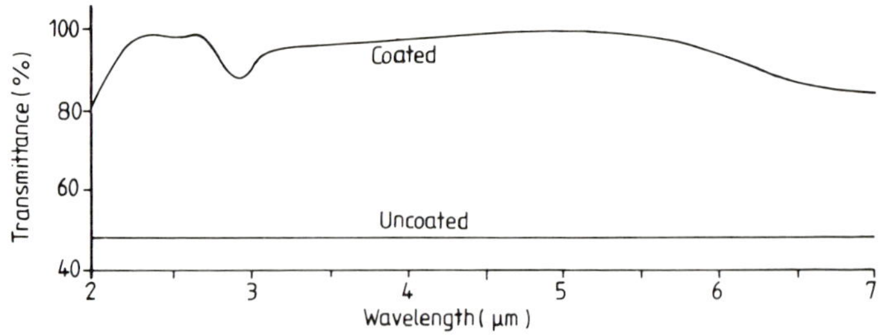

Figure 3.14 Measured transmittance of a germanium plate with coatings consisting of $MgF_2 + CeO_2 + Si$ ($n_1d_1 = n_2d_2 = n_3d_3 = \lambda/4$ at 3.5 μm). (After Cox, Hass and Jacobus[5].)

It is relatively easy to extend the vector method to deal with four layers, where the zeros of reflectance are found at $\frac{5}{8}\lambda_0, \frac{5}{6}\lambda_0, \frac{5}{4}\lambda_0$, and $\frac{5}{2}\lambda_0$: an even broader region than the three-layer coating. Five layers are equally straightforward. Whether or not such coatings are of practical value depends very much on the application. For many purposes the two-layer coating is quite adequate.

The addition of an extra layer makes the exact theory of the three-layer coating very much more involved than that of the two-layer. The number of possible groups of designs is enormous. It therefore becomes profitable to employ techniques which, rather than calculate performance in detail, simply indicate arrangements which are likely to be capable of acceptable performance and eliminate those which are not. Performance can then be accurately calculated by the procedures of chapter 2.

A particularly useful technique of this type has been developed by Musset and Thelen[6]. It is based on Smith's method, that is, the method of effective interfaces. We recall from chapter 2 that this involves the breaking down of the assembly into two subsystems. These we can label a and b. The overall transmittance of the multilayer is then given by

$$T = \left(\frac{T_a T_b}{(1 - R_a^{1/2} R_b^{1/2})^2}\right) \times \left[1 + \frac{4R_a^{1/2} R_b^{1/2}}{(1 - R_a^{1/2} R_b^{1/2})^2} \sin^2\left(\frac{\phi_a + \phi_b - 2\delta}{2}\right)\right]^{-1}. \tag{3.12}$$

We assume that there is no absorption, so that $T_a = 1 - R_a$ and $T_b = 1 - R_b$.

Both of the expressions multiplied together on the right-hand side of equation (3.12) have maximum possible values of unity and for maximum transmittance, therefore, both must be separately maximised. The first expression

$$\frac{T_a T_b}{(1 - R_a^{1/2} R_b^{1/2})^2}$$

will be unity if, and only if, $R_a = R_b$, while the second,

$$\left[1 + \frac{4R_a^{1/2}R_b^{1/2}}{(1 - R_a^{1/2}R_b^{1/2})^2}\sin^2\left(\frac{\phi_a + \phi_b - 2\delta}{2}\right)\right]^{-1}$$

will be unity if, and only if,

$$\sin^2\left(\frac{\phi_a + \phi_b - 2\delta}{2}\right) = 0.$$

The conditions for a perfect antireflection coating are then

$$R_a = R_b$$

called the *amplitude condition* by Musset and Thelen, and

$$\frac{\phi_a + \phi_b - 2\delta}{2} = m\pi \quad (m = 0, \pm 1, \pm 2 \ldots)$$

called the *phase condition*. The amplitude condition is a function of the two subsystems. The phase condition can be satisfied by adjusting the thickness of the spacer layer. The amplitude condition can, using a method devised by Musset and Thelen, be satisfied for all wavelengths, but it is difficult to satisfy the phase condition except at a limited number of discrete wavelengths. At other wavelengths the performance departs from ideal to a variable degree.

The transmittance and reflectance of a multilayer remain constant when the optical admittances are all multiplied by a constant factor or when they are all replaced by their reciprocals, in both cases keeping the optical thicknesses constant. These properties can readily be demonstrated from the structure of the characteristic matrices[7]. They enable the design of pairs of substructures having identical reflectance so that only the phase condition need be satisfied for perfect antireflection. We can, following Musset and Thelen, imagine a multilayer consisting of two subsections a and b, as shown in figure 3.15, with a medium of admittance η_i in between. At this stage we put no restrictions on this medium in terms either of refractive index or thickness but, as we shall see, they will become defined at a later stage. Subsection a is bounded by η_m on one

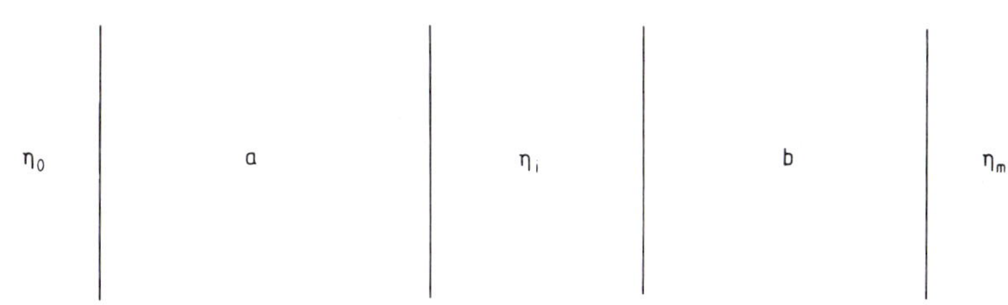

Figure 3.15 Multilayer antireflection coating consisting of two subsystems a and b separated by a central layer.

side and η_i on the other, while b is bounded in the same way by η_i and η_0. We can now apply the appropriate rules for ensuring that the amplitude condition is satisfied. We set up any subsystem a and then convert it into subsystem b by retaining the optical thicknesses and either multiplying the admittances by a constant multiplier, or taking the reciprocals of the admittances and multiplying them by a constant multiplier. Systems derived by the former procedure are classified by Musset and Thelen as of Type I, those by the latter as Type II.

For Type I systems we must have

$$\eta_m f = \eta_i$$
$$\eta_i f = \eta_0$$

so that

$$\eta_i = (\eta_0 \eta_m)^{1/2}$$

and

$$f = (\eta_0/\eta_m)^{1/2}.$$

In this way, any η_a gives a corresponding η_b of $\eta_a(\eta_0/\eta_m)^{1/2}$.

Type II systems, on the other hand, convert so that

$$f/\eta_m = \eta_0$$
$$f/\eta_i = \eta_i$$
$$f/\eta_a = \eta_b$$

i.e.

$$\eta_i = (\eta_0\eta_m)^{1/2} \qquad \text{and} \qquad f = \eta_0\eta_m$$

so that any η_a gives a corresponding η_b of $\eta_0\eta_m/\eta_a$.

There are no restrictions on layer thickness or on the number of layers in each subsystem except that they must be equal in number, and it is simpler if quarter-wave layers are used. Once the individual subsystems a and b are established, the amplitude condition is automatically satisfied at all wavelengths and it remains to satisfy the phase condition. This involves the coupling arrangement. It is impossible to meet the phase condition at all wavelengths and the problem is so complex that it is best to take the easy way out and adopt a layer of admittance η_i with thickness zero, in which case the layer is omitted, or a quarter-wave like the remaining layers of the assembly.

The method can be illustrated by application to the antireflection of germanium at normal incidence. In this case, $n_0 = 1.00$ and $n_m = 4.00$. Hence $n_i = (n_0 n_m)^{1/2} = 2.0$ in both Type I and II systems. First of all we take, for subsystem a, a straightforward single quarter-wave matching the substrate to the coupling medium:

n_1	n_a	n_m
	$(n_i n_m)^{1/2}$	
2.0	2.826	4.0

Subsystem b is then, for both Type I and II systems,

n_0	n_b	n_i
1.0	1.414	2.0

Putting the two subsystems together, we have either a two-layer coating if we permit the thickness of the coupling layer to shrink to zero, or a three-layer coating if the coupling layer is a quarter-wave. In the former case we have the design:

Air	1.414	2.828	Ge
1.0	$0.25\lambda_0$	$0.25\lambda_0$	4.0

and in the latter:

Air	1.414	2.0	2.828	Ge
1.0	$0.25\lambda_0$	$0.25\lambda_0$	$0.25\lambda_0$	4.0

The first design gives a single minimum. The second, which is similar to the three-layer design already obtained by the vector method, has a broad three-minimum characteristic (figure 3.16).

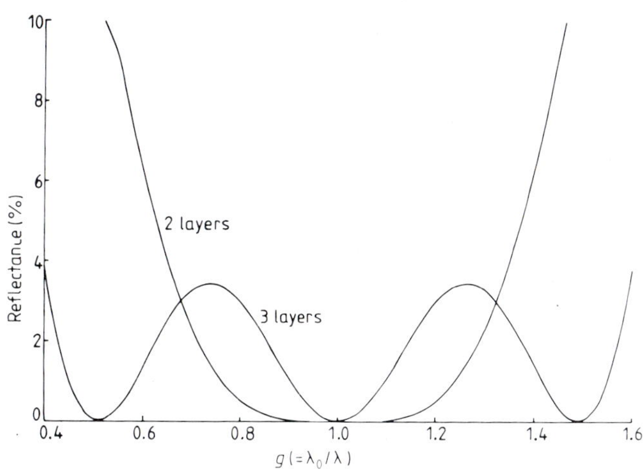

Figure 3.16 Theoretical performance of antireflection coatings on germanium designed by the method of Mussett and Thelen[6].

2 layers:

Air	1.414	2.828	Ge
1.0	$0.25\lambda_0$	$0.25\lambda_0$	4.0

3 layers:

Air	1.414	2.00	2.828	Ge
1.0	$0.25\lambda_0$	$0.25\lambda_0$	$0.25\lambda_0$	4.0

The subsystems need not be perfect matching systems for n_m to n_i and n_i to n_0. We could, for instance, use

$$n_0 = 1.0$$
$$n_b = (1.0 \times 4.0)^{1/3} = 1.587$$
$$n_m = 2.0$$

from the two-layer coating derived by the vector method. This gives complete two- and three-layer coatings, as follows.

Type I

Air	1.587	3.174	Ge
1.0	$0.25\lambda_0$	$0.25\lambda_0$	4.0

Air	1.587	2.0	3.174	Ge
1.0	$0.25\lambda_0$	$0.25\lambda_0$	$0.25\lambda_0$	4.0

Type II

Air	1.587	2.520	Ge
1.0	$0.25\lambda_0$	$0.25\lambda_0$	4.0

Air	1.587	2.0	2.520	Ge
1.0	$0.25\lambda_0$	$0.25\lambda_0$	$0.25\lambda_0$	4.0

The first of the Type II designs is identical with the vector method coating. Performance curves are given in figure 3.17.

Analytical expressions for calculating the positions of the zeros and the residual reflectance maxima of two- and three-layer coatings of the above types are given by Musset and Thelen. The method can be readily extended to four and more layers.

Young[8] has developed alternative techniques for coatings consisting of quarter-wave optical thicknesses based on the correspondence between the theory of thin-film multilayers and that of microwave transmission lines. He gives a useful set of tables for the design of multilayer coatings where all thicknesses are quarter-waves. Given the bandwidth and the maximum permissible reflectance it is possible quickly to derive the coating which meets the specification with the least number of layers. The method, of course, takes no account of the possibility of achieving the given indices in practice, as with many of the other methods we have been discussing, but the optimum solution is a very useful point of departure in the design of coatings using real indices.

ANTIREFLECTION COATINGS ON LOW-INDEX SUBSTRATES

Although the theory developed for antireflection coatings on high-index materials applies equally well to low-index materials, the problem is made

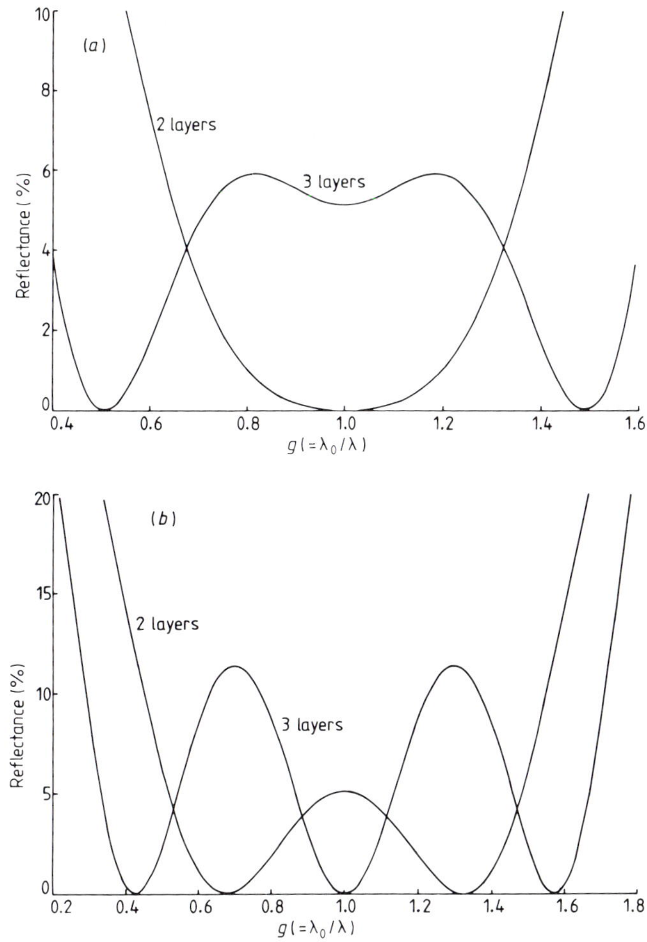

Figure 3.17 (*a*) Theoretical performance of Type I antireflection coatings on germanium designed by the method of Mussett and Thelen[6].

2 layers:

Air	1.587	3.174	Ge
1.0	$0.25\lambda_0$	$0.25\lambda_0$	4.0

3 layers:

Air	1.587	2.00	3.174	Ge
1.0	$0.25\lambda_0$	$0.25\lambda_0$	$0.25\lambda_0$	4.0

(*b*) Theoretical performance of Type II antireflection coatings on germanium designed by the method of Mussett and Thelen[6].

2 layers:

Air	1.587	2.520	Ge
1.0	$0.25\lambda_0$	$0.25\lambda_0$	4.0

3 layers:

Air	1.587	2.00	2.520	Ge
1.0	$0.25\lambda_0$	$0.25\lambda_0$	$0.25\lambda_0$	4.0

much more severe by the lack of any rugged thin-film materials of very low index. Magnesium fluoride, with an index of around 1.38, represents the lowest practical index which can be achieved. This immediately makes the manufacture of designs arrived at by the straightforward application of the techniques so far discussed, largely impossible. Design techniques for antireflection coatings on low-index materials are less well organised and involve much more intuition and trial and error than those for high-index materials.

The commonest low-index material is crown glass, and coatings are most frequently required for the visible region of the spectrum, which extends from around 400 nm to around 700 nm. For the purposes of most of the coatings which we will discuss here, we will assume an index of 1.52, although this varies a little with the particular type of glass and also with wavelength. Although much of what follows is applied directly to the antireflection coating of crown glass, the techniques apply equally well to the coating of other low-index materials. We begin with the simplest coating, a single layer.

The single-layer antireflection coating

We can make use of the expressions already developed for high-index materials.

The optimum single-layer coating is a quarter-wave optical thickness for the central wavelength λ_0 with optical admittance given by

$$\eta_f = (\eta_0 \eta_m)^{1/2}. \tag{3.13}$$

For crown glass in air, this represents

$$\eta_f = (1.0 \times 1.52)^{1/2} = 1.23.$$

As already mentioned, the lowest useful film index which can be obtained at present is that of magnesium fluoride, around 1.38 at 500 nm. While not ideal, this does give a worthwhile improvement. The reflectance at the minimum is given by

$$R = \left(\frac{\eta_0 - \eta_f^2/\eta_m}{\eta_0 + \eta_f^2/\eta_m} \right)^2 \tag{3.14}$$

i.e. 1.3% per surface.

At angles of incidence other than normal, the phase thickness of the layer is reduced, so that for a given layer thickness the wavelength corresponding to the minimum becomes shorter. The optical admittance appropriate to the angle of incidence and the plane of polarisation should also be used in calculating the reflectance. Figure 3.18 indicates the way in which the reflectance of a single layer of magnesium fluoride on a substrate of index 1.52 can be expected to vary with angle of incidence.

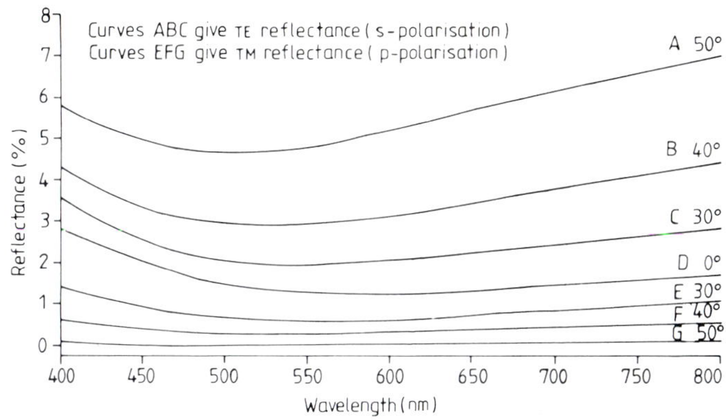

Figure 3.18 The computed reflectance at various angles of incidence of a single surface of glass of index 1.52 coated with a single layer of magnesium fluoride of index 1.38 and optical thickness at normal incidence one quarter-wave at 600 nm.

Two-layer antireflection coatings

The single-layer coating cannot achieve zero reflectance even at the minimum because of the absence of suitable low-index materials. Instinct suggests that a thin layer of high-index material placed next to the substrate might make it appear to have a higher index so that a subsequent layer of magnesium fluoride would be more effective. This proves to be the case. Two-layer coatings have already been considered with regard to high-index substrates and a complete analysis has been derived.

We can study the Schuster diagram (page 81) for coatings on glass of index 1.52, and this is reproduced as figure 3.19. We can assume 1.38 as the lowest possible index, while a realistic upper bound to the range of possible indices is 2.45. Possible solutions are then limited to the shaded area of the diagram. This area is bounded by the lines

$$\eta_1 = 1.38 \qquad \eta_2 = 2.45 \qquad \eta_1 = \eta_2(\eta_0/\eta_m)^{1/2}.$$

Solutions on the line

$$\eta_1 = \eta_2(\eta_0/\eta_m)^{1/2}$$

will consist of two quarter-wave layers. Solutions elsewhere will consist of two layers of unequal thickness, one greater and the other less than a quarter-wave. The thicknesses are given by the expressions

$$\tan^2\delta_1 = \frac{(\eta_m - \eta_0)(\eta_2^2 - \eta_0\eta_m)\eta_1^2}{(\eta_1^2\eta_m - \eta_0\eta_2^2)(\eta_0\eta_m - \eta_1^2)} \tag{3.15}$$

$$\tan^2\delta_2 = \frac{(\eta_m - \eta_0)(\eta_0\eta_m - \eta_1^2)\eta_2^2}{(\eta_1^2\eta_m - \eta_0\eta_2^2)(\eta_2^2 - \eta_0\eta_m)}. \tag{3.16}$$

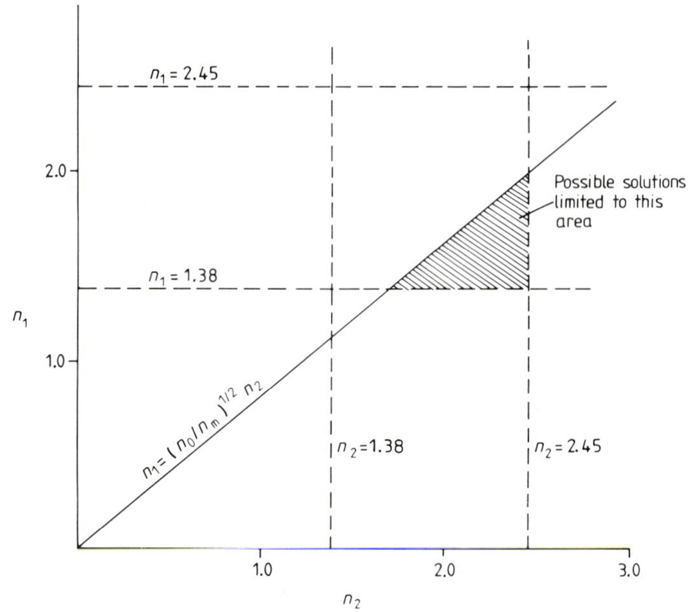

Figure 3.19 A Schuster diagram for two-layer coatings on glass ($n = 1.52$) in air ($n = 1.0$). Possible layer indices are assumed to be limited to the range 1.38–2.45.

As an example, we can take a value of 2.2 for the high-index layer, corresponding to cerium oxide, and of 1.38, corresponding to magnesium fluoride, for the low-index layer. The two possible solutions are then

$$\delta_1/2\pi = 0.3208 \qquad \delta_2/2\pi = 0.05877$$

and

$$\delta_1/2\pi = 0.1792 \qquad \delta_2/2\pi = 0.4412$$

respectively.

These two solutions are plotted in figure 3.20 and it can be clearly seen that the characteristic of the coating is a single minimum with a narrower bandwidth than the single layer and that the broader of the two possible solutions is associated with the thinner high-index layer. The coating is also an effective one for other values of substrate index. The higher the index of the substrate, the thinner need the high-index layer be and the broader is the characteristic of the coating.

We can follow Catalan[2] and plot curves showing how the values of δ_1 and δ_2 vary with the index of the layer next to the substrate. Such curves are shown in figure 3.21 and from them several points of interest emerge. First, as already predicted by the Schuster plot, there is a region in which no solution is possible. Second, and more important, the curves flatten out as the index of the layer increases and changes in refractive index are accompanied by only small

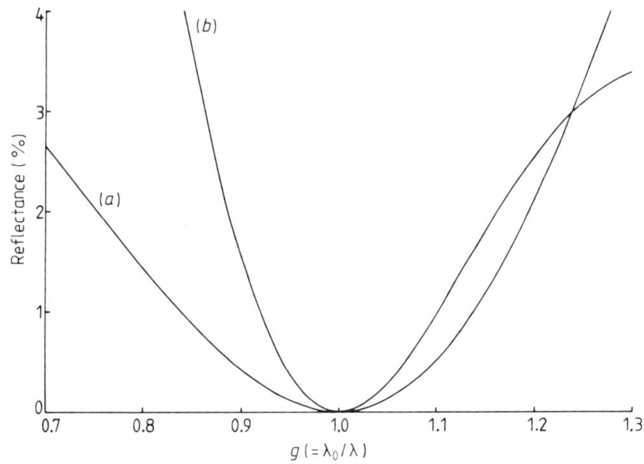

Figure 3.20 Two-layer antireflection coatings for glass.

(*a*)	Air 1.0	1.38 $0.321\lambda_0$	2.20 $0.0588\lambda_0$	Glass 1.52
(*b*)	Air 1.0	1.38 $0.179\lambda_0$	2.20 $0.441\lambda_0$	Glass 1.52

(*a*), the broader characteristic, is usually selected. Because of the characteristic single minimum the coating is often known as a V-coat.

changes in optical thickness. One of the problems in manufacturing coatings is the control of the refractive index of the layers, particularly of the high-index layers, and the curves indicate good stability of the performance of the coating in this respect.

The equations are not limited to normal incidence. Catalan has also computed, for various angles of incidence, values of reflectance of a two-layer coating consisting of bismuth oxide, with index 2.45, and magnesium fluoride, with index 1.38, on glass of index 1.5. Curves showing the variation of reflectance with angle of incidence are given in figures 3.22 and 3.23. The performance is very good up to an angle of incidence of 20° but beyond that it begins to fall off.

It may also be necessary to design coatings for angles of incidence other than normal. Turbadar[9] has considered this problem and published designs for angle of incidence 45°. The materials were once again bismuth oxide and magnesium fluoride, of indices 2.45 and 1.38 respectively, on glass of index 1.5. Four possible solutions were given, which are reproduced as table 3.1 where the bismuth oxide is next to the glass.

A large number of performance curves of the various designs under different conditions, including the effect of errors, were produced. Particularly valuable

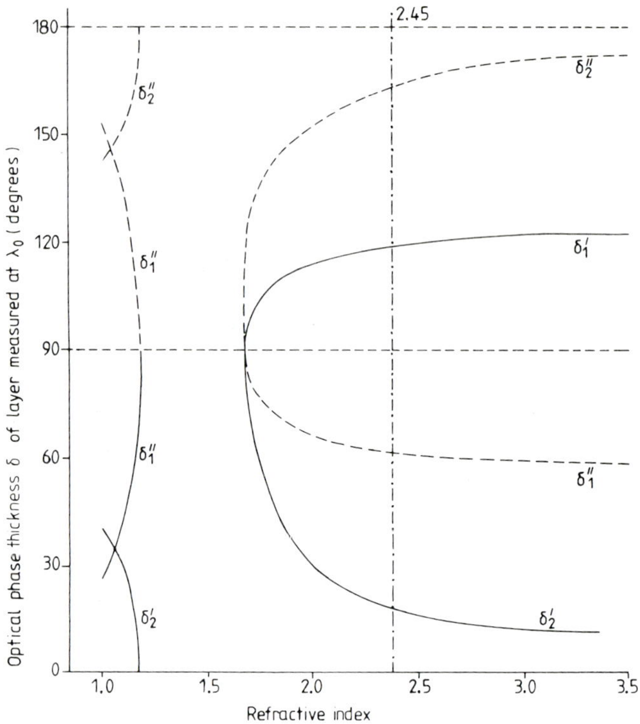

Figure 3.21 Optimum thicknesses of the layers in a double-layer antireflection coating at normal incidence. δ_1 and δ_2, the optical phase thicknesses given by equations (3.7) and (3.8), are plotted against n_2, the refractive index of the high-index layer. The low-index layer is assumed to be magnesium fluoride of index 1.38 and the coating is deposited on glass of index 1.50. Two pairs of solutions of (3.7) and (3.8) are possible for each set of refractive indices and are denoted by δ'_1 and δ'_2, and δ''_1 and δ''_2. The value, 2.45, of refractive index, shown by the broken line, corresponds to bismuth oxide and was used by Catalan in his calculations. (After Catalan[2].)

are plots of equireflectance contours over a grid of angle of incidence against wavelength. These are given in figure 3.24.

It is useful to consider an admittance plot for a two-layer coating, which can be a great help in visualising performance. The plot consist of two circles, the first corresponding to the low-index layer η_1, which passes through the point $(\eta_0, 0)$ if the reflectance is to be zero and which must, therefore, also pass through the point $(\eta_1^2/\eta_0, 0)$. The second circle corresponds to the high-index layer η_2 which must pass through the point $(\eta_m, 0)$ corresponding to the substrate and, therefore, also through the point $(\eta_2^2/\eta_m, 0)$. Provided that these two circles intersect, then a two-layer antireflection coating of this type is possible. Such a plot is shown in figure 3.25. There are two possible arrangements of the admittance circles which will give the required zero

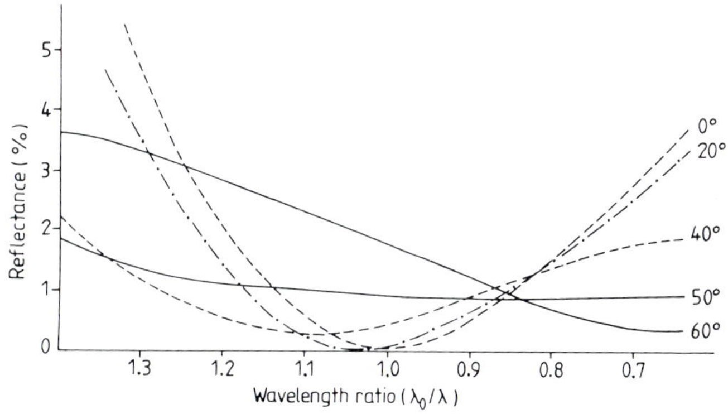

Figure 3.22 Theoretical p-reflectance (TM) as a function of wavelength ratio g $(= \lambda_0/\lambda)$ of a double-layer antireflection coating. $n_0 = 1.00$; $n_1 = 1.38$; $n_2 = 2.45$; $n_{sub} = 1.50$. (After Catalan[2].)

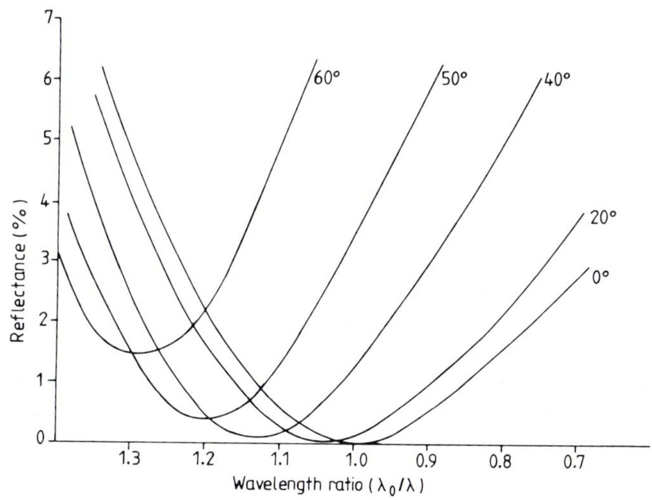

Figure 3.23 Theoretical s-reflectance (TE) as a function of wavelength ratio g $(= \lambda_0/\lambda)$ of a double-layer antireflection coating. $n_0 = 1.00$; $n_1 = 1.38$; $n_2 = 2.45$; $n_{sub} = 1.50$. (After Catalan[2].)

reflectance. If we recall that a semicircle starting and finishing on the real axis corresponds to a quarter-wave, then we can see that either the high-index layer will be thinner than a quarter-wave with the low-index layer thicker, or the reverse, just as we have already established.

The special case where the layers are both quarter-waves can then be seen to occur when the η_2 circle just touches the η_1 circle internally. In that case

$$\eta_1^2/\eta_0 = \eta_2^2/\eta_m$$

Table 3.1

		Bismuth oxide	Magnesium fluoride
s-polarisation (TE wave)	S′	$0.065\ \lambda_0$	$0.376\ \lambda_0$
	S″	$0.457\ \lambda_0$	$0.206\ \lambda_0$
p-polarisation (TM wave)	P′	$0.021\ \lambda_0$	$0.382\ \lambda_0$
	P″	$0.501\ \lambda_0$	$0.201\ \lambda_0$

or

$$\eta_1 = \eta_2(\eta_0/\eta_m)^{1/2}$$

which is the equation of the oblique line in the Schuster plot. The admittance plot for $\lambda = \lambda_0$ and the theoretical performance curve for such a coating are shown in figure 3.26.

All the two-layer coatings considered so far exhibit one single minimum, which can be theoretically zero, at $\lambda = \lambda_0$. On either side of the minimum, the reflectance rises rather more rapidly than for the single-layer coating. An alternative two-layer coating makes use of the broadening effects of a half-wave layer to produce an improvement over the single-layer performance. A half-wave layer of index higher than the substrate is inserted between the substrate and the quarter-wave low-index film. If magnesium fluoride, of index 1.38, is once again chosen for the low-index film, then, for a substrate of index 1.52, the high-index layer should preferably be in the range 1.7–1.9, while, for a substrate of index 1.7, the range should be increased to 1.9–2.1. The way in which the half-wave layer acts to improve the performance can readily be understood by sketching an admittance plot, as in figure 3.27. The opening of the end of the high-index locus as the value of g decreases from 1.0 partially compensates for the shortening of the low-index locus. A similar effect exists as g increases from 1.0, when the lengthening of the low-index locus is compensated by an overlapping with the high-index locus. The half-wave layer must be of an index higher than that of the substrate, otherwise the opening of the half-wave circle would pull the low-index locus even further from the point $g = 1.0$, hence increasing the reflectance further and effectively narrowing the characteristic. The important feature of the arrangement is that at the reference wavelength, the second quarter-wave portion of the half-wave layer and the following quarter-wave layer should have loci on the same side of the real axis.

Multilayer antireflection coatings

There is little further improvement in performance which can be achieved with two-layer coatings, given the limitations which exist in usable film indices. For higher performance, further layers are required.

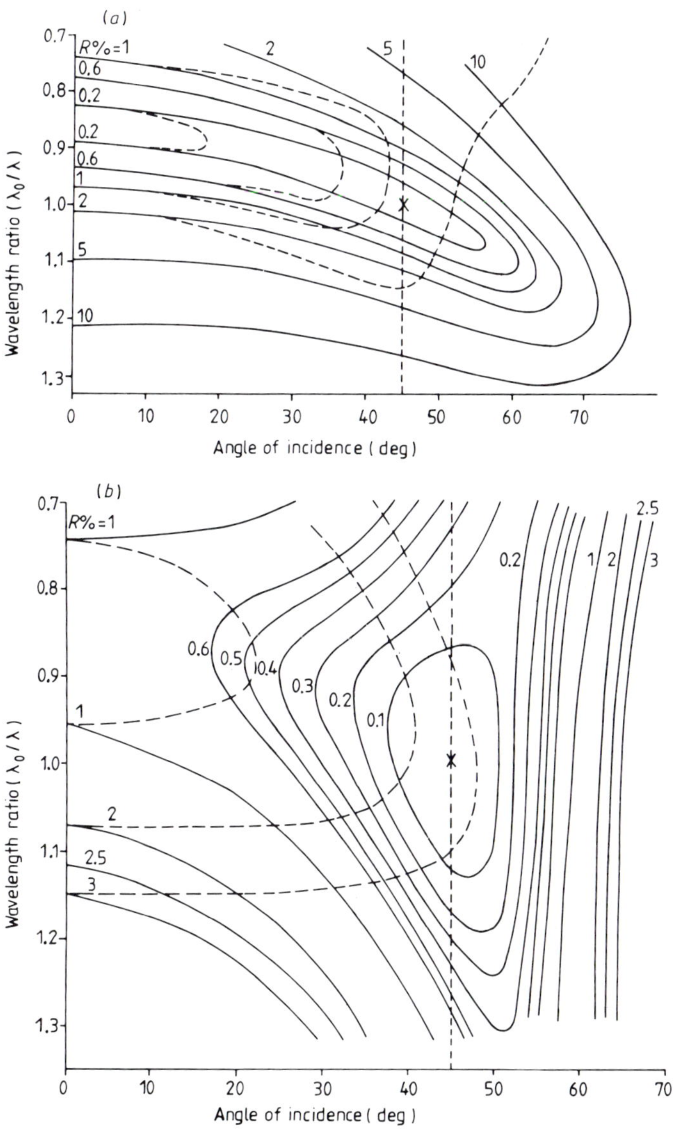

Figure 3.24 (*a*) Equireflectance contours for double-layer antireflection coatings on glass. $n_0 = 1.00$; $n_1 = 1.38$; $n_2 = 2.45$; $n_{sub} = 1.50$, with layer thicknesses optimised for s-polarisation (TE) at 45° angle of incidence, given by S′ in table 3.1. Solid curves s-reflectance (TE); broken curves p-reflectance (TM). (After Turbadar[9].)
(*b*) Equireflectance contours for double-layer antireflection coatings on glass. $n_0 = 1.00$; $n_1 = 1.38$; $n_2 = 2.45$; $n_{sub} = 1.50$, with layer thicknesses optimised for p-polarisation (TM) at 45° angle of incidence, given by P′ in table 3.1. Solid curves p-reflectance (TM); broken curves s-reflectance (TE). (After Turbadar[9].)

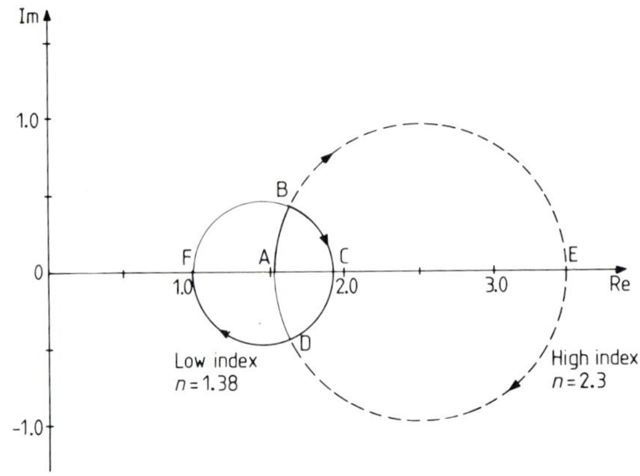

Figure 3.25 Admittance diagram showing the two possible double-layer antireflection coating designs.

Thetford[10] has devised a technique for designing three-layer antireflection coatings where the reflectance is zero at two wavelengths and low over a wider range than in the two-layer coating. The arrangement consists of a layer of intermediate index next to the substrate, followed by a high-index layer and finally by a low-index layer on the outside. The indices are chosen at the outset and the method yields the necessary layer thicknesses. There is an advantage in specifying layer indices rather than thicknesses because of the limited range of materials available.

The technique is based on both the vector method and Smith's method (the method of effective interfaces). We recall that the transmittance of an assembly will be unity if, and only if, the reflectances of the structures on either side of the chosen spacer layer are equal and the thickness of the spacer layer is such that the phase change suffered by a ray of the appropriate wavelength, after having completed a round trip in the layer, being reflected once at each of the boundaries, is zero or an integral multiple of 2π. If the phase thickness of the layer is δ, then this is equivalent to saying that

$$\phi + \phi' - 2\delta = 2s\pi \qquad s = 0, \pm 1, \pm 2 \ldots \tag{3.17}$$

where ϕ and ϕ' are the phases of the amplitude reflection coefficients at the boundaries of the layer. Thetford split the assembly into two parts on either side of the middle layer and then computed the two amplitude reflection coefficients by the vector method, combining the calculations on one diagram. He chose thicknesses for the layers which made the reflectances equal at a reference wavelength. He then found expressions for the change in reflectance with wavelength for each of the two structures, and, from them, a second value of wavelength, shorter than the first, at which the reflectances were again equal. The next step was to compute the thickness of the middle layer to satisfy the

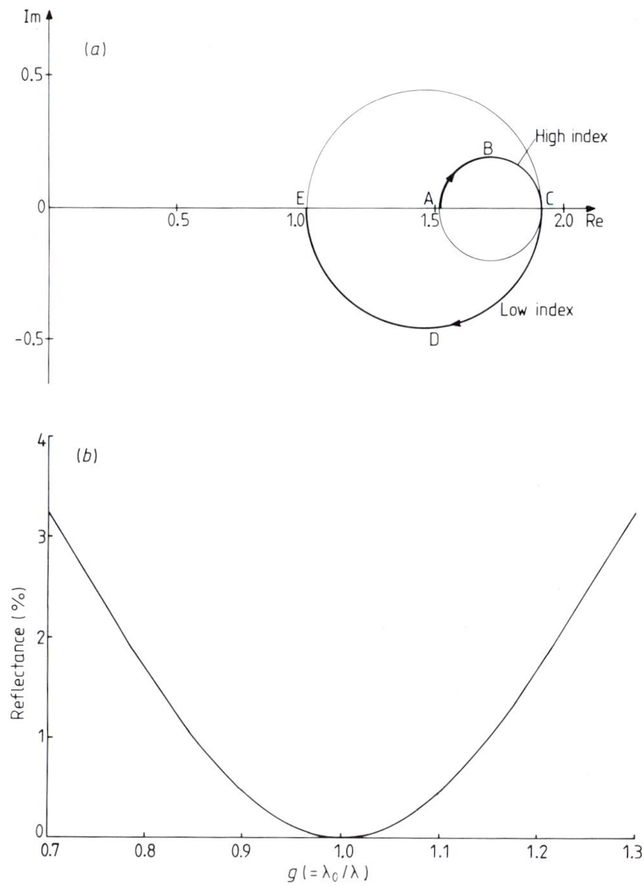

Figure 3.26 Special case of the two-layer antireflection coating where the layers become quarter waves and the two solutions of figure 3.25 merge into one. The design is:

Air	1.38	1.70	Glass
1.0	$0.25\lambda_0$	$0.25\lambda_0$	1.52

(*a*) The admittance locus. (*b*) The theoretical performance curve.

phase condition at the first wavelength and hence to give zero reflectance for the complete coating at that wavelength, and then to check whether or not the phase condition was also satisfied at the second wavelength. If it was, then the reflectance of the complete coating was known to be zero at this wavelength also and the design was complete. If it was not, then the procedure was repeated with slightly different initial conditions at the reference wavelength. This trial-and-error procedure turned out to be a very quick method of

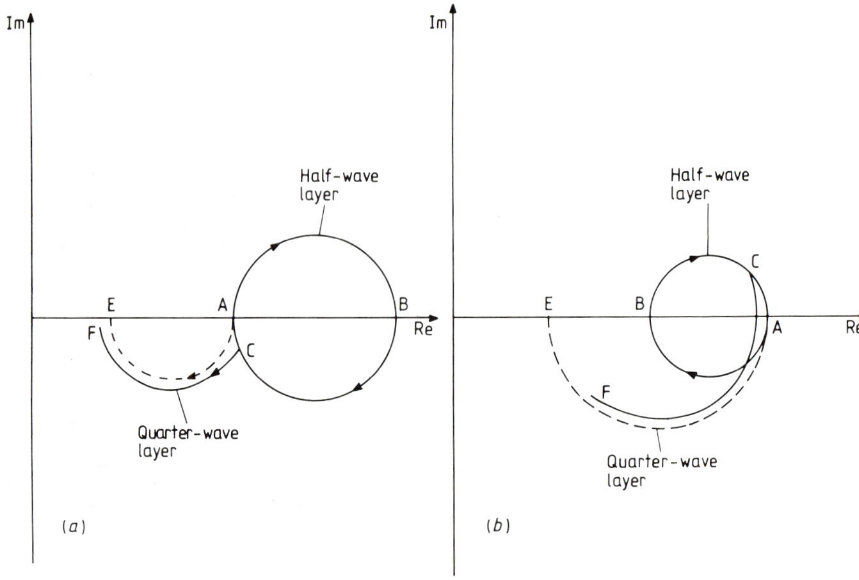

Figure 3.27 The operation of a half-wave flattening layer. The contour AE represents a low-index quarter-wave coating and in ABA a half-wave layer is inserted between it and the substrate. In (*a*) the half-wave is of index higher than the substrate and as g ($= \lambda_0/\lambda$) varies, the action of the half-wave keeps the end of the quarter-wave near the point E and the reflectance remains low. ABCF represents the locus with g somewhat less than unity. g greater than unity would give a similar effect with the point C now above the real axis and the loci slightly longer than full circle and semicircle. (*b*) shows the corresponding diagram for a low-index half-wave. Here the end point is dragged rapidly away from E as g varies and the reflectance rises rapidly. Flattening is therefore effective in (*a*) but not in (*b*). Note that the reflectance curve for another coating with half-wave flattening layer of design:

Air	1.38	1.90	Glass
1.0	$0.25\lambda_0$	$0.5\lambda_0$	1.52

is shown as curve (*a*) of figure 3.31. This latter coating is sometimes called a W-coat because of the shape of the characteristic.

arriving at the final solution. The only step which remained was the accurate calculation of the performance of the design as a check.

The three-layer coating is shown in figure 3.28. Thetford's notation has been altered to fit in with the practice in this book. The vector diagrams for the two structures are shown at (*b*) and (*c*) and then combined at (*d*), with vectors in such a position that the resultant amplitude reflection coefficients ρ and ρ' are equal in length but not necessarily in phase. In the solution shown, both ρ and ρ' are in the fourth quadrant. It is very easy to arrive at this initial condition. All that is required is a circle with centre the origin which cuts both the loci of

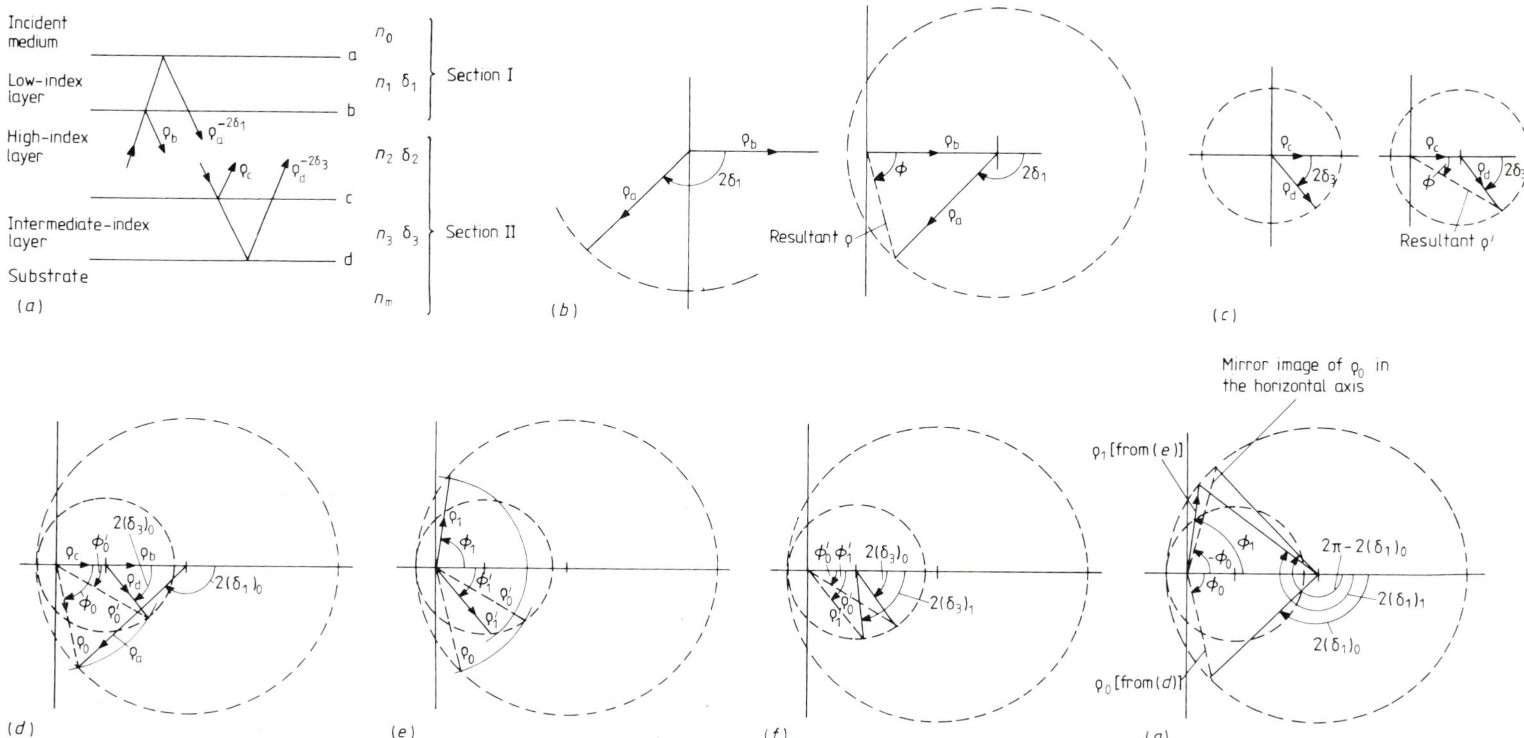

Figure 3.28 Thetford's method for antireflection coating design. (*a*) The three-layer coating split into two sections, I and II, on either side of the central high-index layer. (*b*) The vector diagram of section I. (*c*) The vector diagram of section II. (*d*) The vector diagrams of section I and II superimposed, showing one possible orientation of ρ and ρ' so that they are of equal length. (*e*) An alternative orientation of ρ and ρ', where again they are of equal length. The arrangement in (*d*) is also shown. The vectors ρ_a, ρ_b, ρ_c and ρ_d, have been omitted. (*f*) The two possible orientations of ρ' from (*d*) and (*e*). (*g*) The two possible orientations of ρ from (*d*) and (*e*). Also shown is the mirror image of ρ_0 in the horizontal axis. (After Thetford[10].)

vectors ρ_a and ρ_d. This initial condition we can take as corresponding to our reference wavelength λ_0. Figure 3.28 (*e*) shows a second solution for a shorter wavelength λ_1 plotted on top of the first. The values of δ_1 and δ_3 which correspond to this solution are given by λ_0/λ_1 times the values corresponding to λ_0, and ρ is now in the first quadrant while ρ' remains in the fourth. To find this second solution, Thetford has derived approximate expressions for the change in reflectance with change in wavelength which turn out to give surprisingly accurate results.

The reflectances corresponding to ρ and ρ' are given, from the diagram, by

$$\rho^2 = \rho_a^2 + \rho_b^2 + 2\rho_a\rho_b \cos 2\delta_1 \tag{3.18}$$

and

$$(\rho')^2 = \rho_c^2 + \rho_d^2 + 2\rho_c\rho_d \cos 2\delta_3. \tag{3.19}$$

For a reasonably small change in wavelengths we can find the corresponding change in ρ^2 and $(\rho')^2$ by differentiating equations (3.18) and (3.19), i.e.

$$\Delta(\rho^2) = -4\rho_a\rho_b \sin 2\delta_1 \times \Delta\delta_1$$

$$\Delta[(\rho')^2] = -4\rho_c\rho_d \sin 2\delta_3 \times \Delta\delta_3.$$

Now since the two values of $(\rho')^2$ in which we are interested are in the fourth quadrant, and well clear of any turning values, we can apply this approximate expression directly, giving

$$\Delta[(\rho')^2] = -4\rho_c\rho_d \sin 2(\delta_3)_0 \times \Delta\delta_3$$

$$\Delta\delta_3 = \left(\frac{\lambda_0}{\lambda_1} - 1\right)(\delta_3)_0$$

for the change in $(\rho')^2$ corresponding to the shift in wavelength from λ_0 to λ_1, where $(\delta_3)_0$ is the value at λ_0.

ρ^2, however, is not so simple. It passes through a turning value between the two solutions. Thetford observed that in figure 3.28(*c*) the mirror image of ρ_a in the horizontal axis would also give the same resultant ρ^2 (although with a different phase angle) and that this would be fairly near the desired solution. This new position of ρ_a has angle $2\delta_1$ with value $2\pi - 2(\delta_1)_0$ and a change in this angle of

$$\Delta\delta_1 = \left[\left(1 + \frac{\lambda_0}{\lambda_1}\right)(\delta_1)_0 - \pi\right] \tag{3.20}$$

would swing it round exactly into the correct position. We can therefore find the change in ρ^2 that we want by using the approximate expression, but calculating it as a change of $\Delta\delta_1$ (equation (3.20)) from this fictitious position of ρ_a. $\Delta(\rho^2)$ is then given by

$$\Delta(\rho^2) = -4\rho_a\rho_b \sin[2\pi - 2(\delta_1)_0]\left[\left(1 + \frac{\lambda_0}{\lambda_1}\right)(\delta_1)_0 - \pi\right]$$

$$= 4\rho_a\rho_b \sin 2(\delta_1)_0\left[\left(1 + \frac{\lambda_0}{\lambda_1}\right)(\delta_1)_0 - \pi\right].$$

We must now set $\Delta[(\rho')^2] = \Delta(\rho^2)$, which permits us to solve for λ_1. Next, we investigate the phase condition and the thickness of the middle layer.

From the vector diagram for the first solution we can find the phase angles ϕ_0 and ϕ'_0 associated with ρ and ρ' and λ_0. The necessary phase thickness of the middle layer to satisfy the condition for zero reflectance is given from equation (3.17) by

$$2(\delta_2)_0 = 2\pi + \phi_0 + \phi'_0$$

where we must remember to include the signs of ϕ_0 and ϕ'_0, (both negative in figure 3.28(*d*)) and where we have taken *s* as $+1$ to give the thinnest possible positive value for $(\delta_2)_0$. Next, from the vector diagram we find the values of phase angle ϕ_1 and ϕ'_0 associated with λ_1. If these satisfy the expression

$$2(\delta_2)_0 \frac{\lambda_0}{\lambda_1} = 2\pi + \phi_1 + \phi'_1 \qquad (3.21)$$

then we know we have a valid solution. The phase angles of the layers at λ_0 are then given by $(\delta_1)_0$, $(\delta_2)_0$ and $(\delta_3)_0$ respectively and the optical thicknesses of the layers in terms of a quarter-wave at λ_0 can be found by dividing by $\pi/2$. If, however, the phase condition is not met at λ_1, then it is necessary to go back to the beginning and try a new set of solutions. In fact, a satisfactory solution will be found quickly, especially if the error in equation (3.21) is plotted against, say, $(\delta_1)_0$.

One advantage which Thetford has pointed out for this type of coating is that once the phase condition has been satisfied at both λ_0 and λ_1, it will be approximately satisfied at all wavelengths between them. This means that the design will possess a broad region of low reflectance without any pronounced peaks of high reflectance. Some of Thetford's designs are shown in figure 3.29 which also demonstrates how the characteristic varies with the index of the middle layer. This coating is clearly a considerable improvement over the two-layer coating.

It is not easy to establish analytical expressions for the ranges of n_1, n_2 and n_3 which will give an acceptable reflectance characteristic. Generally, if the Argand diagram is not too far removed in appearance from the form of figure 3.28 where the two positions of ρ are near the minimum which corresponds to $2\delta_1 = \pi$, then a good antireflection coating will be obtained.

If it should be a requirement that only two values of refractive index rather than three be used in the construction of the coating, then it is possible to achieve a similar performance if four layers of alternate high and low index are used. Thetford[11] has used a similar technique for the design of such a coating. He split the coating (which has a high-index layer next to the glass) at the high-index layer nearest the air, so that the high–low combination next to the glass took the place of the intermediate index layer of the three-layer design. If the thicknesses of these two layers are fairly small, then an Argand diagram is obtained which is not too different from that for the three-layer design.

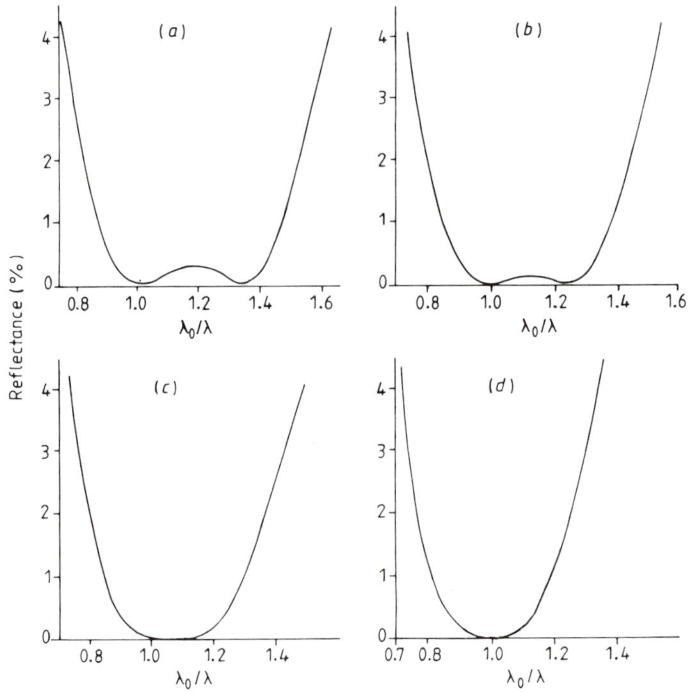

Figure 3.29 Calculated reflectance of some three-layer antireflection coatings designed by Thetford. The designs are as follows:

(*a*) $n_0 = 1.00$, $n_1 = 1.38$, $n_2 = 2.00$, $n_3 = 1.80$, $n_4 = n_s = 1.52$. $n_1d_1 = 0.205\lambda_0$, $n_2d_2 = 0.336\lambda_0$, $n_3d_3 = 0.132\lambda_0$.
(*b*) $n_0 = 1.00$, $n_1 = 1.38$, $n_2 = 2.10$, $n_3 = 1.80$, $n_4 = n_s = 1.52$. $n_1d_1 = 0.225\lambda_0$, $n_2d_2 = 0.359\lambda_0$, $n_3d_3 = 0.152\lambda_0$.
(*c*) $n_0 = 1.00$, $n_1 = 1.38$, $n_2 = 2.20$, $n_3 = 1.80$, $n_4 = n_s = 1.52$. $n_1d_1 = 0.227\lambda_0$, $n_2d_2 = 0.338\lambda_0$, $n_3d_3 = 0.170\lambda_0$.
(*d*) $n_0 = 1.00$, $n_1 = 1.38$, $n_2 = 2.40$, $n_3 = 1.80$, $n_4 = n_s = 1.52$. $n_1d_1 = 0.247\lambda_0$, $n_2d_2 = 0.445\lambda_0$, $n_3d_3 = 0.181\lambda_0$. (After Thetford[10].)

Because the expressions would be much more complicated in this case, Thetford did not attempt an analytical solution, but rather arrived at a design which appeared reasonable, by trial and error. The reflectance characteristic of such a design is shown in figure 3.30. This solution was then refined by C Butler, using a computer technique, to give optimum performance. This improved coating is also shown in figure 3.30.

Another design based on four layers of alternate high and low index has been published by C Reichert Optische Werke AG[12]. Full details of the design method are, unfortunately, not given. The thicknesses and materials of the particular result quoted are given in table 3.2. The published reflectance curve

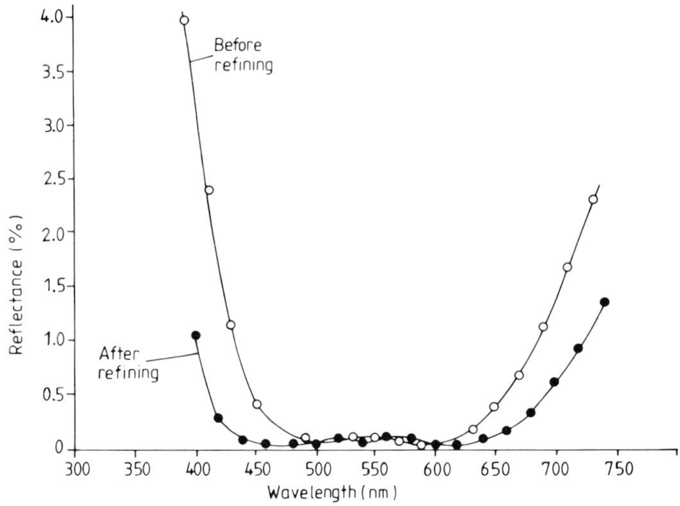

Figure 3.30 Calculated reflectance of four-layer antireflection coatings on glass showing the performance before and after the design was refined by computer. The two designs are: ○ Before refining: $n_0 = 1.00$, $n_1 = n_3 = 1.38$, $n_2 = n_4 = 2.10, n_3 = 1.80$, $n_5 = n_s = 1.52$. $n_1 d_1 = 0.21\lambda_0, n_2 d_2 = 0.37\lambda_0$, $n_3 d_3 = 0.036\lambda_0, n_4 d_4 = 0.070\lambda_0$. ● After refining: $n_0 = 1.00, n_1 = n_3 = 1.38, n_2 = n_4 = 2.10, n_3 = 1.80, n_5 = n_s = 1.52$. $n_1 d_1 = 0.216\lambda_0$, $n_2 d_2 = 0.458\lambda_0$, $n_3 d_3 = 0.072\lambda_0$, $n_4 d_4 = 0.049\lambda_0$. (Communicated by Thetford.)

Table 3.2

Material	Index	Optical thickness (nm)
Glass	1.52	Massive
TiO_2	2.28	54
MgF_2	1.37	56.5
TiO_2	2.28	78.5
MgF_2	1.37	161
Air	1.00	Massive

of this coating is slightly better than the unrefined curve of figure 3.30 but inferior to the refined curve.

There are also many coatings which involve layers of either quarter-wave or half-wave optical thicknesses. A number of these can be looked upon as modifications of some of the two-layer designs already considered.

First, we take the two-layer coating consisting of a half-wave layer next to the substrate followed by a quarter-wave layer. This has a peak reflectance in the centre of the low-reflectance region. This peak corresponds to the minimum reflectance of a single-layer coating because the inner layer, being a half-wave at that wavelength, is an absentee. We can reduce the peak but retain to some extent the flattening effect of the half-wave layer by splitting it into two quarter-waves, only slightly different in index. The first layer we can retain as 1.9, although it is in no way critical, and then if we make the second quarter-wave of slightly higher index, 2.0, say, the design now becoming

Air	1.38	2.0	1.9	Glass
1.0	$0.25\,\lambda_0$	$0.25\,\lambda_0$	$0.25\,\lambda_0$	1.52

we find a reduction in the reflectance at λ_0 from 1.26% to 0.38%. The characteristic remains fairly broad. Increasing the index of the central layer still further, to 2.13, i.e. a design

Air	1.38	2.13	1.9	Glass
1.0	$0.25\,\lambda_0$	$0.25\,\lambda_0$	$0.25\,\lambda_0$	1.52

reduces the reflectance at λ_0 to virtually zero, but the width of the coating becomes much more significantly reduced. The characteristic curves of these two coatings are shown in figure 3.31.

Yet a further increase in the width of the coating can be achieved by adding a

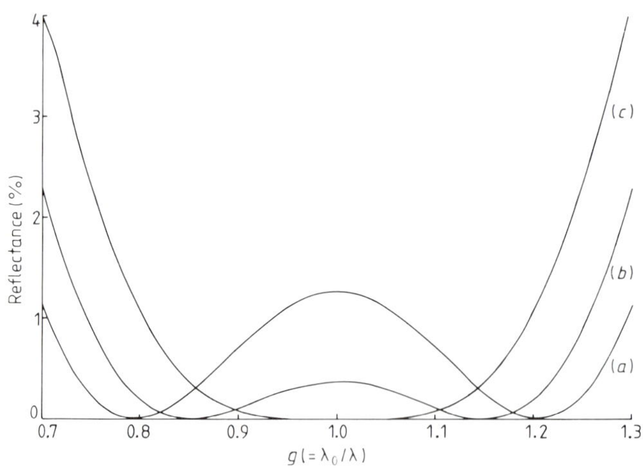

Figure 3.31 Progressive changes in an antireflection coating consisting of three quarter-wave layers. (*a*) The original coating:

Air	1.38	1.90	1.90	Glass
1.00	$0.25\lambda_0$	$0.25\lambda_0$	$0.25\lambda_0$	1.52

The two 1.90 index layers combine to form a single half-wave layer. This is known as a W-coat because of the shape of the characteristic. (*b*) The index of the central layer is increased to 2.00. (*c*) The index of the central layer is increased further to 2.13.

half-wave layer of low index next to the substrate. The admittance plot is shown in figure 3.32 and we see the characteristic shape where the final part of the locus of the half-wave layer and the start of the following layer are on the same side of the real axis. A half-wave layer in the same position with index higher than the substrate would be ineffective. A certain amount of trial and

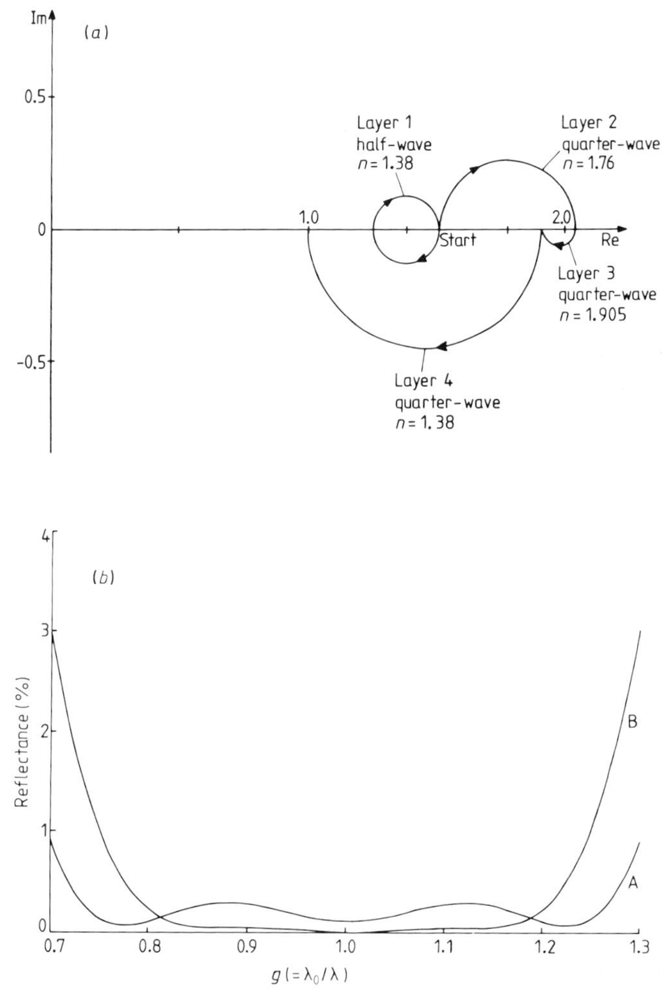

Figure 3.32 (*a*) The admittance locus of the coating:

Air	1.38	1.905	1.76	1.38	Glass
1.00	$0.25\lambda_0$	$0.25\lambda_0$	$0.25\lambda_0$	$0.5\lambda_0$	1.52

(*b*) The characteristics of (A) the coating of 3.32 (*a*) and (B) the coating (*c*) of figure 3.31 with a half-wave flattening layer of index 1.38 added next to the substrate.

error leads to the designs shown in figure 3.32, that is

Air	1.38	1.905	1.76	1.38	Glass
1.0	$0.25\,\lambda_0$	$0.25\,\lambda_0$	$0.25\,\lambda_0$	$0.5\,\lambda_0$	1.52

and

Air	1.38	2.13	1.90	1.38	Glass
1.0	$0.25\,\lambda_0$	$0.25\,\lambda_0$	$0.25\,\lambda_0$	$0.5\,\lambda_0$	1.52.

An alternative approach is to broaden the quarter–quarter design of figure 3.26 by inserting a half-wave layer between the two quarter-waves. In order to achieve the broadening effect it must, of course, be of high index, so that the admittance plot will be of the form shown in figure 3.33. The coating is frequently referred to as the quarter–half–quarter coating. Coatings that fit into this general type date back to the 1940s and were described by Lockhart and King[13]. A systematic design technique explaining the functions of the various layers, however, was not available until the detailed study of Cox, Hass and Thelen[14]. A certain amount of trial and error leads to the characteristics of figure 3.34. However, good results are obtained with values of the index of the half-wave layer in the range 2.0–2.4. Cox *et al* also investigated the effect of varying the indices of the quarter-wave layers and found that for the best results on crown glass, the outermost layer index should be between 1.35 and 1.45, and the innermost layer index between 1.65 and 1.70. The outermost layer is the most critical in the design.

Figure 3.35 also comes from their paper and shows the measured reflectance of an experimental coating consisting of magnesium fluoride, index 1.38, zirconium oxide, index 2.1, and cerium fluoride, which was evaporated rather

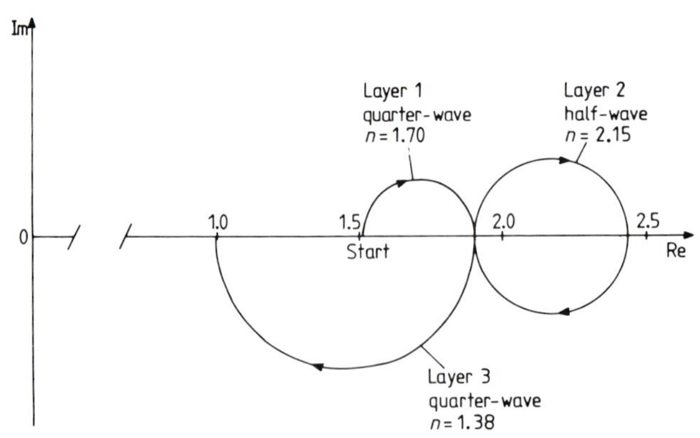

Figure 3.33 Admittance locus of the quarter–half–quarter coating:

Air	1.38	2.15	1.70	Glass
1.00	$0.25\lambda_0$	$0.5\lambda_0$	$0.25\lambda_0$	1.52

The half-wave layer acts to flatten the performance of the two quarter-waves.

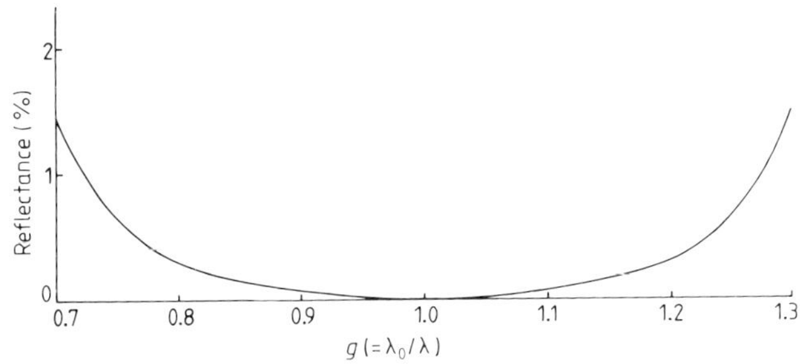

Figure 3.34 The calculated reflectance of the quarter–half–quarter coating shown in figure 3.33.

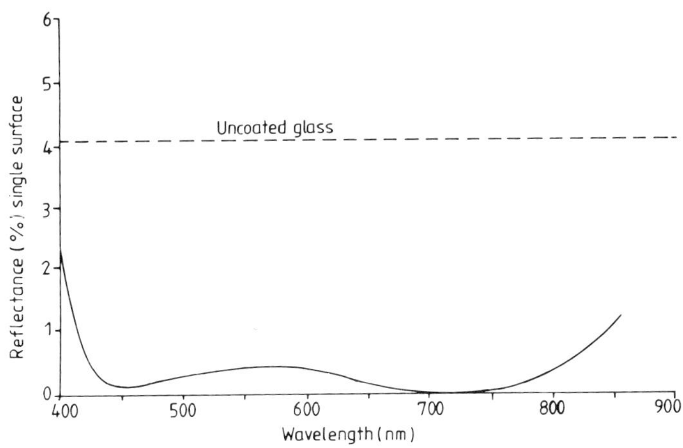

Figure 3.35 Measured reflectance of a quarter–half–quarter antireflection coating of $MgF_2 + ZrO_2 + CeF_3$ on crown glass. $\lambda_0 = 550$ nm. (After Cox, Hass and Thelen[14].)

too slowly and had an index of 1.63, which accounts for the slight rise in the middle of the range. Otherwise, the coating is an excellent practical confirmation of the theory.

The effect of variations in angle of incidence has also been examined. Cox *et al*'s results for tilts up to 50° of a coating designed for normal incidence are shown in figure 3.36. The performance of the coating is excellent up to 20° but begins to fall off beyond 30°. The coatings can, of course, be designed for use at angles of incidence other than normal, and Turbadar[15] has published a full account of a design for use at 45°. The particular design depends on whether the light is s- or p-polarised and figure 3.37 shows sets of equireflectance contours for both designs.

The quarter–half–quarter coating is certainly the most significant of the

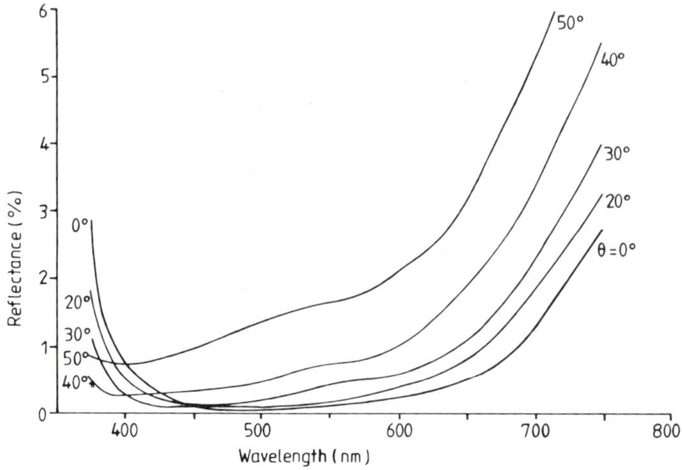

Figure 3.36 Calculated reflectance as a function of wavelength for quarter–half–quarter antireflection coatings on glass at various angles of incidence. $n_0 = 1.00$, $n_1 = 1.38$, $n_2 = 2.2$, $n_3 = 1.70$, $n_{sub} = 1.51$. (After Cox, Hass and Thelen[14].)

multilayer coatings for low-index glass and it has had considerable influence on the development of the field.

The success of the broadening effect of the half-wave layer on the quarter–quarter coating prompts us to consider inserting a similar half-wave in the two-layer coating of figure 3.25. In this case, there is an advantage in using a layer of the same index as that next to the substrate. Here we cannot split the coating at the interface between the high- and the low-index layers, because the admittance plot would not show the correct broadening configuration. Instead, we must split the coating at the point where the low-index locus cuts the real axis so that the plot appears as in figure 3.38. The design of the coating is then

Air	1.38	2.30	1.38	2.30	Glass
1.0	$0.25\,\lambda_0$	$0.5\,\lambda_0$	$0.0734\,\lambda_0$	$0.0522\,\lambda_0$	1.52

where this time we have used a value of 2.30 for the high index, and the performance is shown in figure 3.39. There is a considerable resemblance of the admittance plot to that of the quarter–half–quarter design, and this particular arrangement is due to Vermeulen[16] who arrived at it in a completely different way. There is a difficulty in achieving the correct value for the intermediate index in the quarter–half–quarter design in practice and Vermeulen realised that the deposition of a low-index layer over a high-index layer of less than a quarter-wave would lead to a maximum turning value in reflectance rather lower than would have been achieved with a quarter-wave of high index on its own. He therefore designed a two-layer high–low combination to give an identical turning value to that which should be obtained with the 1.70 index

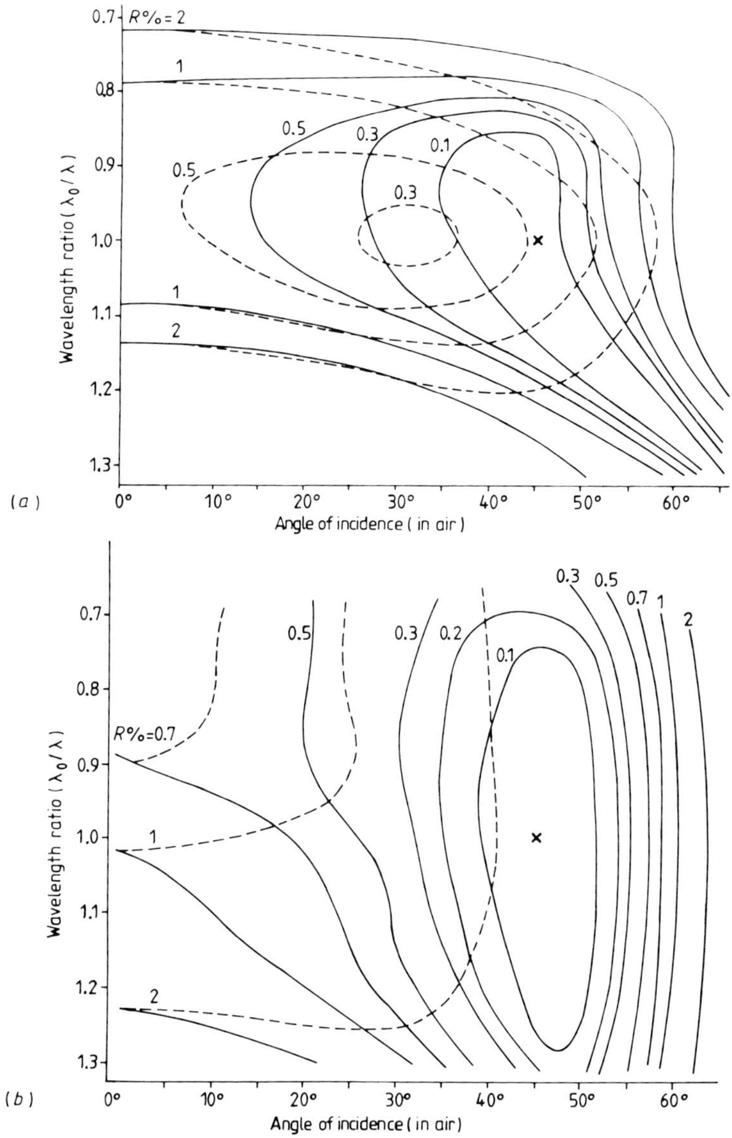

Figure 3.37 (*a*) Equireflectance contours for a quarter–half–quarter antireflection coating designed for use at 45° on crown glass. The indices are chosen for best performance with s-polarisation (TE). $n_0 = 1.00$, $n_1 = 1.35$, $n_2 = 2.45$, $n_3 = 1.70$, $n_{sub} = 1.50$. Solid curves s-polarisation (TE); broken curves p-polarisation (TM). (After Turbadar[15].)

(*b*) Equireflectance contours for a quarter–half–quarter antireflection coating designed for use at 45° on crown glass. The indices are chosen for best performance with p-polarisation (TE). $n_0 = 1.00$, $n_1 = 1.40$, $n_2 = 1.75$, $n_3 = 1.58$, $n_{sub} = 1.50$. Solid curves p-polarisation (TM); broken curves s-polarisation (TE). (After Turbadar[15].)

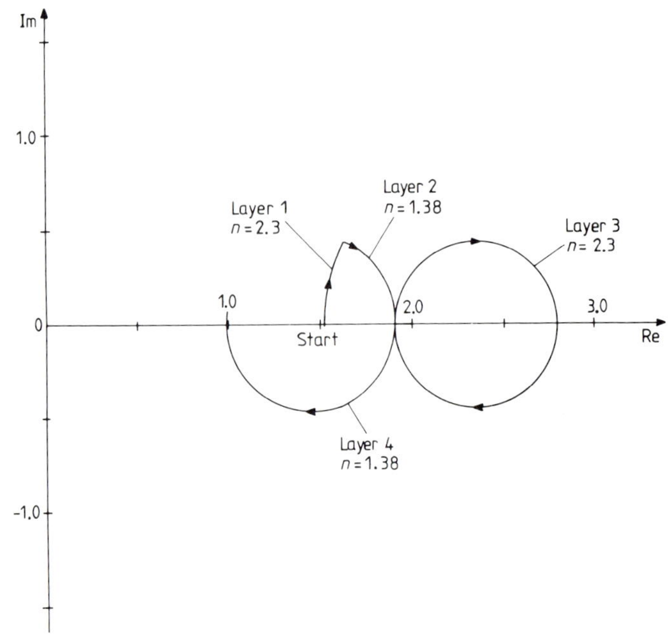

Figure 3.38 The two-layer coating of figure 3.25 with the low-index layer split where it intersects the real axis and a high-index flattening layer inserted.

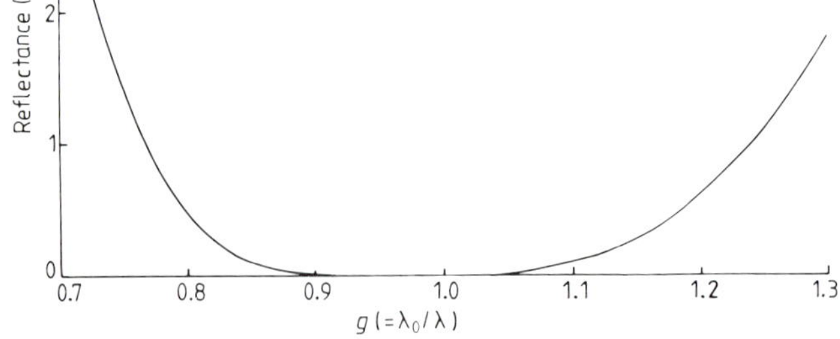

Figure 3.39 The performance of the coating of figure 3.38. Although arrived at by way of the admittance plot of figure 3.38, the design is virtually identical to one published by Vermeulen[16] whose design technique was quite different (see text).

layer of the quarter–half–quarter coating, and he discovered that good performance was maintained. The turning value in reflectance must, of course, correspond to the intersection of the locus with the real axis, and the rest follows. We shall return to this coating later.

The quarter–half–quarter coating can be further improved by replacing the layer of intermediate index by two quarter-wave layers. The layer next to the substrate should have an index lower than that of the substrate. A practical coating of this general type is shown in figure 3.40[17]. Trial and error leads to a design

Air	1.38	2.05	1.60	1.45	Glass
1.0	$0.25\,\lambda_0$	$0.5\,\lambda_0$	$0.25\,\lambda_0$	$0.25\,\lambda_0$	1.52

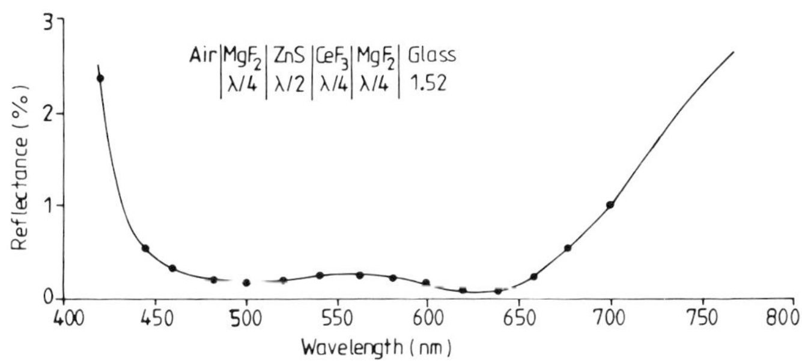

Figure 3.40 Measured reflectance of a four-layer antireflection coating on crown glass. The results are for a single surface. (After Shadbolt[17].)

the theoretical performance of which is shown in figure 3.41. Similar designs with slightly different index values are given by Cox and Hass[18] and by Musset and Thelen[6]. Ward[19] has published a particularly useful version of this coating with indices chosen to match those of available materials rather than to achieve optimum performance. Examples of four-layer coatings for substrates of indices other than 1.52 are also given by Ward[19] and by Musset and Thelen[6].

Yet a further four-layer design can be obtained by splitting the half-wave layer of the quarter–half–quarter coating into two quarter-waves and adjusting the indices to improve the performance. A five-layer design derived in a similar way from the design of figure 3.41:

Air	1.38	1.86	1.94	1.65	1.47	Glass
1.0	$0.25\,\lambda_0$	$0.25\,\lambda_0$	$0.25\,\lambda_0$	$0.25\,\lambda_0$	$0.25\,\lambda_0$	1.52

is shown in figure 3.42.

The possibilities are clearly enormous and problems are found much more in the construction of the coatings because not all the required indices are readily available. One solution is discussed in the next section.

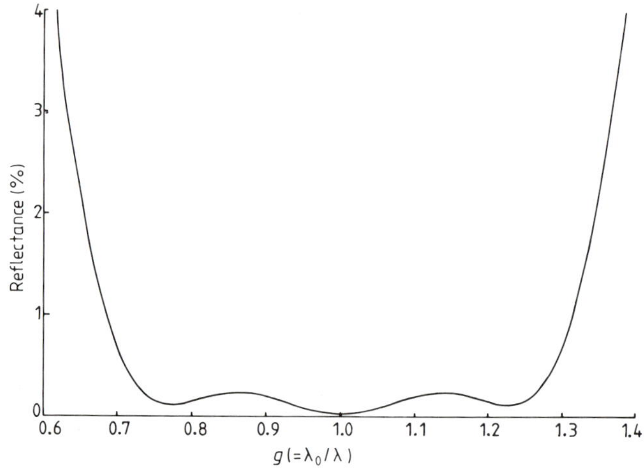

Figure 3.41 The performance of the four-layer coating of design:

Air	1.38	2.05	1.60	1.45	Glass
1.00	$0.25\lambda_0$	$0.5\lambda_0$	$0.25\lambda_0$	$0.25\lambda_0$	1.52

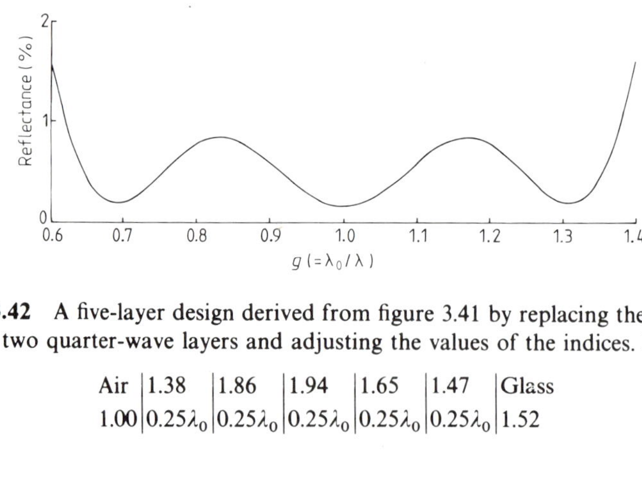

Figure 3.42 A five-layer design derived from figure 3.41 by replacing the half-wave layer by two quarter-wave layers and adjusting the values of the indices. Design:

Air	1.38	1.86	1.94	1.65	1.47	Glass
1.00	$0.25\lambda_0$	$0.25\lambda_0$	$0.25\lambda_0$	$0.25\lambda_0$	$0.25\lambda_0$	1.52

Equivalent layers

There are great advantages in using a series of quarter-waves or multiples of quarter-waves in the first stages of the design of antireflection coatings because the characteristic curves of such coatings are symmetrical about $g = 1.0$. However, problems are presented in construction because the indices which are specified in this way do not often correspond exactly with indices which are readily available. Using mixtures of materials of higher and lower indices to produce a layer of intermediate index is a technique which has been used successfully (see chapter 9), but a more straightforward technique is to replace the layers by equivalent combinations which involve only two materials, one of high index and one of low index. These two materials can be well tried stable materials, the characteristics of which have been established over many production runs in the plant which will be used for, and under the conditions which will apply to, the production of the coatings. To illustrate the method, we assume two materials of index 2.30 and 1.38, corresponding approximately to titanium dioxide and magnesium fluoride, respectively.

The first technique to mention is that of Vermeulen[16] which has already been referred to. It involves the replacing of a quarter-wave by a two-layer equivalent. The analysis is exactly that already given for the two-layer antireflection coating and it is assumed that the quarter-wave to be replaced has a locus which starts and terminates at predetermined points on the real axis. The replacement is, therefore, valid for the particular starting and terminating points used in its derivation only, and for that single wavelength for which the original layer is a quarter-wave. Under conditions which are increasingly remote from these ideal ones, the two-layer replacement becomes increasingly less satisfactory. It is advisable, when calculating the parameters of the layers, to sketch a rough admittance plot because otherwise there is a real danger of picking incorrect values of layer thickness. In the particular case we are considering, the starting admittance is 1.52 on the real axis and the terminating admittance is 1.9044, which will ensure that the outermost 1.38 index quarter-wave layer will terminate at the point 1.00 on the real axis. Clearly the high-index layer should be next to the substrate. The thicknesses are then, using equations (3.5) and (3.6) and selecting the appropriate pair of solutions, 0.05217 and 0.07339 full waves for the high- and low-index layers respectively. We complete the design by adding a half-wave of index 2.30 and a quarter-wave of index 1.38. The characteristic curve of this coating is shown in figure 3.39 which, we recall, was arrived at in a completely different way.

To obtain a better replacement for a quarter-wave, we turn to a technique originated by Epstein[20] involving the symmetrical periods and the Herpin admittance mentioned briefly in chapter 2. We recall that any symmetrical combination of layers acts as a single layer with an equivalent phase thickness and equivalent optical admittance. In this particular application we consider combinations of the form ABA only. We choose for the indices of A and B those of the two materials from which the coating is to be constructed and then for each quarter-wave layer of the coating we construct a three-layer symmetrical period which has an equivalent thickness of one quarter-wave and an equivalent admittance equal to that required from the original layer.

To proceed further, we need expressions for the equivalent thickness and admittance of a symmetrical period. These are derived later in chapter 6. Since the symmetrical period is of the form ABA, then

$$\eta_E = \eta_A \times \left(\frac{\sin 2\delta_A \cos \delta_B + \frac{1}{2}[(\eta_B/\eta_A) + (\eta_A/\eta_B)] \cos 2\delta_A \sin \delta_B + \frac{1}{2}[(\eta_B/\eta_A) - (\eta_A/\eta_B)] \sin \delta_B}{\sin 2\delta_A \cos \delta_B + \frac{1}{2}[(\eta_B/\eta_A) + (\eta_A/\eta_B)] \cos 2\delta_A \sin \delta_B - \frac{1}{2}[(\eta_B/\eta_A) - (\eta_A/\eta_B)] \sin \delta_B} \right)^{1/2} \quad (3.21)$$

$$\cos\gamma = \cos 2\delta_A \cos \delta_B - \tfrac{1}{2}[(\eta_B/\eta_A) + (\eta_A/\eta_B)] \sin 2\delta_A \sin \delta_B \quad (3.22)$$

where η_E is the equivalent optical admittance and γ is the equivalent phase thickness. The important feature of the symmetrical combination is that it behaves as a single layer of phase thickness γ and admittance η_E regardless of the starting point for the admittance locus.

In our particular case, the equivalent thickness of the combination should be a quarter-wave. That is

$$\begin{aligned}\cos\gamma &= \cos(\pi/2) = 0 \\ &= \cos 2\delta_A \cos \delta_B - \tfrac{1}{2}[(\eta_B/\eta_A) + (\eta_A/\eta_B)] \sin 2\delta_A \sin \delta_B\end{aligned}$$

which gives

$$\tan 2\delta_A \tan \delta_B = \frac{2\eta_A\eta_B}{\eta_A^2 + \eta_B^2}. \quad (3.23)$$

Substituting in equation (3.21) and manipulating the expression we have

$$\eta_E = \eta_A \left(\frac{1 + [(\eta_B^2 - \eta_A^2)/(\eta_B^2 + \eta_A^2)] \cos 2\delta_A}{1 - [(\eta_B^2 - \eta_A^2)/(\eta_B^2 + \eta_A^2)] \cos 2\delta_A} \right)^{1/2} \quad (3.24)$$

which yields

$$\cos 2\delta_A = \frac{(\eta_B^2 + \eta_A^2)(\eta_E^2 - \eta_A^2)}{(\eta_B^2 - \eta_A^2)(\eta_E^2 + \eta_A^2)}. \quad (3.25)$$

δ_B is given by equation (3.23), i.e.

$$\tan \delta_B = \frac{2\eta_A\eta_B}{\eta_A^2 + \eta_B^2} \frac{1}{\tan 2\delta_A} \quad (3.26)$$

and the optical thicknesses are then

$$\frac{n_A d_A}{\lambda_0} = \frac{\delta_A}{2\pi} \quad \text{full waves at } \lambda_0$$

$$\frac{n_B d_B}{\lambda_0} = \frac{\delta_B}{2\pi} \quad \text{full waves at } \lambda_0.$$

If an equivalent combination for a half-wave layer is required, then it is considered as two quarter-waves in series.

As an example of the application of this technique we take the four-layer coating of figure 3.32:

Air	1.38	2.13	1.90	1.38	Glass
1.0	$0.25\,\lambda_0$	$0.25\,\lambda_0$	$0.25\,\lambda_0$	$0.5\,\lambda_0$	1.52

The layers which must be replaced are the quarter waves with indices 2.13 and 1.90. There are two possible combinations, *HLH* or *LHL*, for each of these layers.

2.13	1.38	2.30	1.38
$0.25\,\lambda_0$ →	$0.04128\,\lambda_0$	$0.15861\,\lambda_0$	$0.04128\,\lambda_0$
	2.30	1.38	2.30
	$0.11198\,\lambda_0$	$0.02302\,\lambda_0$	$0.11198\,\lambda_0$

1.90	1.38	2.30	1.38
$0.25\,\lambda_0$ →	$0.06793\,\lambda_0$	$0.10438\,\lambda_0$	$0.06793\,\lambda_0$
	2.30	1.38	2.30
	$0.09216\,\lambda_0$	$0.05868\,\lambda_0$	$0.09216\,\lambda_0$

As an indication of the closeness of fit between the symmetrical periods and the layers they replace, the variation, with g, of equivalent admittance and equivalent optical thickness is plotted in figure 3.43.

We can now replace the layers in the actual design of antireflection coating. There are two possible replacements for each of the relevant layers, but where *HLH* and *LHL* combinations are mixed, there is a tendency towards an excessive number of layers in the final design, and so we consider two possibilities only, one based on *HLH* periods and one on *LHL*. These are shown in table 3.3.

The spectral characteristics of these coatings along with the original design are shown in figure 3.44. The replacements have a slightly inferior performance due to the effective dispersion that can be seen in figure 3.43. The process of design need not stop at this point, however, because the designs are excellent starting points for refinement. Figure 3.45 shows the performance of a refined version of one of the coatings. In practice, the refinement will include an allowance for the dispersion of the indices of the materials and there will be a certain amount of adjustment of the coating during the production trials.

If performance over a much wider region is required, then the apparent dispersion of the equivalent periods may become a problem. This dispersion can be reduced by using equivalent periods of $\frac{1}{8}$-wave thickness instead of quarter-wave. Each quarter-wave in the original design is then replaced by two symmetrical periods in series. This adds considerably to the number of layers and, in addition, the solution of the appropriate equations is no longer simple.

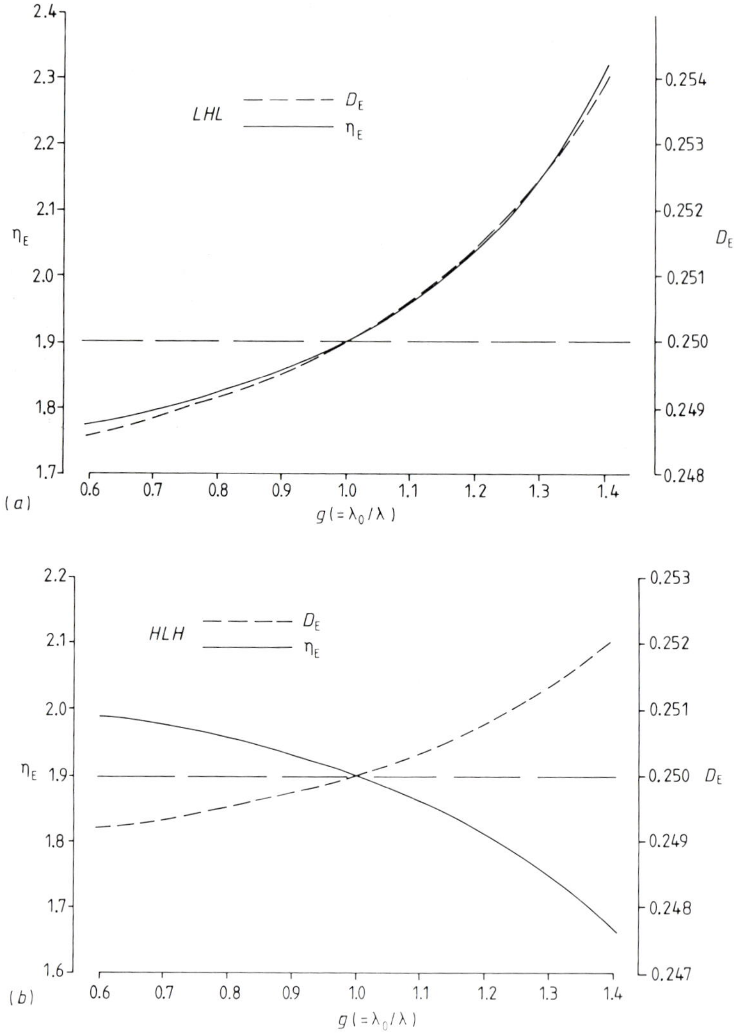

Figure 3.43 The equivalent admittances and optical thickness as a function of $g\ (=\lambda_0/\lambda)$ of symmetrical period replacements for a single quarter-wave of index 1.90. The indices used in the symmetrical replacement are 2.30 for the high index and 1.38 for the low index. (*a*) *LHL* combination. (*b*) *HLH* combination. For a perfect match D_E and η_E should both be constant at $0.25\lambda_0$ and 1.9 respectively, whatever the value of g.

Antireflection coatings for two zeros

There are occasional applications where antireflection coatings are required which have zeros at certain well defined wavelengths rather than over a wide spectral region. One of the most frequent of these applications is frequency

Table 3.3

Layer number	Design based on *LHL* periods		Design based on *HLH* periods	
	Index	Thickness	Index	Thickness
0	1.0	Incident medium	1.0	Incident medium
1	1.38	0.29128 λ_0	1.38	0.25 λ_0
2	2.30	0.15861 λ_0	2.30	0.11198 λ_0
3	1.38	0.10921 λ_0	1.38	0.02302 λ_0
4	2.30	0.10438 λ_0	2.30	0.20414 λ_0
5	1.38	0.56793λ_0	1.38	0.05868 λ_0
6	1.52	Substrate	2.30	0.09216 λ_0
7			1.38	0.5 λ_0
8			1.52	Substrate

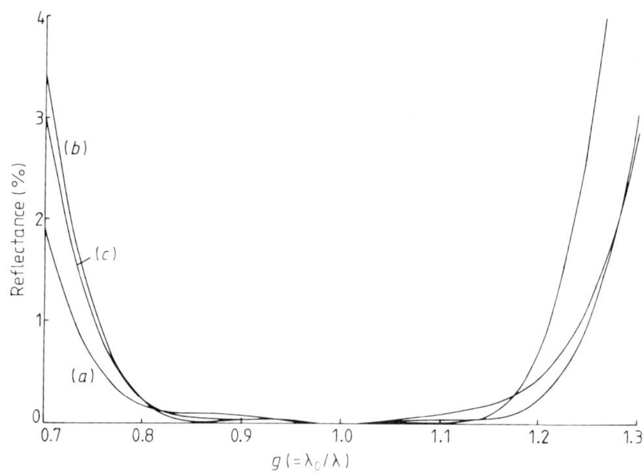

Figure 3.44 The performance of the designs of table 3.3: (*a*) Five-layer design based on *LHL* periods. (*b*) Seven-layer design based on *HLH* periods. (*c*) The original four-layer design from which (*a*) and (*b*) were derived.

doubling, where antireflection is required at two wavelengths, one of which is twice the other.

The simplest coating that will satisfy this requirement is the quarter–quarter that has already been considered. We recall that the coating has two zeros at $\lambda = 3\lambda_0/4$ and $\lambda = 3\lambda_0/2$, just what is required. The conditions are

$$n_1 = (n_0^2 n_s)^{1/2}$$
$$n_2 = (n_0 n_s^2)^{1/2}. \tag{3.27}$$

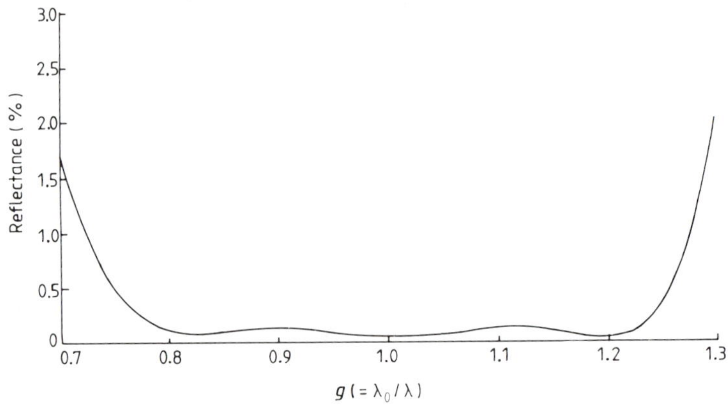

Figure 3.45 Refined version of the five-layer design of figure 3.44 and table 3.2. Design:

Air	1.38	2.30	1.38	2.30	1.38	Glass
1.00	$0.2973\lambda_0$	$0.1252\lambda_0$	$0.1244\lambda_0$	$0.0874\lambda_0$	$0.5597\lambda_0$	1.52

The principal problem with this coating is once again the low-index substrate. With an index of 1.38 as the lowest value for n_1 the lowest value of substrate index that can be accommodated, from equation (3.27), is $1.38^3 = 2.63$. Thus the coating is suitable only for high-index substrates.

A common material that requires antireflection coatings at λ and 2λ is lithium niobate, which has an index of around 2.25. The quarter–quarter coating should have indices of 1.310 and 1.717. Indices of 1.38 and 1.717 give a reflection loss of 0.2%, which will probably be adequate for many applications, and indeed similar performance is obtained with any index between 1.7 and 1.8 for the high-index layer.

Should this performance be inadequate, then an additional layer can be added. Provided we keep to quarter-waves and multiples of quarter-waves, we retain the symmetry about $g = 1$ and therefore have to consider the performance at $g = \frac{2}{3}$ only since that at $g = \frac{4}{3}$ will be automatically equivalent. From the point of view of the vector diagram, the problem with the quarter–quarter coating is ρ_a, the amplitude reflection coefficient from the first interface, which is too large. The vectors are inclined at 120° to each other and for zero reflectance they should be of equal length so that they form an equilateral triangle. If an extra quarter-wave n_3 is added, there will be four vectors and the fourth, ρ_0, will be along the same direction as ρ_a. If ρ_d is made to be of opposite sense to ρ_a, that is if $n_3 > n_m$, then it is possible to reduce the resultant of the two vectors to the same length as the other two. This can be achieved by the design

Air	1.38	1.808	2.368	Lithium niobate
1.0	$0.25\,\lambda_0$	$0.25\,\lambda_0$	$0.25\,\lambda_0$	2.25

We can take 2.35, the index of zinc sulphide, for n_3, and then any index in the range 1.75–1.85 for n_2 to keep the minimum reflectance at $g = \frac{2}{3}$ to below 0.1 %.

There are many other possible arrangements. A coating with the first layer a half-wave, instead of a quarter-wave, can give a similar improvement, this time through a combination with ρ_c which means that $n_2 > n_3$. Here the ideal design is

Air	1.38	1.81	1.72	Lithium niobate
1.0	$0.5\,\lambda_0$	$0.25\,\lambda_0$	$0.25\,\lambda_0$	2.25

and once again there is reasonable flexibility in the values of n_2 and n_3 if the aim is simply a reflectance of less than 0.1 %. It is interesting to note the similarity between this coating and the quarter–quarter. The quarter–quarter has another zero at $g = \frac{8}{3}$. If the inner quarter-waves in the above design were merged into a single half-wave of index around 1.75, then the coating would be identical with the quarter–quarter used at $g = \frac{4}{3}$ and $g = \frac{8}{3}$. Figure 3.46 shows the performance of these coatings.

This idea of using the fourth vector to trim the length of one of the other

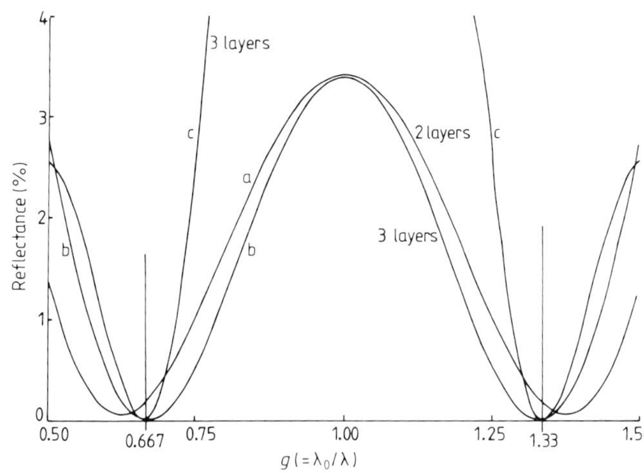

Figure 3.46 The performance of various two-zero 2:1 antireflection coatings on a high-index substrate such as lithium niobate with $n = 2.25$. The ideal positions for the two zeros are $g = 0.667$ and $g = 1.333$.

(*a*)

Air	1.38	1.72	Lithium niobate
1.0	$0.25\lambda_0$	$0.25\lambda_0$	2.25

(*b*)

Air	1.38	1.808	2.368	Lithium niobate
1.0	$0.25\lambda_0$	$0.25\lambda_0$	$0.25\lambda_0$	2.25

(*c*)

Air	1.38	1.81	1.72	Lithium niobate
1.0	$0.5\lambda_0$	$0.25\lambda_0$	$0.25\lambda_0$	2.25

three so that a low reflectance is obtained can be extended to low-index substrates. The coating now, of course, departs considerably from the original quarter–quarter coating. A quarter–quarter–quarter design based on this approach is

Air	1.38	1.585	1.82	Glass
1.0	$0.25\,\lambda_0$	$0.25\,\lambda_0$	$0.25\,\lambda_0$	1.52

and its performance is shown in figure 3.47 where the monitoring wavelength has been assumed to be 707 nm and the two zeros are situated at 530 nm and 1.06 μm.

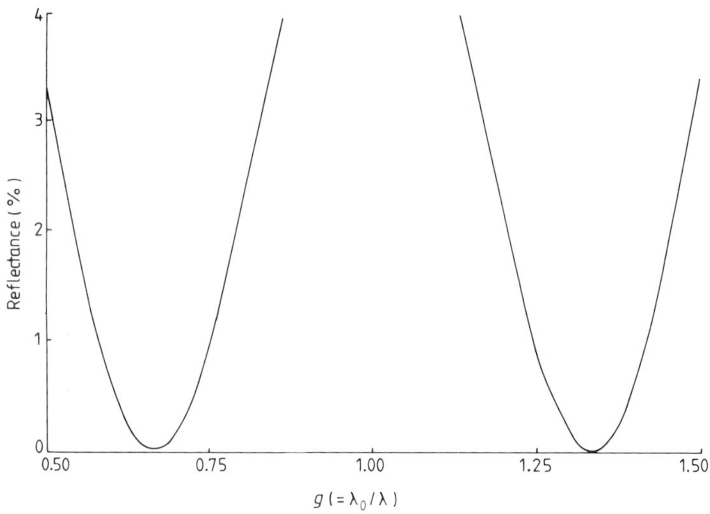

Figure 3.47 A three-layer two-zero 2:1 antireflection coating for a low-index substrate. Design ($\lambda_0 = 707$ nm):

Air	1.38	1.585	1.82	Glass
1.0	$0.25\lambda_0$	$0.25\lambda_0$	$0.25\lambda_0$	1.52

The method can be extended to four and even more quarter-waves, although the derivation of the final designs is very much more of a trial and error process because of the rather cumbersome expressions that cannot be reduced to explicit formulae for the various indices. Indeed, there are now too many parameters for there to be just one solution and the surplus can be used in an optimising process for broadening the reflectance minima. A number of interesting designs is given by Baumeister[21].

Mouchart[22] too, has considered the derivation of antireflection coatings intended to eliminate reflection at two wavelengths. In coatings where all layers have thicknesses that are specified in advance to be multiples of a quarter-wave at $g = 1$ it is possible arbitrarily to choose the indices of all the layers except the final two, which can then be calculated from the values given to the others. The calculation involves the solution of an eighth-order equation that can be set up using expressions derived by Mouchart. The values of $\partial^2 R/\partial\lambda^2$ at the antireflection wavelength, which is inversely related to the bandwidth of the coating, can be used to assist in choosing the more promising designs from the enormous number that can be produced. Mouchart considers three-layer coatings of this type in some detail.

Antireflection coatings for the visible region and an infrared wavelength

There are frequent requirements for coatings that span the visible region and also reduce the reflectance at an infrared wavelength corresponding to a laser line. Such coatings are required in instruments where visual information and laser light share common elements, such as surgical instruments, surveying devices and the like. There are very many designs for such coatings and manufacturers seldom publish them. Design is largely a process of trial and error and frequently the final operation is to replace the unobtainable or difficult indices by symmetrical combinations of better behaved materials and to refine the design so obtained to take account of the dispersion of the optical constants of real materials and to compensate for the apparent dispersion that occurs in connection with the symmetrical periods. In this section we consider the fundamental design process only, neglecting dispersion and in most cases retaining the ideal values of index. We assume that the substrate is always glass of index 1.52 and that, as usual, the incident medium is air of index 1.0.

The simplest type of coating that has low reflectance in the visible region and at a wavelength in the near infrared is a single layer of low-index material of thickness three quarter-waves. This has low reflectance at both λ_0 and $3\lambda_0$. Unfortunately, the lowest index, of 1.38, corresponding to magnesium fluoride, gives a residual reflectance of 1.25% at the minima and the performance in the visible region is rather narrower than that for the single quarter-wave coating since the layer is three times thicker. The magnesium fluoride layer could be considered as an outer quarter-wave over an inner half-wave and a high-index half-wave flattening layer, of index 1.8, could be introduced between them giving the design

$$\text{Air}\,|\,LHHLL\,|\,\text{Glass}.$$

Unfortunately, the half-wave layer, while it flattens the performance in the visible region, destroys the performance in the infrared at $3\lambda_0$, where it is two thirds of a quarter-wave thick. The solution is to make the layer three half-waves thick in the visible, so that it is still a half-wave, and therefore an absentee, at $3\lambda_0$. The design then becomes

$$\text{Air}\,|\,L6H2L\,|\,\text{Glass}$$

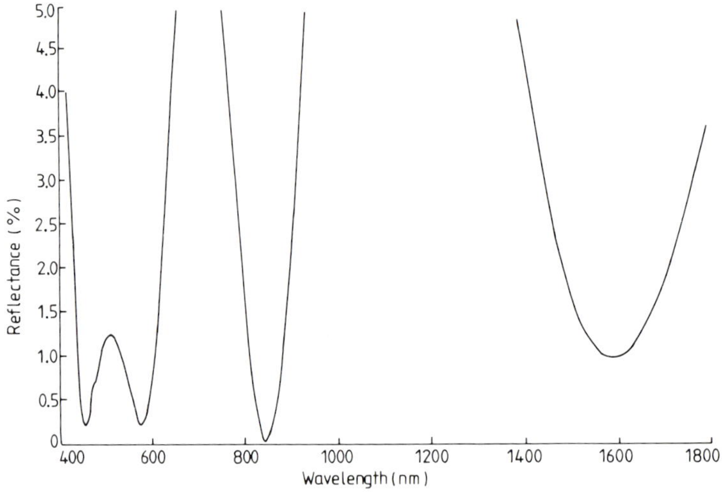

Figure 3.48 The performance of the coating:

$$\text{Air}(1.0)|L\ 6H\ 2L|\text{Glass}\ (1.52)$$

with L a quarter-wave of index 1.38 and H of 1.8. λ_0 is 510 nm.

and the performance is shown in figure 3.48, where the reference wavelength is 510 nm. The performance in the visible region is indeed flattened in the normal way, although, because the flattening layer is three times thicker than normal, the characteristic rises sharply in the blue and red regions. The minimum in the infrared around 1.53 μm is still present, although slightly skewed because of the half-wave layer. However, perhaps the most surprising feature is the appearance of a third and very deep minimum at 840 nm. We use the admittance diagram to help in understanding the origin of this dip.

Figure 3.49 shows the admittance diagram for the coating at the wavelength 840 nm. Layer 2, the 1.8 index layer, is almost two half-waves thick at this wavelength and so describes almost two complete revolutions, linking the ends of the loci of the two 1.38 index layers in such a way that almost zero reflectance is obtained. The loci of the two low-index layers are not very sensitive to changes in wavelength and therefore the position of the dip is fixed almost entirely by the high-index layer. Changes in its thickness will change the position of the dip. Making it thinner, 1.0 full waves instead of 1.5 for example, will move the dip to a longer wavelength. The performance characteristic of a coating of design

$$\text{Air}\,|\,L4H2L\,|\,\text{Glass}$$

is shown in figure 3.50. The dip is now fairly near the desired wavelength of 1.06 μm.

A coating that gives good performance over the visible region but has high

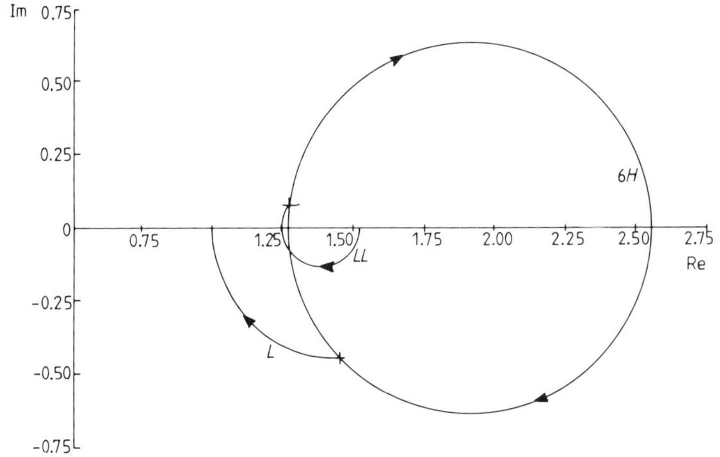

Figure 3.49 The admittance diagram for the coating of figure 3.48 at 840 nm, corresponding to the unexpected sharp zero, explains the occurrence of the dip.

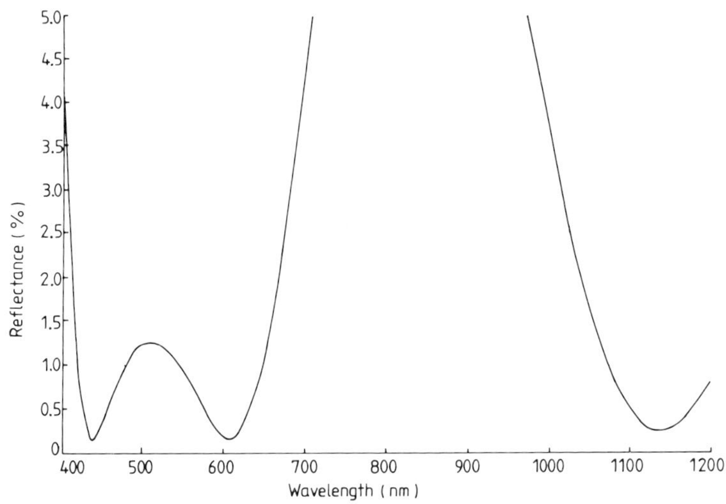

Figure 3.50 The performance of the coating:

Air (1.0)| L 4H 2L| Glass (1.52)

with L a quarter-wave of index 1.38, H of 1.8 and reference wavelength, λ_0, 510 nm. Note that the dip has moved to a longer wavelength than in figure 3.48.

reflectance at 1.06 μm is the quarter–half–quarter coating. The admittance diagram at λ_0 for such a coating is shown in figure 3.33. The locus intersects or crosses the real axis at the points 1.9 and 2.45. It is possible to insert layers of index 1.9 or 2.45 respectively at these points in the design without any effect on the performance at λ_0 at all. The loci of these layers, whatever their thicknesses, would simply be points. Such layers are known as 'buffer layers' and were devised by Mouchart[23]. At the reference wavelength they exert no influence whatsoever but at other wavelengths, where the starting points of their loci move away from their reference wavelength positions, the loci appear in the normal way and can have important effects on performance. They are similar in some respects to half-wave layers, that by virtue of their precise thickness are absentees at λ_0, but which have considerable influence on other wavelengths. The index can be chosen to sharpen or flatten a characteristic. The buffer layer has a precise value of index, but can have any thickness, which can be chosen to adjust performance at wavelengths other than λ_0. Here we attempt to use buffer layers to alter the performance at 1.06 μm. One buffer layer is not sufficient and we need to insert the two possible 1.9 index layers so that the design becomes:

$$\text{Air} \mid LB'HHB''M \mid \text{Glass}$$

where $y_L = 1.38$, $y_H = 2.15$ and $y_M = 1.70$. B' and B'' are buffer layers of admittance 1.9. Trial and error establishes thicknesses for B' of $0.342\,\lambda_0$ and for B'' of $0.084\,\lambda_0$. However, although the reflectance at 1.06 μm is reduced considerably, the buffer layers do distort the performance characteristic somewhat in the visible region (figure 3.51) and only by refining the design

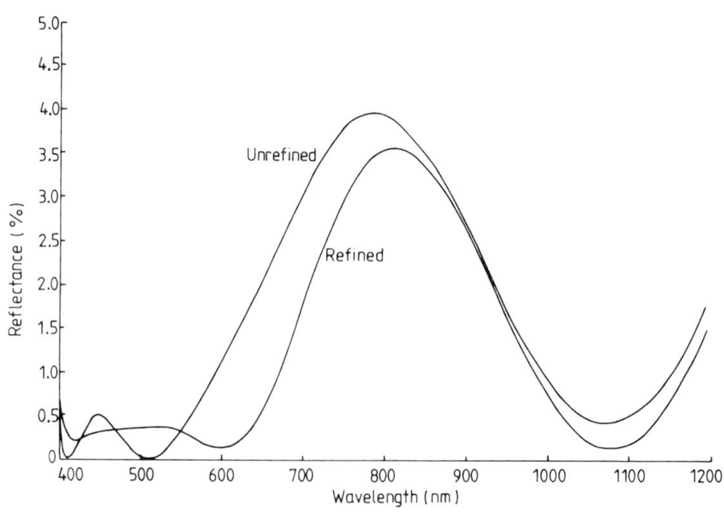

Figure 3.51 The performance of the design:

$$\text{Air (1.0)} \mid L\,B'\,HH\,B''\,M \mid \text{Glass (1.52)}$$

with L, H, M quarter-waves of indices 1.38, 2.15 and 1.70 respectively. B' and B'' are buffer layers of index 1.9 (see text) and thicknesses $0.342\lambda_0$ and $0.084\lambda_0$ respectively. λ_0 is 510 nm. The design has also been refined to yield the second performance curve. The refined design is given in the text.

is a completely satisfactory performance obtained. The final design, also illustrated in figure 3.51, is:

Air	1.38	1.90	2.15	1.90	1.70	Glass
1.00	$0.2667\,\lambda_0$	$0.3085\,\lambda_0$	$0.5395\,\lambda_0$	$0.1316\,\lambda_0$	$0.1796\,\lambda_0$	1.52.

Many of the designs currently used for the visible and 1.06 μm involve just two materials of high and low index. Designs of this type can be arrived at in a number of ways. The arrangements above that use ideal layers can be replaced by symmetrical periods in the way already discussed. This type of design is seldom immediately acceptable because the very wide wavelength range makes it difficult exactly to match the layers with symmetrical periods and they are therefore usually refined by computer.

Figure 3.52 shows the performance of a six-layer design arrived at by computer synthesis:

Air	1.38	2.25	1.38	2.25	1.38	2.25	Glass
1.00	$0.3003\,\lambda_0$	$0.1281\,\lambda_0$	$0.0657\,\lambda_0$	$0.6789\,\lambda_0$	$0.0718\,\lambda_0$	$0.0840\,\lambda_0$	1.52.

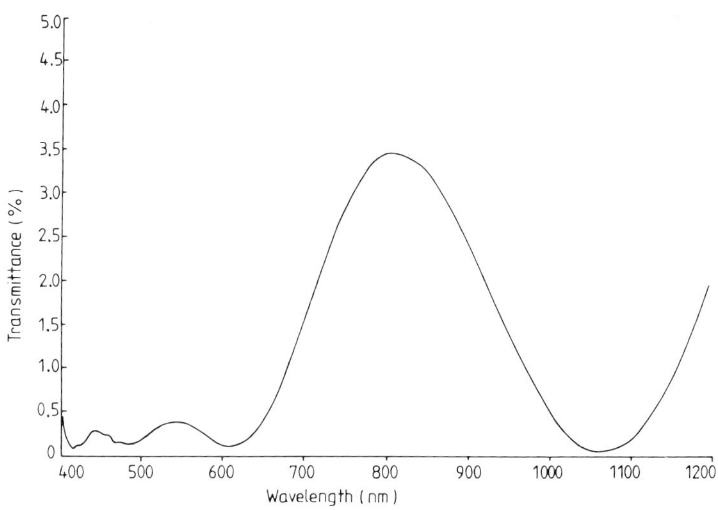

Figure 3.52 The performance of a 6-layer design of antireflection coating for the visible region and 1.06 μm, arrived at purely by computer synthesis. The reference wavelength is 510 nm and the design is given in the text.

INHOMOGENEOUS LAYERS

Inhomogeneous layers are ones in which the refractive index varies through the thickness of the layer. As we shall see in chapter 9, many of the thin-film materials which are commonly used give films which are inhomogeneous. This inhomogeneity is often quite small and the layers can safely be treated as if they were homogeneous in all but the most precise and exacting coatings. There are, however, a number of films which show sufficient inhomogeneity to affect the performance of an antireflection coating perceptibly. If such a layer is used instead of a homogeneous one in a well corrected antireflection coating then a reduction in performance is the normal result. Provided the inhomogeneity is not large, an adjustment of the indices of the other layers is usually sufficient correction and as Ogura[24] has pointed out, a slightly decreasing index with thickness associated with the high-index layer in the quarter–half–quarter coating can actually broaden the characteristic. Zirconium oxide is a much used material which exhibits an index which increases with film thickness when deposited at room temperature, but decreasing with thickness when deposited at substrate temperatures above 200°C. Vermeulen[25] has considered the effect of the inhomogeneity of zirconium oxide on the quarter–half–quarter coating and has shown how it is possible to correct for the inhomogeneity by varying the index of the intermediate-index layer which for virtually complete compensation should be of the two-layer composite type[16] already referred to in this chapter. This type of inhomogeneity is one which is intrinsic and relatively small. By arranging for the evaporation of mixtures of composition varying with film thickness it is possible to produce layers which show an enormous degree of inhomogeneity and which permit the construction of entirely new types of antireflection coating.

Accurate calculation techniques for such layers are reviewed by Jacobson[26] and by Knittl[27]. The simplest method involves the splitting of the inhomogeneous layer into a very large number of thin sub-layers. Each sub-layer is then replaced by a homogeneous layer of the same thickness and mean refractive index so that the smoothly-varying index of the inhomogeneous layer is represented by a series of small steps. Computation can then be carried out as for a multilayer of homogeneous layers. There is no difficulty with modern computers in accommodating very large numbers of sub-layers so that, although an approximation, the method can be made to yield results identical for all practical purposes with those which would have been obtained by exact calculation (in cases where exact calculation techniques exist).

For our purposes, we can approach the theory of such coatings from the starting point of the multilayer antireflection coating for high-index substrates. As more and more layers are added to the coating, the performance, both from the bandwidth and the maximum reflectance in the low-reflectance region, steadily improves. In the limit, there will be an infinite number of layers with infinitesimal steps in optical admittance from one layer to the next. If, as layers are added, the total optical thickness of the multilayer is kept constant, the thickness of the individual layers will tend to zero and the multilayers will become indistinguishable from a single layer of identical optical thickness, but with optical admittance varying smoothly from that of the substrate to that of the incident medium.

If there are n layers in the multilayer, then the total optical thickness of the coating will be $n\lambda_0/4$ which may be denoted by T. There will be n zeros of reflectance extending from a shortwave limit

$$\lambda_S = \frac{(n+1)}{n}\frac{\lambda_0}{2}$$

to a longwave limit

$$\lambda_L = (n+1)\frac{\lambda_0}{2}.$$

In terms of T, the total optical thickness, these limits are

$$\lambda_S = \frac{2(n+1)}{n^2}T$$

$$\lambda_L = \frac{2(n+1)}{n}T.$$

At wavelengths of $2\lambda_L$ or longer, the arrows in the vector diagram are confined to the third and fourth quadrant so that the antireflection coating is no longer effective.

If now n tends to infinity but T remains finite, the multilayer tends to a single inhomogeneous layer, λ_S tends to zero, and λ_L tends to $2T$. For all wavelengths between these limits the reflectance of the assembly is zero. Thus the inhomogeneous film with smoothly varying refractive index is a perfect antireflection coating for all wavelengths shorter than twice the optical thickness of the film. At wavelengths longer than this limit the performance falls off, and at the wavelength given by four times the optical thickness of the film, the coating is no longer effective.

Of course, in practice there is no useful thin-film material with refractive index as low as unity and any inhomogeneous thin film must terminate with an index of around 1.35, say, which, in the infrared, is the index of magnesium fluoride. The reflectance of such a coated component will be equal to that of a plate of magnesium fluoride, 2.2% per surface.

Inhomogeneous coatings received much attention at the Institute of Optical Research in Stockholm, and Jacobsson and Martensson actually produced an inhomogeneous antireflection coating of this type on a germanium plate[28]. The films were manufactured by the simultaneous evaporation of germanium and magnesium fluoride, the relative proportions of which were varied throughout the deposition to give a smooth transition between the indices of the two materials. An example of the performance attained is shown in figure 3.53. For this particular coating the physical thickness is quoted as 1.2 μm. To find the optical thickness we assume that the variation of refractive index with physical thickness is linear (mainly because any other assumed law of variation would lead to very difficult calculations, although possibly more accurate results) so that the optical thickness is given by the physical thickness times the mean of the two terminal indices. For this present film, starting with an index of 4.0 and

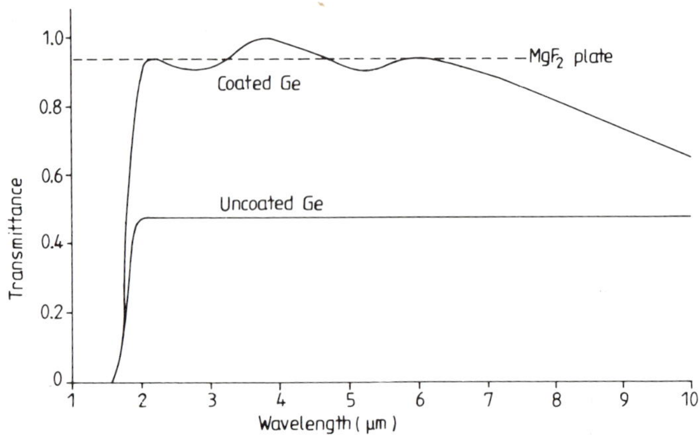

Figure 3.53 Measured transmittance of a germanium plate, coated on both sides with an inhomogeneous Ge–MgF_2 film with geometrical thickness 1.2 μm. (After Jacobsson and Martensson[28].)

finishing with 1.35, the mean is 2.68 and the optical thickness, therefore, 2.68 × 1.2 μm, i.e. 3.2 μm. This implies that the coating should give excellent antireflection for wavelengths out to 6.4 μm, after which it should show a gradually reducing transmission until a wavelength of 4 × 3.2 μm, i.e. 12.8 μm. The curve of the coated component in figure 3.53 shows that this is indeed the case.

Berning[29] has suggested the use of the Herpin index concept for the design of antireflection coatings which are composed of homogeneous layers of two materials, one of high index and the other of low index, which are step approximations to the inhomogeneous layer and which, because they involve homogeneous layers of well understood and stable materials, might be easier to manufacture than the ideal inhomogeneous layers. He has suggested designs for the antireflection coating of germanium consisting of up to 39 alternate layers of germanium and magnesium fluoride equivalent to twenty quarter-waves of gradually decreasing index.

As with coatings consisting of homogeneous layers, the most serious limitation is the lack of low-index materials. A single inhomogeneous layer to match a substrate to air must terminate at an index of around 1.38 which means that the best that can be done with such a layer is a residual reflectance of 2.5 %. This limits their direct use to high-index substrates. For low-index substrates it is likely that their role will remain the improving of the performance of designs incorporating homogeneous materials.

FURTHER INFORMATION

It has not been possible in a single chapter in this book to cover completely the field of antireflection coatings. Further information will be found in Cox and

Hass[18] and Musset and Thelen[6]. There is also a very useful account of antireflection coatings in Knittl[27] which contains some alternative techniques.

REFERENCES

1 Cox J T and Hass G 1958 Antireflection coatings for germanium and silicon in the infrared *J. Opt. Soc. Am.* **48** 677–80

2 Catalan L A 1962 Some computed optical properties of antireflection coatings *J. Opt. Soc. Am.* **52** 437–40

3 Schuster K 1949 Anwendung der Vierpoltheorie auf die Probleme der optischen Reflexionsminderung, Reflexionsverstärkung, und der interferenzfilter *Ann. Phys., Lpz.* **4** 352–6

4 Cox J T 1961 Special type of double-layer antireflection coating for infrared optical materials with high refractive indices *J. Opt. Soc. Am.* **51** 1406–8

5 Cox J T, Hass G and Jacobus G F 1961 Infrared filters of antireflected Si, Ge, InAs and InSb *J. Opt. Soc. Am.* **51** 714–18

6 Musset A and Thelen A 1966 Multilayer antireflection coatings *Progress in Optics* vol 8 ed. E Wolf (Amsterdam: North-Holland) pp 201–37

7 Thelen A 1969 Design of multilayer interference filters *Physics of Thin Films* vol 5 ed. G Hass and R E Thun (New York: Academic) pp 47–86

8 Young L 1961 Synthesis of multiple antireflection films over a prescribed frequency band *J. Opt. Soc. Am.* **51** 967–74

9 Turbadar T 1964 Equireflectance contours of double layer antireflection coatings *Opt. Acta* **11** 159–70

10 Thetford A 1969 A method of designing three-layer antireflection coatings *Opt. Acta* **16** 37–44

11 Thetford A 1968 Private communication

12 C Reichert Optische Werke AG 1962 Improvements in or relating to optical components having reflection-reducing coatings *UK Patent Specification* 991 635

13 Lockhart L B and King P 1947 Three-layered reflection-reducing coatings *J. Opt. Soc. Am.* **37** 689–94

14 Cox J T, Hass G and Thelen A 1962 Triple-layer antireflection coatings on glass for the visible and near infrared *J. Opt. Soc. Am.* **52** 965–9

15 Turbadar T 1964 Equireflectance contours of triple-layer antireflection coatings *Opt. Acta* **11** 195–205

16 Vermeulen A J 1971 Some phenomena connected with the optical monitoring of thin-film deposition and their application to optical coatings *Opt. Acta* **18** 531–8

17 Shadbolt M J 1967 Private communication (Sira Institute, South Hill, Chislehurst, Kent BR7 5RQ, England)

18 Cox J T and Hass G 1964 Antireflection coatings *Physics of Thin films* vol 2 ed. G Hass and R E Thun (New York: Academic) pp 239–304

19 Ward J 1972 Towards invisible glass *Vacuum* **22** 369–75

20 Epstein L I 1952 The design of optical filters *J. Opt. Soc. Am.* **42** 806–10

21 Baumeister P W, Moore R and Walsh K 1977 Application of linear programming to antireflection coating design *J. Opt. Soc. Am.* **67** 1039–45

22 Mouchart J 1978 Thin film optical coatings. 6: Design method for two given wavelength antireflection coatings *Appl. Opt.* **17** 1458–65

23 Mouchart J 1978 Thin film optical coatings. 5: Buffer layer theory *Appl. Opt.* **17** 72–5
24 Ogura S 1975 Some features of the behaviour of optical thin films *PhD Thesis* Newcastle upon Tyne Polytechnic
25 Vermeulen A J 1976 Influence of inhomogeneous refractive indices in multilayer anti-reflection coatings *Opt. Acta* **23** 71–9
26 Jacobsson R 1975 Inhomogeneous and coevaporated homogeneous films for optical applications *Physics of Thin Films* vol 8 ed. G Hass, M Francombe and R W Hoffman (New York: Academic) pp 51–98
27 Knittl Z 1976 *Optics of Thin Films* (London: Wiley)
28 Jacobsson R and Martensson J O 1966 Evaporated inhomogeneous thin films *Appl. Opt.* **5** 29–34
29 Berning P H 1962 Use of equivalent films in the design of infrared multilayer antireflection coatings *J. Opt. Soc. Am.* **52** 431–6

4 Neutral mirrors and beam splitters

HIGH-REFLECTANCE MIRROR COATINGS

Almost as important as the transmitting optical components of the previous chapter are those whose function is to reflect a major portion of the incident light. In the vast majority of cases the sole requirement is that the specular reflectance should be as high as conveniently possible, although, as we shall see, there are specialised applications where not only should the reflectance be high, but also the absorption should be extremely low. For mirrors in optical instruments, simple metallic layers usually give adequate performance and these will be examined first. For some applications where the reflectance must be higher than can be achieved with simple metallic layers, their reflectance can be boosted by the addition of extra dielectric layers. Multilayer all-dielectric reflectors, which combine maximum reflectance with minimum absorption, and which transmit the energy which they do not reflect, are reserved for the next chapter.

Metallic layers

The performance of the commonest metals used as reflecting coatings is shown[1] in figure 4.1.

Aluminium is easy to evaporate and has good ultraviolet, visible and infrared reflectance, together with the additional advantage of adhering strongly to most substances, including plastics. As a result it is the most frequently used film material for the production of reflecting coatings. The reflectance of an aluminium coating does drop gradually in use, although the thin oxide layer, which always forms on the surface very quickly after coating, helps to protect it from further corrosion. In use, especially if the mirror is at all exposed, dust and dirt invariably collect on the surface and cause a fall in reflectance. The performance of most instruments is not seriously affected by a slight drop in reflectance, but in some cases where it is important to collect the maximum amount of light, as it is difficult to clean the coatings without

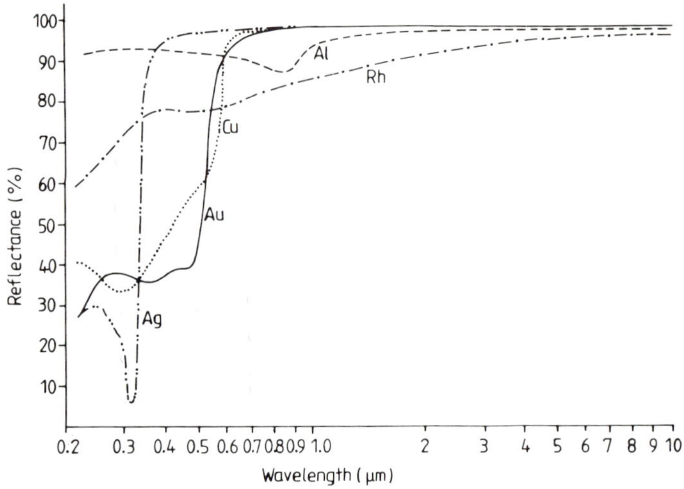

Figure 4.1 Reflectance of freshly deposited films of aluminium, copper, gold, rhodium and silver as a function of wavelength from 0.2–10 µm. (After Hass[1].)

damaging them, the components are recoated periodically. This applies particularly to the mirrors of large astronomical reflecting telescopes. The primary mirrors of these are recoated with aluminium usually around once a year in coating plants which are installed in the observatories for this purpose. Because the primaries are very large and heavy (for example, the 98 inch primary of the Isaac Newton Memorial Telescope of the Royal Greenwich Observatory weighs some 9000 lb), it is not usual to rotate them during coating and the uniformity of coating is achieved through the use of multiple sources.

Silver was once the most popular material of all. It does tarnish when exposed to the atmosphere, owing mainly to the formation of silver sulphide, but the initial high reflectance and the extreme ease of evaporation still make it a common choice for components used only for a short period of time. Silver is also often used where it is necessary to coat temporarily a component, such as an interferometer plate, for a test of flatness.

Gold is probably the best material for infrared reflecting coatings. Its reflectance drops off rapidly in the visible region and it is really useful only beyond 700 nm. On glass, gold tends to form rather soft, easily damaged films, but it adheres strongly to a film of chromium or Nichrome, and this is often used as an underlayer between the gold and the glass substrate.

The reflectance of rhodium and platinum is much less than that of the other metals mentioned and these metals are used only where stable films very resistant to corrosion are required. Both materials adhere very strongly to glass.

Protection of metal films

Most metal films are rather softer than hard dielectric films and can be scratched easily. Unprotected evaporated aluminium layers, for example, can be badly damaged if wiped with a cloth, while gold and silver films are even softer. This is a serious disadvantage, especially when periodic cleaning of the mirrors is necessary. One solution, as we have seen, is periodic recoating. An alternative, which improves the ruggedness of the coatings and also protects them from atmospheric corrosion, is overcoating with an additional dielectric layer. The behaviour of a single dielectric layer on a metal is a useful illustration of the calculation techniques of chapter 2. We shall also require some related results later and so it is useful to spend a little time on the problem.

First of all, the admittance diagram (figure 4.2) gives us a qualitative picture of the behaviour of the system as the dielectric layer is added. The metal layer will normally be thick enough for the optical admittance at its front surface to be simply that of the metal, the substrate optical constants having no effect. The optical admittance of the metal will always be in the fourth quadrant and so, as a dielectric layer is added, the reflectance must fall until the locus of the admittance of the assembly crosses the real axis. (We recall that the reflectance associated with the locus of a dielectric layer always falls as the locus is traced out in the fourth quadrant and always rises in the first—figure 2.17(a)). This minimum of reflectance will occur at a dielectric layer thickness of less than a quarter-wave. For layer thicknesses of up to twice this figure, therefore, the

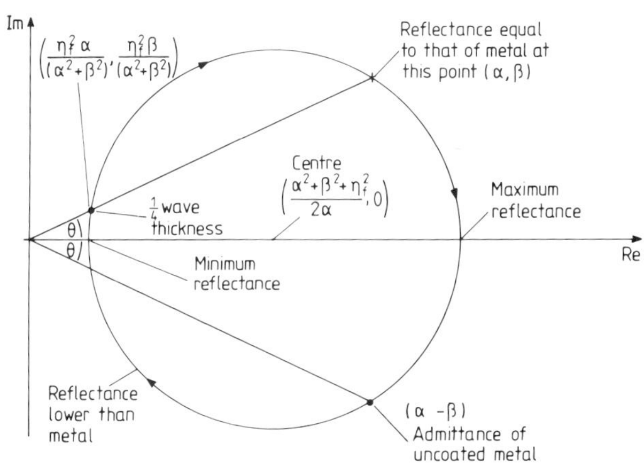

Figure 4.2 Admittance diagram of a dielectric layer deposited over a metal. The metal admittance would usually be much closer to the imaginary axis but has been moved for greater clarity in the diagram. The dielectric locus starts at the admittance of the uncoated metal. The construction to find the quarter-wave point is explained in the text, as are the other parameters.

reflectance of the protected metal film will be reduced. The reduction in reflectance depends very much on the particular metal and the index of the dielectric film.

We can mark the position of the quarter-wave dielectric layer thickness by a simple construction. We draw the line from the origin to the starting point of the dielectric locus, that is the metal admittance (α, $-\beta$) which lies in the fourth quadrant. This line makes an angle θ with the real axis. Then, also through the origin, we draw a line in the first quadrant making the same angle θ with the real axis. This cuts the dielectric locus in two points. One is the point (α, β), the image of the starting point in the real axis, and at this point the reflectance of the assembly is identical to that of the uncoated metal. The second point of intersection is

$$\left(\frac{\eta_f^2\alpha}{(\alpha^2+\beta^2)}, \frac{\eta_f^2\beta}{(\alpha^2+\beta^2)}\right) \qquad \text{i.e.} \qquad \frac{\eta_f^2}{\alpha-\mathrm{i}\beta}$$

and at this point the layer is one quarter-wave thick.

We can derive straightforward analytical expressions for the various parameters, and, in particular, the points of intersection of the locus with the real axis, which we know correspond to the points of maximum and minimum reflectance.

The characteristic matrix is given by

$$\begin{bmatrix} B \\ C \end{bmatrix} = \begin{bmatrix} \cos\delta_f & \mathrm{i}(\sin\delta_f/\eta_f) \\ \mathrm{i}\eta_f\sin\delta_f & \cos\delta_f \end{bmatrix} \begin{bmatrix} 1 \\ \alpha-\mathrm{i}\beta \end{bmatrix} \tag{4.1}$$

where $\alpha-\mathrm{i}\beta$ is the characteristic admittance of the metal, i.e. $\mathscr{Y}(n_m-\mathrm{i}k_m)$ at normal incidence, $\delta_f = 2\pi n_f d_f \cos\theta_f/\lambda$, and η_f is the characteristic admittance of the film material. Then

$$\begin{bmatrix} B \\ C \end{bmatrix} = \begin{bmatrix} \cos\delta_f + (\beta\sin\delta_f)/\eta_f + \mathrm{i}(\alpha\sin\delta_f)/\eta_f \\ \alpha\cos\delta_f + \mathrm{i}(\eta_f\sin\delta_f - \beta\cos\delta_f) \end{bmatrix}.$$

Now, at the points of intersection of the locus with the real axis, we must have that the admittance, which we can denote by μ, is real. But

$$\mu = C/B$$

and, equating real and imaginary parts,

$$\alpha\cos\delta_f = \mu[\cos\delta_f + (\beta\sin\delta_f)/\eta_f] \tag{4.2}$$

$$\eta_f\sin\delta_f - \beta\cos\delta_f = \mu(\alpha\sin\delta_f)/\eta_f. \tag{4.3}$$

Hence, first eliminating μ,

$$(\alpha\cos\delta_f)(\alpha\sin\delta_f)/\eta_f = (\eta_f\sin\delta_f - \beta\cos\delta_f)[\cos\delta_f + (\beta\sin\delta_f)/\eta_f]$$

i.e.

$$[(\alpha^2+\beta^2-\eta_f^2)/(2\eta_f)]\sin(2\delta_f) = -\beta\cos(2\delta_f).$$

Thus

$$\tan(2\delta_f) = 2\beta\eta_f/(\eta_f^2-\alpha^2-\beta^2)$$

so that

$$\delta_f = \tfrac{1}{2}\tan^{-1}[2\beta\eta_f/(\eta_f^2-\alpha^2-\beta^2)] + \frac{m\pi}{2} \qquad m = 0, 1, 2, 3\ldots \tag{4.4}$$

or, in full waves,

$$D_f/\lambda_0 = (1/4\pi)\tan^{-1}[2\beta\eta_f/(\eta_f^2-\alpha^2-\beta^2)] + m/4 \tag{4.5}$$

where the arctangent is to be taken in either the first or second quadrant so that δ_f for $m = 0$ is positive and represents the first intersection with the real axis where the film is less than, or at the very most, equal to a quarter-wave. A similar result has been derived by Park[2] using a slightly different technique.

The value of μ can be found by rearranging equations (4.2) and (4.3) slightly:

$$(\mu-\alpha)\cos\delta_f + (\beta\mu/\eta_f)\sin\delta_f = 0$$
$$\beta\cos\delta_f + [(\mu\alpha/\eta_f)-\eta_f]\sin\delta_f = 0$$

and, eliminating δ_f,

$$(\mu-\alpha)[(\mu\alpha/\eta_f)-\eta_f] - \beta(\beta\mu/\eta_f) = 0.$$

The two solutions are

$$\mu = [(\alpha^2+\beta^2+\eta_f^2)/2\alpha] \pm \{[(\alpha^2+\beta^2+\eta_f^2)/4\alpha^2]-\eta_f^2\}^{1/2}$$

but this is not the best form for calculation. We know that the two solutions μ_1 and μ_2 are related by $\mu_1\mu_2 = \eta_f^2$ and so we write

$$\mu_1 = 2\alpha\eta_f^2/\{(\alpha^2+\beta^2+\eta_f^2)+[(\alpha^2+\beta^2+\eta_f^2)^2-4\alpha^2\eta_f^2]^{1/2}\} \tag{4.6}$$

$$\mu_2 = [(\alpha^2+\beta^2+\eta_f^2)/2\alpha] + \{[(\alpha^2+\beta^2+\eta_f^2)/4\alpha^2]-\eta_f^2\}^{1/2} \tag{4.7}$$

and the value which corresponds to the first intersection ($m = 0$ in equation (4.4)) is

$$\mu_1 = 2\alpha\eta_f^2/\{(\alpha^2+\beta^2+\eta_f^2)+[(\alpha^2+\beta^2+\eta_f^2)^2-4\alpha^2\eta_f^2]^{1/2}\}. \tag{4.6}$$

Often

$$(\alpha^2+\beta^2+\eta_f^2)^2 \gg 4\alpha^2\eta_f^2$$

and in that case

$$\mu_1 = \alpha\eta_f^2/(\alpha^2+\beta^2+\eta_f^2) \tag{4.8}$$

$$\mu_2 = (\alpha^2+\beta^2+\eta_f^2)/\alpha. \tag{4.9}$$

The limits of reflectance are given by

$$R_{\text{minimum}} = [(\eta_0-\mu_1)/(\eta_0+\mu_1)]^2 \tag{4.10}$$

$$R_{\text{maximum}} = [(\eta_0-\mu_2)/(\eta_0+\mu_2)]^2. \tag{4.11}$$

The higher the index of the dielectric film, the greater is the fall in reflectance at the minimum. The reflectance rises above that of the bare metal at the maximum, but, for the metals commonly used as reflectors, the increase is not great, and so the lower-index films are to be preferred as protecting layers. As an example, we can consider aluminium, which has a refractive index of 0.82 − i5.99 at 546 nm,[3] with protecting layers of quartz of index 1.45 or a high-index layer, 2.3, such as cerium oxide. The results in table 4.1 were calculated from equations (4.5)–(4.7), (4.10) and (4.11). Clearly, if high-index films are used for protecting metal layers, then the monitoring of layer thickness must be accurate, otherwise there is a risk of a sharp drop in reflectance.

Table 4.1

Aluminium (0.82–i5.99)	$R_{uncoated}$ (%)	R_{min} (%)	D_{min} (Full waves)	R_{max} (%)	D_{max} (Full waves)
Quartz (1.45)	91.63	83.64	0.2128	91.86	0.4628
CeO_2 (2.30)	91.63	65.90	0.1925	92.44	0.4425

Aluminium is probably the commonest mirror coating material for the visible region, and, in addition to the quartz and cerium oxide mentioned above, there is a large number of materials which can be used for protecting it. Silicon oxide, SiO, for example, is also a very effective protecting material, but it has strong absorption at the blue end of the spectrum, where it causes the reflectance of the composite coating to be rather low. Another useful coating is sapphire, Al_2O_3. This can be vacuum deposited, or the aluminium at the surface of the coating can be anodised by an electrolytic technique[1], forming a very hard layer of aluminium oxide. Gold and silver are more difficult to protect because of the difficulty of getting films to stick to them. However, it has been found that aluminium oxide sticks very well to silver[4,5]. Aluminium oxide does not appear to be a very effective barrier against moisture and so it has been used principally as a bonding layer between the silver and a layer of silicon oxide which affords good moisture resistance and which, although it adheres only weakly to silver, adheres strongly to the aluminium oxide. Further details of the coating are given by Hass and his colleagues[4]. To reduce the absorption at the blue end of the spectrum, the silicon oxide should be reactively deposited (see chapter 9) when the actual oxide which is produced lies between SiO and SiO_2. With such a coating it is possible to achieve a reflectance greater than 95% over the visible and infrared from 0.45–20 μm.

Aluminium oxide and silicon oxide are absorbing at wavelengths longer than 8 μm and it has been discovered by Pellicori[6] and confirmed theoretically by Cox *et al*[5] that reflectors protected by these materials exhibit a sharp dip in reflectance at high angles of incidence, that is, 45° and above. The dip can be avoided by the use of a protecting material which does not absorb in this region. Magnesium fluoride is such a material, but it must be deposited on a hot substrate (temperatures in excess of 200°C) if it is to be robust. The metals have their best performance if deposited at room temperature and thus the substrates should only be heated after they have been coated with the metal.

Overall system performance, boosted reflectance

In optical instruments of any degree of complexity there will be a number of reflecting components in series, and the overall transmission of the system will be given by the product of the reflectances of the various elements. Figure 4.3 gives the overall transmission of any system with a number of components in series, with identical values of reflectance. It is obvious from the diagram that even with the best metal coatings, the performance with ten elements, say, is low. If the instrument is to be used over a wide range there is little that can be done to alleviate the situation. Most spectrometers, for instance, have ten or more reflections with a consequent severe drop in transmission, but are

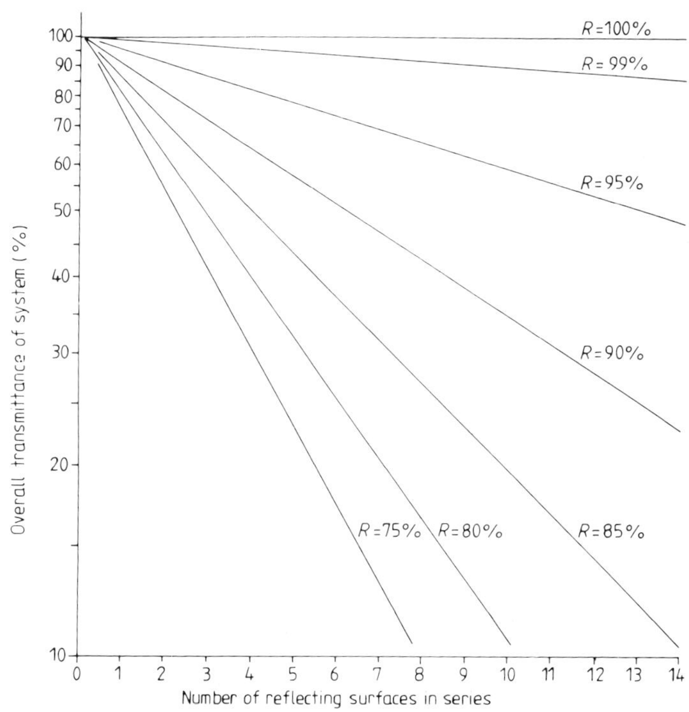

Figure 4.3 Overall transmittance of an optical system which has a number of reflecting elements in series.

required to work over a wide region—possibly as much as a 25:1 variation in wavelength. The spectrometer designer normally just accepts this loss and designs the rest of the instrument accordingly.

In cases where the wavelength range is rather more limited, say, to the visible region or to a single wavelength, it is possible to increase the reflectance of a simple metal layer by boosting it with extra dielectric layers.

The characteristic admittance of a metal can be written $n-ik$ and the reflectance in air at normal incidence is

$$R=\left|\frac{1-(n-ik)}{1+(n-ik)}\right|^2=\frac{(1-n)^2+k^2}{(1+n)^2+k^2}=\frac{1-[2n/(1+n^2+k^2)]}{1+[2n/(1+n^2+k^2)]}. \tag{4.1}$$

On p 46 it was shown that the optical admittance of an assembly Y becomes n^2/Y when a quarter-wave optical thickness of index n, that is admittance in free space units, is added.

If the metal is overcoated with two quarter-waves of material of indices n_1 and n_2, n_2 being next to the metal, then the optical admittance at normal incidence is

$$\left(\frac{n_1}{n_2}\right)^2(n-ik)$$

and the reflectance in air, also at normal incidence,

$$R=\left|\frac{1-(n_1/n_2)^2(n-ik)}{1+(n_1/n_2)^2(n-ik)}\right|^2$$

i.e.

$$R=\frac{[1-(n_1/n_2)^2n]^2+(n_1/n_2)^4k^2}{[1+(n_1/n_2)^2n]^2+(n_1/n_2)^4k^2}$$

$$=\frac{1-[2(n_1/n_2)^2n]/[1+(n_1/n_2)^4(n^2+k^2)]}{1+[2(n_1/n_2)^2n]/[1+(n_1/n_2)^4(n^2+k^2)]}. \tag{4.2}$$

This will be greater than the reflectance of the bare metal, given by equation (4.1), if

$$\frac{2(n_1/n_2)^2n}{1+(n_1/n_2)^4(n^2+k^2)}<\frac{2n}{1+n^2+k^2} \tag{4.3}$$

which is satisfied by either

$$\left(\frac{n_1}{n_2}\right)^2>1$$

or

$$\left(\frac{n_1}{n_2}\right)^2<\frac{1}{n^2+k^2} \tag{4.4}$$

assuming that $n^2+k^2\geqslant 1$.

The first solution is of greater practical value than the second, which can be ignored. This shows that the reflectance of any metal can be boosted by a pair of quarter-wave layers for which $(n_1/n_2) > 1$, n_1 being on the outside and n_2 next to the metal. The higher this ratio, the greater the increase in reflectance. As an example, consider aluminium at 550 nm with $n - ik = 0.82 - i5.99$. From equation (4.1), the untreated reflectance of this is approximately 91.6 %.

If the aluminium is covered by two quarter-waves consisting of magnesium fluoride of index 1.38, next to the aluminium, followed by zinc sulphide of index 2.35, then $(n_1/n_2)^2 = 2.9$ and, from equation (4.2), the reflectance jumps to 96.9 %.

An approximate result can be obtained very quickly using $A = (1 - R)$. When the two layers are added, A is reduced roughly to $A/(n_1/n_2)^2$. Inserting the above figures, for aluminium, A is 8.4 % initially, and on addition of the layers drops to 2.9 %, corresponding to a boosted reflectance of 97.1 % (instead of the more accurate figure of 96.9 %).

A second similar pair of dielectric layers will boost the reflectance even higher—to approximately 99 %, and greater numbers of quarter-wave pairs may be used to give an even higher reflectance.

Unfortunately, the region over which the reflectance is boosted is limited. Outside this zone the reflectance is less than it would be for the bare metal. Jenkins[7] has measured the reflectance of an aluminium layer overcoated with six quarter-wave layers of cryolite, of index 1.35, and zinc sulphide of index 2.35. With layers monitored at 550 nm, the reflectance of the boosted aluminium was greater than 95 % over a region 280 nm wide, and greater than 99 % over the major part.

More robust coatings can be obtained using magnesium fluoride, silicon dioxide or aluminium oxide as the low-index layers, and cerium oxide or titanium oxide as the high-index layers. To attain maximum toughness, the dielectric layers should be deposited on a hot substrate. Aluminium, however, if deposited hot, tends to scatter badly and so the substrates should be heated only after deposition of the aluminium is complete. Figure 4.4 shows the reflectance of aluminium boosted by four quarter-wave layers, which enhanced the reflectance over the visible region.

We have already considered more exactly the behaviour of a single dielectric layer on a metal, and have shown, as did Park[2], that the thickness of the dielectric layer for minimum reflectance should be

$$D = \{\tan^{-1}[2\beta\eta_f/(\eta_f^2 - \alpha^2 - \beta^2)]\}[\lambda_0/(4\pi)]$$

where $(\alpha - i\beta)$ is the admittance of the metal and the angle is in the first or second quadrant. This is the thickness which the low-index layer next to the metal should have if the maximum possible increase in reflectance is to be achieved. A moment's consideration of the admittance diagram will show that this is indeed the case. Layers other than that next to the metal will, of course, retain their quarter-wave thicknesses.

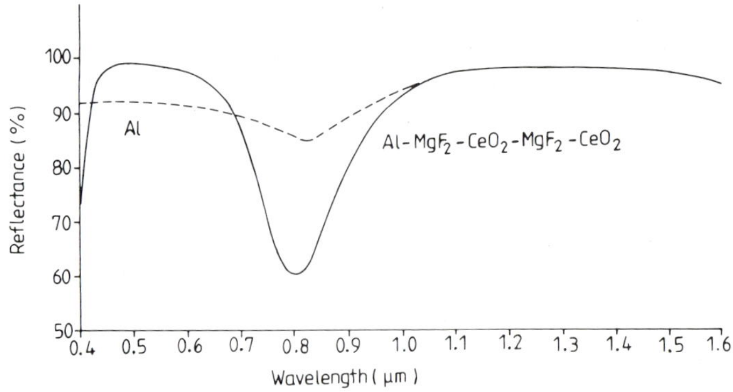

Figure 4.4 Reflectance of evaporated aluminium with (solid curve) and without (broken) two reflectance-increasing film pairs of MgF_2 and CeO_2 as a function of wavelength from 0.4–1.6 μm. (After Hass[1].)

Reflecting coatings for the ultraviolet

The production of high-reflectance coatings for the ultraviolet is a much more exacting task than for the visible and infrared. A very full review of the topic is given by Madden[8], supplemented in great detail by a later account of Hass and Hunter[9]. The following is a very brief summary.

The most suitable material known for the production of reflecting coatings for the ultraviolet out to around 100 nm is aluminium. To achieve the best results, the aluminium should be evaporated at a very high rate, 40 nm s^{-1} or more if possible, on to a cold substrate, the temperature of which should not be permitted to exceed 50°C, and at pressures of 10^{-6} torr or lower. The aluminium should be of the purest grade. Hass and Tousey[10] have quoted results which show that there is a significant improvement (as high as 10% at 150 nm) in the ultraviolet reflectance of aluminium films if 99.99% pure aluminium is used in preference to 99.5% pure. Aluminium should, in theory, have a much higher reflectance than is usually achieved in practice, particularly at the shortwave end of the range. This has been found to be due to the formation of a thin oxide layer on the surface, and as we have already shown, such a layer must, unless it is very thick, lead to a reduction in reflectance. This oxidation takes place even at partial pressures of oxygen below 10^{-6} torr. Unprotected aluminium films, therefore, inevitably show a rapid fall in reflectance with time when exposed to the atmosphere. The reflectance stabilises when the layer is of sufficient thickness to inhibit further oxidation, but this occurs only when the reflectance at short wavelengths has fallen catastrophically.

Attempts have been made to find suitable protecting material for aluminium to prevent oxidation, and very promising results have been obtained with magnesium fluoride (very robust coatings) and lithium fluoride (less robust), which in crystal form are very useful window materials for the ultraviolet. Figures 4.5 and 4.6 show the effect of an extra protecting layer of magnesium fluoride[11] or lithium fluoride[12] on the reflectance of aluminium. The increase in reflectance is partly due to the lack of oxide layer, but also to interference effects.

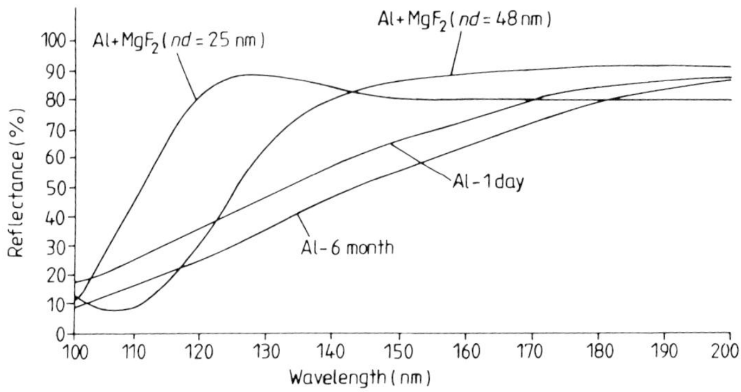

Figure 4.5 Reflectance of evaporated aluminium from 100–200 nm with and without protective layers of MgF_2 of two different thicknesses. (After Canfield, Hass, and Waylonis[11].)

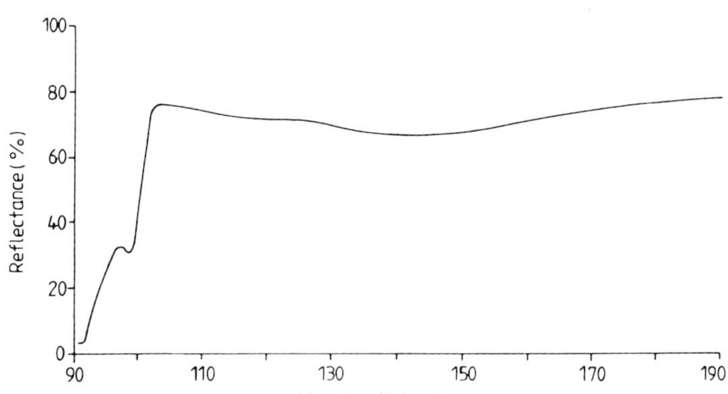

Figure 4.6 Reflectance of an evaporated aluminium film with a 14 nm thick LiF overcoating in the region from 90–190 nm. Measurements were begun 10 minutes after the evaporation was completed. (After Cox, Hass, and Waylonis[12].)

It is necessary to evaporate the protecting layer immediately after the aluminium in order that the minimum amount of oxidation should be allowed to take place. This is usually achieved by running the two sources simultaneously and arranging for the shutter which covers the aluminium source at the end of the deposition of the aluminium layer to uncover at the same time the magnesium or lithium fluoride source. The use of magnesium fluoride overcoated aluminium as a reflecting coating for the ultraviolet is now becoming standard practice.

The aluminium and magnesium fluoride coating is examined in some detail by Canfield *et al*[11]. Amongst other results they show that provided the magnesium fluoride is thicker than 10 nm the coatings will withstand, without deterioration, exposure to ultraviolet radiation and to electrons (up to 10^{16}, 1 MeV electrons/cm^2) and protons (up to 10^{12}, 5 MeV protons/cm^2).

NEUTRAL BEAM SPLITTERS

A device which divides a beam of light into two parts is known as a beam splitter. The functional part of a beam splitter generally consists of a plane surface coated to have a specified reflectance and transmittance over a certain wavelength range. The incident light is split into a transmitted and a reflected portion at the surface, which is usually tilted so that the incident and reflected beams are separated. The ideal values of reflectance and transmittance may vary from one application to another. The beam splitters considered in this section are known as neutral beam splitters, because reflectance and transmittance should ideally be constant over the wavelength range concerned.

Neutral beam splitters are usually specified by the ideal values of transmittance and reflectance expressed as a percentage and written T/R. 50/50 beam splitters are probably the most common.

Beam splitters using metallic layers

Apart from a single uncoated surface, which is sometimes used, the simplest type of beam splitter consists of a metal layer deposited on a glass plate. Silver, which has least absorption of all the common metals used in the visible region, is traditionally the most popular material for this. 50/50 beam splitters are frequently referred to as being 'half-silvered', although commercial beam splitters nowadays are usually constructed from metals such as chromium which are less prone to damage by abrasion and corrosion.

All metallic beam splitters suffer from absorption. The transmission of a metal film is the same, regardless of the direction in which it is measured. This is not so for reflectance, and that measured at the air side is slightly higher than that measured at the glass side. This effect does not appear with a transparent film. Since $T+A+R=1$, the reduction in reflectance at the substrate side means that the absorption from that side must always be higher. Figure 4.7 shows curves for platinum demonstrating this behaviour[13]. Because of this difference in reflection, metallic beam splitters should always be used in the manner shown in figure 4.8 if the highest efficiency is to be achieved.

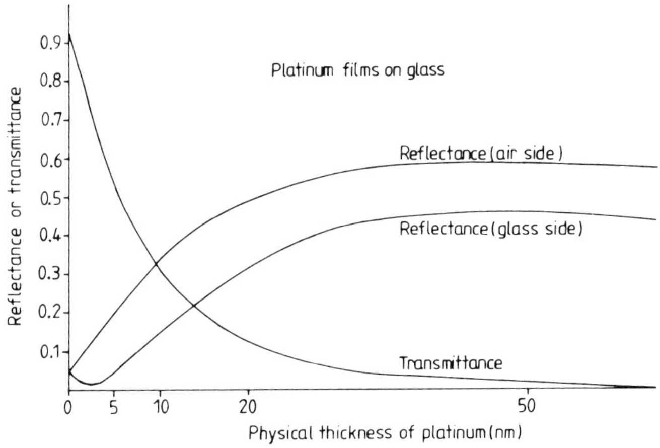

Figure 4.7 Reflectance and transmittance curves for a platinum film on glass, calculated from the optical constants of the bulk metal. (After Heavens[13].)

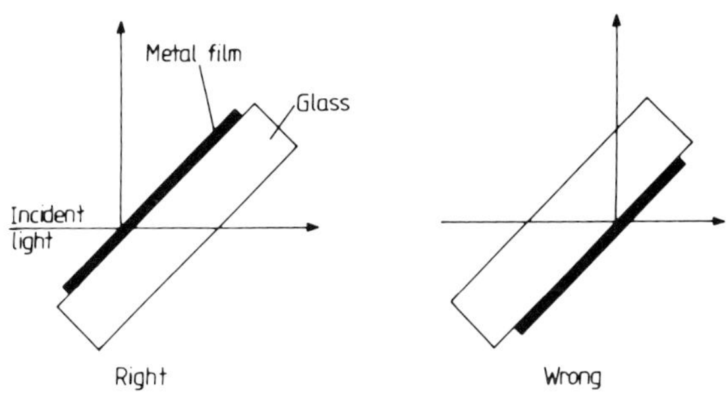

Figure 4.8 Correct use of a metallic beam splitter.

It is possible to decrease the absorption in metallic beam splitters by adding an extra dielectric layer. The method has been applied to chromium films by Pohlack[14] and figure 4.9 gives some of the measurements made.

The first pair of results is for a simple chromium film on glass of index 1.52 measured both from the air side and the glass side. The second pair of results shows how the absorption in the chromium can be reduced by the presence of a quarter-wave layer of high refractive index material (zinc sulphide of index approximately 2.4 in this case) between the metal and the glass. This layer forms an antireflection coating on the rear surface of the metal, and the effect can be seen particularly strongly in the results for reflectance and transmission from the glass side. There, the transmission remains exactly as before, but the reflectance is considerably reduced. Results are also given for a chromium layer protected by a cover glass cemented on the front surface with and without the

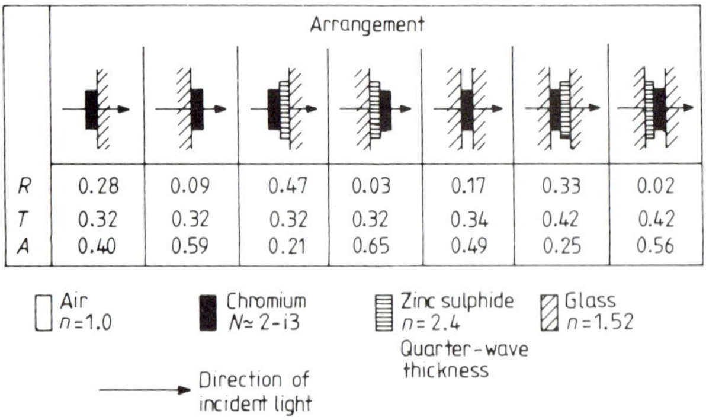

R	0.28	0.09	0.47	0.03	0.17	0.33	0.02
T	0.32	0.32	0.32	0.32	0.34	0.42	0.42
A	0.40	0.59	0.21	0.65	0.49	0.25	0.56

Figure 4.9 Values of reflectance, transmittance and absorptance at 550 nm and normal incidence for semi-reflecting films of chromium on glass showing the effect of adding a quarter-wave layer of zinc sulphide (after Pohlack[11]).

antireflecting layer. The metallic absorption again is very much less when the antireflection layer is on the side of the metal remote from the incident light.

Shkliarevskii and Avdeenko[15] increased the transparency and decreased the absorption in metallic coatings using an antireflection coating in a similar manner. The antireflection coating in this case, instead of being dielectric, was a thin metallic layer. They found that a layer of silver deposited on a substrate heated to around 300°C increased the transparency of an aluminium coating, deposited on top of the silver at room temperature, by a factor as high as 3.5 at 1 μm and 2.5 at 700 nm without any decrease in reflectance at the aluminium–air interface.

If the beam splitter is used correctly, the reduction in reflectance at the glass–film interface can be useful in reducing the stray light derived from reflection, first from the back surface of the glass blank and then from the glass–film interface.

One complication found with beam splitters is a difference in the values of reflectance for the two planes of polarisation when the beam splitter is tilted. The TE (or s-) reflectance is higher than the TM (or p-) reflectance. In calculating the efficiency of a beam splitter this must be taken into account. Anders[16] describes a method for calculating efficiency and stray light performance.

It is not always possible to use the flat plate beam splitter in some optical systems. Reflections from the rear surface can be a problem in spite of the antireflection layer behind the metal film, and in applications where the light passing through the plate is not collimated, aberrations are introduced. To overcome these difficulties a beam-splitting cube, as shown in figure 4.10, can be used, although the absorption in the metal is greater in this configuration because both surfaces, instead of just one, are now in contact with a medium whose index is greater than unity. Since the cemented assembly protects the

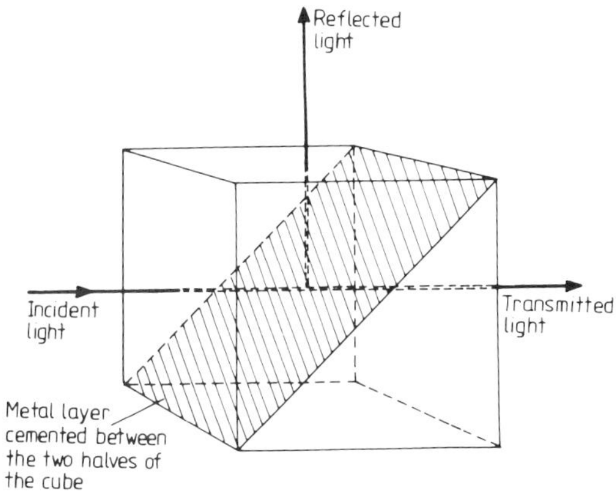

Figure 4.10 A cube beam splitter.

metal layers the choice of materials is wide. Silver is probably most frequently used, although chromium, aluminium and gold are also popular.

Chromium gives almost neutral beam splitting over the visible region, with an absorption of approximately 0.55 for both planes of polarisation, the TE reflectance being approximately 0.30 and the TM 0.15. Silver varies more with wavelength, the reflectance falling towards the blue end of the spectrum, but the absorption is rather less than for chromium, around 0.15 at 550 nm, with TE reflectance 0.50 and TM 0.30. Curves of the performance of several different metallic beam splitters are given by Anders[16].

Beam splitters using dielectric layers

There are many optical instruments where the light undergoes a transmission followed by a reflection, or vice versa, both at the same, or at the same type of, beam splitter. In two-beam interferometers, for example, the beams are first of all separated by one pass through a beam splitter and then combined again by a further pass either through the same beam splitter, as in the Michelson interferometer, or through a second beam splitter, as in the Mach–Zehnder interferometer. The effective transmittance of the instrument is given by the product of the transmission and the reflectance of the beam splitter, taking into account the particular polarisation involved. For a perfect beam splitter, TR would be 0.25; for most metallic beam splitters it is around 0.08 or 0.10. The absorption in the film is the primary source of loss.

A beam splitter of improved performance, as far as the TR product is concerned, can be obtained by replacing the metallic layer with a transparent high-index quarter-wave. At normal incidence the reflectance of a quarter-wave is given by

$$R = \left(\frac{1 - n_1^2/n_2}{1 + n_1^2/n_2}\right)^2.$$

At 45° angle of incidence in air the position of the peak is shifted to a shorter wavelength, and the appropriate optical admittances must be used in calculating peak reflectance.

$$R = \left(\frac{\eta_0 - (\eta_1^2/\eta_2)}{\eta_0 - (\eta_1^2/\eta_2)}\right)^2$$

and since η varies with the plane of polarisation, R will have two values, R_{TE} and R_{TM}.

Figure 4.11 shows the peak reflectance of a quarter-wave of index between 1.0 and 3.0 on glass of index 1.52 for both 45° incidence and normal incidence. At 45°, the peak reflectance for unpolarised light, $\frac{1}{2}(R_{TE} + R_{TM})$, is within 1.5% of the peak value for normal incidence.

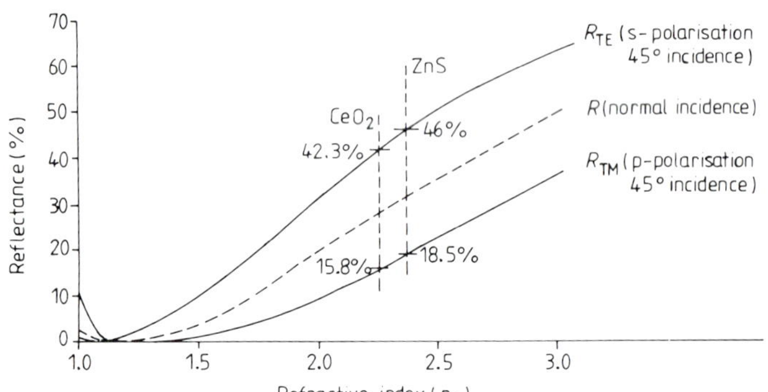

Figure 4.11 Peak reflectance in air of a quarter-wave of index n_1 on glass of index 1.52 at normal and 45° incidence.

Zinc sulphide, with index 2.35, is a popular material for beam splitters. At 45° we have

$$(TR)_{TE} = (0.46 \times 0.54) = 0.248$$
$$(TR)_{TM} = (0.185 \times 0.815) = 0.151$$

and

$$(TR)_{unpolarised} = \tfrac{1}{2}(0.248 + 0.151) = 0.200.$$

($(TR)_{unpolarised}$ cannot be calculated using $T_{mean}\ R_{mean}$ (= 0.219) because the light, after having undergone one reflection or transmission, is then partly polarised.)

If a more robust film is required, cerium oxide, with an index approximately 2.25, is a good choice. Here

$$(TR)_{TE} = (0.423 \times 0.577) = 0.244$$
$$(TR)_{TM} = (0.158 \times 0.842) = 0.133$$
$$(TR)_{unpolarised} = 0.189.$$

Clearly the dielectric beam splitter, even if it does tend to have characteristics which more nearly correspond to 70/30 rather than 50/50, has a considerably better performance than the metallic beam splitter. The reflectance curve of a typical 70/30 beam splitter in figure 4.12 shows how the reflectance varies on either side of the peak.

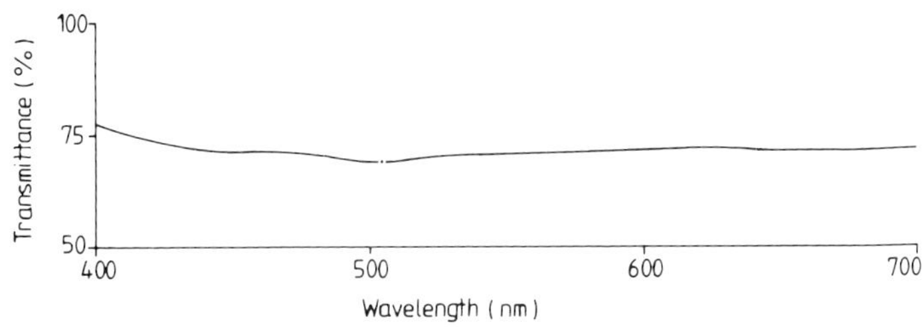

Figure 4.12 Measured transmittance curve of a dielectric 70/30 beam splitter at 45° angle of incidence. (Courtesy of Sir Howard Grubb, Parsons & Co Ltd.)

Beam splitters with 55/45 characteristics can be made by evaporating pure titanium in a good vacuum and subsequently oxidising it to TiO_2 by heating at 420°C in air at atmospheric pressure. The titanium oxide thus formed has rutile structure and a refractive index of 2.8. Titanium films produced in a poor vacuum oxidise subsequently to the anatase form, having rather lower refractive index. The production of very large beam splitters, of this type, 17 × 13 inches, is described in a paper by Holland *et al*[17].

The single layer beam splitter suffers from a fall in reflectance on either side of the central wavelength. In the same way that single-layer antireflection coatings can be broadened by adding a half-wave layer, so the single quarter-wave beam splitter can be broadened. The same basic pattern of admittance circles can be achieved either by a low-index half-wave layer between the high-index quarter-wave and the glass substrate or an even higher index half-wave deposited over the quarter-wave. Since no suitable materials for the latter solution exist in practice, the low-index half-wave is the only feasible approach. The admittance diagram is shown in figure 4.13 and the performance in figure 4.14.

The technique is effective also for multilayer systems to give a higher reflectance. Approximately 50% reflectance can be achieved by a four-layer coating, Air $|LHLH|$ Glass, and this can be flattened by an additional low index half-wave at the glass end of the multilayer, that is, Air $|LHLHLL|$ Glass. Figure 4.14 shows the performance calculated for this design of beam splitter.

A detailed discussion of the role of half-wave layers is given by Knittl[18].

As mentioned above, beam-splitting cubes must be used in some applications where plate beam splitters are unsuitable. Unfortunately, the main problem connected with dielectric beam splitters, the low reflectance for TM waves, becomes even worse with cube beam splitters. The reason for this is

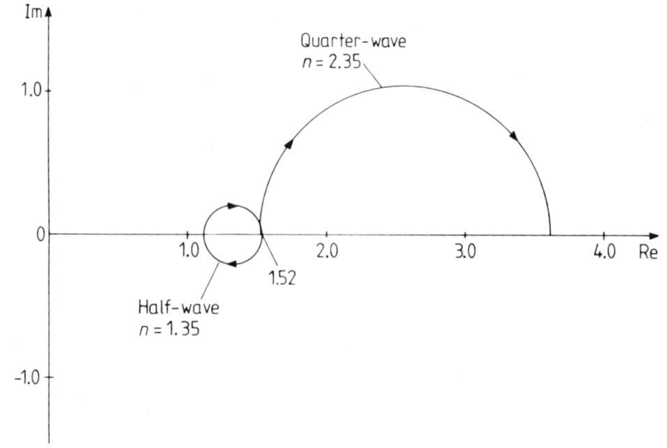

Figure 4.13 Admittance diagram at λ_0 of a two-layer beam splitter. The high-index quarter-wave layer gives the required high reflectance. The low-index half-wave layer flattens the performance over the visible region.

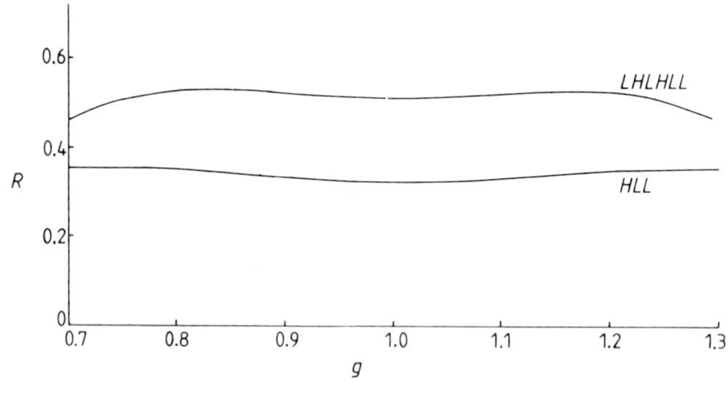

Figure 4.14(*a*) The performance of the beam splitter shown in figure 4.13. Design: Air (1.00) |*HLL*| Glass (1.52) with *L* a quarter-wave of index 1.35 and *H* of 2.35. (*b*) The performance of a beam splitter of design: Air(1.00) |*LHLHLL*| Glass (1.52) with indices as for (*a*).

simply that 45° incidence in glass is effectively a much greater angle of incidence than 45° in air. Consequently, the polarisation splitting is even greater and the TM performance becomes so poor that the beam splitter is unusable in most applications. Metal layers are, therefore, the only ones which can be used in the straightforward cube beam splitter and combiner. This disadvantage of the dielectric layer can, however, be turned to advantage in the construction of polarisers as we shall see in chapter 8.

NEUTRAL-DENSITY FILTERS

A filter which is intended to reduce the intensity of an incident beam of light evenly over a wide spectral region is known as a neutral-density filter.

The performance of neutral-density filters is usually defined in terms of the optical density, D:

$$D = \log_{10} (I_0/I_T)$$

where I_0 is the incident intensity and I_T is the transmitted intensity measured either at one particular wavelength or integrated over a region.

Absorption and absorptance are terms which are not correctly used of neutral-density filters because they represent the fraction of energy which is actually absorbed in the film, and in neutral-density filters a proportion of the incident energy is removed by reflection.

The advantage of using the logarithmic term is that the effect of placing two or more neutral-density filters in series is easily calculated. The overall density is simply the sum of the individual densities (provided that multiple reflections are not permitted to occur between the individual filters, which would affect the result in the way shown in chapter 2, p 68, equation (2.93)).

Thin-film neutral-density filters consist of single metallic layers with thicknesses chosen to give the correct transmission values. Rhodium, palladium, tungsten, chromium, as well as other metals, are all used to some extent, but the best performance is obtained by the evaporation of a nickel chromium alloy, approximately 80 % nickel and 20 % chromium. Chromel A or Nichrome are standard resistance wires which have this composition and can be readily obtained. The method is described by Banning[19]. The Chromel or Nichrome should be evaporated at 10^{-4} torr or better from a thick tungsten spiral. Neutral films, having densities up to around 1.5, corresponding to a

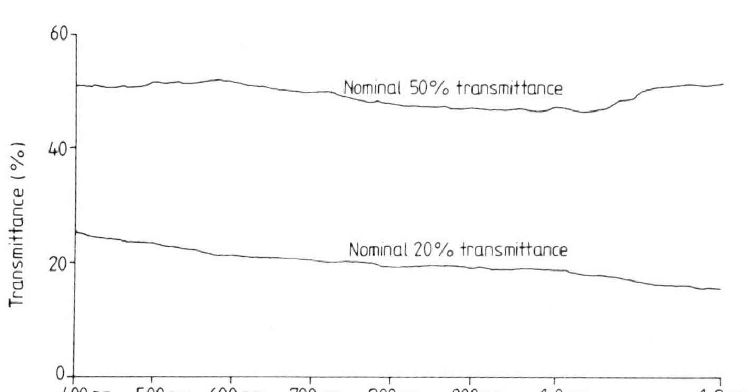

Figure 4.15 Measured transmittance curves of neutral density filters consisting of Nichrome films on glass substrates (courtesy of Sir Howard Grubb, Parsons & Co Ltd).

transmission of 3%, can be manufactured in this way. If the films are made thicker, they are not as neutral and tend to have a higher transmission in the red, owing to excess chromium. The films are very robust and do not need any protection, especially if they are heated to around 200°C after evaporation.

Figure 4.15 shows some response curves of neutral-density filters made from Nichrome on glass. The filters are reasonably neutral over the visible and near infrared out to 2 μm. In fact, if quartz substrates are used the filters will be good over the range 0.24–2 μm.

REFERENCES

1 Hass G 1955 Filmed surfaces for reflecting optics *J. Opt. Soc. Am.* **45** 945–52

2 Park K C 1964 The extreme values of reflectivity and the conditions for zero reflection from thin dielectric films on metal *Appl. Opt.* **3** 877–81

3 Hass G 1972 Optical constants of metals *American Institute of Physics Handbook* ed. D E Gray (New York: MacGraw-Hill) pp 6-124–6-156. The value used for aluminium, 0.82–i5.99 at 546 nm, is quoted on p 6-125.

4 Hass G, Heany J B, Herzig H, Osantowski J F and Triolo J J 1975 Reflectance and durability of Ag mirrors coated with thin layers of Al_2O_3 plus reactively deposited silicon oxide *Appl. Opt.* **14** 2639–44.

5 Cox J T, Hass G and Hunter W R 1975 Infrared reflectance of silicon oxide and magnesium fluoride protected aluminium mirrors at various angles of incidence from 8 μm to 12 μm *Appl. Opt.* **14** 1247–50

6 Pellicori S F 1974 *Private communication* (Santa Barbara Research Center, Goleta, California) see reference 5

7 Jenkins F A 1958 Extension du domaine spectral de pouvoir réflecteur élevé des couches multiples diélectriques *J. Phys. Radium* **19** 301–6

8 Madden R P 1963 Preparation and measurement of reflecting coatings for the vacuum ultraviolet *Physics of Thin Films* vol 1 ed. G Hass (New York: Academic) pp 123–86

9 Hass G and Hunter W R 1978 The use of evaporated films for space applications—extreme ultraviolet astronomy and temperature control of satellites *Physics of Thin Films* vol 10 ed. G Hass and M H Francombe (New York: Academic) pp 71–166

10 Hass G and Tousey R 1959 Reflecting coatings for the extreme ultraviolet *J. Opt. Soc. Am.* **49** 593–602

11 Canfield L R, Hass G and Waylonis J E 1966 Further studies on MgF_2-overcoated aluminium mirrors with highest reflectance in the vacuum ultraviolet *Appl. Opt.* **5** 45–50

12 Cox J T , Hass G and Waylonis J E 1968 Further studies on LiF overcoated aluminium mirrors with highest reflectance in the vacuum ultra-violet *Appl. Opt.* **7** 1535–9

13 Heavens O S 1955 *Optical properties of thin solid films* (London: Butterworth) figure 6.5, p 162

14 Pohlack H 1953 Beitrag zur Optik dünnster Metallschichten *Jenaer Jahrbuch* pp 241–5 (Jena: Zeiss)

15 Shkliarevskii I N and Avdeenko A A 1959 Increasing the transparency of metallic coatings *Opt. Spectrosc.* **6** 439–43
16 Anders H 1965 *Dunne Schichten für die Optik* (Stuttgart: Wissenschaftliche Verlagsgesellschaft) pp 82–91
17 Holland L, Hacking K and Putner T 1953 The preparation of titanium dioxide beam-splitters of large surface area *Vacuum* **3** 159–61
18 Knittl Z 1976 *Optics of Thin Films* (London: Wiley)
19 Banning M 1947 Neutral density filters of Chromel A *J. Opt. Soc. Am.* **37** 686–7

5 Multilayer high-reflectance coatings

The metal reflecting layers of the previous chapter suffer from a considerable absorption loss which, although unfortunate, still permits a high level of performance in most simple systems. There are applications where the absorption in metal layers is too high and the reflectance too low. These include multiple-beam interferometers and resonators, where the large number of successive reflections magnifies the effects of absorption, and high-power systems where the energy absorbed can be sufficient to damage the coating. One way of increasing the reflectance of an opaque metal coating, as we have seen, is to boost the reflectance by adding dielectric layers. This also reduces the absorptance, but the transmittance remains effectively zero. For high-reflecting coatings which must transmit what they do not reflect, all-dielectric multilayers are required. The description which follows is built around the most successful of the multiple-beam interferometers, the Fabry–Perot interferometer. As we shall see later, this interferometer is also of considerable importance in the development of thin-film band-pass filters, and this is a further reason for dealing with it in some detail here.

THE FABRY–PEROT INTERFEROMETER

First described in 1899 by Fabry and Perot,[1] the interferometer known by their names has profoundly influenced the development of thin-film optics. It belongs to the class of interferometers known as multiple-beam interferometers because a large number of beams is involved in the interference. The theory of each of the various types of multiple-beam interferometer is similar. They differ mainly in physical form. Their common feature is that their fringes are much sharper than those in two-beam interferometers, thus improving both measuring accuracy and resolution. Multiple-beam interferometers are described in almost all textbooks on optics, for example that by Born and Wolf.[2]

A Fabry–Perot interferometer consists of two flat plates separated by a distance d_s and aligned so that they are parallel to a very high degree of accuracy. The separation is usually maintained by a spacer ring made of invar or quartz, and the assembly of two plates and a spacer is known as an etalon. The inner surfaces of the two plates are usually coated to enhance their reflectance.

Figure 5.1 shows an etalon in diagrammatic form. The amplitude reflection and transmission coefficients are defined as shown. The basic theory has already been given in chapter 2 (p 54), where it was shown that the transmission for a plane wave is given by

$$T = \frac{T_a T_b}{[1-(R_a^- R_b^+)^{1/2}]^2}\left[1 + \frac{4R_a^- R_b^+}{[1-(R_a^- R_b^+)^{1/2}]^2}\sin^2\left(\frac{\phi_a+\phi_b}{2} - \delta\right)\right]^{-1} \tag{5.1}$$

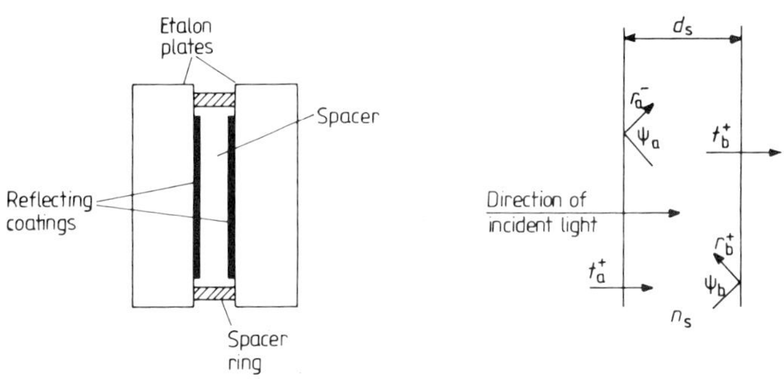

Figure 5.1 A Fabry–Perot etalon. The amplitude coefficients in the diagram are converted to the intensity coefficients of equation (5.1) as shown on page 54.

where $\delta = (2\pi n_s d_s \cos\theta_s)/\lambda$, d_s and n_s being the physical thickness and refractive index of the spacer layer. This is similar to (2.81) except that δ has been modified to include oblique incidence θ_s. In order to simplify the discussion, let the reflectances and transmittances of the two surfaces be equal, let there be no phase change on reflection, i.e. let $\phi_a = \phi_b = 0$, and let n_s be unity, i.e. an air spacer. Then

$$T = \frac{T_s^2}{(1-R_s)^2}\frac{1}{1+[4R_s/(1-R_s)^2]\sin^2\delta} \tag{5.2}$$

and, writing

$$F = \frac{4R_s}{(1-R_s)^2} \tag{5.3}$$

then

$$T = \frac{T_s^2}{(1-R_s)^2}\frac{1}{1+F\sin^2\delta}. \tag{5.4}$$

If there is no absorption in the reflecting layers, then

$$1 - R_s = T_s$$

and

$$T = \frac{1}{1 + F \sin^2 \delta}. \tag{5.5}$$

The form of this function is given in figure 5.2 where T is plotted against δ. T is a maximum for $\delta = m\pi$, where $m = 0, \pm 1, \pm 2, \ldots$, and a minimum halfway between these values. The successive peaks of T are known as fringes and m is known as the order of the appropriate fringe. As F increases, the widths of the fringes become very much narrower. The ratio of the separation of adjacent fringes to the halfwidth (the fringe width measured at half the peak transmission) is called the 'finesse' of the interferometer and is written $\mathscr{F}$. From equation (5.5), the value of δ corresponding to a transmission of half the peak value is given by

$$0.5 = \frac{1}{1 + F \sin^2 \delta}$$

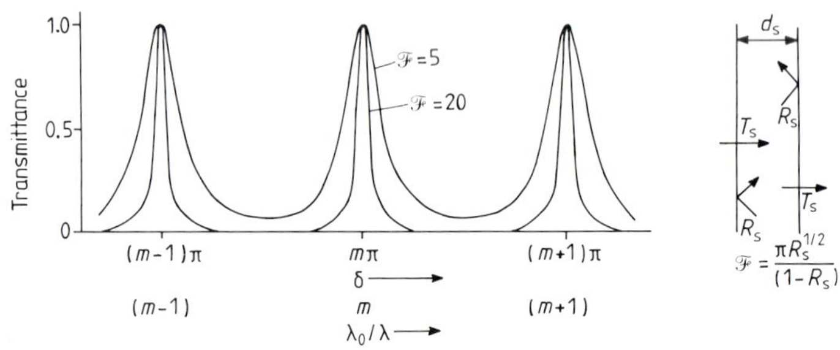

Figure 5.2 Fabry–Perot fringes.

and if δ is sufficiently small so that we can replace $\sin^2 \delta$ by δ^2, then

$$0.5 = \frac{1}{1 + F\delta^2}$$

i.e.

$$\delta = \frac{1}{F^{1/2}}$$

which is *half* the width of the fringe. The separation between values of δ representing successive fringes is π, so that

$$\mathscr{F} = \frac{\pi F^{1/2}}{2}$$

or

$$\mathscr{F} = \frac{\pi R_s^{1/2}}{(1 - R_s)}. \tag{5.6}$$

The Fabry–Perot interferometer is used principally for the examination of the fine structure of spectral lines. The fringes are produced by passing light from the source in question through the interferometer. Measurement of the fringe pattern as a function of the physical parameters of the etalon can yield very precise values of the wavelengths of the various components of the line. The two most common arrangements are either to have the incident light highly collimated and incident normally, or at some constant angle, when the fringes can be scanned by varying the spacer thickness, or it is possible to keep the spacer thickness constant and scan the fringes by varying θ_s, the angle of incidence. Possible arrangements corresponding to these two methods are shown in figure 5.3.

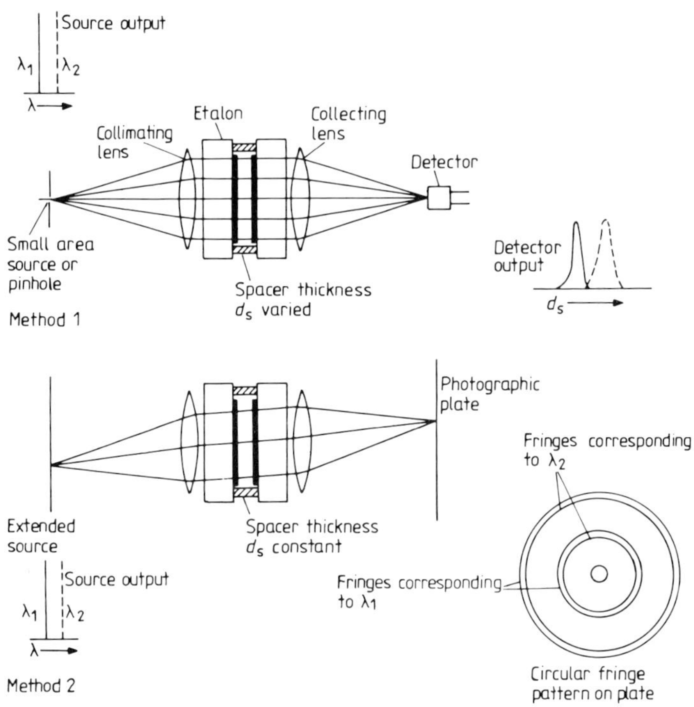

Figure 5.3 Two possible arrangements of a Fabry–Perot interferometer.

Practical considerations limit the achievable finesse to a maximum normally of around 25, or perhaps 50 in exceptional cases. This is due mainly to imperfections in the plates themselves. It is extremely difficult to manufacture a plate with flatness better than $\lambda/100$ at, say, 546 nm. Variations in flatness of the plates give rise to local variations of d_s and hence δ, causing the fringes to shift. These variations should not be greater than the fringe width, otherwise the luminosity of the instrument will suffer. Chabbal[3] has considered this problem in great detail, but for our present purpose it is sufficient to assume that, for a pair of $\lambda/100$ plates (i.e. having errors not greater than $\pm\lambda/200$ about the mean), the variation in thickness of the spacer layer will be of the order of $\pm\lambda/100$ about the mean. This will occur when the defects in the plates are in the form of either spherical depressions in both plates or else protrusions. This in turn means a change in δ of $\pm 2\pi/100$ corresponding to a total excursion of $2\pi/50$. Any decrease in fringe width below this will not increase the resolution of the system but merely reduce the overall luminosity, so that $2\pi/50$ represents a lower limit on the fringe width. Since the interval between fringes is π, this condition is equivalent to an upper limit on finesse of $\pi/(2\pi/50)$, i.e. 25. In more general terms, if the plates are good enough to limit the total thickness variation in the spacer to λ/p (not quite the same as saying that each plate is good to λ/p), then the finesse should be not greater than $p/2$.

The resolution of an optical instrument is normally determined by the Rayleigh criterion, which is particularly concerned with intensity distributions of the form

$$I(\delta) = \left(\frac{\sin \delta/2}{\delta/2}\right)^2 I_{\max}$$

which are of a type produced by diffraction rather than interference effects. Two wavelengths are considered just resolved by the instrument if the intensity maximum of one component falls exactly over the first intensity zero of the other component. This implies that if the two components are of equal intensity, then, in the combined fringe pattern, the minimum which will exist between the two maxima will be of intensity $8/\pi^2$ times that at either of them. In the Fabry–Perot interferometer the fringes are of rather different form, and the pattern of zeros and successively weaker maxima associated with the $[(\sin \delta/2)/(\delta/2)]^2$ function is missing. The Rayleigh criterion cannot, therefore, be applied directly. Born and Wolf[4] suggest that a suitable alternative form of the criterion, which could be applied in this case, might be that two equally intense lines are just resolved when the resultant intensity between the peaks in the combined fringe pattern is $8/\pi^2$ that at either peak. On this basis they have shown that the resolving power of the Fabry–Perot interferometer is

$$\frac{\lambda}{\Delta\lambda} = 0.97m\mathscr{F}$$

which is virtually indistinguishable from

$$\frac{\lambda}{\Delta\lambda} = m\mathscr{F}$$

and which is the ratio of the peak wavelength of the appropriate order to the halfwidth of the fringe. Thus the halfwidth of the fringe is a most useful parameter because it is directly related to the resolution of the instrument in a most simple manner. We shall make much use of the concept of halfwidth in chapter 7 when we discuss band-pass filters.

Since resolution is the product of finesse and order number, a low finesse does not necessarily mean low resolution, but it does mean that to achieve high resolution the interferometer must be used in high order. This in its turn means that the separation of neighbouring orders in terms of wavelength is small—in high order this is given approximately by λ/m. If steps are not taken to limit the range of wavelengths accepted by the interferometer then the interpretation of the fringe patterns becomes impossible. This limiting of the range can be achieved by using some sort of filter in series with the etalon. This filter could be a thin-film filter of a type discussed in chapter 7. Another method is to use, in series with the etalon, other etalons of lower order, and hence resolution, arranged so that the fringes coincide only at the wavelength of interest and at wavelengths very far removed. The wide fringe interval or, as it is also called, free spectral range, of the low-order, low-resolution instrument is thus combined with the high resolution and narrow free spectral range of the high-order instrument. A simpler and more convenient method, which is probably that most often employed, involves a spectrograph and is generally used in conjunction with the second method of scanning the interferometer: variation of θ_s keeping d_s constant. The resolution of the spectrograph need not be high and the entrance slit can be quite broad. It is usually placed where the photographic plate is in figure 5.3, so that it accepts a broad strip down the centre of the circular fringe pattern. The plate from the spectrograph then shows a low-resolution spectrum with a fringe pattern along each line corresponding to the fine-structure components within the line.

So far in our examination of the Fabry–Perot interferometer we have neglected to consider absorption in the reflecting coatings. Equation (5.4) contains the information we need.

$$T = \frac{T_s^2}{(1-R_s)^2}\,\frac{1}{1+F\sin^2\delta}\,. \tag{5.4}$$

Let A_s be the absorptance of the coatings; then

$$1 = R_s + T_s + A_s$$

and equation (5.4) becomes

$$T = \frac{T_s^2}{(T_s+A_s)^2}\,\frac{1}{1+F\sin^2\delta}$$

i.e.

$$T = \frac{1}{(1+A_s/T_s)^2}\,\frac{1}{1+F\sin^2\delta}\,. \tag{5.7}$$

Clearly the all-important parameter is A_s/T_s.

Curves are shown in figure 5.4 which connect the transmission of the etalon with finesse, given the absorption of the coatings. It is possible on this diagram

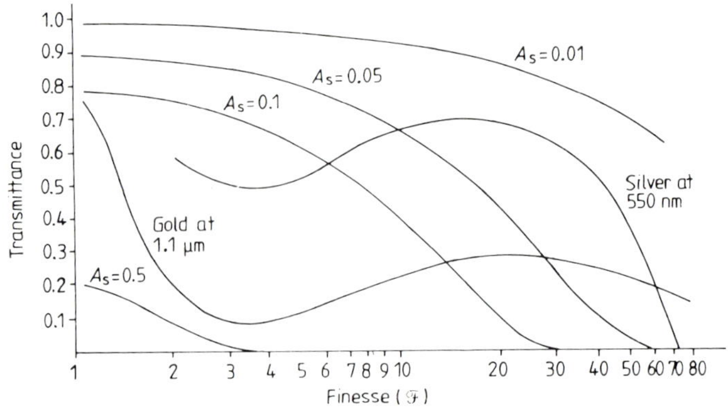

Figure 5.4 Etalon transmittance against finesse for various values of absorptance of the coatings.

to plot the performance of any type of coating if the way in which R_s, T_s and A_s vary is known. This has been done for silver layers at 550 nm and gold at 1.1 μm. The figures from which these curves were plotted were taken from Mayer.[5] Other sources of information, particularly on silver films, are available[6, 7] and results may differ from those plotted in some respects. However, the curves are adequate for their primary purpose, which is to show that the performance of silver, the best metal of all for the visible and near infrared, begins to fall off rapidly beyond a finesse of 20 and is inadequate for the very best interferometer plates. An enormous improvement is possible with all-dielectric multilayer coatings.

MULTILAYER DIELECTRIC COATINGS

In chapter 1 it was mentioned that a high reflectance can be obtained from a stack of quarter-wave dielectric layers of alternate high and low index. This is because the beams reflected from all the interfaces in the assembly are of equal phase when they reach the front surface, where they combine constructively. An expression is given on p 46 for the optical admittance of a series of quarter waves. If n_H and n_L are the indices of the high- and low-index layers and if the stack is arranged so that the high-index layers are outermost at both sides, then

$$Y = \left(\frac{n_H}{n_L}\right)^{2p} \frac{n_H^2}{n_s} \tag{5.8}$$

where n_s is the index of the substrate and $(2p + 1)$ the number of layers in the stack.

The reflectance in air or free space is then

$$R = \left(\frac{1 - (n_H/n_L)^{2p}(n_H^2/n_s)}{1 + (n_H/n_L)^{2p}(n_H^2/n_s)}\right)^2. \tag{5.9}$$

The greater the number of layers the greater the reflectance. Maximum reflectance for a given odd number of layers is always obtained with the high-index layers outermost.

If

$$\left(\frac{n_H}{n_L}\right)^{2p}\frac{n_H^2}{n_s} > 1$$

then

$$R \simeq 1 - 4\left(\frac{n_L}{n_H}\right)^{2p}\frac{n_s}{n_H^2}$$

and

$$T = 1 - R \simeq 4\left(\frac{n_L}{n_H}\right)^{2p}\frac{n_s}{n_H^2} \tag{5.10}$$

which shows that when reflectance is high, then the addition of two extra layers reduces the transmission by a factor of $(n_L/n_H)^2$.

Provided the materials which are used are transparent, the absorption in a multilayer stack can be made very small indeed. We shall return later to this topic, but we can note here that in the visible region of the spectrum the absorptance can be less than 0.01 %.

Dielectric multilayers, however, suffer from two defects. The first, which is more of a complication than a fault, is that there is a variable change in phase associated with the reflection. The second, which is more serious, is that the high reflectance is obtained over a limited range of wavelengths.

We can see, qualitatively, how the phase shift varies, using the admittance diagram. If, as is usual, the multilayer consists of an odd number of layers with high-index layers on the outside, then at the outer surface of the final layer the admittance will be on the real axis with a high positive value. This is shown diagrammatically in figure 5.5. The quadrants are marked on the figure with reference to figure 2.17(*b*). Clearly the phase shift associated with the coating is π, for the reference wavelength for which all the layers are quarter-waves. For slightly longer wavelengths, the circles shrink slightly from the semicircles associated with the quarter-waves and so the terminal point of the locus moves upwards into the region associated with the third quadrant. If the wavelength decreases, the terminal point moves into the second quadrant. The phase shift, therefore, increases with wavelength. If, on the other hand, the coating ends with a quarter-wave of low-index material so that at the reference wavelength the admittance is real, but less than unity, then the phase shift on reflection will be zero, moving into the first quadrant as the wavelength increases or into the fourth as it decreases.

To investigate the effect of the phase change, and also of the dispersion of

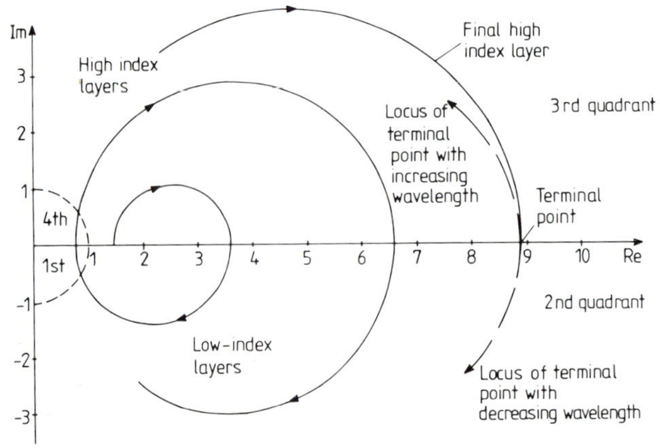

Figure 5.5 Admittance diagram for a quarter-wave stack ending with a high-index layer. The quadrants for the phase shift on reflection ϕ are marked on the diagram and correspond to those in figure 2.17(*b*). For decreasing wavelength the terminal point moves into the region associated with values of ϕ in the second quadrant while for increasing wavelength ϕ moves into the third quadrant.

phase change, on the operation of the interferometer, we return to the original formula, equation (5.1). In our analysis we made the assumption that the phase change on reflection was zero and concluded that transmission peaks would be obtained at wavelengths given by

$$\delta = m\pi$$

where $m = 0, \pm 1, \pm 2, \ldots$. If we now permit ϕ_a and ϕ_b to be non-zero, then the positions of the transmission peaks will be given by

$$\frac{\phi_a + \phi_b - 2\delta}{2} = q\pi$$

where $q = 0, \pm 1, \pm 2, \ldots$. The effect of the phase changes ϕ_a and ϕ_b is simply to shift the positions of the peak wavelengths. If the order is fairly high (and as we have seen most interferometers are used in high order), the shift is quite small. The effect of the phase change, and of any phase dispersion, can be completely eliminated from the determination of wavelength with the interferometer, by a method described by Stanley and Andrew[8] which involves the use of two spacers of different thickness.

The behaviour of a typical quarter-wave stack is shown in figure 5.6. The high-reflection zone can be seen to be limited in extent. On either side of a plateau, the reflectance falls abruptly to a low, oscillatory value. The addition of extra layers does not affect the width of the zone of high reflectance, but increases the reflectance within it and the number of oscillations outside.

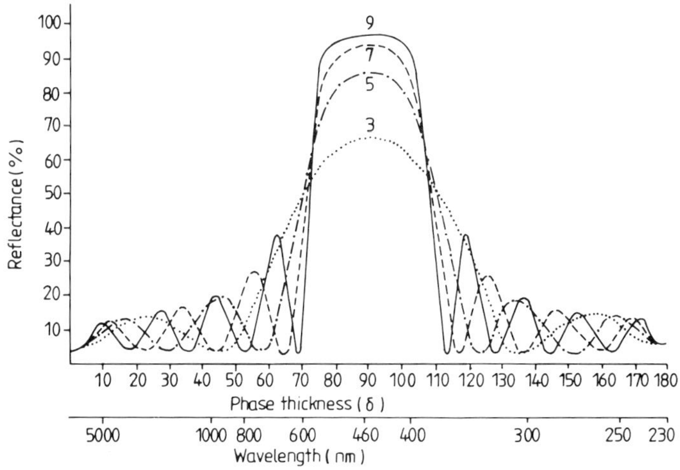

Figure 5.6 Reflectance R for normal incidence of alternating $\lambda_0/4$ layers of high- ($n_H = 1.52$) and low-index ($n_L = 1.38$) dielectric materials on a transparent substrate ($n_s = 1.52$) as a function of the phase thickness $\delta = 2\pi nd/\lambda$ (upper scale) or the wavelength λ for $\lambda_0 = 460$ nm (lower scale). The number of layers is shown as a parameter on the curves. (After Penselin and Steudel[12].)

The width of the high-reflectance zone can be computed using the following method. If a multilayer consists of n repetitions of a fundamental period consisting of two, three or indeed any number of layers, then the characteristic matrix of the multilayer is given by

$$[\mathscr{M}] = [M]^n$$

where $[M]$ is the matrix of the fundamental period. Let $[M]$ be written

$$\begin{bmatrix} M_{11} & M_{12} \\ M_{21} & M_{22} \end{bmatrix}.$$

Then it can be shown that for wavelengths which satisfy

$$\left| \frac{M_{11} + M_{22}}{2} \right| \geqslant 1 \qquad (5.11)$$

the reflectance increases steadily with increasing number of periods. This is therefore the condition that a high-reflectance zone should exist and the boundaries are given by

$$\left| \frac{M_{11} + M_{22}}{2} \right| = 1. \qquad (5.12)$$

A rigorous proof of this result is somewhat involved. One version is given by Born and Wolf[9] and another by Welford.[10] A justification of the result, rather than a proof, was given by Epstein[11] and it is his method which is followed here.

If the characteristic matrix of a thin-film assembly on a substrate of admittance η_{n+1} is given by

$$\begin{bmatrix} B \\ C \end{bmatrix}$$

then, if η_{n+1} is real, equation (2.67) shows that

$$T = \frac{4\eta_0\eta_{n+1}}{(\eta_0 B + C)(\eta_0 B + C)^*} = \frac{4\eta_0\eta_{n+1}}{|\eta_0 B + C|^2}$$

where η_0 is the admittance of the incident medium. Let the characteristic matrix of the assembly of thin films be, as above,

$$[\mathcal{M}] = \begin{bmatrix} \mathcal{M}_{11} & \mathcal{M}_{12} \\ \mathcal{M}_{21} & \mathcal{M}_{22} \end{bmatrix}.$$

Then

$$\begin{bmatrix} B \\ C \end{bmatrix} = \begin{bmatrix} \mathcal{M}_{11} & \mathcal{M}_{12} \\ \mathcal{M}_{21} & \mathcal{M}_{22} \end{bmatrix} \begin{bmatrix} 1 \\ \eta_{n+1} \end{bmatrix} = \begin{bmatrix} \mathcal{M}_{11} + \eta_{n+1}\mathcal{M}_{12} \\ \eta_{n+1}\mathcal{M}_{22} + \mathcal{M}_{21} \end{bmatrix}$$

where $[\mathcal{M}] = [M]^n$ as before and we have

$$T = \frac{4\eta_0\eta_{n+1}}{|\eta_0(\mathcal{M}_{11} + \eta_{n+1}\mathcal{M}_{12}) + \eta_{n+1}\mathcal{M}_{22} + \mathcal{M}_{21}|^2}.$$

If there is no absorption, $\mathcal{M}_{11}$ and $\mathcal{M}_{22}$ are real, and $\mathcal{M}_{12}$ and $\mathcal{M}_{21}$ are imaginary. Then

$$T = \frac{4\eta_0\eta_{n+1}}{|\eta_0\mathcal{M}_{11} + \eta_{n+1}\mathcal{M}_{22}|^2 + |\eta_0\eta_{n+1}\mathcal{M}_{12} + \mathcal{M}_{21}|^2}. \tag{5.13}$$

In the absence of the multilayer, the transmission of the substrate will be

$$T_{sub} = \frac{4\eta_0\eta_{n+1}}{(\eta_0 + \eta_{n+1})^2}. \tag{5.14}$$

To simplify the discussion, let $\eta_0 = \eta_{n+1}$. Then, from equations (5.13) and (5.14), T will be less than T_{sub} if

$$\frac{|\mathcal{M}_{11} + \mathcal{M}_{22}|}{2} \geqslant 1$$

regardless of the values of $\mathcal{M}_{12}$ and $\mathcal{M}_{21}$. Now, if

$$\frac{|M_{11} + M_{22}|}{2} > 1$$

where $[M]$ is the matrix of the fundamental period in the multilayer, then, generally, as the number of periods increases, that is, as n tends to infinity,

$$\frac{|\mathcal{M}_{11}+\mathcal{M}_{22}|}{2} \to \infty .$$

That this is plausible may be seen by first of all squaring $[M]$, whence, writing M'_{pq} for the terms in $[M]^2$,

$$M'_{11}+M'_{22} = (M_{11})^2 + 2M_{12}M_{21} + (M_{22})^2.$$

Since $\det[M] = 1$,

$$2M_{12}M_{21} = 2M_{11}M_{22} - 2$$

so that

$$M'_{11}+M'_{22} = (M_{11}+M_{22})^2 - 2.$$

If

$$\frac{|M_{11}+M_{22}|}{2} = 1+\delta$$

when δ is positive, then

$$M'_{11}+M'_{22} = (2+2\delta)^2 - 2 = 2 + 8\delta + 4\delta^2$$

so that by squaring $[M']$ and resquaring the result and so on, it can be seen that

$$\frac{|\mathcal{M}_{11}+\mathcal{M}_{22}|}{2} \to \infty \qquad \text{as} \qquad n \to \infty .$$

The quarter-wave stack, which we have so far been considering, consists of a number of two-layer periods, together with one extra high-index layer. Each period has a characteristic matrix:

$$[M] = \begin{bmatrix} \cos\delta & (\mathrm{i}\sin\delta)/n_L \\ \mathrm{i}n_L\sin\delta & \cos\delta \end{bmatrix} \begin{bmatrix} \cos\delta & (\mathrm{i}\sin\delta)/n_H \\ \mathrm{i}n_H\sin\delta & \cos\delta \end{bmatrix}.$$

Since the two layers are of equal optical thickness, δ without any suffix has been used for phase thickness.

$$\frac{M_{11}+M_{22}}{2} = \cos^2\delta - \tfrac{1}{2}\left(\frac{n_H}{n_L}+\frac{n_L}{n_H}\right)\sin^2\delta.$$

The right-hand side of this expression cannot be greater than $+1$, and so to find the boundaries of the high-reflectance zone we must set

$$-1 = \cos^2\delta_e - \tfrac{1}{2}\left(\frac{n_H}{n_L}+\frac{n_L}{n_H}\right)\sin^2\delta_e$$

which, with some rearrangement, gives

$$\left(\frac{n_H-n_L}{n_H+n_L}\right)^2 = \cos^2\delta_e .$$

Now,

$$\delta = \frac{\pi}{2}\frac{\lambda_0}{\lambda}$$

where λ_0 is, as usual, the wavelength for which the layers have quarter-wave optical thickness. We can also write this as

$$\delta = \frac{\pi}{2}g$$

where

$$g = \frac{\lambda_0}{\lambda}.$$

Let the edges of the high-reflectance zone be given by

$$\delta_e = \frac{\pi}{2}g_e = \frac{\pi}{2}(1 \pm \Delta g)$$

so that

$$\cos^2 \delta_e = \sin^2\left(\pm\frac{\pi\Delta g}{2}\right)$$

and the width of the zone is $2\Delta g$. Then

$$\Delta g = \frac{2}{\pi}\sin^{-1}\left(\frac{n_H - n_L}{n_H + n_L}\right). \tag{5.15}$$

This shows that the width of the zone is a function only of the indices of the two materials used in the construction of the multilayer. The higher the ratio, the greater the width of the zone. Figure 5.7 shows Δg plotted against the ratio of refractive indices.

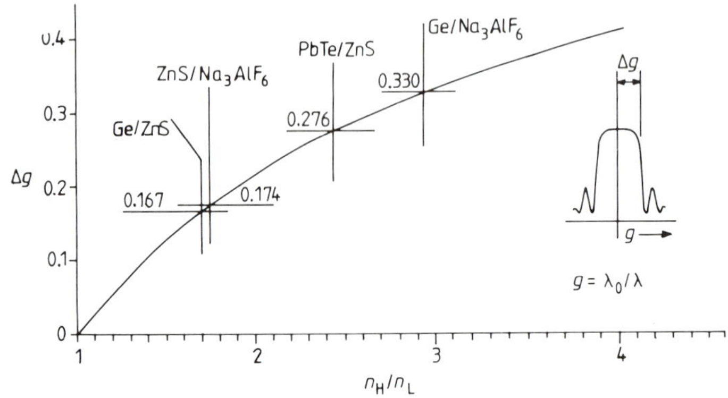

Figure 5.7 The width of the high-reflectance zone of a quarter-wave stack plotted against the ratio of the refractive indices, n_H/n_L.

So far we have considered only the fundamental reflectance zone for which all the layers are one quarter of a wavelength thick. It is obvious that high-reflectance zones will exist at all wavelengths for which the layers are an odd number of quarter wavelengths thick. That is, if the centre wavelength of the fundamental zone is λ_0, then there will also be high-reflectance zones with centre wavelengths $\lambda_0/3$, $\lambda_0/5$, $\lambda_0/7$, $\lambda_0/9$, and so on.

At wavelengths where the layers have optical thickness equivalent to an even number of quarter-waves, which is the same as an integral number of half-waves, the layers will all be absentee layers and the reflectance will be that of the bare substrate.

The analysis determining Δg for the fundamental zone is valid also for all higher-order zones so that the boundaries are given by

$$g_0 \pm \Delta g, \qquad 3g_0 \pm \Delta g, \qquad 5g_0 \pm \Delta g$$

and so on. Higher-order reflectance curves are shown in figure 5.8.

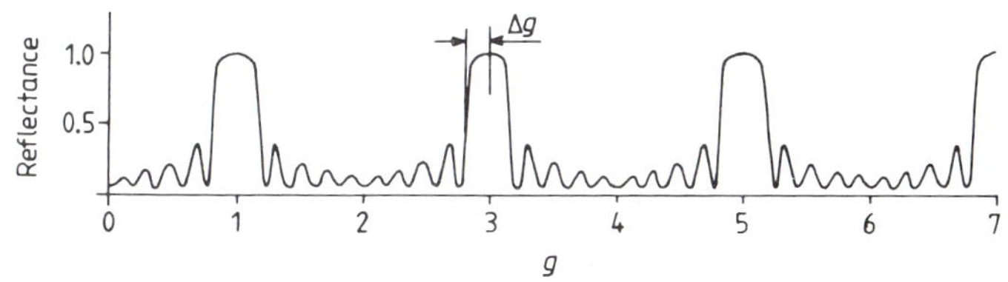

Figure 5.8 Reflectance of a nine-layer stack of zinc sulphide ($n_H = 2.35$) and cryolite ($n_L = 1.35$) on glass ($n = 1.52$) showing the high-reflectance bands.

For the visible region, the most common coating materials are zinc sulphide and cryolite. Absorption levels less than 0.5 % can be achieved with ease, 0.1 % with extra care and 0.001 % with minute attention to detail. Neither material in thin-film form is particularly hard, but they are both easy to evaporate and give high optical performance even when evaporated onto a cold substrate. This means that the risk of distortion of very accurate interferometer plates through heating is eliminated. The layers are rather susceptible to attack by moisture and care should be taken to avoid any condensation, such as might happen when cold plates are exposed to a warmer atmosphere; otherwise, the coatings will be ruined. Touching by fingers is also to be avoided at all costs. The softness of the coatings can, however, be turned to advantage. Etalon plates are extremely expensive and if the coatings are easily removable, the plates can be recoated for use at other wavelengths. Prolonged soaking in warm water is often sufficient to bring zinc sulphide and cryolite coatings off. In cases where the coatings are not completely removed in this way, the addition of two or three drops of hydrochloric acid to the water will quickly complete the operation. This should obviously be done with great care and the plates immediately rinsed in running water to avoid any risk of surface damage.

Where substrates are worked to somewhat lower tolerances, harder materials can be used. Oxide layers, such as titanium dioxide, zirconium dioxide or cerium dioxide, are all useful high-index materials with indices in the region of 2.2. Magnesium fluoride evaporated on to a hot substrate with an index of 1.38, or quartz, with index 1.45, or silicon oxide, with an index around 1.5, are all useful low-index layers. Such combinations will withstand handling, humidity and abrasion.

For the ultraviolet, a good combination for the 300–400 nm region is antimony trioxide with cryolite, evaporated on to a cold substrate. They should be handled as carefully as zinc sulphide and cryolite.

For the infrared, germanium for the region 1.8–20 μm with an index of 4.0, or lead telluride for the region 3.5–40 μm, with an index of 5.5, are good high-index materials. Zinc sulphide, with an index of 2.35, is a useful low-index material out to 20 μm. In the near infrared, silicon monoxide, calcium fluoride, magnesium fluoride, cerium fluoride, or thorium fluoride are all good low-index materials. More details of these and of all the other materials mentioned in this chapter will be found in chapter 8.

The losses experienced in the coatings are as much a function of the technique used as of the materials themselves. Great care in preparing the plant and substrates is needed. Everything should be scrupulously clean. Two papers which will be found useful if the maximum performance is required are by Perry[13] and Heitmann[14]. Both these authors are concerned with laser mirrors, where losses must be of an even lower order than in the case of the Fabry–Perot interferometer.

All-dielectric multilayers with extended high-reflectance zones

The limited range over which high reflectance can be achieved with a quarter-wave stack is a difficulty in some applications, and a number of attempts have been made to extend the range by altering the design. Most of these have involved the staggering of the thicknesses of successive layers throughout the stack to form a regular progression, the aim being to ensure that at any wavelength in a fairly wide range, enough of the layers in the stack have optical thickness sufficiently near a quarter-wave to give high reflectance.

Penselin and Steudel[12] were probably the first workers to try this method. They produced a number of multilayers where the layer thicknesses were in a harmonic progression. The best thirteen-layer results which they published were obtained with the scheme in table 5.1. See also figure 5.9.

Heavens and Liddell[15] used a similar approach. They computed a large number of reflection curves for assemblies of layers for which the thicknesses were in either arithmetic or geometric progression. With the same number of layers the geometric progression gave very slightly broader reflection zones. In the computations the high index was assumed to be 2.36 (zinc sulphide), the

Table 5.1

Number of layers	Material	Index	Wavelength for which layer is a quarter-wave (nm)
	Quartz substrate	1.45	Massive
1	$PbCl_2$	2.20	330
2	MgF_2	1.38	344
3	$PbCl_2$	2.20	360
4	MgF_2	1.38	377
5	ZnS	2.35	396
6	Na_3AlF_6	1.35	417
7	ZnS	2.35	440
8	Na_3AlF_6	1.35	466
9	ZnS	2.35	495
10	Na_3AlF_6	1.35	528
11	ZnS	2.35	566
12	Na_3AlF_6	1.35	609
13	ZnS	2.35	660
	Air	1.00	Massive

The performance is shown as curve B in figure 5.9.

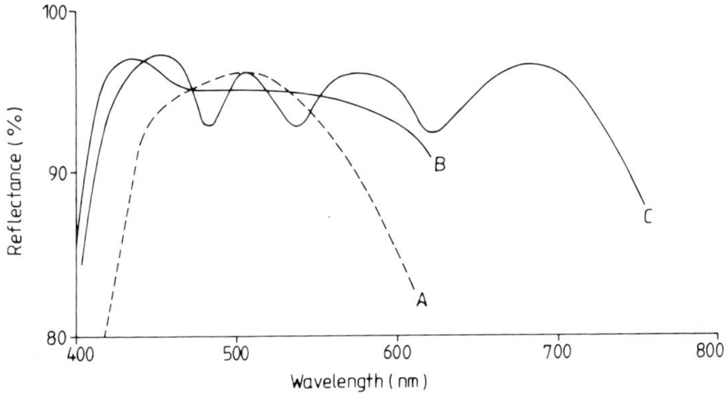

Figure 5.9 Broadband multilayer reflectors. A, computed curve for a seven-layer quarter-wave stack. B, measured reflectance of a broadband design (Penselin and Steudel[12]). C, measured reflectance of an alternative design (Baumeister and Stone[16]).

low index 1.39 (magnesium fluoride) and the substrate index 1.53 (glass). Values of common difference for the arithmetic progression ranged from −0.05 to +0.05, and for the common ratio of the geometric progression from 0.95 to 1.05. Their results for −0.02 and 0.97 respectively are summarised in table 5.2.

Table 5.2

	Number of layers	High-reflectance region (nm)	Wavelength of first-layer quarter-wave (nm)
Arithmetic filters	15	419–625	600
	25	418–725	700
	35	330–840	800
Geometric filters	15	394–625	600
	25	342–730	700
	35	300–826	800

The monitoring wavelengths for which each layer is a quarter-wave are given for the arithmetic filters by

$$t, t(1+k), \ldots, t[1+(q-2)k], t[1+(q-1)k]$$

and for the geometric filters by

$$t, kt, \ldots, k^{q-2}t, k^{q-1}t$$

where q is the number of layers, t the monitoring wavelength for the first layer, and k the common difference or common ratio respectively. A 35-layer geometric curve is shown in figure 5.10.

As in the case of antireflection coatings, computer refinement can be used to improve an initial, less satisfactory performance. Baumeister and Stone[16, 17]

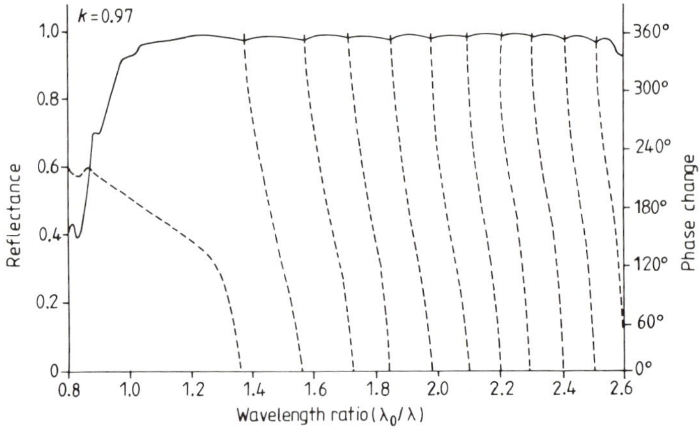

Figure 5.10 Reflectance of a 35-layer geometric stack on glass. Reflectance (full curve) and phase change on reflection (broke curve); $n_0 = 1.00$, $n_H = 2.36$, $n_L = 1.39$, $n_s = 1.53$, common difference $k = 0.97$. (After Heavens and Liddell[15].)

pioneered the use of this technique in optical thin films. By trial and error they arrived at a preliminary 15-layer design with high reflectance over an extended range but with unacceptably large dips. The aim was to produce a reflectance of around 95% using zinc sulphide ($n = 2.3$) and cryolite ($n = 1.35$) and the final result is shown as curve C of figure 5.9 with design details listed in table 5.3. Computer limitations forced the use of a very coarse net for the relaxation—only five points were involved—and in addition, arbitrary relationships between the various layers were used to reduce the number of independent variables to five. This was in 1956. Since then, advances in the technique have kept pace with the increasing power of computers. The detailed methods are outside the scope of this book. They are considered in depth by Liddell[18]. As an illustration of what is possible, figure 5.11 shows the calculated performance of a 21-layer design giving greater than 97% reflectance over the region 400–800 nm. Dispersion of the indices of zinc sulphide and cryolite, the materials used, have been included both in the design procedure and in the performance calculation[19].

Table 5.3

Number of layers	Substance	Index	Wavelength for which layer is a quarter wave (nm)
	Glass substrate		
1	ZnS	2.30	690.8
2	Na_3AlF_6	1.35	690.8
3	ZnS	2.30	690.8
4	Na_3AlF_6	1.35	666.7
5	ZnS	2.30	575.7
6	Na_3AlF_6	1.35	701.3
7	ZnS	2.30	626.2
8	Na_3AlF_6	1.35	517
9	ZnS	2.30	520.5
10	Na_3AlF_6	1.35	463.7
11	ZnS	2.30	463.7
12	Na_3AlF_6	1.35	434.8
13	ZnS	2.30	414
14	Na_3AlF_6	1.35	414
15	ZnS	2.30	414
	Air		

Possibly the simplest method of all is to place a quarter-wave stack for one wavelength on top of another for a different wavelength. This process has been considered in detail by Turner and Baumeister[20]. Unfortunately, if each stack consists of an odd number of layers with outermost layers of the same index,

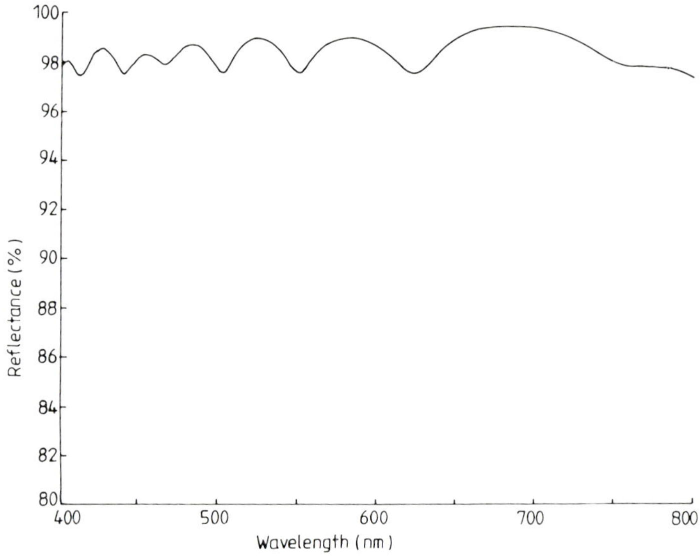

Layer no	Material	Geometrical thickness (nm)	Layer no	Material	Geometrical thickness (nm)
0	Air	Medium	12	Na_3AlF_6	120.4
1	ZnS	41.6	13	ZnS	77.6
2	Na_3AlF_6	76.8	14	Na_3AlF_6	129.9
3	ZnS	51.4	15	ZnS	69.1
4	Na_3AlF_6	94.3	16	Na_3AlF_6	153.0
5	ZnS	49.0	17	ZnS	65.4
6	Na_3AlF_6	94.0	18	Na_3AlF_6	155.7
7	ZnS	47.9	19	ZnS	69.6
8	Na_3AlF_6	95.2	20	Na_3AlF_6	179.1
9	ZnS	58.6	21	ZnS	105.3
10	Na_3AlF_6	147.3	22	SiO_2	Substrate
11	ZnS	62.2			

Figure 5.11 The calculated performance and the design of a 21-layer high-reflectance coating for the visible and near infrared. Dispersion of the indices of the materials has been taken into account in both design by refinement and in performance calculation. (After Pelletier *et al*[19].)

then a peak of transmission is found in the centre of the high-reflectance zone. This peak arises because the two stacks act in much the same way as Fabry–Perot reflectors. In a Fabry–Perot interferometer, as we have seen, provided the reflectances and transmittances of the structures on either side of the spacer layer are equal in magnitude, then the transmittance of the assembly will be unity for

$$\frac{\phi_a + \phi_b - 2\delta}{2} = q\pi$$

where $q = 0, \pm 1, \pm 2, \ldots$

The situation is sketched in figure 5.12. The assembly of the two stacks is divided at the boundary between them and spaced apart leaving a layer of free space forming a spacer layer. The phase angle ϕ associated with each reflection coefficient is also shown. At one wavelength, given by the mean of the centre wavelengths of the stacks, it can be seen that

$$\phi_a + \phi_b = 2\pi.$$

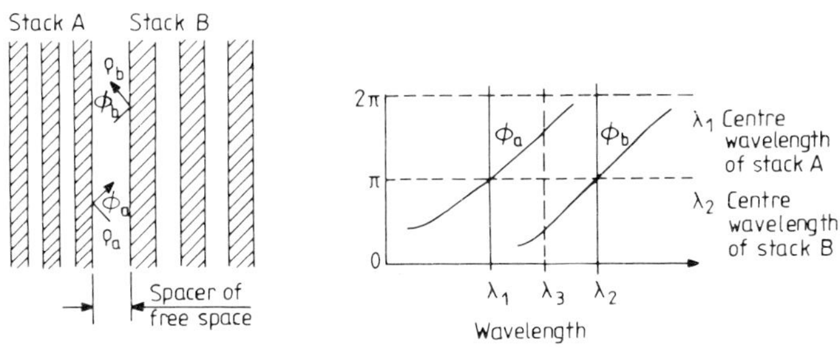

Figure 5.12 At λ_3, $(\phi_a + \phi_b)/2 = \pi$. Also, by symmetry, at λ_3, $(\lambda_2/\lambda_3) - 1 = 1 - (\lambda_1/\lambda_3)$, i.e. $\lambda_3 = \frac{1}{2}(\lambda_1 + \lambda_2)$.

Also by symmetry, at this wavelength the reflectances of both stacks are equal and, therefore, the condition for unity transmittance will be completely satisfied if $2\delta = 0$, that is if the spacer layer of free space is allowed to shrink until it vanishes completely. A peak of transmission will always exist, therefore, if two stacks are deposited so that they are overlapping at the mean of the two monitoring wavelengths. This is shown in figure 5.13, which is reproduced from Turner and Baumeister[20]. Curves A and B are measured reflectance of two high-reflectance quarter-wave stacks, each with the same odd number of layers, starting and finishing with a high-index layer. Curve C shows the measured reflectance of a coating made by combining the two stacks. The peak of transmission can be clearly seen as a dip in the reflectance curve. Experimental errors, either in monitoring or measurement, prevent its reaching the theoretical minimum.

The dip can be removed by destroying the relationship

$$\frac{\phi_a + \phi_b - 2\delta}{2} = q\pi$$

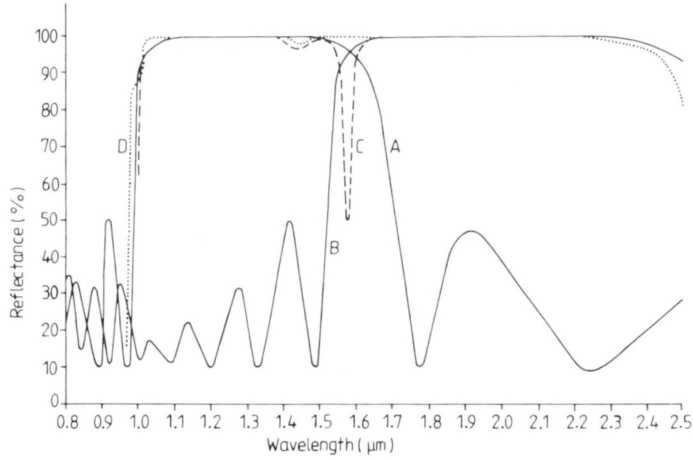

Figure 5.13 Measured reflectances of two quarter-wave stacks with slightly overlapping high-reflectance bands. Individual stacks, full curves: Curve A: A $|0.8(HLHLHLHLH)|$ G. Curve B: A $|1.2(HLHLHLHLH)|$ G. When these are combined in a single coating, there is a minimum in the overlap region resulting from the condition in figure 5.12: Curve C (broken): A $|0.8(HLHLHLHLH)$ $1.2(HLHLHLHLH)|$ G. An inserted L layer eliminates the minimum by destroying the π phase shift. Curve D (dotted): A $|0.8(HLHLHLHLH)\ L\ 1.2(HLHLHLHLH)|$ G. G denotes the glass substrate ($n = 1.52$), A the air incident medium ($n = 1.00$), H the stibnite high-index films and L the chiolite low-index films. H and L are quarter-wave thicknesses at the reference wavelength, λ_0, of 1.6 µm (after Turner and Baumeister[20]).

in the region where both stacks have high reflectance. Turner and Baumeister achieved the result quite simply by adding a low-index layer, one quarter-wave thick at the mean wavelength, in between the stacks. This gave a value for δ of $\pi/2$ and for $(\phi_a + \phi_b - 2\delta)/2$ of $\pi/2$, which corresponds to minimum possible transmission and maximum reflectance. This is illustrated by curve D. The dip has disappeared completely, leaving a broad flat-topped reflectance curve.

Turner and Baumeister have also considered the design of broadband reflectors from a slightly different point of view and achieved similar results to the above, although the reasoning is completely different. If a stack is made up of a number of symmetrical periods such as

$$\frac{H}{2} L \frac{H}{2} \quad \text{or} \quad \frac{L}{2} H \frac{L}{2}$$

it can be represented mathematically by a single layer of thickness similar to the actual thickness of the multilayer and with a real optical admittance. This relationship holds good for all regions except the zones of high reflectance where the thickness and optical admittance are both imaginary. This result has already been referred to on p 48 and will be examined in much greater detail in the following two chapters. For our present purpose it is sufficient to note that the relationship does exist. If a single layer of real refractive index is deposited on top of a 100 % reflector, no interference maxima and minima can possibly exist. For reflectors falling short of the 100 % condition, maxima and minima can exist, but are very weak. Thus, in the region where the overlapping stack has a real refractive index, the high reflectance of the lower stack remains virtually unchanged, provided enough layers are used. The high-reflectance zones can either just touch without overlapping, in which case no reflectance minima will exist, or overlap, in which case the minima will be suppressed because the central layer, composed of an eighth-wave from each stack, is a quarter wavelength thick at the mean of the two monitoring wavelengths, and, as has been shown above, this effectively removes any reflectance minima. Figure 5.14(a) shows the measured reflectance of two stacks,

$$\left(\frac{L}{2}H\frac{L}{2}\right)^4$$

on a barium fluoride substrate together with the measured reflectance of two similar stacks superimposed on the same substrate in such a way that the high-reflectance zones just touch.

Coating uniformity requirements

One feature of the broadband reflectors which we have been considering is that the change in phase on reflection varies very rapidly with wavelength, much more rapidly than in the case of the simple quarter-wave stack. The difficulty which this could cause if such coatings were used in the determination of wavelength in a Fabry–Perot interferometer has frequently been mentioned. Actually, the method proposed by Stanley and Andrew[8], which uses two spacers, completely eliminates the effect of even the most rapid phase change with wavelength, but there is another effect which is the subject of a dramatic report by Ramsay and Ciddor[21]. They used a thirteen-layer coating of a design similar to that of Baumeister and Stone. Their scheme is given in table 5.4.

The coating was deposited with layer uniformity in the region of 1–2 nm from centre to edge of the 75 mm diameter plates. When tested, however, after coating, the plates appeared to be $\lambda/60$ concave at 546 nm, very uniform at 588 nm and $\lambda/10$ convex at 644 nm. This curvature is, of course, only apparent. Tests on the plates using silver layers showed that they were probably $\lambda/60$ concave. The apparent curvature results from changes both in the thickness of the coatings and in the phase change on reflection.

In fact, a theory sufficient to explain the effect was published, together with some estimates of required uniformity, by Giacomo[22] in 1958. He obtained the result that the apparent variation of spacer thickness (measured in units of phase) was equal to the error in uniformity of the coating (measured as the variation in physical thickness) times a factor

$$\left(\frac{\nu}{e}\frac{\partial\phi}{\partial\nu}+4\pi\nu\right)$$

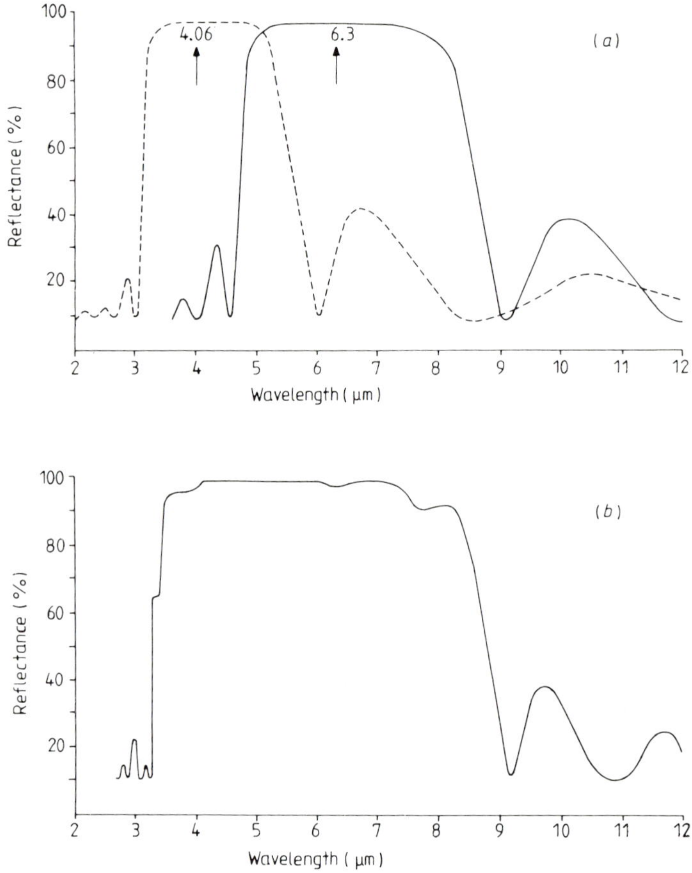

Figure 5.14 (*a*) Measured reflectances of two stacks A $|(0.5L\ H\ 0.5L)^4|$ G on BaF_2 substrates. G denotes the BaF_2 and A air; *H* and *L* are films of stibnite and chiolite a quarter-wave thick at reference wavelengths $\lambda_0 = 4.06$ µm (broken curve) or 6.3 µm (full curve) (after Turner and Baumeister[20]). (*b*) Measured reflectance of the two stacks of (*a*) superimposed in a single coating for an extended high-reflectance region. (After Turner and Baumeister[20].)

where *e* is the total thickness of the coating (physical thickness), $\nu = 1/\lambda$ is the wavenumber and ϕ is the phase change on reflection at the surface of the coating. Another way of stating the result is to take $\Delta\rho_m$ as the maximum allowable error in spacer thickness (measured in units of phase) due to this cause, and then the uniformity in coating must be better than

$$\frac{\Delta e}{e} = \frac{\Delta\rho_m}{[(\partial\phi/\partial\nu) + 4\pi e]\nu}.$$

Table 5.4

Number of layers	Material	Wavelength for which layer is a quarter-wave (nm)
	Fused silica substrate	
1	ZnS	589
2	Na_3AlF_6	671
3	ZnS	720
4	Na_3AlF_6	594
5	ZnS	562
6	Na_3AlF_6	573
7	ZnS	539
8	Na_3AlF_6	535
9	ZnS	571
10	Na_3AlF_6	392
11	ZnS	385
12	Na_3AlF_6	355
13	ZnS	454

Giacomo showed that the two terms in the expression, $\partial\phi/\partial\nu$ (which is generally negative) and $4\pi e$, could cancel, or partially cancel, so that some designs of coating would be more sensitive to uniformity errors than others. Ramsay and Ciddor carried this further by pointing out that the two terms in the expression vary in magnitude throughout the high-reflectance zone of the coating, and, although the cancellation or partial cancellation does occur, in addition, the varying magnitudes mean that it is possible in some cases for the apparent curvature due to uniformity errors to vary from concave to convex or vice versa throughout the range. This is so for the particular coating they considered, and it is this change in apparent curvature which is particularly awkward, implying that the interferometer must be tested for flatness over the entire working range, not, as is normal, at one convenient wavelength.

For the conventional quarter-wave coating, the magnitude of $\partial\phi/\partial\nu$ falls far short of $4\pi e$; for example, in the case of a seven-layer coating of zinc sulphide and cryolite, for the visible region $\partial\phi/\partial\nu$ is only -1.5 μm compared with $4\pi e$ of around $+21.5$ μm, and the uniformity which is required can readily be calculated from the finesse requirement and the physical thickness of the coating, neglecting the effect of the variations in phase angle altogether. In the case of the broadband multilayer however, the magnitude of $\partial\phi/\partial\nu$ is very much greater, and at some wavelengths will exceed the value of $4\pi e$. For example, Giacomo quotes a case where $\partial\phi/\partial\nu$ reached -125 μm, completely swamping the thickness effect, $4\pi e$. Heavens and Liddell, in their paper, quote values of $\partial\phi/\partial\nu$ varying from 10 to 26 μm for the staggered multilayers. The change in apparent curvature can therefore occur with these staggered systems, and it is dangerous to attempt to calculate the required uniformity simply from the coating thickness and the finesse requirement. An analysis which is very similar in certain respects, especially in the end result, has been carried out for random errors in the layers of certain types of band-pass filters, and is considered in chapter 7. One point which does arise is the possibility of designing a coating where the two terms cancel almost completely throughout the entire working range. This is mentioned by Ramsay and Ciddor. Since then, Ciddor[23] has carried this a stage further and has now produced several possible designs. Particularly successful is a design for a reflector to give approximately 75 % reflectance over the major part of the visible, which is apparoximately three times less sensitive to thickness variations than would be the case with a reflector exhibiting no phase change at all with change in thickness. The design is intended for film indices of 2.30 and 1.35 on a substrate of index 1.46, corresponding to zinc sulphide and cryolite on fused silica. The thicknesses are given in table 5.5. The reflectance is constant within perhaps ±2 % over the region 650 nm to 400 nm and an interferometer plate with such a coating would behave as if it were much flatter than the purely geometrical lack of uniformity of the coating would suggest.

Table 5.5

Number of layers	Index	Wavelength for which the layer is one quarter-wave thick (nm)
0	1.00	Massive—incident medium
1	1.35	309
2	2.30	866
3	1.35	969
4	2.30	436
5	1.35	521
6	2.30	369
7	1.35	484
8	2.30	441
9	1.35	795
10	2.30	768
11	1.46	Massive—substrate

LOSSES

If lossless materials are used, then the reflectance which can be attained by a quarter-wave stack depends solely on the number of layers. If the reflectance is high then the addition of a further pair of layers reduces the transmittance by a factor $(n_L/n_H)^2$. In practice, the reflectance which can be ultimately achieved is limited by losses in the layers. These losses can be scattering or absorption. Scattering losses are principally due to defects such as dust in the layers or to surface roughness, and techniques for reducing them are considered in chapter 10. Absorption losses are a property of the material, which may be intrinsic or due to impurities or to composition or to structure. Absorption losses are related to the extinction coefficient of the material, and it is useful to consider the absorption losses of a quarter-wave stack composed of weakly absorbing layers having small but non-zero extinction coefficients. Expressions for this have been derived by a number of workers. The technique we use here is adapted from an approach devised by Hemingway and Lissberger[24].

We use the concept of potential transmittance introduced in chapter 2. We split the multilayer into subassemblies of single layers each with its own value of potential transmittance. The potential transmittance of the assembly is then the product of the individual transmittances.

For the entire multilayer we can write

$$\psi = \frac{T}{(1-R)}.$$

Then, if A is the absorptance

$$1-\psi = \frac{1-R-T}{(1-R)} = \frac{A}{(1-R)}$$

and

$$A = (1-R)(1-\psi).$$

Now $0 \leqslant \psi \leqslant 1$ and so we can introduce a quantity $\mathscr{A}_f$ and write

$$\psi_f = 1 - \mathscr{A}_f$$

for each individual layer, and, since we are considering only weak absorption, the potential transmittance will be very near unity and so $\mathscr{A}_f$ will be very small. Then the potential transmittance of the entire assembly will be given by

$$\psi = \prod_{f=1}^{p} \psi_f = \prod_{f=1}^{p} (1-\mathscr{A}_f)$$

$$= 1 - \sum_{f=1}^{p} \mathscr{A}_f + \ldots$$

so that, neglecting higher powers of $\mathscr{A}_f$,

$$A = (1-R)(1-\psi) = (1-R)\sum_{f=1}^{p} \mathscr{A}_f.$$

Now let us consider one single layer. The relevant parameters are contained in

$$\begin{bmatrix} B \\ C \end{bmatrix} = \begin{bmatrix} \cos\phi_f & \mathrm{i}(\sin\phi_f)/\eta_f \\ \mathrm{i}\eta_f \sin\phi_f & \cos\phi_f \end{bmatrix} \begin{bmatrix} 1 \\ y_\mathrm{e} \end{bmatrix} \quad (5.16)$$

and

$$\psi_f = \frac{\text{Re}\,(y_e)}{\text{Re}\,(BC^*)}$$

from equation (2.69). Also

$$\eta_f = n_f - \mathrm{i}k_f \qquad \text{(in free space units)}$$

$$\begin{aligned}\phi_f &= 2\pi(n_f - \mathrm{i}k_f)d_f/\lambda \\ &= 2\pi n_f d_f/\lambda - \mathrm{i}\,2\pi k_f d_f/\lambda \\ &= \alpha - \mathrm{i}\beta\end{aligned}$$

where k_f, and hence β, is small.

If we consider layers which are approximately quarter-waves, we can set

$$\alpha = (\pi/2 + \varepsilon)$$

where ε is small. Then

$$\cos\phi_f \simeq (-\varepsilon + \mathrm{i}\beta)$$
$$\sin\phi_f \simeq 1$$

and the matrix expression becomes

$$\begin{bmatrix} B \\ C \end{bmatrix} = \begin{bmatrix} (-\varepsilon + \mathrm{i}\beta) & \mathrm{i}/(n - \mathrm{i}k) \\ \mathrm{i}(n - \mathrm{i}k) & (-\varepsilon + \mathrm{i}\beta) \end{bmatrix} \begin{bmatrix} 1 \\ y_e \end{bmatrix}$$

whence

$$\begin{bmatrix} B \\ C \end{bmatrix} = \begin{bmatrix} (-\varepsilon + \mathrm{i}\beta) + \mathrm{i}y_e/(n - \mathrm{i}k) \\ \mathrm{i}(n - \mathrm{i}k) + y_e(-\varepsilon + \mathrm{i}\beta) \end{bmatrix}$$

so that

$$BC^* = [(-\varepsilon + \mathrm{i}\beta) + \mathrm{i}y_e/(n - \mathrm{i}k)]\,[\mathrm{i}(n - \mathrm{i}k) + y_e(-\varepsilon + \mathrm{i}\beta)]^*$$

and, neglecting terms of second order and above in k, β and ε

$$\text{Re}\,(BC^*) = (\beta n + y_e + y_e^2\beta/n)$$

i.e.

$$\psi_f = \frac{y_e}{(\beta n + y_e + y_e^2\beta/n)} = \frac{1}{1 + \beta[(n/y_e) + (y_e/n)]}$$

and, since β is small,

$$\psi_f = 1 - \beta\left(\frac{n}{y_e} + \frac{y_e}{n}\right)$$

so that

$$\mathscr{A}_f = 1 - \psi_f = \beta\left(\frac{n}{y_e} + \frac{y_e}{n}\right).$$

Next we must find

$$(1 - R)\sum \mathscr{A}_f.$$

For this we need the value of y_e at each interface. Let the stack of quarter-wave layers end with a high-index layer. Then the admittance of the whole assembly will be Y, where Y is large. If we denote the admittance of the incident medium by n_0 then

$$R=\left[\frac{n_0-Y}{n_0+Y}\right]^2$$

where n_0 and Y are real.

If Y is sufficiently large,

$$R=1-4n_0/Y$$

or

$$(1-R)=4n_0/Y.$$

Further, since Y is the terminating admittance and the layers are all quarter-waves, the admittances at each of the interfaces follow the scheme:

$$\begin{array}{ccccccccccccccc} & Y & & \dfrac{n_H^2}{Y} & & \dfrac{n_L^2 Y}{n_H^2} & & \dfrac{n_H^4}{n_L^2 Y} & & \dfrac{n_L^4 Y}{n_H^4} & & \dfrac{n_H^6}{n_L^4 Y} & & \dfrac{n_L^6 Y}{n_H^6} & \\ n_0 & | & n_H & | & n_L & | & n_H & | & n_L & | & n_H & | & n_L & | & \cdots \end{array}$$

Then

$$\begin{aligned} A&=(1-R)\sum_{f=1}^{p}\mathscr{A}_f\\ &=\frac{4n_0}{Y}\left[\left(\frac{n_H}{n_H^2/Y}+\frac{n_H^2/Y}{n_H}\right)\beta_H+\left(\frac{n_L}{n_L^2Y/n_H^2}+\frac{n_L^2Y/n_H^2}{n_L}\right)\beta_L\right.\\ &\quad\left.+\left(\frac{n_H}{n_H^4/n_L^2Y}+\frac{n_H^4/n_L^2Y}{n_H}\right)\beta_H+\ldots\right] \end{aligned}$$

i.e.

$$A=4n_0\left[\left(\frac{1}{n_H}+\frac{n_H}{Y^2}\right)\beta_H+\left(\frac{n_L}{n_H^2}+\frac{n_H^2}{n_LY^2}\right)\beta_L+\left(\frac{n_L^2}{n_H^3}+\frac{n_H^3}{n_L^2Y^2}\right)\beta_H+\ldots\right].$$

Since β_H and β_L are small and Y is large, we can neglect terms in β/Y^2 and the absorptance is then given by

$$A=4n_0\left[\left(\frac{1}{n_H}+\frac{n_L^2}{n_H^3}+\frac{n_L^4}{n_H^5}+\ldots\right)\beta_H+\left(\frac{n_L}{n_H^2}+\frac{n_L^3}{n_H^4}+\frac{n_L^5}{n_H^6}+\ldots\right)\beta_L\right].$$

$(n_L/n_H)^2$ is less than unity and, although the series are not infinite, we can assume that they have a sufficiently large number of terms so that any error which is involved in assuming that they are in fact infinite is very small.

Thus

$$A=4n_0\left(\frac{\beta_H/n_H}{1-(n_L/n_H)^2}+\frac{n_L\beta_L/n_H^2}{1-(n_L/n_H)^2}\right)=\frac{4n_0(n_H\beta_H+n_L\beta_L)}{(n_H^2-n_L^2)}.$$

Now

$$n\beta = n\frac{2\pi kd}{\lambda} = \frac{2\pi nd}{\lambda}k$$

and, since the layers are quarter-waves,

$$\frac{2\pi nd}{\lambda} = \frac{\pi}{2}$$

so that

$$A = \frac{2\pi n_0(k_H + k_L)}{(n_H^2 - n_L^2)} \qquad \text{(final layer of high index).}$$

The case of a multilayer terminating with a low-index layer can be dealt with in the same way. The final low-index layer acts to reduce the reflectance and so increase the absorption, which is given by

$$A = \frac{2\pi}{n_0}\frac{(n_H^2 k_L + n_L^2 k_H)}{(n_H^2 - n_L^2)} \qquad \text{(final layer of low index).}$$

As an example, we can consider a multilayer with $k_H = k_L = 0.0001$, $n_H = 2.35$ and $n_L = 1.35$, in air, i.e. $n_0 = 1.0$.

$A = 0.03\%$ (high-index layer outermost)

$A = 0.12\%$ (low-index layer outermost).

In fact, in the red part of the spectrum, the losses in a zinc sulphide and cryolite stack can be less than 0.001 %, indicating that the value of k must be less than 6×10^{-6} assuming that the loss is entirely in one material.

In absolute terms, the absorption loss affects the reflectance more than the transmittance in any given quarter-wave stack. Giacomo[25] has shown that $\Delta T/T$ and $\Delta R/R$ are of the same order, and therefore, since $R \gg T$ then $\Delta R \gg \Delta T$. We will return to this question of loss later.

REFERENCES

1 Fabry C and Perot A 1899 Théorie et applications d'une nouvelle méthode de spectroscopie interférentielle *Ann. Chim. Phys., Paris* **16** 115–44

2 Born M and Wolf E 1975 *Principles of Optics* 5th edn (London: Pergamon)

3 Chabbal R 1953 Recherche des meilleures conditions d'utilisation d'un spectrométre photoélectrique Fabry–Perot *J. Rech. CNRS* **24** 138–85

4 Born M and Wolf E *ibid.* (pp 333–5 in the 3rd edition, 1965)

5 Mayer H 1950 *Physik dünner Schichten* (Stuttgart: Wissenschaftliche Verlagsgesellschaft)

6 Kuhn H and Wilson B A 1950 Reflectivity of thin silver films and their use in interferometry *Proc. Phys. Soc.* **B63** 745–55

7 Oppenheim U 1956 Semi-reflecting silver films for infrared interferometry *J. Opt. Soc. Am.* **46** 628–33
8 Stanley R W and Andrew K L 1964 Use of dielectric coatings in absolute wavelength measurements with a Fabry–Perot interferometer *J. Opt. Soc. Am.* **54** 625–7
9 Born M and Wolf E *ibid.* (pp 66–69 in the 3rd edition, 1965)
10 Welford W (writing as W Weinstein) 1954 Computations in thin film optics *Vacuum* **4** 3–19. (The proof is on page 10)
11 Epstein L I 1955 Improvements in heat-reflecting filters *J. Opt. Soc. Am* **45** 360–2
12 Penselin S and Steudel A 1955 Fabry–Perot Interferometerverspiegelungen aus dielektrischen Vielfachschichten *Z. Phys.* **142** 21–41
13 Perry D L 1965 Low loss multilayer dielectric mirrors *Appl. Opt.* **4** 987–91
14 Heitmann W 1966 Extrem hochreflektierende dielektrische Spiegelschichten mit Zincselenid *Z. Angew. Phys.* **21** 503–8
15 Heavens O S and Liddell H M 1966 Staggered broad-band reflecting multilayers *Appl. Opt.* **5** 373–6
16 Baumeister P W and Stone J M 1956 Broad-band multilayer film for Fabry–Perot interferometers *J. Opt. Soc. Am.* **46** 228–9 (More information about this design technique is given in reference 17)
17 Baumeister P W 1958 Design of multilayer filters by successive approximations *J. Opt. Soc. Am.* **48** 955–8
18 Liddell H M 1981 *Computer-aided techniques for the design of multilayer filters* (Bristol: Adam Hilger)
19 Pelletier E, Klapisch M and Giacomo P 1971 Synthèse d'empilements de couches minces *Nouv. Rev. Opt. Appl.* **2** 247–54
20 Turner A F and Baumeister P W 1966 Multilayer mirrors with high reflectance over an extended spectral region *Appl. Opt.* **5** 69–76
21 Ramsay J V and Ciddor P E 1967 Apparent shape of broad-band, multilayer reflecting surfaces *Appl. Opt.* **6** 2003–4
22 Giacomo P 1958 Propriétés chromatiques des couches réfléchissantes multidiélectriques *J. Phys. Rad.* **19** 307–11
23 Ciddor P E 1968 Minimization of the apparent curvature of multilayer reflecting surfaces *Appl. Opt.* **7** 2328–9
24 Hemingway D J and Lissberger P H 1973 Properties of weakly absorbing multilayer systems in terms of the concept of potential transmittance *Opt. Acta.* **20** 85–96
25 Giacomo P 1956 Les couches réfléchissantes multidiélectriques appliquées a l'interféromètre de Fabry–Perot. Etude théorique et experimentale des couches réelles *Rev. Opt.* **35** 317–54, 442–67

6 Edge filters

Filters in which the primary characteristic is an abrupt change between a region of rejection and a region of transmission are known as edge filters. Edge filters are divided into two main groups, longwave-pass and shortwave-pass. The operation may depend on many different mechanisms and the construction may take a number of different forms. The following account is limited to thin-film edge filters. These rely for their operation on absorption or interference or both.

THIN-FILM ABSORPTION FILTERS

A thin-film absorption filter consists of a thin film of material which has an absorption edge at the required wavelength and is usually longwave-pass in character. Semiconductors which exhibit a very rapid transition from opacity to transparency at the intrinsic edge are particularly useful in this respect, making excellent longwave-pass filters. The only complication which usually exists is a reflection loss in the pass region due to the high refractive index of the film. Germanium, for example, with an edge at 1.65 μm, has an index of 4.0, and, as the thickness of germanium necessary to achieve useful rejection will be at least several quarter-waves, there will be prominent interference fringes in the pass zone showing variations from substrate level, at the half-wave positions, to a reflectance of 68% (in the case of a glass substrate) at the quarter-wave position. The problem can be readily solved by placing antireflection coatings between the substrate and the germanium layer, and between the germanium layer and the air. Single quarter-wave antireflection coatings are usually quite adequate. For optimum matching the values required for the indices of the antireflecting layers are 2.46 between glass and germanium, and 2.0 between germanium and air. The index of zinc sulphide, 2.35, is sufficiently near to both values and, with it, the reflectance near the peak of the quarter-wave coatings will oscillate between

$$\left(\frac{1-(2.35^4)/(4^2\times 1.52)}{1+(2.35^4)/(4^2\times 1.52)}\right)^2 = 1.3\%$$

for wavelengths where the germanium layer is equal to an integral odd number of quarter-waves, and 4 %, that is the reflectance of the bare substrate, where the germanium layer is an integral number of half-waves thick (for at such a wavelength the germanium layer acts as an absentee layer and the two zinc sulphide layers combine also to form a half-wave and, therefore, an absentee layer).

Other materials used to form single-layer absorption filters in this way include cerium dioxide, giving an ultraviolet rejection–visible transmitting filter, silicon, giving a longwave-pass filter with an edge at 1 μm, and lead telluride, giving a longwave-pass filter at 3.4 μm.

A practical lead telluride filter characteristic is shown in figure 6.1, which also gives the design. The two zinc sulphide layers were arranged to be quarter-waves at 3.0 μm. Better results would probably have been obtained if the thicknesses had been increased to quarter-waves at 4.5 μm.

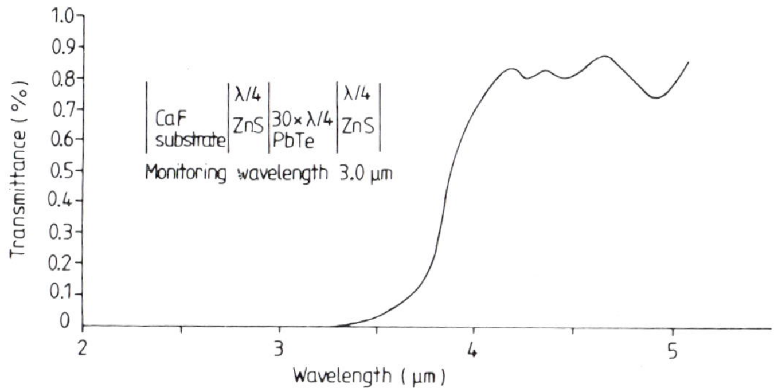

Figure 6.1 The measured characteristic of a lead telluride filter. The small dip at 4.25 μm is probably due to atmospheric CO_2 causing a slight unbalance of the measuring spectrometer. (Courtesy of Sir Howard Grubb, Parsons & Co Ltd.)

INTERFERENCE EDGE FILTERS

The quarter-wave stack

The basic type of interference edge filter is the quarter-wave stack of the previous chapter. As was explained there, the principal characteristic of the optical transmission curve plotted as a function of wavelength is a series of high-reflection zones, i.e. low transmission, separated by regions of high transmission. The shape of the transmission curve of a quarter-wave stack is

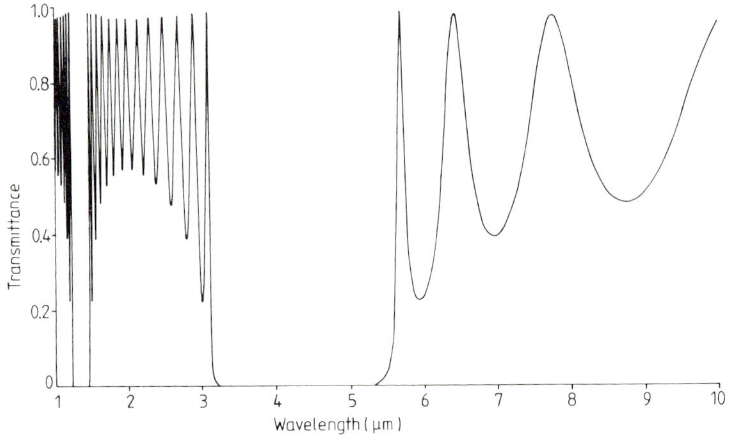

Figure 6.2 Computed characteristic of a 13-layer quarter-wave stack of germanium (index 4.0) and silicon monoxide (index 1.70) on a substrate of index 1.42. The reference wavelength, λ_0, is 4.0 μm.

shown in figure 6.2. The particular combination of materials shown is useful in the infrared beyond 2 μm, but the curve is typical of any pair of materials having a reasonably high ratio of refractive indices.

The system of figure 6.2 can be used either as a longwave-pass filter with an edge at 5.0 μm or a shortwave-pass filter with an edge at 3.3 μm. These wavelengths can be altered at will by changing the monitoring wavelength.

It sometimes happens that the width of the rejection zone is adequate for the particular application, as, for example, where light of a particularly narrow spectral region only is to be eliminated, or where the detector itself is insensitive to wavelength beyond the opposite edge of the rejection zone. In most cases, however, it is desirable to eliminate all wavelengths shorter than, or longer than, a particular value. The rejection zone, shown in figure 6.2, must somehow be extended. This is usually done by coupling the interference filter with one of the absorption type.

Absorption filters usually have very high rejection in the stop region, but, as they depend on the fundamental optical properties of the basic materials, they are inflexible in character and the edge positions are fixed. Using interference and absorption filters together combines the best properties of both, the deep rejection of the absorption filter with the flexibility of the interference filter. The interference layers can be deposited on an absorption filter, which acts as the substrate, or the interference section can sometimes be made from material which itself has an absorption edge within the interference rejection zone. Within the absorption region the filter behaves in much the same way as the single layers of the previous section.

Other methods of improving the width of the rejection zone will be dealt with shortly, but now we must turn our attention to the more difficult problem created by the magnitude of the ripple in transmission in the pass region. As the curve of figure 6.2 shows, the ripple is severe and the performance of the filter would be very much improved if somehow the ripple could be reduced.

Before we can reduce the ripple we must first investigate the reason for its appearance, and this is not an easy task, because of the complexity of the mathematics. A paper published by Epstein[1] in 1952 is of immense importance, in that it lays the foundation of a method which gives the necessary insight into the problem to enable the performance to be not only predicted but also improved.

Symmetrical multilayers and the Herpin index

The paper written by Epstein[1] in 1952 dealt with the mathematical equivalence of a symmetrical combination of films and a single layer, and was the beginning of what has become the most powerful design method to date for thin-film filters.

Any thin-film combination is known as symmetrical if each half is a mirror image of the other half. The simplest example of this is a three-layer combination in which a central layer is sandwiched between two identical outer layers. If a multilayer can be split into a number of equal symmetrical periods, then it can be shown that it is equivalent in performance to a single layer having a thickness similar to that of the multilayer and an optical admittance that can be calculated. This is a most important result. Unfortunately, the accurate calculation of the equivalent optical admittance is rather involved, but the basic form of the result can be established relatively easily and used as a qualitative guide. Once the basic form of a filter has been established, computer techniques can be used to finalise the design.

Consider first a symmetrical three-layer period *pqp*, made up of dielectric materials free from absorption. The characteristic matrix of the combination is given by

$$\begin{bmatrix} M_{11} & M_{12} \\ M_{21} & M_{22} \end{bmatrix} = \begin{bmatrix} \cos\delta_p & (\mathrm{i}\sin\delta_p)/\eta_p \\ \mathrm{i}\eta_p\sin\delta_p & \cos\delta_p \end{bmatrix} \begin{bmatrix} \cos\delta_q & (\mathrm{i}\sin\delta_q)/\eta_q \\ \mathrm{i}\eta_q\sin\delta_q & \cos\delta_q \end{bmatrix}$$

$$\times \begin{bmatrix} \cos\delta_p & (\mathrm{i}\sin\delta_p)/\eta_p \\ \mathrm{i}\eta_p\sin\delta_p & \cos\delta_p \end{bmatrix} \tag{6.1}$$

(where we have used the more general optical admittance η rather than the refractive index n). By performing the multiplication we find:

$$M_{11} = \cos 2\delta_p \cos\delta_q - \tfrac{1}{2}\left(\frac{\eta_q}{\eta_p} + \frac{\eta_p}{\eta_q}\right)\sin 2\delta_p \sin\delta_q \tag{6.2a}$$

$$M_{12}=\frac{\mathrm{i}}{\eta_p}\left[\sin 2\delta_p\cos\delta_q+\tfrac{1}{2}\left(\frac{\eta_q}{\eta_p}+\frac{\eta_p}{\eta_q}\right)\cos 2\delta_p\sin\delta_q\right.$$
$$\left.+\tfrac{1}{2}\left(\frac{\eta_p}{\eta_q}-\frac{\eta_q}{\eta_p}\right)\sin\delta_q\right] \qquad (6.2b)$$

$$M_{21}=\mathrm{i}\eta_p\left[\sin 2\delta_p\cos\delta_q+\tfrac{1}{2}\left(\frac{\eta_p}{\eta_q}+\frac{\eta_q}{\eta_p}\right)\cos 2\delta_p\sin\delta_q\right.$$
$$\left.-\tfrac{1}{2}\left(\frac{\eta_p}{\eta_q}-\frac{\eta_q}{\eta_p}\right)\sin\delta_q\right] \qquad (6.2c)$$

and

$$M_{22}=M_{11}. \qquad (6.2d)$$

It is this last relationship which permits the next step.

Now, let

$$M_{11}=\cos\gamma=M_{22} \qquad (6.3)$$

and if we set

$$M_{12}=\frac{\mathrm{i}\sin\gamma}{E} \qquad (6.4)$$

then, since $M_{11}M_{22}-M_{12}M_{21}=1$

$$M_{21}=\mathrm{i}E\sin\gamma. \qquad (6.5)$$

These quantities have exactly the same form as a single layer of phase thickness γ and admittance E. The equations can be solved for γ and E, choosing the particular value of γ which is nearest to the total phase thickness of the period. γ is then the equivalent phase thickness of the three-layer combination and E is the equivalent optical admittance, also known sometimes as the Herpin index. M_{11} does not equal M_{22} in an unsymmetrical arrangement and such a combination cannot, therefore, be replaced by a single layer.

It can be easily shown that this result can be extended to cover any symmetrical period consisting of any number of layers. First the central three layers which, by definition, will form a symmetrical assembly on their own can be replaced by a single layer. This equivalent layer can then be taken along with the next layers on either side as a second symmetrical three-layer combination, which can, in its turn, be replaced by a single layer. The process can be repeated until all the layers have been replaced and a single equivalent layer found.

The importance of this result lies both in the ease of interpretation (the properties of a single layer can be visualised much more readily than those of a multilayer) and in the ease with which the result for a single period may be extended to that for a multilayer consisting of many periods.

If a multilayer is made up of, say, S identical symmetrical periods, each of which has an equivalent phase thickness γ and equivalent admittance E, then physical considerations show that the multilayer will be equivalent to a single layer of thickness $S\gamma$ and admittance E. This result also follows because of an easily derived result:

$$\begin{bmatrix} \cos\gamma & \mathrm{i}\sin\gamma/E \\ \mathrm{i}E\sin\gamma & \cos\gamma \end{bmatrix}^S = \begin{bmatrix} \cos S\gamma & \mathrm{i}\sin S\gamma/E \\ \mathrm{i}E\sin S\gamma & \cos S\gamma \end{bmatrix}. \tag{6.6}$$

It should be noted that the equivalent single layer is not an exact replacement for the symmetrical combination in every respect physically. It is merely a mathematical expression of the product of a number of matrices. The effect of changes in angle of incidence, for instance, cannot be estimated by converting the multilayer to a single layer in this way.

In any practical case when the matrix elements are computed it will be found that there are regions where $M_{11} < -1$, i.e. $\cos\gamma < -1$. This expression cannot be solved for real γ, and in this region γ and E are both imaginary. The physical significance of this was explained in the previous chapter, where it was shown that as the number of basic periods is increased the reflectance of a multilayer tends to unity in regions where $|M_{11} + M_{22}|/2 > 1$, M_{11} and M_{22} being elements of the matrix of the basic period. In the present symmetrical case this is equivalent to

$$|M_{11}| = |M_{22}| > 1$$

which therefore denotes a region of high reflectance, i.e. a stop band. Inside the stop band, the equivalent phase thickness and the equivalent admittance are both imaginary. Outside the stop band the phase thickness and admittance are real and these regions are known as pass regions or pass bands. The edges of the pass bands and stop bands are given by $M_{11} = -1$.

Application of the Herpin index to the quarter-wave stack

Returning for the moment to our quarter-wave stack, we see that it is possible to apply the above results directly if a simple alteration to the design is made. This is simply to add a pair of eighth-wave layers to the stack, one at each end. Low-index layers are required if the basic stack begins and ends with quarter-wave high-index layers and vice versa. The two possibilities are

$$\frac{H}{2}LHLHLH\ldots HL\frac{H}{2}$$

and

$$\frac{L}{2}HLHLHL\ldots LH\frac{L}{2}$$

These arrangements we can replace immediately by

$$\frac{H}{2}L\frac{H}{2}\,\frac{H}{2}L\frac{H}{2}\,\frac{H}{2}L\frac{H}{2}\,\frac{H}{2}L\frac{H}{2}\,\frac{H}{2}L\frac{H}{2}\ldots\frac{H}{2}L\frac{H}{2}$$

and

$$\frac{L}{2}H\frac{L}{2}\frac{L}{2}H\frac{L}{2}\frac{L}{2}H\frac{L}{2}\frac{L}{2}H\frac{L}{2}\frac{L}{2}H\frac{L}{2}\ldots\frac{L}{2}H\frac{L}{2}$$

respectively which can then be written as

$$\left[\frac{H}{2}L\frac{H}{2}\right]^{S} \quad \text{and} \quad \left[\frac{L}{2}H\frac{L}{2}\right]^{S}$$

$(H/2)L(H/2)$ and $(L/2)H(L/2)$ being the basic periods in each case. The results in equations (6.1)–(6.6) can then be used to replace both the above stacks by single layers making the performance in the pass bands and also the extent of the stop bands easily calculable. We shall examine first the width of the stop bands. As mentioned above, the edges of the stop bands are given by $M_{11} = -1$. Using equation (6.2*a*) this is equivalent to

$$\cos^2\delta_{qe} - \tfrac{1}{2}\left(\frac{\eta_q}{\eta_p} + \frac{\eta_p}{\eta_q}\right)\sin^2\delta_{qe} = -1$$

which is exactly the same expression as was obtained in the previous chapter for the width of the unaltered quarter-wave stack. There, δ was replaced by $(\pi/2)g$, where $g = \lambda_0/\lambda$ (or ν/ν_0, where ν is the wavenumber), and the edges of the stop band were defined by

$$\delta_e = \frac{\pi}{2}(1 \pm \Delta g).$$

The width is therefore

$$2\Delta g = 2\Delta\left(\frac{\lambda_0}{\lambda}\right)$$

where, if $\eta_p < \eta_q$,

$$\Delta g = \frac{2}{\pi}\sin^{-1}\left(\frac{\eta_q - \eta_p}{\eta_q + \eta_p}\right) \tag{6.7}$$

or, if $\eta_q < \eta_p$,

$$\Delta g = \frac{2}{\pi}\sin^{-1}\left(\frac{\eta_p - \eta_q}{\eta_p + \eta_q}\right). \tag{6.8}$$

These expressions are plotted in figure 5.7. The width of the stop band is therefore exactly the same regardless of whether the basic period is $(H/2)L(H/2)$, or $(L/2)H(L/2)$. Of course, it is possible to have other three-layer combinations where the width of the central layer is not equal to twice the thickness of the two outer layers, and some of the other possible arrangements will be examined, both in this chapter and the next, as they have some interesting properties, but, as far as the width of the stop band is concerned, it has been shown by Vera[2] that the maximum width for a three-layer symmetrical period is obtained when the central layer is a quarter-wave and the outer layers an eighth-wave each.

Let us now turn our attention to the pass band; first the equivalent admittance and then the equivalent optical thickness. The expression for the equivalent admittance in the pass band is quite a complicated one. From equations (6.2*b*), (6.2*c*), (6.4) and (6.5)

$$E = +\left(\frac{M_{21}}{M_{12}}\right)^{1/2}$$

$$= +\left(\frac{\eta_p^2[\sin 2\delta_p \cos \delta_q + \frac{1}{2}(\eta_p/\eta_q + \eta_q/\eta_p)\cos 2\delta_p \sin \delta_q - \frac{1}{2}(\eta_p/\eta_q - \eta_q/\eta_p)\sin \delta_q]}{\sin 2\delta_p \cos \delta_q + \frac{1}{2}(\eta_p/\eta_q + \eta_q/\eta_p)\cos 2\delta_p \sin \delta_q + \frac{1}{2}(\eta_p/\eta_q - \eta_q/\eta_p)\sin \delta_q}\right)^{1/2}. \tag{6.9}$$

This is not a particularly easy expression to handle analytically, but evaluation is straightforward, either by computer or even a programmable calculator. Figure 6.3 shows the equivalent admittance and optical thickness of combinations of zinc sulphide and cryolite. The form of this curve is quite

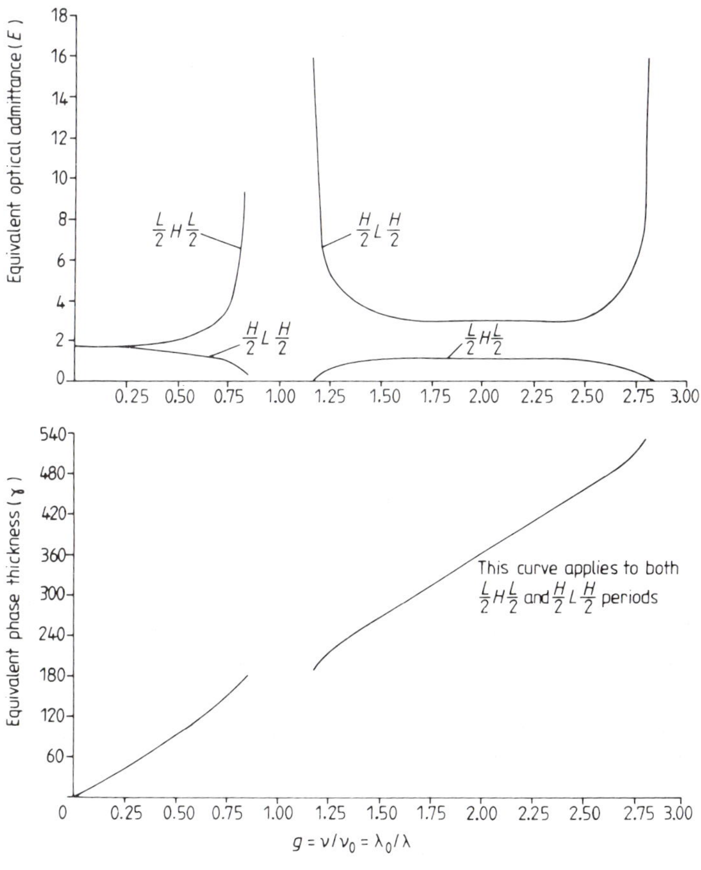

Figure 6.3 Equivalent optical admittance, E, and phase thickness, γ, of a symmetrical period of zinc sulphide ($n = 2.35$) and cryolite ($n = 1.35$) at normal incidence.

typical of such periods. Once the equivalent admittance and thickness have been evaluated, the calculation of the performance of the filter in the pass region, and its subsequent improvement, become much more straightforward. They are dealt with in greater detail later in this chapter. First we shall examine some of the properties of the expression for the equivalent optical admittance.

We can normalise expression (6.9) by dividing both sides by η_p. E/η_p is then solely a function of δ_p, δ_q and the ratio η_p/η_q. Next, we can make the further simplification, which we have not used so far, that $2\delta_p = \delta_q$. The expression for E/η_p then becomes

$$\frac{E}{\eta_p} = +\left(\frac{\{1+\frac{1}{2}[\rho+(1/\rho)]\}\cos\delta_q\sin\delta_q - \frac{1}{2}[\rho-(1/\rho)]\sin\delta_q}{\{1+\frac{1}{2}[\rho+(1/\rho)]\}\cos\delta_q\sin\delta_q + \frac{1}{2}[\rho-(1/\rho)]\sin\delta_q}\right)^{1/2} \tag{6.10}$$

where $\rho = \eta_p/\eta_q$.

It is is now easy to see that the following relationships are true. We write $(E/\eta_p)\,(\rho, \delta_q)$ to indicate that it is a function of the variables ρ and δ_q.

$$\frac{E}{\eta_p}(\rho, \pi-\delta_q) = \frac{1}{(E/\eta_p)\,(\rho, \delta_q)} \tag{6.11}$$

$$\frac{E}{\eta_p}\left(\frac{1}{\rho}, \delta_q\right) = \frac{1}{(E/\eta_p)\,(\rho, \delta_q)}. \tag{6.12}$$

These relationships are, in fact, true for all symmetrical periods, even ones which involve inhomogeneous layers, and general statements and proofs of these and other theorems are given by Thelen[3].

Thelen has shown how these relationships may be used to reduce the labour in calculating the equivalent admittance over a wide range. Figure 6.4 shows a set of curves giving the equivalent admittance for various values of the ratio of admittances. The vertical scale has been made logarithmic which has the advantage of making the various sections of the curve repetitions of the first section. This follows directly from the relationships (6.11) and (6.12). The values of the ratios of optical admittances which have been used are all greater than unity. Values less than unity can be derived from the plotted curves using relation (6.12). Again the logarithmic scale means that it is necessary only to reorient the curve for $\eta_p/\eta_q = k$ to give that for $\eta_p/\eta_q = 1/k$. All the information necessary to plot the curves is therefore given in the enlarged version of the first section of figure 6.4 which is reproduced in figure 6.5. Figures 6.4 and 6.5 are both taken from the paper by Thelen[3].

It is also useful to note the limiting values of E:

$$\begin{aligned} &E \text{ tends to } (\eta_p\eta_q)^{1/2} \quad \text{as} \quad \delta_q \text{ tends to zero} \\ &\text{and} \\ &E \text{ tends to } \eta_p(\eta_p/\eta_q)^{1/2} \quad \text{as} \quad \delta_q \text{ tends to } \pi. \end{aligned} \tag{6.13}$$

The equivalent phase thickness of the period is given by (6.2*a*) and (6.3) as

$$\gamma = \cos^{-1}\left[\cos 2\delta_p \cos\delta_q - \frac{1}{2}\left(\frac{\eta_p}{\eta_q}+\frac{\eta_q}{\eta_p}\right)\sin 2\delta_p \sin\delta_q\right]. \tag{6.14}$$

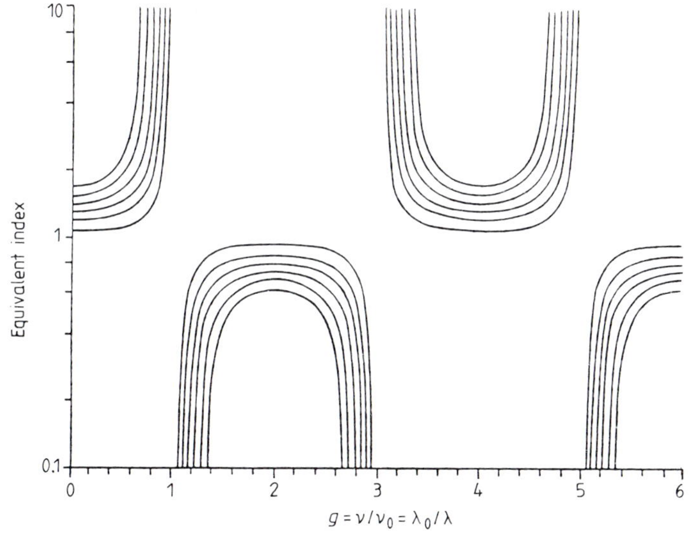

Figure 6.4 Equivalent admittance for the system $(L/2)H(L/2)$. $n_L = 1.00$ and n_H/n_L is a parameter with values 1.23, 1.50, 1.75, 2.0, 2.5, 3.0. The curves with the wider stop bands have the higher n_H/n_L values. (After Thelen[3].)

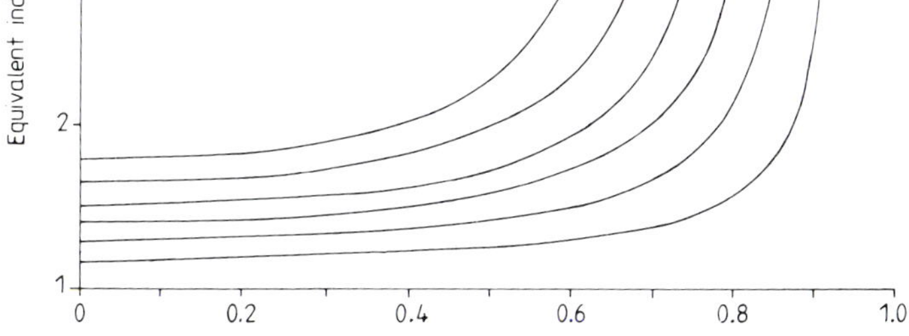

Figure 6.5 Enlarged first part of figure 6.4 (after Thelen[3]).

This expression for γ is multivalued, and the value chosen is that nearest to $2\delta_p + \delta_q$, the actual sum of the individual phase thicknesses, which is the most easily interpreted value. It is clear from the expression for γ that it does not matter whether the ratio of the admittances is greater or less than unity. The phase thickness for ρ is the same as that for $1/\rho$. Figure 6.6, which is also taken from Thelen's paper, shows the phase thickness of the combinations in figures 6.4 and 6.5. Because of the obvious symmetries, all the information necessary for the complete curve of the equivalent phase thickness is given in this diagram. The equivalent thickness departs significantly from the true thickness only near the edge of the high-reflectance zone. At any other point in the pass bands the equivalent phase thickness is almost exactly equal to the actual phase thickness of the combination.

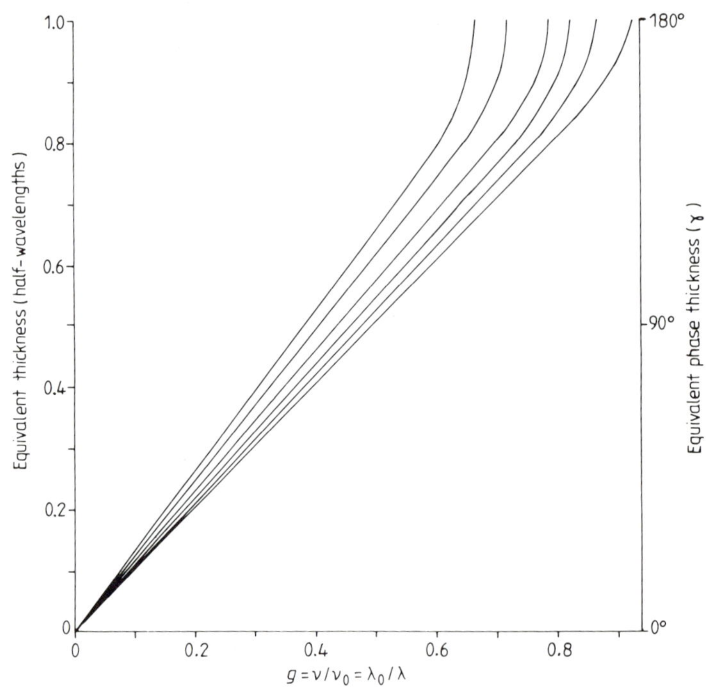

Figure 6.6 Equivalent thickness of the system described in figure 6.4 (after Thelen[3]).

Application of the Herpin index to multilayers of other than quarter-waves

All the curves shown so far are for |eighth-wave| quarter-wave |eighth-wave| periods. If the relative thicknesses of the layers are varied from this arrangement then the equivalent admittance is altered. It has already been mentioned that the reflectance zones for a combination other than the above must be narrower. Some idea of the way in which the equivalent admittance alters can be obtained from the value as $g \to 0$. Let $2\delta_p/\delta_q = \psi$. Then, from equation (6.9)

$$E = +\eta_p^2\left[\frac{\sin 2\delta_p}{\sin 2\delta_q}\cos\delta_q + \tfrac{1}{2}\left(\frac{\eta_p}{\eta_q}+\frac{\eta_q}{\eta_p}\right)\cos 2\delta_p - \tfrac{1}{2}\left(\frac{\eta_p}{\eta_q}-\frac{\eta_q}{\eta_p}\right)\right]^{1/2}$$

$$\times\left[\frac{\sin 2\delta_p}{\sin\delta_q}\cos\delta_q + \tfrac{1}{2}\left(\frac{\eta_p}{\eta_q}+\frac{\eta_q}{\eta_p}\right)\cos 2\delta_p + \tfrac{1}{2}\left(\frac{\eta_p}{\eta_q}-\frac{\eta_q}{\eta_p}\right)\right]^{-1/2}. \quad (6.15)$$

Now $\sin 2\delta_p/\sin\delta_q \to \psi$ as $g \to 0$, since $\delta_q \to 0$, $\delta_p \to 0$, i.e.

$$E \to \eta_p\left[\psi + \tfrac{1}{2}\left(\frac{\eta_p}{\eta_q}+\frac{\eta_q}{\eta_p}\right) - \tfrac{1}{2}\left(\frac{\eta_p}{\eta_q}-\frac{\eta_q}{\eta_p}\right)\right]^{1/2}$$

$$\times\left[\psi + \tfrac{1}{2}\left(\frac{\eta_p}{\eta_q}+\frac{\eta_q}{\eta_p}\right) + \tfrac{1}{2}\left(\frac{\eta_p}{\eta_q}-\frac{\eta_q}{\eta_p}\right)\right]^{-1/2}.$$

Rearranging this we obtain

$$\frac{E}{\eta_p} \to \left(\frac{\psi + (\eta_q/\eta_p)}{\psi + (\eta_p/\eta_q)}\right)^{1/2}. \quad (6.16)$$

This result shows that, for small g, it is possible to vary the equivalent admittance throughout the range of values between η_p and η_q but not outside that range. This result has already been referred to in the chapter on antireflection coatings, where it was shown how to use the concept of equivalent admittance to create replacements for layers having indices difficult to reproduce.

Epstein[1] has considered in more detail the variation of equivalent admittance by altering the thickness ratio and gives tables of results for zinc sulphide/cryolite multilayers.

Ufford and Baumeister[4] give sets of curves which assist in the use of equivalent admittance in a wide range of design problems.

Some results which are at first sight rather surprising are obtained when the value of the equivalent admittance around $g = 2$ is investigated. As $g \to 2$, $2\delta_p \to \pi$ and $\delta_q \to \pi$ so that, from equation (6.15)

$$\frac{E}{\eta_p} \to \left(\frac{-1-\tfrac{1}{2}[(\eta_p/\eta_q)+(\eta_q/\eta_p)]-\tfrac{1}{2}[(\eta_p/\eta_q)-(\eta_q/\eta_p)]}{-1-\tfrac{1}{2}[(\eta_p/\eta_q)+(\eta_q/\eta_p)]+\tfrac{1}{2}[(\eta_p/\eta_q)-(\eta_q/\eta_p)]}\right)^{1/2} = \left(\frac{\eta_p}{\eta_q}\right)^{1/2}. \quad (6.17)$$

This is quite a straightforward result. Now let $2\delta_p/\delta_q = \psi$, as in the case just considered where $g \to 0$. Let $g \to 2$ so that

$$2\delta_p + \delta_q \to 2\pi.$$

(This is really how, in this case, we define $g = \lambda_0/\lambda$, by defining λ_0 as that wavelength which makes $2\delta_p + \delta_q = \pi$.)

We have, as $g \to 2$,

$$\cos 2\delta_p \to \cos(2\pi - \delta_q) = \cos\delta_q$$
$$\sin 2\delta_p \to -\sin(2\pi - \delta_q) = -\sin\delta_q$$

and $\delta_q \to 2\pi/(1+\psi)$ so that

$$\begin{aligned}
\frac{E}{\eta_p} \to & \left[-\sin\delta_q\cos\delta_q + \tfrac{1}{2}\left(\frac{\eta_p}{\eta_q} + \frac{\eta_q}{\eta_p}\right)\cos\delta_q\sin\delta_q - \tfrac{1}{2}\left(\frac{\eta_p}{\eta_q} - \frac{\eta_q}{\eta_p}\right)\sin\delta_q\right]^{1/2} \\
& \times\left[-\sin\delta_q\cos\delta_q + \tfrac{1}{2}\left(\frac{\eta_p}{\eta_q} + \frac{\eta_q}{\eta_p}\right)\cos\delta_q\sin\delta_q + \tfrac{1}{2}\left(\frac{\eta_p}{\eta_q} - \frac{\eta_q}{\eta_p}\right)\sin\delta_q\right]^{-1/2} \\
= & \left\{-\cos\delta_q\left[1 - \tfrac{1}{2}\left(\frac{\eta_p}{\eta_q} + \frac{\eta_q}{\eta_p}\right)\right] - \tfrac{1}{2}\left(\frac{\eta_p}{\eta_q} - \frac{\eta_q}{\eta_p}\right)\right\}^{1/2} \\
& \times\left\{-\cos\delta_q\left[1 - \tfrac{1}{2}\left(\frac{\eta_p}{\eta_q} + \frac{\eta_q}{\eta_p}\right)\right] + \tfrac{1}{2}\left(\frac{\eta_p}{\eta_q} - \frac{\eta_q}{\eta_p}\right)\right\}^{-1/2}
\end{aligned} \tag{6.18}$$

where $\cos\delta_q = \cos[2\pi/(1+\psi)]$.

Whatever the value of ψ, the quantities within the square root brackets have opposite sign, which means that the equivalent admittance is imaginary. Even as $\psi \to 1$, where one would expect the limit to coincide with the result in equation (6.17), the admittance is still imaginary.

The explanation of this apparent paradox is as follows. An imaginary equivalent admittance, as we have seen, indicates a zone of high reflectance. Consider first the ideal eighth-wave | quarter-wave | eighth-wave stack of equation (6.17). At the wavelength corresponding to $g = 2$, the straightforward theory predicts that the reflectance of the substrate shall not be altered by the presence of the multilayer, because each period of the multilayer is acting as a full wave of real admittance and is therefore an absentee layer. Looking more closely at the structure of the multilayer we can see that this can also be explained by the fact that all the individual layers are a half-wavelength thick. If the ratio of the thicknesses is altered, the layers are no longer a half-wavelength thick and cannot act as absentees. In fact, the theory of the above result shows that a zone of high reflectance occurs.

The transmission of a shortwave-pass filter at the wavelength corresponding to $g = 2$ is therefore very sensitive to errors in the relative thicknesses of the layers. Even a small error leads to a peak of reflection. The width of this spurious high-reflectance zone is quite narrow if the error is small. Thus the appearance of a pronounced narrow dip in the transmission curve of a shortwave-pass filter is quite a common feature and is difficult to eliminate. The dip is referred to sometimes as a 'half-wave hole'.

Performance calculations

We are now in a position to make some performance calculations.

Transmission at the edge of a stop band

The transmission in the high-reflectance region, or stop band, is an important parameter of the filter. Thelen[3] gives a useful method for calculating this at the edges of the band. His analysis is as follows:

Let the multilayer be made up of S fundamental periods so that the characteristic matrix of the multilayer is

$$[M]^S = \begin{bmatrix} \cos\gamma & (\mathrm{i}\sin\gamma)/E \\ \mathrm{i}E\sin\gamma & \cos\gamma \end{bmatrix}^S = \begin{bmatrix} \cos S\gamma & (\mathrm{i}\sin S\gamma)/E \\ \mathrm{i}E\sin S\gamma & \cos S\gamma \end{bmatrix}.$$

At the edges of the stop band we know that $\cos S\gamma \to 1$, $\sin S\gamma \to 0$, and $E \to 0$ or ∞ depending on the particular combination of layers. Now,

$$\frac{\sin S\gamma}{\sin\gamma} \to S \qquad \text{as} \qquad \sin\gamma \to 0$$

so that the matrix tends to

$$\begin{bmatrix} 1 & (\mathrm{i}S\sin\gamma)/E \\ \mathrm{i}ES\sin\gamma & 1 \end{bmatrix} = \begin{bmatrix} 1 & SM_{12} \\ SM_{21} & 1 \end{bmatrix}$$

at the stop band limits. Either M_{12} or M_{21} will also tend to zero because

$$M_{11}M_{22} - M_{12}M_{21} = 1$$

and, depending on which tends to zero, we have either

$$\begin{bmatrix} 1 & SM_{12} \\ 0 & 1 \end{bmatrix} \qquad \text{or} \qquad \begin{bmatrix} 1 & 0 \\ SM_{21} & 1 \end{bmatrix}$$

for the matrix.

If η_0 is the admittance of the incident medium and η_m of the substrate, then the transmittance of the multilayer at the edge of the stop band is given by equation (2.67):

$$T = \frac{4\eta_0\eta_\mathrm{m}}{(\eta_0 B + C)(\eta_0 B + C)^*}$$

where

$$\begin{bmatrix} B \\ C \end{bmatrix} = \begin{bmatrix} 1 & SM_{12} \\ 0 & 1 \end{bmatrix}\begin{bmatrix} 1 \\ \eta_\mathrm{s} \end{bmatrix} \qquad \text{if} \qquad M_{21} = 0$$

or

$$\begin{bmatrix} 1 & 0 \\ SM_{21} & 1 \end{bmatrix} \begin{bmatrix} 1 \\ \eta_m \end{bmatrix} \quad \text{if} \quad M_{12} = 0$$

i.e.

$$\begin{bmatrix} B \\ C \end{bmatrix} = \begin{bmatrix} 1 + S\eta_m M_{12} \\ \eta_m \end{bmatrix} \quad \text{or} \quad \begin{bmatrix} 1 \\ \eta_m + SM_{21} \end{bmatrix}$$

so that, if there is no absorption,

$$T = \frac{4\eta_0\eta_m}{(\eta_0 + \eta_m)^2 + (S\eta_m\eta_0|M_{12}|)^2} \quad \text{when} \quad M_{21} = 0 \tag{6.19}$$

or

$$T = \frac{4\eta_0\eta_m}{(\eta_0 + \eta_m)^2 + (S|M_{21}|)^2} \quad \text{when} \quad M_{12} = 0 \tag{6.20}$$

(since M_{12} and M_{21} are imaginary in the absence of absorption). For M_{12} or M_{21} to be zero requires that

$$\sin 2\delta_p \cos \delta_q + \tfrac{1}{2}\left(\frac{\eta_p}{\eta_q} + \frac{\eta_q}{\eta_p}\right)\cos 2\delta_p \sin \delta_q = \mp\tfrac{1}{2}\left(\frac{\eta_p}{\eta_q} - \frac{\eta_q}{\eta_p}\right)\sin \delta_q.$$

If M_{12} is zero we can deduce that

$$|M_{21}| = \left|\eta_p\left(\frac{\eta_p}{\eta_q} - \frac{\eta_q}{\eta_p}\right)\sin \delta_q\right| \tag{6.21}$$

or, if M_{21} is zero, that

$$|M_{12}| = \left|\frac{1}{\eta_p}\left(\frac{\eta_p}{\eta_q} - \frac{\eta_q}{\eta_p}\right)\sin \delta_q\right|. \tag{6.22}$$

At the limits of the high-reflectance zone we have already seen that

$$\cos^2 \delta = \left(\frac{\eta_q - \eta_p}{\eta_q + \eta_p}\right)^2$$

i.e.

$$\sin^2\delta = 1 - \cos^2 \delta = \frac{4\eta_p\eta_q}{(\eta_q + \eta_p)^2}.$$

Substituting this in the expressions (6.21) and (6.22) for $|M_{21}|$ and $|M_{12}|$ we find

$$|M_{21}|^2 = \left|\frac{4\eta_p(\eta_p - \eta_q)^2}{\eta_q}\right| \quad \text{for} \quad M_{12} = 0. \tag{6.23}$$

$$|M_{12}|^2 = \left|\frac{4(\eta_p - \eta_q)^2}{\eta_p^3\eta_q}\right| \quad \text{for} \quad M_{21} = 0. \tag{6.24}$$

To give the transmittance at the edges of the high-reflectance zone, these expressions should be used in equations (6.19) and (6.20) according to the rule:

If E, the equivalent admittance, is zero, then M_{21} is zero.
If E, the equivalent admittance, is ∞, then M_{12} is zero.

Transmission in the centre of a stop band

For the simple quarter-wave stack an expression for transmittance at the centre of the high-reflectance zone has already been given in chapter 5. For the present multilayer, the transmittance is of a similar order of magnitude but the eighth-wave layers at the outer edges of the stack complicate matters. The stack may be represented by

$$\frac{p}{2}\,q\,\frac{p}{2}\,\frac{p}{2}\,q\,\frac{p}{2}\,\ldots\,\frac{p}{2}\,q\,\frac{p}{2}$$

which is

$$\frac{p}{2}\,qpqpqp\ldots q\,\frac{p}{2}\,.$$

If there are S periods, then the layer q appears S times in this expression. At the centre of the high-reflectance zone, the matrix product becomes:

$$\begin{bmatrix} 1/\sqrt{2} & \mathrm{i}/(\eta_p\sqrt{2}) \\ \mathrm{i}\eta_p/\sqrt{2} & 1/\sqrt{2} \end{bmatrix}\begin{bmatrix} 0 & \mathrm{i}/\eta_q \\ \mathrm{i}\eta_p & 0 \end{bmatrix}\begin{bmatrix} 0 & \mathrm{i}/\eta_p \\ \mathrm{i}\eta_p & 0 \end{bmatrix}\cdots\begin{bmatrix} 0 & \mathrm{i}/\eta_q \\ \mathrm{i}\eta_q & 0 \end{bmatrix}\begin{bmatrix} 1/\sqrt{2} & \mathrm{i}/(\eta_p\sqrt{2}) \\ \mathrm{i}\eta_p/\sqrt{2} & 1/\sqrt{2} \end{bmatrix}$$

$$=\begin{bmatrix} 1/\sqrt{2} & \mathrm{i}/(\eta_p\sqrt{2}) \\ \mathrm{i}\eta_p/\sqrt{2} & 1/\sqrt{2} \end{bmatrix}\begin{bmatrix} 0 & \mathrm{i}/\eta_q \\ \mathrm{i}\eta_q & 0 \end{bmatrix}\begin{bmatrix} -\eta_q/\eta_p & 0 \\ 0 & -\eta_p/\eta_q \end{bmatrix}^{S-1}\begin{bmatrix} 1/\sqrt{2} & \mathrm{i}/(\eta_p\sqrt{2}) \\ \mathrm{i}\eta_p/\sqrt{2} & 1/\sqrt{2} \end{bmatrix}$$

$$=\tfrac{1}{2}\begin{bmatrix} (-\eta_q/\eta_p)^S+(-\eta_p/\eta_q)^S & (\mathrm{i}/\eta_p)[(-\eta_q/\eta_p)^S-(-\eta_p/\eta_q)^S] \\ \mathrm{i}\eta_p[(-\eta_p/\eta_q)^S-(-\eta_q/\eta_p)^S] & (-\eta_q/\eta_p)^S+(-\eta_q/\eta_p)^S \end{bmatrix}. \tag{6.25}$$

Let η_m be the admittance of the substrate. Then

$$\begin{bmatrix} B \\ C \end{bmatrix}=\frac{1}{2}\begin{bmatrix} \left(-\dfrac{\eta_p}{\eta_q}\right)^S+\left(-\dfrac{\eta_q}{\eta_p}\right)^S+\dfrac{\mathrm{i}\eta_\mathrm{m}}{\eta_p}\left[\left(-\dfrac{\eta_q}{\eta_p}\right)^S-\left(-\dfrac{\eta_p}{\eta_q}\right)^S\right] \\ \eta_\mathrm{m}\left[\left(-\dfrac{\eta_p}{\eta_q}\right)^S+\left(-\dfrac{\eta_q}{\eta_p}\right)^S\right]+\mathrm{i}\eta_p\left[\left(-\dfrac{\eta_p}{\eta_q}\right)^S-\left(-\dfrac{\eta_q}{\eta_p}\right)\right] \end{bmatrix}. \tag{6.26}$$

Equation (2.67) gives

$$\begin{aligned} T&=\frac{4\eta_0\eta_\mathrm{m}}{(\eta_0 B+C)(\eta_0 B+C)^*} \\ &=\frac{16\eta_0\eta_\mathrm{m}}{\{(\eta_0+\eta_\mathrm{m})[(-\eta_q/\eta_p)^S+(-\eta_p/\eta_q)^S]\}^2} \\ &\quad+\frac{16\eta_0\eta_\mathrm{m}}{\{[(\eta_0\eta_\mathrm{m}/\eta_p)-\eta_p][(-\eta_q/\eta_p)^S-(-\eta_p/\eta_q)^S]\}^2}. \end{aligned} \tag{6.27}$$

If S is sufficiently large so that

$$\left(\frac{\eta_H}{\eta_L}\right)^S \gg \left(\frac{\eta_L}{\eta_H}\right)^S$$

which will usually be the case, this expression reduces to

$$T=\frac{16\eta_0\eta_\mathrm{m}}{(\eta_H/\eta_L)^{2S}\{(\eta_0+\eta_\mathrm{m})^2+[(\eta_0\eta_\mathrm{m}/\eta_p)-\eta_p]^2\}}. \tag{6.28}$$

Transmission in the pass band

In the pass band, the multilayer behaves as if it were a single layer of slightly variable optical thickness and admittance. Let us consider the case of $[(L/2)H(L/2)]^S$. Figure 6.7 shows part of the curve of equivalent admittance E for $[(L/2)H(L/2)]$. γ, the equivalent phase thickness, is also shown.

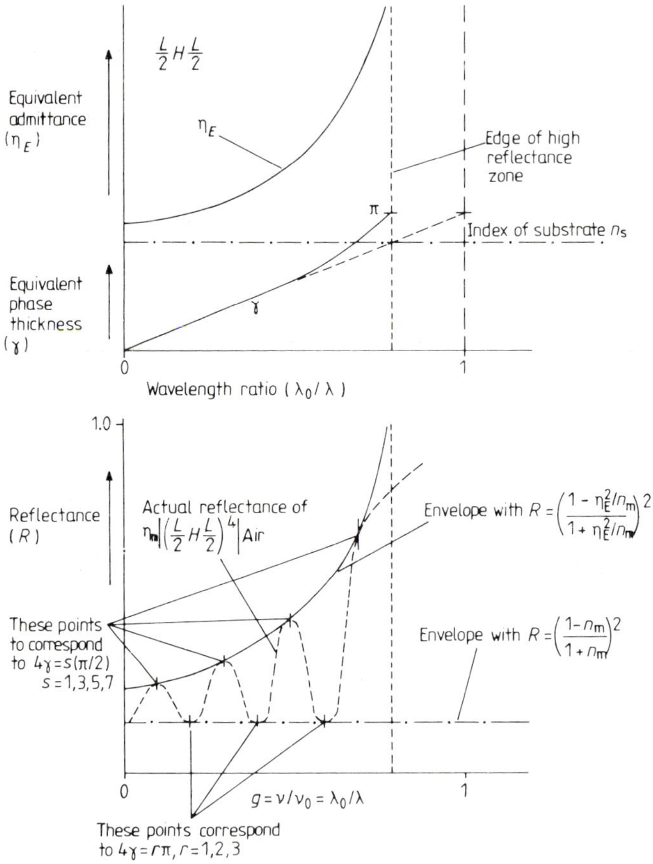

Figure 6.7 Diagram explaining the origin of the ripple in the pass band of an edge filter.

In the case of a real single transparent layer on a transparent substrate the reflectance oscillates between two limiting values which correspond to layer thicknesses of an integral number of quarter-waves. When the layer is equivalent to an even number of quarter-waves, that is a whole number of half-waves, it is an absentee layer and behaves as if it did not exist, so that the reflectance is that of the bare substrate. When the layer is equivalent to an odd number of quarter-waves, then, according to whether the index is higher or lower than that of the substrate, the reflectance will either be a maximum or a minimum. Thus if η_f is the admittance of the film, η_m of the substrate and η_0 of the incident medium, the reflectance will be $[(\eta_0-\eta_m)/(\eta_0+\eta_m)]^2$, corresponding to an even number of quarter-waves, and

$$\left(\frac{\eta_0-(\eta_f^2/\eta_m)}{\eta_0+(\eta_f^2/\eta_m)}\right)^2$$

corresponding to an odd number of quarter-waves. Regardless of the actual thickness of the film, we can draw two lines

$$R=\left(\frac{\eta_0-\eta_m}{\eta_0+\eta_m}\right)^2 \tag{6.29}$$

and

$$R=\left(\frac{\eta_0-(\eta_f^2/\eta_m)}{\eta_0+(\eta_f^2/\eta_m)}\right)^2 \tag{6.30}$$

which are the loci of maximum and minimum reflectance values, that is, the envelope of the reflectance curve of the film. If the optical thickness of the film is D, then the actual positions of the turning values will be given by

$$D=2n\,\lambda/4 \qquad n=0,\ 1,\ 2,\ 3,\ 4,\ \ldots$$

for those in equation (6.29), and by

$$D=(2n+1)\lambda/4$$

for those in equation (6.30). That is at wavelengths given by

$$\lambda=4D/2n=2D/n$$

and

$$\lambda=4D/(2n+1)$$

respectively.

We can now return to our multilayer. Since the multilayer can be replaced by a single film, the reflectance will oscillate between two values: the reflectance of the bare substrate

$$R=\left(\frac{\eta_0-\eta_m}{\eta_0+\eta_m}\right)^2 \tag{6.31}$$

and that given by

$$R=\frac{[\eta_0-(E^2/\eta_m)]^2}{[\eta_0+(E^2/\eta_m)]^2} \tag{6.32}$$

where we have replaced η_f in equation (6.29) by E, the equivalent admittance of the period. Equation (6.32) now represents a curve, since E is variable, rather than a line. To find the positions of the maxima and minima we look for values of $g=\lambda_0/\lambda$ for which the total thickness of the multilayer is a whole number of quarter-waves, which is the same as saying that the total equivalent phase thickness of the multilayer must be a whole number times $\pi/2$; an odd number corresponds to equation (6.32) and an even number to equation (6.31). If there are n periods in the multilayer, then the equivalent phase thickness will be $n\gamma$, which will be a multiple of $\pi/2$ when the equivalent phase thickness of a single period, γ, is a multiple of $\pi/2n$, i.e.

$$\gamma = s\pi/2n \qquad s = 1, 3, 5, 7, \ldots \qquad \text{corresponding to (6.32)}$$

and

$$\gamma = r\pi/n \qquad r = 1, 2, 3, 4, \ldots \qquad \text{corresponding to (6.31).}$$

At the very edge of the pass band, the equivalent phase thickness is π and so we might expect that the multilayer should act as an absentee layer. However, the equivalent admittance at that point is either zero or infinite and so the multilayer cannot be treated in this way, and, in fact, we apply the expressions (6.21)–(6.24), which we have already derived.

Figure 6.7 illustrates the situation where a four-period multilayer has been taken as an example. The important point, however, is that the envelopes of the reflectance curve do not vary with the number of periods.

The reason for the excessive ripple in the pass band of a filter is now clear. It is due to mismatching of the equivalent admittances of the substrate, multilayer stack, and medium. To reduce the ripple, better matching is required.

Reduction of pass band ripple

There are a number of different approaches for reducing ripple. The simplest approach is to choose a combination which has an equivalent admittance similar to that of the substrate. Provided the reflection loss due to the bare substrate is not too great, this method should yield an adequate result. Figure 6.3 shows that the combination $[(H/2)L(H/2)]$ where $\eta_H = 2.35$, $\eta_L = 1.35$, should give a reasonable performance as a longwave-pass filter on glass, and this is indeed the case. The performance of such a filter is shown in figure 6.8. For a shortwave-pass filter, the combination $[(L/2)H(L/2)]$ is better and this is also shown in figure 6.8. Often, however, the materials which are available do not yield a suitable equivalent admittance and other measures to reduce ripple must be adopted.

One method which is very straightforward has been suggested by Welford[5] but does not seem to have been much used. This is simply to vary the thicknesses of the films in the basic period so that the equivalent admittance is altered to bring it nearer to the desired value. For this method to be successful, the reflectance from the bare substrate must be kept low and the substrate should have a low index. Glass in the visible region is quite satisfactory, but the method could not be used with, for example, silicon and germanium in the infrared without modification.

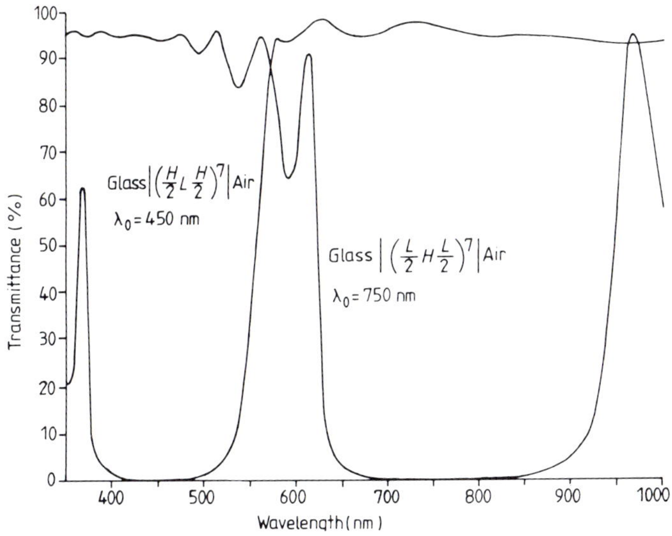

Figure 6.8 Computed transmittance of a 15-layer longwave-pass filter and a 15-layer shortwave-pass filter.

The more usual approach is to add matching layers at either side of the multilayer to match it to the substrate and to the medium. If a quarter-wave layer of admittance η_3 is inserted between the multilayer and substrate, and a quarter-wave layer of admittance η_1 between the multilayer and medium, then good matching will be obtained if

$$\eta_3 = (\eta_m E)^{1/2} \qquad \text{and} \qquad \eta_1 = (\eta_0 E)^{1/2}. \tag{6.33}$$

The layers are simply acting as antireflection layers between the multilayer and its surroundings. As a quick check that this does give the required performance we can compute the behaviour of the multilayer, considering just those wavelengths where the multilayer is equivalent either to an odd or to an even number of quarter-waves and to plot as before the envelope of the reflectance curve. At wavelengths where the multilayer acts like a quarter-wave, the equivalent admittance of the assembly is just

$$Y = \frac{\eta_1^2\eta_3^2}{E^2\eta_m}$$

so that the reflectance is

$$R = \left(\frac{\eta_0 - (\eta_1^2\eta_3^2/E^2\eta_m)}{\eta_0 + (\eta_1^2\eta_3^2/E^2\eta_m)}\right)^2 \tag{6.34}$$

which will be zero for

$$\eta_1^2\eta_3^2 = E^2\eta_m\eta_0. \tag{6.35}$$

When the multilayer acts like a half-wave it is an absentee, and the reflectance is

$$R = \left(\frac{\eta_0 - (\eta_1^2 \eta_m / \eta_3^2)}{\eta_0 + (\eta_1^2 \eta_m / \eta_3^2)} \right)^2 \tag{6.36}$$

which is zero if

$$\frac{\eta_1^2}{\eta_3^2} = \frac{\eta_0}{\eta_m}. \tag{6.37}$$

Solving equations (6.35) and (6.37) for η_1 and η_3 gives equation (6.33), as we expected.

If ideal matching layers do not exist, the suitability of any available materials can quickly be checked by substituting the appropriate values in equations (6.34) and (6.36).

Figure 6.9 shows a shortwave-pass filter before and after the matching layers

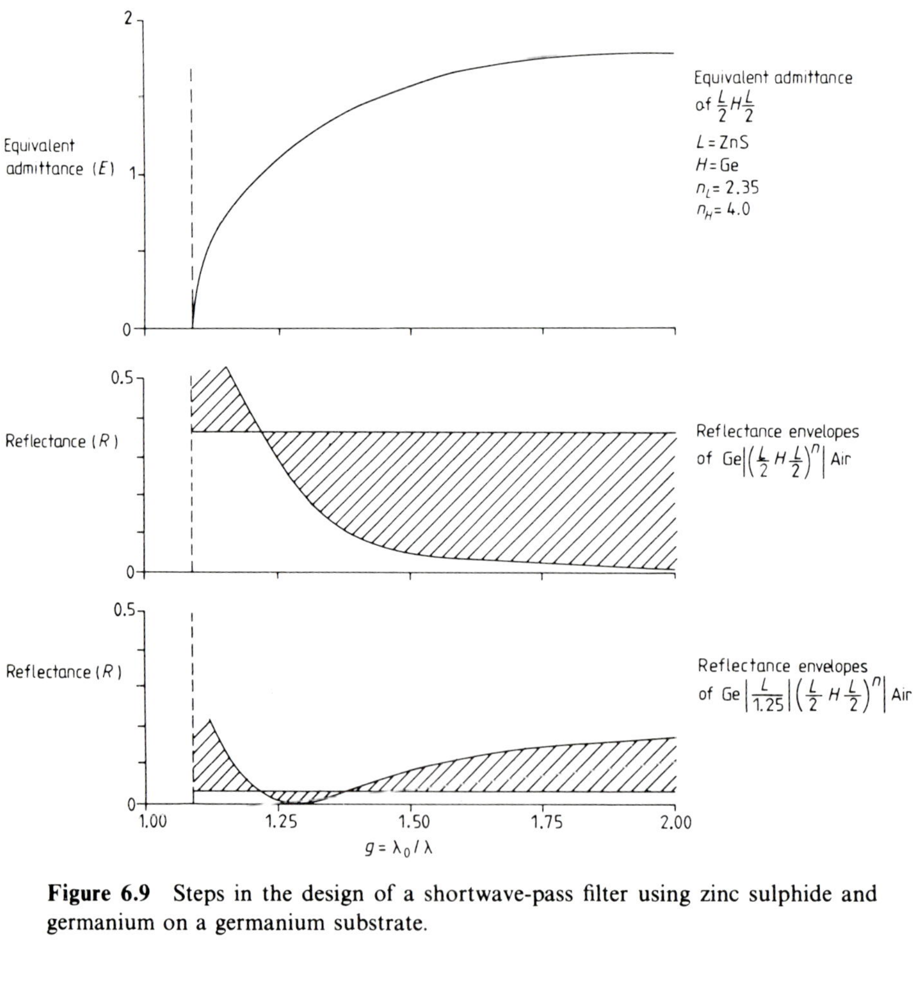

Figure 6.9 Steps in the design of a shortwave-pass filter using zinc sulphide and germanium on a germanium substrate.

have been added. The final reflectance envelopes are given by equations (6.34) and (6.36). The computed performance of the filter is shown in figure 6.10. As the value of g increases from 1.25, the ripple becomes a little greater than that predicted by the envelopes. This is because the envelopes were calculated on the basis of quarter-wave matching layers, and this is strictly true for $g = 1.25$ only.

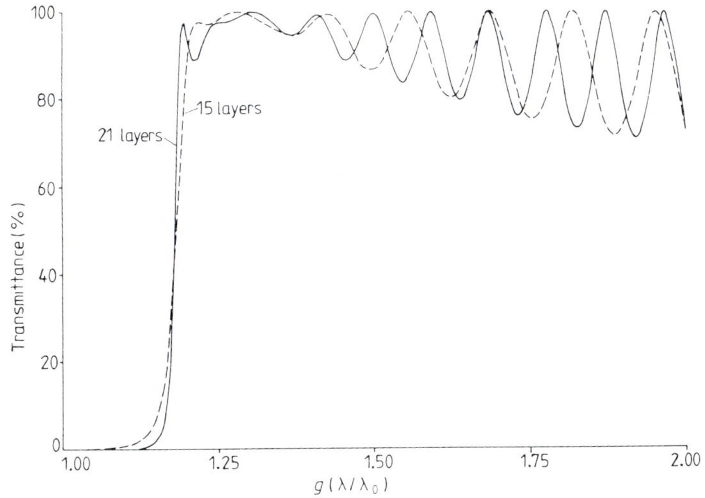

Figure 6.10 The calculated performance of filters designed according to figure 6.9 with design:

$$\text{Air}\,|\,(0.5L\ H\ 0.5L)^q L/1.25\,|\,\text{Ge}$$

with $n_L = 2.35$, $n_H = 4.0$, $n_{Ge} = 4.0$, and $n_{Air} = 1.00$. (*a*) $q = 7$ (*b*) $q = 10$.

Summary of design procedure so far

We have now established a simple design procedure for edge filters. First, two materials of different refractive index which are transparent in the region where transmission is required are chosen and used to form a multilayer of the form $[(L/2)H(L/2)]^S$ or $[(H/2)L(H/2)]^S$. Generally, it is better to choose as high a ratio of refractive indices as possible to give the widest rejection zone and also the maximum rejection for a given number of periods. The width of the rejection zone is given by equation (6.7) or (6.8) and is plotted in figure 5.6. The level of rejection at the edges of the zone is given by equations (6.19), (6.20), (6.23) and (6.24) and at the centre of the zone by equation (6.28). Next, the equivalent admittance of the stack must be calculated. This can be done either by a computer or by using the design curves given in figure 6.5. The formulae given in equations (6.13) for E/η_p at $g = 0$ and $g = 2$ will be found useful as a guide to interpolating curves. The reflectance envelopes can now be drawn using the formulae (6.31) and (6.32). This will immediately give some idea of the likely ripple. The positions of the peaks and troughs of the ripple can, if necessary, be found using the curves of γ in figure 6.6 and the method given on p 204. If this ripple is adequate the next step can be omitted and the design can proceed to the final step. If the ripple is not adequate then matching layers between multilayer and substrate, and multilayer and medium should be inserted. These should be quarter wavelength films at the most important wavelength and should have admittances as nearly as possible given by

$$\eta_1 = (\eta_0 E)^{1/2} \qquad \eta_3 = (\eta_{\mathrm{m}} E)^{1/2} \tag{6.33}$$

where η_1 is between the multilayer and medium and η_3 between the multilayer and substrate. Generally materials with the exact values will not be available and a compromise must be made. To test the effectiveness of the compromise the new reflectance envelope curves can be calculated using equations (6.34) and (6.36). If this is satisfactory, the next step is to calculate the actual performance on a computer. This is advisable because the quarter-wave matching layers are effective over a narrower region than assumed in equations (6.34) and (6.36). From the curve produced by the computer, the monitoring wavelength and thicknesses of the layers to position the characteristic at the correct wavelength can be calculated. The method is illustrated by the design of a shortwave-pass filter made from germanium and zinc sulphide on a germanium substrate as shown in figures 6.9 and 6.10.

A longwave-pass filter, designed by this method, with construction Air | 1.488 $L[(L/2)H(L/2)]^7$ 1.488 H|Ge (H = PbTe with $n_H = 5.3$, L = ZnS with $n_L = 2.35$), is shown in figure 10.10.

More advanced procedures for eliminating ripple

At the present time, probably the most common technique for eliminating ripple, apart from that already discussed, is computer refinement. This was introduced into optical coating design by Baumeister[6] who programmed a computer to eliminate the effects of slight changes in the thicknesses of the individual layers on a merit function representing the deviation of the performance of the coating from the ideal. An initial design, not too far from ideal, was adopted and the thicknesses of the layers modified, successively, gradually to improve the performance. This is still the basis of the technique. The optimum thickness of any one layer is not independent of the thicknesses of the other layers so that the changes in thickness at each iteration cannot be large without running the risk of instability. Computer speed and capacity has increased considerably since the early work of Baumeister, but the essentials of the method are still the same. Rather than change the layers successively, it is more usual to estimate changes which should be made in all the layers. These changes are then made simultaneously and the new function of merit computed. New changes are then estimated and the process repeated. The way in which the changes to be made are assessed is the principal difference between the techniques in frequent use. If the function of merit is considered as a surface in $(p+1)$-dimensional space with p independent variables being layer thicknesses, then a common method involves determining the direction of greatest slope of the merit surface and then altering the layer thicknesses so as to move along it, computing the new figure of merit and repeating the process. A battery of techniques for ensuring rapid convergence exists, and for further details the book by Liddell[7] should be consulted.

Less usual is complete design synthesis with no starting solution. This is still very much a research area and at the time of writing the most impressive results are those of Dobrowolski[8].

Computer refinement is a very powerful design aid but it can only function with an initial design. It then finds a modified design with an improved performance and repeats the process until stopped or until the performance reaches a maximum. This maximum will normally be simply a local maximum rather than the best possible performance, and the most useful way of ensuring that the maximum reached will be sufficiently high is to start from an initial design which is sufficiently good. The better the performance required, the better must be the initial design. Thus the existence of efficient computer refinement techniques does not in any way imply that the analytical design methods are obsolete and can be discarded. Refinement should be looked upon as a way of making a good design better. Applied to a poor design, computer refinement techniques usually yield disappointing results. For this reason, we continue with our examination of analytical techniques. It should always be remembered, however, that the manufacture of edge filters is not altogether an easy task, and unless the design performance of the simple design is being achieved in manufacture, there is little point in attempting anything more complicated until the sources of error have been eliminated.

The first and obvious method for improving the design is to improve the efficiency of the matching layers. In the chapter on antireflection coatings there were many multilayer coatings discussed which gave a rather better performance than the single layer. Any of these coatings can be used to eliminate the ripple. The ultimate performance is obtained with an inhomogeneous layer, but, as we have seen, the difficulty with inhomogeneous layers is that, in all practical cases, it is impossible to manufacture a layer with a graded index terminating in an index below 1.35, which means that there is always some small residual ripple. Jacobsson[9] has, however, considered briefly the matching of a multilayer longwave-pass filter $[(H/2)L(H/2)]^6$, consisting of germanium with an index of 4.0 and silicon monoxide with an index of 1.80, to a germanium substrate by means of an inhomogeneous layer. His paper shows the three curves reproduced in figure 6.11. The first curve 1 is the multilayer on a glass substrate of index 1.52. Since, in the pass band, the equivalent admittance of the multilayer falls gradually from $(1.8 \times 4.0)^{1/2} = 2.7$ to zero as the wavelength approaches the edge, it will be a value not too different from the

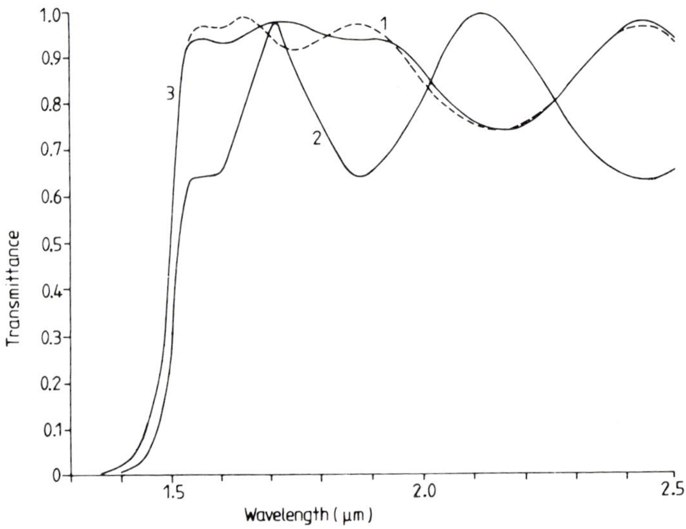

Figure 6.11 Reflectance versus wavelength of a multilayer on a substrate with index $n_{sub} = 1.52$ (curve 1), $n_{sub} = 4.00$ (curve 2) and on a substrate with $n_{sub} = 4.00$ with an inhomogeneous layer between substrate and multilayer (curve 3). (After Jacobsson[9].)

index of the substrate in the vicinity of the edge. The transmission near the edge is, therefore, high, as we might expect. When, as in curve 2, the same multilayer is deposited on a germanium substrate of index 4.0, the severe mismatching causes a very large ripple to appear. With an inhomogeneous layer between the germanium substrate and the multilayer and with the index varying from that of germanium next to the substrate to 1.52 next to the multilayer, the performance achieved, curve 3, is almost exactly that of the original multilayer on the glass substrate.

One of the examples examined by Baumeister was a shortwave-pass filter, and the design that he eventually obtained suggested a new approach to Young and Cristal[10]. It was mentioned in chapter 3 that Young had devised a method for designing antireflection coatings based on the quarter-wave transformer used in microwave filters. The antireflection coating takes the form of a series of quarter-waves with refractive indices in steady progression from the index of one medium to the index of the other. Young has given a series of tables enabling antireflection coatings of given bandwidth and ripple to be designed.

In their paper, Young and Cristal explain that they examined Baumeister's filter, and realised that the design might be written as a series of symmetrical periods with thicknesses increasing steadily from the middle of the stack to the outside, and they were struck by the resemblance which this bore to an antireflection coating in which each layer had been replaced by a symmetrical period. They then designed a coating, by microwave techniques, to match the admittance at the centre of the filter, which they arbitrarily took as 0.6, to air, with admittance 1.0, at the outside, each layer being replaced by an equivalent period. The scheme is shown as filter B in table 6.1, where the thicknesses given by Young and Cristal for one of their filters have been broken down into their

Table 6.1

Layer number	Filter B		Filter D		
	Layer thickness	Periods	Layer thickness	Periods	Period
1 Na_3AlF_6	47.50°	47.50°	48.5°	48.5°	1
2 ZnS	95.00°	95.00°	97.0°	97.0°	1
		47.50°		48.5°	1
3 Na_3AlF_6	93.25°		94.5°		
		45.75°		46.0°	2
4 ZnS	91.50°	91.50°	92.0°	92.0°	2
		45.75°		46.0°	2
5 Na_3AlF_6	90.00°		90.25°		
		44.25°		44.25°	3
6 ZnS	88.50°	88.50°	88.5°	88.5°	3
		44.25°		44.25°	3
7 Na_3AlF_6	87.50°		86.63°		
		43.25°		42.38°	4
8 ZnS	86.50°	86.50°	84.75°	84.75°	4
		43.25°		42.38°	4
9 Na_3AlF_6	86.50°		84.75°		
		43.25°		42.38°	5
10 ZnS	86.50°	86.50°	84.75°	84.75°	5
		43.25°		42.38°	5
11 Na_3AlF_6	87.50°		86.63°		
		44.25°		44.25°	6
12 ZnS	88.50°	88.50°	88.5°	88.5°	6
		44.25°		44.25°	6
13 Na_3AlF_6	90.00°		90.25°		
		45.75°		46.0°	7
14 ZnS	91.50°	91.50°	92.0°	92.0°	7
		45.75°		46.0°	7
15 Na_3AlF_6	93.25°		94.5°		
		47.50°		48.5°	8
16 ZnS	95.00°	95.00°	97.0°	97.0°	8
17 Na_3AlF_6	47.50°	47.50°	48.5°	48.5°	8

The second column in each case gives the filter split into its component periods.

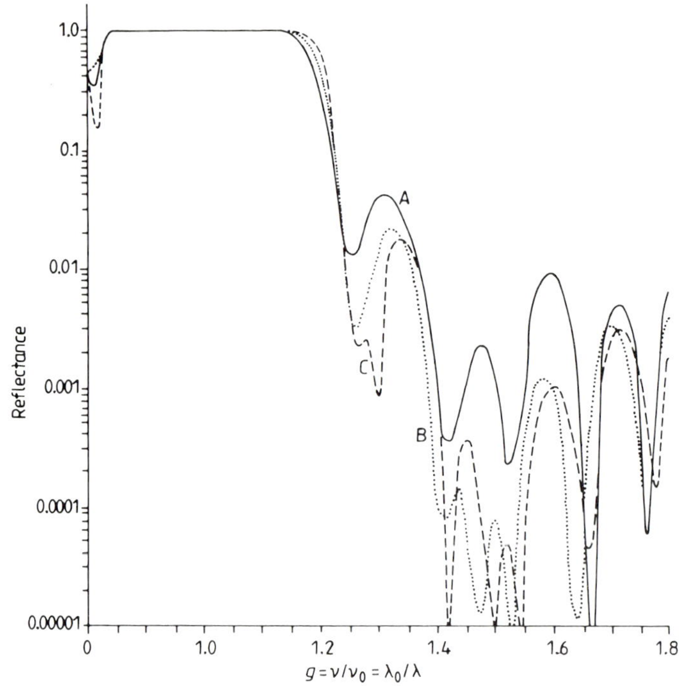

Figure 6.12 Reflectance of the three shortwave-pass filter designs, A, B and C, having unequal layer thickness. (After Young and Cristal[10].)

symmetrical periods. The performance of the filter is shown in figure 6.12 along with one other filter of their design and Baumeister's original design. The thicknesses are all shown in table 6.2. To simplify the discussion, Young and Cristal designed the filter to match with air on both sides of the multilayer, instead of, as is more usual, glass on one side and air on the other.

Young and Cristal do not discuss their design procedure in detail, but, from the final design of filter, it is possible to deduce it. First, the equivalent admittance of a single period was plotted, as in figure 6.13. The wavelength corresponding to 240° was chosen for optimising. From the value of equivalent admittance at 240° the value of 0.6 was probably selected intuitively as the value to use for the centre of the stack. An antireflection coating consisting of four layers, each three-quarter wavelengths thick, was designed to match this value to air, and the admittances of the layers computed. The admittances were then matched by that of three-layer symmetrical periods by altering thicknesses of each period, following the scheme shown in figure 6.13. This meant that the admittances were ideal but the thicknesses were not. However, the antireflection coating is not very susceptible to errors in layer thickness, and, as can be seen from the curve in figure 6.12, the performance achieved is excellent.

Table 6.2

Number of layers	Thickness (degrees)		
	Filter A	Filter B	Filter C
1	46.00	47.50	46.60
2	96.00	95.00	93.20
3	93.20	93.25	91.70
4	91.70	91.50	90.20
5	91.10	90.00	89.15
6	89.75	88.50	88.10
7	87.50	87.50	87.30
8	86.05	86.50	86.50
9	86.70	86.50	86.50
10	86.05	86.50	86.50
11	87.50	87.50	87.30
12	89.75	88.50	88.10
13	91.10	90.00	89.15
14	91.70	91.50	90.20
15	93.20	93.25	91.70
16	96.00	95.00	93.20
17	46.00	47.50	46.60

Filter A: The half of Baumeister's filter on the air side repeated symmetrically. (The design is referred to as design IX in Baumeister's paper.)
Filter B: New design based on a prototype transformer with a fractional bandwidth of 1.5
Filter C: New design based on a prototype transformer with a fractional bandwidth of 1.6

A similar approach is to use one of the multilayer antireflection coatings mentioned in chapter 3. Since the equivalent admittance of a symmetrical period varies with wavelength, any optimising at one wavelength is strictly correct over only a narrow range, and a simple approach, such as this, is probably as good as a more complicated one. Taking 240° as corresponding to the design wavelength, we find the value for equivalent admittance of the single period to be 0.8. We want the periods in the final design to be symmetrically placed around this period, so we find the starting admittance at the centre of the stack by assuming that this period should be able to act as a $3\lambda/4$ antireflection coating between the centre and the outside air. The admittance at the centre of the filter should therefore be $0.8^2 = 0.64$. Next, we design a four-layer antireflection coating to replace this basic period, using the formulae:

$$\eta_1 = \eta_0(\eta_s/\eta_0)^{1/5} \qquad \eta_3 = \eta_0(\eta_s/\eta_0)^{3/5}$$
$$\eta_2 = \eta_0(\eta_s/\eta_0)^{2/5} \qquad \eta_4 = \eta_0(\eta_s/\eta_0)^{4/5}$$

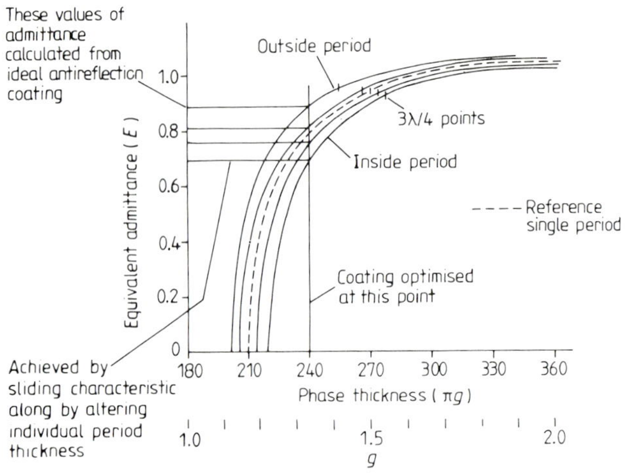

Figure 6.13 The admittance of the ideal four-layer antireflection coating to match air to the admittance of 0.6 are marked along the equivalent admittance axis. The reference single period is shown dotted and the values marked on the g axis refer to this period. By altering the total thickness of each period relative to this reference the four displaced solid line curves are obtained in such a way that the four symmetrical periods have the desired admittances at the wavelengths that correspond to a reference phase thickness of 240°.

where η_0 is air and η_s the admittance at the centre. Taking $\eta_0 = 1.0$ and $\eta_s = 0.64$, these admittances are then

$$\eta_1 = 0.91 \qquad \eta_2 = 0.84 \qquad \eta_3 = 0.76 \qquad \eta_4 = 0.70.$$

The values of total phase thickness πg at which the single period has equivalent admittance corresponding to these values are

$$\pi g_1 = 259^\circ \qquad \pi g_2 = 245^\circ \qquad \pi g_3 = 234^\circ \qquad \pi g_4 = 226^\circ.$$

For each period to have the appropriate admittance at the design wavelength, the phase thicknesses of the layers measured at the monitoring wavelength are given by

$$\text{Period 1}\begin{cases} \dfrac{L}{2}\ 45^\circ \times \dfrac{\pi g_1}{240^\circ} \\[2ex] H\ 90^\circ \times \dfrac{\pi g_1}{240^\circ} \\[2ex] \dfrac{L}{2}\ 45^\circ \times \dfrac{\pi g_1}{240^\circ} \end{cases} \qquad \text{Period 2}\begin{cases} \dfrac{L}{2}\ 45^\circ \times \dfrac{\pi g_2}{240^\circ} \\[2ex] H\ 90^\circ \times \dfrac{\pi g_2}{240^\circ} \\[2ex] \dfrac{L}{2}\ 45^\circ \times \dfrac{\pi g_2}{240^\circ} \end{cases}$$

and so on. The results are shown in table 6.1, filter D. The transmission of filter D is shown in figure 6.14.

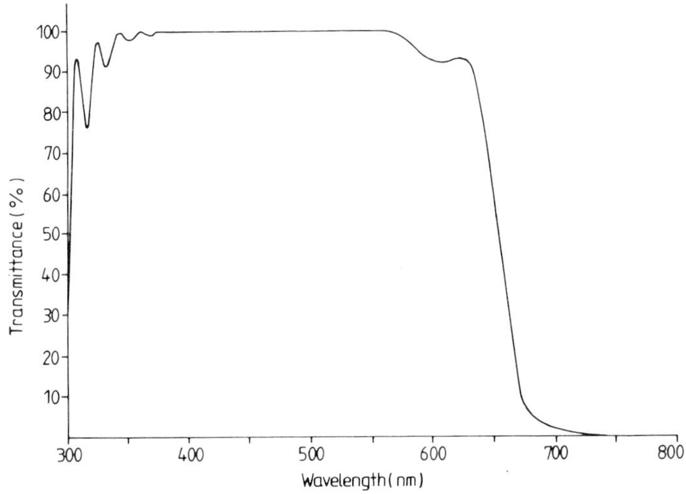

Figure 6.14 The computer transmittance of the shortwave-pass filter of design D of table 6.1. The reference wavelength, λ_0, is 800 nm.

Thelen[3] has pointed out that the rapid variation of equivalent admittance near the edge of the filter is the major source of difficulty in edge filter design. It is a simple matter to match the multilayer to the substrate where the equivalent admittance curve is flat, some distance from the edge, but the variations near the edge usually give rise, with simple designs, to a pronounced dip in the transmission curve. Thelen has devised an ingenious method of dealing with this dip, involving the equivalent of a single-layer antireflection coating. Between the main or primary multilayer, which consists of a number of equal basic periods, Thelen places a secondary multilayer, similar to the first but shifted in thickness so that, in the centre of the steep portion of the admittance curve, the equivalent admittance of the secondary is made equal to the square root of the equivalent admittance of the primary times the admittance of the substrate. The number of secondary periods is chosen to make the thickness at this point an odd number of quarter-waves and to satisfy completely the antireflection condition. Figure 6.15 shows the performance he achieved.

Seeley[11] has developed a different method of adapting results obtained in the synthesis of lumped electrical circuits for use in thin-film optical filters. One of the features of Young's method is that the refractive indices cannot be specified in advance, and as the range of available indices is limited this can lead to difficulties. In certain cases this can be avoided, as we have seen, by constructing three-layer periods with the appropriate equivalent indices, but even this has its limitations. Seeley, therefore, searched for another method which would permit the designer to specify the indices right from the start and to achieve the final performance by varying the thicknesses of the various

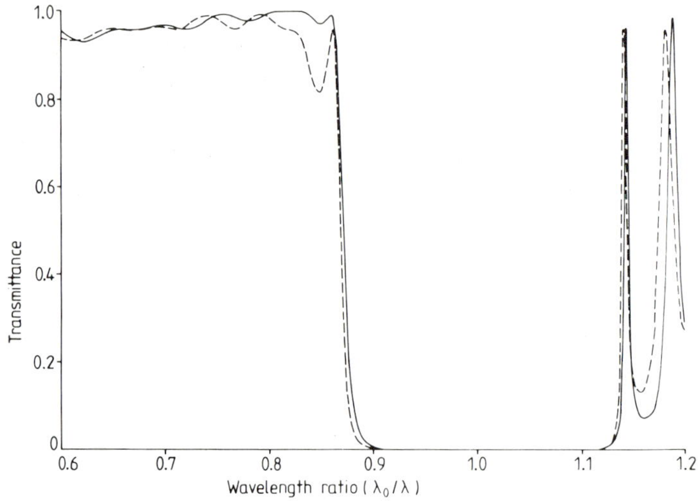

Figure 6.15 Comparison of the computed performance of the filters:

$$1.00|(0.5H\ L\ 0.5H)^{15}|1.52 \quad \text{(broken line)}$$

and

$$1.00|(0.5H\ L\ 0.5H)^{12}\ [(1/1.05)(0.5H\ L\ 0.5H)^3]|1.52 \quad \text{(solid line)}$$

with $n_H = 2.3$, $n_L = 1.56$. (After Thelen[3].)

layers. In a lumped electrical filter, consisting of inductances and capacitances, one parameter only is specified, the admittance. In the thin-film filter there are two parameters for each layer, the refractive index and the thickness. Thus it is possible for the optical designer to fix the values of the refractive indices of the multilayer filter in advance and then to compute the layer thickness by analogy with the lumped filter. As Welford[5] has pointed out, the analogy between thin-film assemblies and lumped electric filters is not exact. Thin films behave, in fact, in the same manner as lengths of waveguides. Seeley, however, devised a way of making the analogy exact, although only at one frequency. At all other frequencies, the analogy is only approximate. If the frequency chosen for exact correspondence is made the cut-off point of the filter, then the performance of the optical filter is found to be sufficiently close to that of the electrical filter over the usual working range. The techniques for optimising the performance of electrical filters are well established.

Seeley's method starts with an electrical filter of the desired type—longwave-pass, shortwave-pass or band-pass—whose performance is known to be optimum. The elements of the electrical filter are then converted by a step-by-step process into an equivalent circuit which is an exact analogue of the thin-film multilayer at one frequency. The process is shown in figure 6.16. In his design work, Seeley usually chooses electrical filters which have been designed using the Tchebyshev equal ripple polynomial. This polynomial allows the best fit to a square pass band when both edge steepness and ripple in the pass band are taken into account. From this, Seeley and Smith[12] have given simple rules for longwave-pass filters:

1 The optical admittance of the substrate η_m should lie between η_H and η_L, the admittances of the high- and low-index layers of the multilayer. If this is not satisfied, then a matching layer or combination of layers will be necessary between the substrate and the multilayer.

2 The first layer at the substrate should be high if $\eta_H/\eta_m > \eta_m/\eta_L$, and low if $\eta_m/\eta_L > \eta_H/\eta_m$.

3 The fractional ripple in the pass band will be

$$\left(\frac{\eta_H}{\eta_m}-\frac{\eta_m}{\eta_L}\right)^2\left(\frac{\eta_H}{\eta_m}+\frac{\eta_m}{\eta_L}\right)^{-2}.$$

4 For filters on germanium substrates using as layer materials lead telluride and zinc sulphide, the phase thicknesses should be in the proportions shown in table 6.3. The first layer at the substrate and all other odd layers, including the antireflection layer, are ZnS ($n = 2.2$). The remaining (even) layers are PbTe ($n = 5.1$). The substrate, germanium, has an index of 4.0.

5 Since the low-index material is usually good for matching the substrate to air, the front layer of the multilayer section of the filter should have a high index.

The computed transmittances of the designs given in table 6.3 are given in figure 6.17. The method is described in greater detail by Seeley *et al*[11].

Practical filters

Because the stop band of the multilayer edge filter is limited in extent, it is usually necessary for practical filters to consist of a multilayer filter together with additional filters which give the broad rejection region that is almost always required. These additional filters may be multilayer and some methods of broadening the stop band in this way are mentioned in the following section. Usually they are absorption filters having wide rejection regions but inflexible characteristics. These absorption filters may be combined with the multilayer filters in a number of different ways. They may simply be placed in series with the substrates carrying the multilayers, the substrates may themselves be the absorption filters or the multilayer materials may also act as thin-film absorption filters.

In the visible and near ultraviolet regions there is available a wide range of glass filters which solve most of the problems, particularly those connected with longwave-pass filters. In the infrared, the position is rather more difficult,

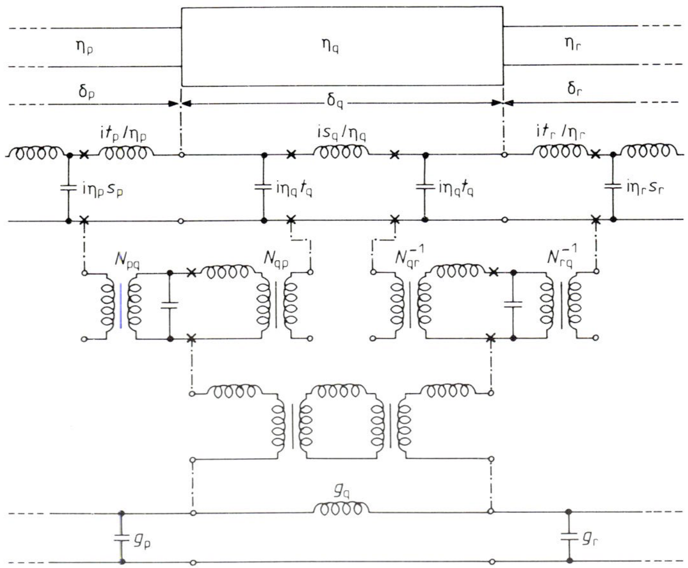

The conversion used by Seeley in diagrammatic form. High-index layers are first replaced by a T circuit and low-index ones by a Π circuit.

Figure 6.16(*a*) The step-by-step process by which Seeley converts a multilayer thin-film filter into a lumped electric filter in such a way that the elements of the electric filter can be identified with the optical thickness of the films, the indices of the films being specified completely independently. (Courtesy of Dr J S Seeley.)

(*Opposite page*) The manipulation takes place at the cut-off frequency of the lumped circuit and all variable quantities are normalised to that frequency. The scheme leads to a fairly complicated set of equations for . . . δ_p, δ_q, δ_r . . . in terms of . . . g_p, g_q, g_r . . ., which cannot be solved analytically but require iteration. Approximate solutions have been derived and are as follows:

$$\text{High-index layers:} \quad \sin\delta_p \simeq \frac{g_p}{(\eta_H/\eta_\mathrm{m}) + (\eta_L/\eta_\mathrm{m})}$$

$$\text{Low-index layers:} \quad \sin\delta_q \simeq \frac{g_q}{(\eta_\mathrm{m}/\eta_L) + (\eta_\mathrm{m}/\eta_H)}$$

δ being between 0 and $\pi/2$ for longwave-pass filters and $\pi/2$ and π for shortwave-pass filters.

The admittance levels in the derivation of these two expressions have been normalised to the terminating admittance (of the substrate), so that for η_p we have written η_H/η_m and for η_q, η_L/η_m, η_H and η_L being the admittances of the high- and low-index layers respectively.

High index — Low index — High index

$$\begin{bmatrix} \cos\delta_p & (\mathrm{i}\sin\delta_p)/\eta_p \\ \mathrm{i}\eta_p\sin\delta_p & \cos\delta_p \end{bmatrix} \times \begin{bmatrix} \cos\delta_q & (\mathrm{i}\sin\delta_q)/\eta_q \\ \mathrm{i}\eta_q\sin\delta_q & \cos\delta_q \end{bmatrix} \times \begin{bmatrix} \cos\delta_r & (\mathrm{i}\sin\delta_r)/\eta_r \\ \mathrm{i}\eta_r\sin\delta_r & \cos\delta_r \end{bmatrix}$$

$$\begin{bmatrix} 1 & \mathrm{i}t_p/\eta_p \\ 0 & 1 \end{bmatrix}\begin{bmatrix} 1 & 0 \\ \mathrm{i}\eta_p s_p & 1 \end{bmatrix}\begin{bmatrix} 1 & \mathrm{i}t_p/\eta_p \\ 0 & 1 \end{bmatrix} \times \begin{bmatrix} 1 & 0 \\ \mathrm{i}\eta_q t_q & 1 \end{bmatrix}\begin{bmatrix} 1 & \mathrm{i}s_q/\eta_q \\ 0 & 1 \end{bmatrix}\begin{bmatrix} 1 & 0 \\ \mathrm{i}\eta_q t_q & 1 \end{bmatrix} \times \begin{bmatrix} 1 & \mathrm{i}t_r/\eta_r \\ 0 & 1 \end{bmatrix}\begin{bmatrix} 1 & 0 \\ \mathrm{i}\eta_r s_r & 1 \end{bmatrix}\begin{bmatrix} 1 & \mathrm{i}t_r/\eta_r \\ 0 & 1 \end{bmatrix}$$

$$\begin{bmatrix} 1-(\eta_q t_p t_q/\eta_p) & \mathrm{i}t_p/\eta_p \\ \mathrm{i}\eta_q t_q & 1 \end{bmatrix} \qquad \begin{bmatrix} 1 & \mathrm{i}t_r/\eta_r \\ \mathrm{i}\eta_q t_q & 1-(\eta_q t_q t_r/\eta_r) \end{bmatrix}$$

$$\begin{bmatrix} N_{pq} & 0 \\ 0 & N_{pq}^{-1} \end{bmatrix}\begin{bmatrix} 1 & \mathrm{i}N_{qp}t_p/N_{pq}\eta_p \\ \mathrm{i}N_{pq}\eta_q t_q/N_{qp} & 1-(\eta_q t_q t_p/\eta_p) \end{bmatrix}\begin{bmatrix} N_{qp} & 0 \\ 0 & N_{qp}^{-1} \end{bmatrix} \qquad \begin{bmatrix} N_{qr}^{-1} & 0 \\ 0 & N_{qr} \end{bmatrix}\begin{bmatrix} 1-(\eta_q t_q t_r/\eta_r) & \mathrm{i}N_{qr}t_r/N_{rq}\eta_r \\ \mathrm{i}N_{rq}\eta_q t_q/N_{qr} & 1 \end{bmatrix}\begin{bmatrix} N_{rq}^{-1} & 0 \\ 0 & N_{rq} \end{bmatrix}$$

$$\begin{bmatrix} 1 & 0 \\ \mathrm{i}N_{pq}\eta_q t_q/N_{qp} & 1 \end{bmatrix} \times \begin{bmatrix} 1 & \mathrm{i}N_{qp}t_p/N_{pq}\eta_p \\ 0 & 1 \end{bmatrix}\begin{bmatrix} N_{qp} & 0 \\ 0 & N_{qp}^{-1} \end{bmatrix}\begin{bmatrix} 1 & \mathrm{i}s_q/\eta_q \\ 0 & 1 \end{bmatrix}\begin{bmatrix} N_{qr}^{-1} & 0 \\ 0 & N_{qr} \end{bmatrix}\begin{bmatrix} 1 & \mathrm{i}N_{qr}t_r/N_{rq}\eta_r \\ 0 & 1 \end{bmatrix} \times \begin{bmatrix} 1 & 0 \\ \mathrm{i}N_{rq}\eta_q t_q/N_{qr} & 1 \end{bmatrix}$$

$$\begin{bmatrix} N_{qp}/N_{qr} & \mathrm{i}[(N_{qp}N_{qr}S_q/\eta_q) + (N_{qp}t_r/N_{rq}\eta_r) + (N_{qr}t_p/N_{pq}\eta_p)] \\ 0 & N_{qr}/N_{qp} \end{bmatrix}$$

$$\begin{bmatrix} 1 & 0 \\ \mathrm{i}g_p & 1 \end{bmatrix} \times \begin{bmatrix} 1 & \mathrm{i}g_q \\ 0 & 1 \end{bmatrix} \times \begin{bmatrix} 1 & 0 \\ \mathrm{i}g_r & 1 \end{bmatrix}$$

$s = \sin\delta$ $\qquad$ $t = \tan\delta/2$

Figure 6.16(*b*) Matrix manipulations corresponding to figure 6.16(*a*).

Table 6.3

Layer number	Relative thickness	
	Longwave-pass	Shortwave-pass
1 and 14	0.55	1.25
2 and 13	0.82	1.11
3 and 12	0.92	1.05
4 and 11	0.96	1.025
5 and 10	0.98	1.015
6 and 9	0.99	1.01
7 and 8	1.0	1.0
15 (antireflection)	2.0	0.5

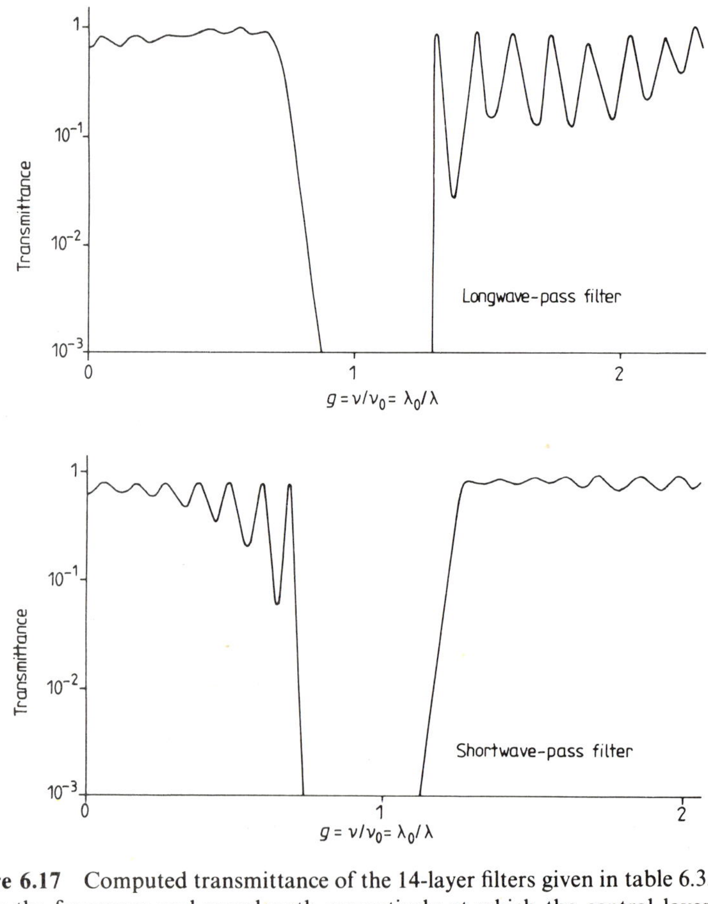

Figure 6.17 Computed transmittance of the 14-layer filters given in table 6.3. ν_0 and λ_0 are the frequency and wavelength respectively at which the central layers are a quarter-wave in thickness. (After Seeley and Smith[12].)

and often the complete filter consists of several multilayers which are necessary to connect the edge of the stop band to the nearest suitable absorption filter. Figure 6.18 shows a longwave-pass filter for the infrared. Figure 6.19 gives some of the infrared absorption filters which have shortwave-pass characteristics. For longwave-pass characteristics, semiconductors such as silicon, with an edge at 1 μm, and germanium, with an edge at 1.65 μm, are the most suitable. Indium arsenide, with an edge at 3.4 μm, and indium antimonide, with edge at 7.2 μm, are also useful, but because of the rather higher absorption they can only be used in very thin slices, around 0.013 cm for indium antimonide and only a little thicker for indium arsenide. This means that they

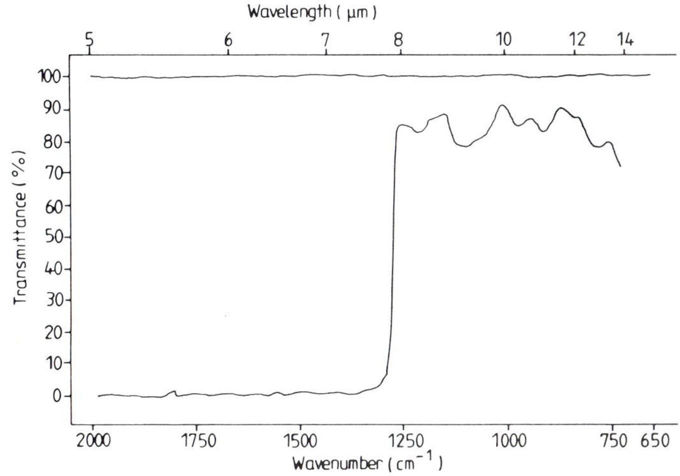

Figure 6.18 Measured transmittance of a practical longwave-pass filter with edge at 1250 cm^{-1} (8 μm). (Courtesy of OCLI Optical Coatings Ltd.)

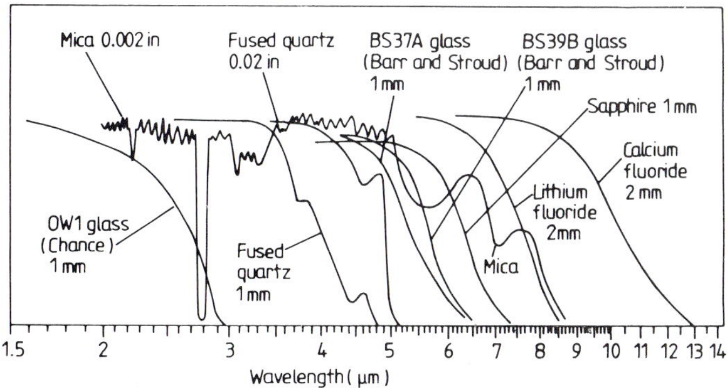

Figure 6.19 A selection of infrared materials which can be used as shortwave-pass absorption filters. (Courtesy of Sir Howard Grubb, Parsons & Co Ltd.)

tend to be extremely fragile and can only be produced in a circular shape of rather limited diameter, not usually greater than 2.0 cm.

The measured transmittance for a longwave-pass filter consisting of an edge filter together with an absorption filter is given in figure 6.20. This filter was originally designed to be used as a shortwave blocking filter with narrowband filters at 15 μm. It consists of two components, a multilayer filter made from a lead telluride and zinc sulphide multilayer on a germanium substrate and placed in series with an indium antimonide filter. The very high rejection achieved can be seen from the logarithmic plot.

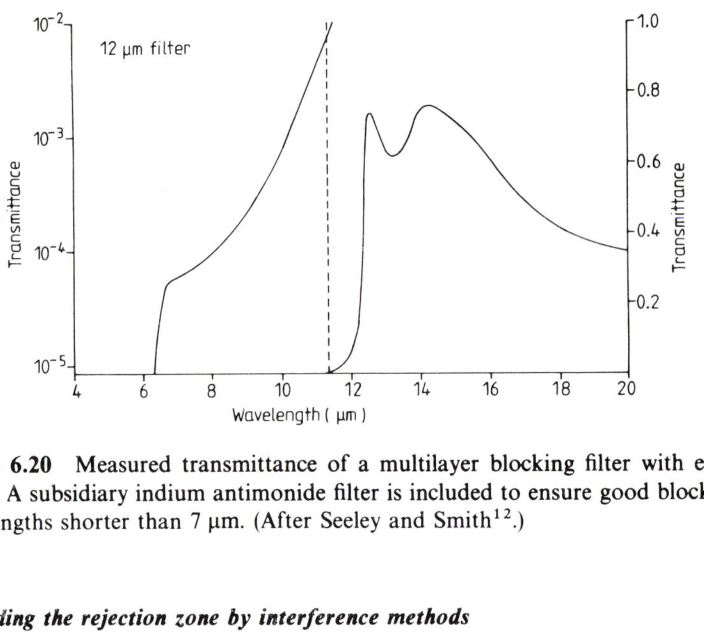

Figure 6.20 Measured transmittance of a multilayer blocking filter with edge at 12 μm. A subsidiary indium antimonide filter is included to ensure good blocking at wavelengths shorter than 7 μm. (After Seeley and Smith[12].)

Extending the rejection zone by interference methods

The most convenient and straightforward way of extending the reflectance zone is to place a second stack in series with the first and to ensure that their rejection zones overlap. The second stack is best placed either on a second substrate or on the opposite side of the substrate from the first stack. Provided that the substrate is reasonably thick or slightly wedged, the transmission of the assembly is then given by equation (2.95)

$$T = \frac{1}{(1/T_a) + (1/T_b) - 1} \qquad (6.38)$$

and a nomogram for calculating this is given in figure 2.21.

Occasionally it may happen that it is impossible to place the stacks on separate surfaces, and one stack must be deposited directly on top of the other. In this case it is necessary to take precautions to avoid the creation of transmission maxima. The problem has already been dealt with in chapter 5 where the extension of the high-reflectance zone of a quarter-wave stack was discussed (pp 175–9).

If we consider the assembly split into two separate multilayers, as shown in figure 5.12, then a transmission maximum will occur at any wavelength for which $(\phi_a + \phi_b)/2 = n\pi$, where $n = 0, \pm 1, \pm 2, \ldots$. The height of this maximum is given by

$$T = \frac{|\tau_a^+|^2 |\tau_b^+|^2}{(1 - |\rho_a^-||\rho_b^+|)^2} = \frac{T_a T_b}{[1 - (R_a R_b)^{1/2}]^2}.$$

If there is no absorption, this expression implies that, for low transmission at the maxima, R_a and R_b should be as dissimilar as possible. This can be achieved by using many layers to keep the reflectance of one multilayer as high as possible in the pass region of the other.

In slightly more quantitative terms, from the reflectance envelope, which does not vary with the number of periods, we can find the highest reflectance in the pass region of either multilayer making up the composite filter. If we denote this reflectance by R_p, then we can be certain that the design will be acceptable if we choose a sufficiently high number of periods to make R_s, the lowest reflectance in the stop band of the other multilayer, sufficiently high to ensure that

$$\frac{(1 - R_p)(1 - R_s)}{[1 - (R_p R_s)^{1/2}]^2} \leqslant T_c \tag{6.39}$$

where T_c is some acceptable level for the transmission in the rejection zone of the complete filter. This formula will give a pessimistic result; the actual transmission achieved in practice will depend on the phase change as well as the reflectance.

The only other danger area is the region where the two high-reflectance bands are overlapping. There, it must be arranged that on no account is $(\phi_a + \phi_b)/2 = n\pi$. The method for dealing with this was described in the previous chapter where a layer of intermediate thickness was placed between the two quarter-wave stacks. The result is equivalent to placing two similar multilayers, both of the form $[(L/2)H(L/2)]^n$ or $[(H/2)L(H/2)]^n$, together.

Equation (6.39) also implies that some of the sections of the composite filter should have more periods than others. In the reduction of the ripple in the pass band of the basic multilayer, the ripple on the other side of the stop band is invariably increased. Thus, in the combination of, say, two multilayers, the rejection zone of one stack will overlap a region of high ripple, while the rejection zone of the other stack will overlap a region of relatively low ripple. Since high ripple means that R_p is high, the former stack should have more periods than the latter if the same level of rejection is required throughout the combined rejection region. Figure 6.21 shows two component edge filters which are combined in a single filter in figure 6.22. The severe ripple which

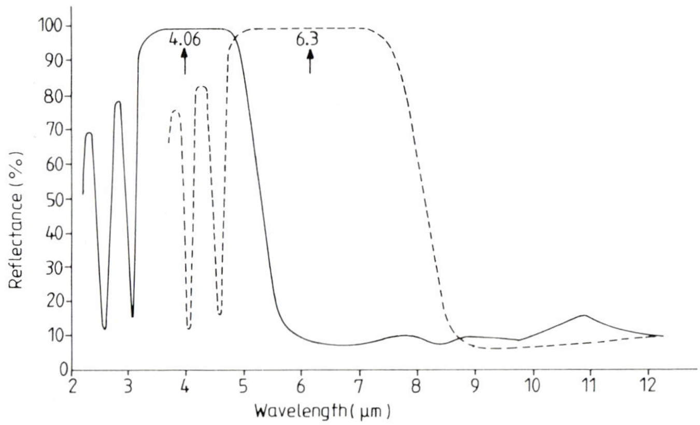

Figure 6.21 Measured reflectance of two longwave-pass stacks:

$$A|(0.5H\ L\ 0.5H)^4|BaF_2.$$

H and L are films of stibnite and chiolite a quarter-wave thick at $\lambda_0 = 4.06$ μm or 6.3 μm. A is air and the substrate is barium fluoride. (After Turner and Baumeister[13].)

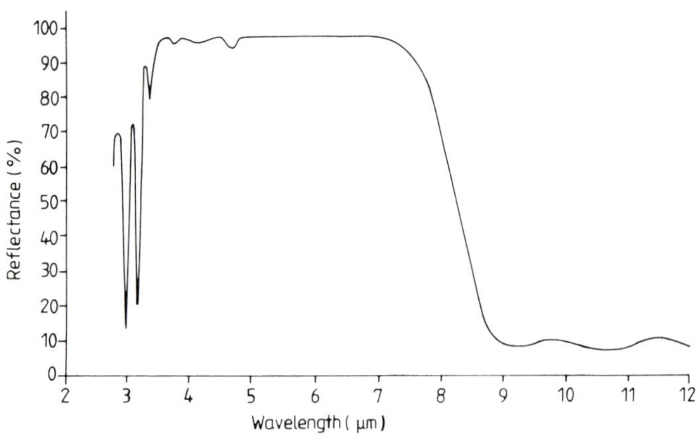

Figure 6.22 Measured reflectance of the two longwave-pass stacks of figure 6.21 superimposed in a single coating for an extended high-reflectance region. (After Turner and Baumeister[13].)

occurs in one of the multilayers can be seen reflected in the rejection zone of the composite filter. This ripple is limited to part of the rejection zone only, and in order to reduce the effect, more periods are necessary in the appropriate multilayer.

Extending the transmission zone

The shortwave-pass filter, as it has been described so far, possesses a limited pass band because of the higher order stop bands. These are not always particulary embarrassing, but occasionally, as for example with some types of heat reflecting filters, a much wider pass band is required. The problem was first considered by Epstein[14] and was studied more extensively by Thelen.[15]

Epstein's analysis was as follows: let the multilayer be represented by S periods each of the form

$$M = \begin{bmatrix} M_{11} & M_{12} \\ M_{21} & M_{22} \end{bmatrix}.$$

If a single period is considered as if it were immersed in a medium of admittance η, then the transmission coefficient of the period is given by

$$t = \frac{2\eta}{\eta\{(M_{11}+M_{22})+[\eta M_{12}+(M_{21}/\eta)]\}}.$$

Let $t = |t|e^{i\tau}$; then

$$\tfrac{1}{2}\{(M_{11}+M_{22})+[\eta M_{12}+(M_{21}/\eta)]\} = \frac{\cos\tau - i\sin\tau}{|t|}.$$

If the period is transparent, equating real parts gives

$$\tfrac{1}{2}(M_{11}+M_{22}) = \frac{\cos\tau}{|t|}.$$

Now, if light which has suffered two or more reflections at interfaces within the period is ignored, then

$$\tau \simeq \sum\delta$$

the total phase thickness of the period.

When $\Sigma\delta = n\pi$, $\cos\tau = \pm 1$, and, if $|t| < 1$, then

$$|\tfrac{1}{2}(M_{11}+M_{22})| > 1$$

and a high-reflectance zone results. If, however, $|t| = 1$, then

$$|\tfrac{1}{2}(M_{11}+M_{22})| = 1$$

and the high-reflectance zone is suppressed. In the simple form of stack,

$$[(L/2)H(L/2)]^S \quad \text{or} \quad [(H/2)L(H/2)]^S$$
$$|t| = 1 \quad \text{for} \quad \tau = 2r\pi \quad r = 1, 2, 3, 4, \ldots$$

and the even-order high-reflectance zones are therefore suppressed. As noted earlier, only a slight change in the relative thicknesses of the layers is enough to reduce t and turn the band into a high-reflectance zone.

Putting this result in another way, a zone of high reflectance potentially exists whenever the total optical thickness of an individual period of the multilayer is an integral number of half-waves, and the high-reflectance zone is prevented from appearing if, and only if, $|\tau| = 1$. This result has been used by Epstein in his paper to design a multilayer in which the fourth- and fifth-order reflectance bands were suppressed. Thelen has extended Epstein's analysis to deal with cases where any two and any three successive orders are suppressed and it is this method which we shall follow.

Following Epstein, Thelen[15] assumed a five-layer form, *ABCBA*, which involves three materials, for the basic period of the multilayer, and noted that if the period is thought of as immersed in a medium *M*, the combination *AB* becomes an antireflection coating for *C* in *M* at the wavelengths where suppression is required. In the construction of the final multilayer, the medium *M* can be considered first to exist between successive periods and then to suffer a progressive decrease in thickness until it just vanishes. The shrinking procedure leaves unchanged the suppression of the various orders which has been arranged. *M* can therefore be chosen quite arbitrarily during the design procedure to be discarded later. The antireflection coating *AB* is of a type studied originally by Muchmore[16] and Thelen adapted his results as follows:

The various parameters of the layers are denoted by the usual symbols with the appropriate suffixes *A*, *B*, *C* and *M*.

Let layers *A* and *B* be of equal optical thickness, i.e.

$$\delta_A = \delta_B \tag{6.40}$$

and let

$$\eta_A \eta_B = \eta_C \eta_M . \tag{6.41}$$

Then the wavelengths for which unity transmittance will be achieved will be given by

$$\tan^2 \delta'_A = \frac{\eta_A \eta_B - \eta_C^2}{\eta_B^2 - (\eta_A \eta_C^2 / \eta_B)} . \tag{6.42}$$

(This result can be derived from equations (3.3) and (3.4). If we replace, in these equations, suffixes 1, 2, m and 0 by *A*, *B*, *C* and *M* respectively, then the condition for $\delta_A = \delta_B$ is, from equation (3.4): $\eta_A \eta_B = \eta_C \eta_M$ and equation (6.42) then follows immediately from equation (3.3).)

Two solutions given by equation (6.42), δ'_A and $(\pi - \delta'_A)$, are possible. We can specify that δ'_A corresponds to λ_1 and $(\pi - \delta'_A)$ to λ_2 where λ_1 and λ_2 are the two wavelengths where suppression is to be obtained. Solving these two equations for δ'_A gives

$$\delta'_A = \frac{\pi}{1 + (\lambda_1 / \lambda_2)} \tag{6.43}$$

which can be entered in equation (6.42), whence

$$\tan^2 \frac{\pi}{1 + (\lambda_1 / \lambda_2)} = \frac{\eta_A \eta_B - \eta_C^2}{\eta_B^2 - (\eta_A \eta_C^2) / \eta_B} . \tag{6.44}$$

This determines the complete design of the coating. The optical thickness of the layer A can be found from equation (6.43) to be

$$\frac{\lambda_1\lambda_2}{2(\lambda_1+\lambda_2)}. \tag{6.45}$$

The only other quantity to be found is the optical thickness of layer C and we note first that the total optical thickness of the period is $\lambda_0/2$, where λ_0 is the wavelength of the first high-reflectance zone. The optical thicknesses of layers A and B have already been defined as equal, so that the optical thickness of layer C is

$$\frac{\lambda_0}{2}-\frac{2\lambda_1\lambda_2}{2(\lambda_1+\lambda_2)}. \tag{6.46}$$

The medium M, which was introduced as an artificial aid to calculation, disappears and does not figure at all in the results. Any two of the optical admittances η_A, η_B and η_C can be chosen at will. The third one is then found from equation (6.44).

Thelen, in his paper, gives a large number of examples of multilayers with various zones suppressed. Particularly useful is a multilayer with the second- and third-order zones suppressed. For this,

$$\lambda_1=\lambda_0/2 \qquad \lambda_2=\lambda_0/3$$

and all the layers are found to be of equal optical thickness $\lambda_0/10$. Two of the refractive indices of the layers are then chosen and equation (6.44) solved for the remaining one. For rapid calculation Thelen gives a nomogram connecting the three quantities. The transmittance of a multilayer with the second and third orders suppressed is given in figure 6.23.

Thelen also considered a multilayer in which the second, third and fourth orders were all suppressed and found the conditions to be:

Layer thicknesses:

$$A\colon \lambda_0/12$$
$$B\colon \lambda_0/12$$
$$C\colon \lambda_0/6.$$

The indices are given by

$$\eta_B=(\eta_A\eta_C)^{1/2}.$$

Figure 6.24 shows the transmittance of a multilayer where the second, third and fourth orders have been suppressed in this way.

A heat reflecting filter using a combination of stacks in which the second and third, and second, third and fourth orders have been suppressed together with the normal quarter-wave stacks, has been designed. The calculated transmittance spectrum is shown in figure 6.25. The production of such a coating would indeed be a formidable task.

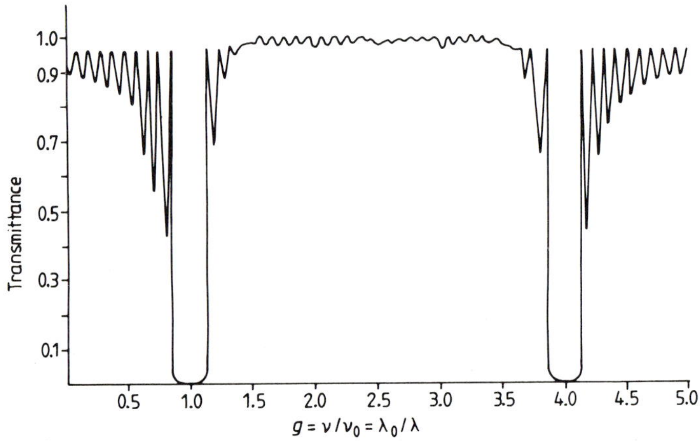

Figure 6.23 Calculated transmittance as a function of g of the design:

$$M\,|\,(ABCBA)^{10}A\,|\,S$$

with $n_S = 1.50$, $n_M = 1.00$, $n_A = 1.38$, $n_B = 1.90$ and $n_C = 2.30$. (After Thelen[15]).

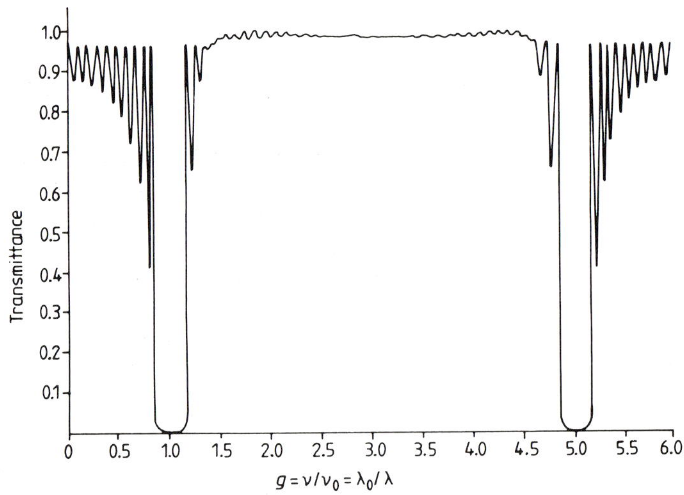

Figure 6.24 Calculated transmittance as a function of g of the design:

$$M\,|(AB2CBA)^{10}A|\,S$$

with $n_S = 1.50$, $n_M = 1.00$, $n_A = 1.38$, $n_B = 1.781$ and $n_C = 2.30$. (After Thelen[15].)

Reducing the transmission zone

The simple quarter-wave multilayer has the even-order high-reflectance bands missing. Sometimes it is useful to have these high-reflectance bands present. The method of the previous section can also be applied to this problem and the

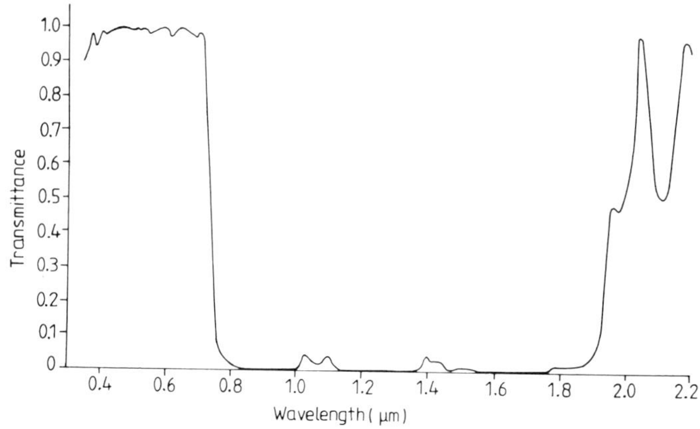

Figure 6.25 Calculated transmittance of a triple-stack heat reflector. Design:

$$M\,|\,[1.1(\tfrac{1}{2}AC\tfrac{1}{2}A)]\,(\tfrac{1}{2}AC\tfrac{1}{2}A)^5[1.25(\tfrac{1}{2}AC\tfrac{1}{2}A)]$$

$$-\,[0.57(ADCDA)]^8\,[0.642(AB2CBA)]^8\tfrac{1}{2}A\,|\,S$$

with $\lambda_0 = 860$ nm, $n_S = 1.50$, $n_M = 1.00$, $n_A = 1.38$, $n_B = 1.781$, $n_C = 2.30$ and $n_D = 1.90$. (After Thelen[15].)

enahancement of the reflectance at the even orders is a relatively simple business.

Because it makes the analysis simpler, we assume that the basic period is of the form AB rather than $(A/2)B(A/2)$. Once the basic result is established, it can easily be converted to the form $(A/2)B(A/2)$ if required. The reason that the even-order peaks are suppressed in the ordinary quarter-wave stack is that each of the layers is an integral number of half-waves thick and so $|t| = 1$ for the basic period. All that is required for a reflectance peak to appear is the destruction of this condition. To achieve this, the thickness of one of the layers must be increased and the other decreased, keeping the overall optical thickness constant. The greater the departure from the half-wave condition, the more pronounced the reflectance peak.

Consider the case where reflectance bands are required at λ_0, $\lambda_0/2$, and $\lambda_0/3$, but not necessarily at $\lambda_0/4$. This will be satisfied by making $n_A d_A = n_B d_B/3$ and $n_A d_A = \lambda_0/8$ so that the basic stack becomes either

$$\frac{H}{2}\,\frac{3L}{2}\,\frac{H}{2}\,\frac{3L}{2}\cdots\frac{3L}{2}$$

or

$$\frac{L}{2}\,\frac{3H}{2}\,\frac{L}{2}\,\frac{3H}{2}\cdots\frac{3H}{2}.$$

The reflectance peak at $\lambda_0/4$ will be suppressed because the layers at that wavelength have integral half-wave thicknesses.

The method can be used to produce any number of high-reflectance zones. However, it should be noted that the further the thicknesses depart from ideal quarter-waves at λ_0, the narrower will be the first-order reflectance band.

Edge steepness

In long- and shortwave-pass filters, the steepness of edge is not usually a parameter of critical importance. The number of layers necessary to produce the required rejection in the stop band of the filter will generally produce an edge steepness which is quite acceptable.

If, however, an exceptional degree of edge steepness is required, then the easiest way of improving it is to use still more layers. Increasing the number of layers will cause an apparent increase in the ripple in the pass band, because the first minimum in the pass band will be brought nearer to the edge, and usually will be on a part of the reflectance envelope which is increasing in width towards the edge. If the increase in number of layers is considerable, then it will probably be advisable to use one of the more advanced techniques for reducing ripple.

An alternative method for increasing the steepness of edge without major alterations to the basic design concept is the use of higher-order stacks. The steepness of edge for a given number of layers will increase in proportion with the order. There are two snags here. The first is that the rejection zone width varies inversely with the order number. This can be dealt with by adding a further first-order stack to extend the rejection zone. The second snag is more serious. The permissible errors in layer thickness are also reduced in inverse proportion with the order number. This is because the performance does not depend directly on the phase thickness of the layers but rather on the sines and cosines of the layer thicknesses, and in the case of the fifth order, for example, these are layer thicknesses greater than 2π. Thus, while for a first-order edge filter, as we shall see in chapter 9, the random errors in layer thickness which can be tolerated are of the order of 5 % or even 10 %, those which are tolerable in the fifth order are of the order of 1 % or possibly 2 %. A possible further practical difficulty with higher-order filters is that considerably more material is required for each layer.

REFERENCES

1 Epstein L I 1952 The design of optical filters *J. Opt. Soc. Am.* **42** 806–10

2 Vera J J 1964 Some properties of multilayer films with periodic structure *Opt. Acta* **11** 315–31

3 Thelen A 1966 Equivalent layers in multilayer filters *J. Opt. Soc. Am.* **56** 1533–8

4 Ufford C and Baumeister P W 1974 Graphical aids in the use of equivalent index in multilayer-filter design *J. Opt. Soc. Am.* **64** 329–34

5 Welford W T (writing as W Weinstein) 1954 Computations in thin film optics *Vacuum* **4** 3–19

6 Baumeister P W 1958 Design of multilayer filters by successive approximations *J. Opt. Soc. Am.* **48** 955–8

7 Liddell H M 1981 Computer-aided techniques for the design of multilayer filters (Bristol: Adam Hilger)

8 Dobrowolski J A and Lowe D 1978 Optical thin film synthesis program based on the use of Fourier transforms *Appl. Opt.* **17** 3039–50

9 Jacobsson R 1964 Matching a multilayer stack to a high-refractive-index substrate by means of an inhomogeneous layer *J. Opt. Soc. Am.* **54** 422–3

10 Young L and Cristal E G 1966 On a dielectric fiber by Baumeister *Appl. Opt.* **5** 77–80

11 Seeley J S, Liddell H M and Chen T C 1973 Extraction of Tschebysheff design data for the lowpass dielectric multilayer *Opt. Acta.* **20** 641–61

12 Seeley J S and Smith S D 1966 High performance blocking filters for the region 1 to 20 microns *Appl. Opt.* **5** 81–5

13 Turner A F and Baumeister P W 1966 Multilayer mirrors with high reflectance over an extended spectral region *Appl. Opt.* **5** 69–76

14 Epstein L I 1955 Improvements in heat reflecting filters *J. Opt. Soc. Am.* **45** 1360–2

15 Thelen A 1963 Multilayer filters with wide transmittance bands *J. Opt. Soc. Am.* **53** 1266–70

16 Muchmore R B 1948 Optimum band width for two layer anti-reflection films *J. Opt. Soc. Am.* **38** 20–26

7 Band-pass filters

A filter which possesses a region of transmission bounded on either side by regions of rejection is known as a band-pass filter. For the broadest band-pass filters, the most suitable construction is a combination of longwave-pass and shortwave-pass filters, which we discussed in chapter 6. For narrower filters, however, this method is not very successful because of difficulties associated with obtaining both the required precision in positioning and the steepness of edges. Other methods are therefore used, involving a single assembly of thin films to produce simultaneously the pass and rejection bands. The simplest of these is the thin-film Fabry–Perot filter, a development of the interferometer already described in chapter 5. The thin-film Fabry–Perot filter has a pass band shape which is triangular and it has been found possible to improve this by coupling simple filters in series in much the same way as tuned circuits. These coupled arrangements are known as multiple cavity filters or multiple half-wave filters. If two simple Fabry–Perot filters are combined, the resultant becomes a double cavity or double half-wave filter, abbreviated to DHW filter, while, if three Fabry–Perot filters are involved, we have a triple cavity filter, abbreviated normally to THW for triple half-wave. In the earlier part of this chapter, we consider single cavity filters. First of all, we examine combinations of edge filters.

BROADBAND-PASS FILTERS

Band-pass filters can be very roughly divided into broadband-pass filters and narrowband-pass filters. There is no definite boundary between the two types and the description of one particular filter usually depends on the application and the filters with which it is being compared. For the purposes of the present work, by broadband filters we mean filters with bandwidths of perhaps 20 % or more which are made by combining longwave-pass and shortwave-pass filters. The best arrangement is probably to deposit the two components on opposite sides of a single substrate. To give maximum possible transmission, each edge filter should be designed to match the substrate into the surrounding medium, a procedure already examined in chapter 6. Such a filter is shown in figure 7.1.

It is also possible, however, to deposit both components on the same side of the substrate. This was a problem which Epstein[1] examined in his early paper on symmetrical periods. The main difficulty is the combining of the two stacks so that the transmission in the pass band is a maximum and also so that one stack does not produce transmission peaks in the rejection zone of the other. The transmission in the pass band will depend on the matching of the first stack to the substrate, the matching of the second stack to the first, and the matching of the second stack to the surrounding medium. Depending on the equivalent admittances of the various stacks it may be necessary to insert quarter-wave matching layers or to adopt any of the more involved matching techniques.

In the visible region, with materials such as zinc sulphide and cryolite, the combination $[(H/2)L(H/2)]^S$ acts as a good longwave-pass filter with an equivalent admittance at normal incidence and at wavelengths in the pass region not too far removed from the edge of near unity. This can therefore be used next to the air without mismatch. The combination $[(L/2)H(L/2)]^S$ acts as a shortwave-pass filter, with equivalent admittance only a little lower than the first section, and can be placed next to it, between it and the substrate, without any matching layers. The mismatch between this second section and the substrate, which in the visible region will be glass of index 1.52, is sufficiently large to require a matching layer. Happily, the $[(H/2)L(H/2)]$ combination with a total phase thickness of 270°, i.e. effectively three quarter-waves, has an admittance exactly correct for this. The transmission of the final design is shown in figure 7.2(*b*) with the appropriate admittances of the two sections in figure 7.2(*a*). Curve A refers to a $[(L/2)H(L/2)]^4$ shortwave-pass section and B to a $[(H/2)L(H/2)]^4$ longwave-pass. The complete design is shown in table 7.1. The edges of the two sections have been chosen quite arbitrarily and could be moved as required.

To avoid the appearance of transmission peaks in the rejection zones of either component, it is safest to deposit them so that high-reflectance zones do not overlap. The complete rejection band of the shortwave-pass section will always lie over a pass region of the longwave-pass filter, but the higher-order bands should be positioned, if at all possible, clear of the rejection zone of the longwave-pass section. The combination of edge filters of the same type has already been investigated in chapter 6 and the principles discussed there apply to this present situation. It should also be remembered that, although in the normal shortwave-pass filter the second-order reflection peak is missing, a small peak can appear if any thickness errors are present. This can, if superimposed on a rejection zone of the other section, cause the appearance of a transmission peak if the errors are sufficiently pronounced. The expression

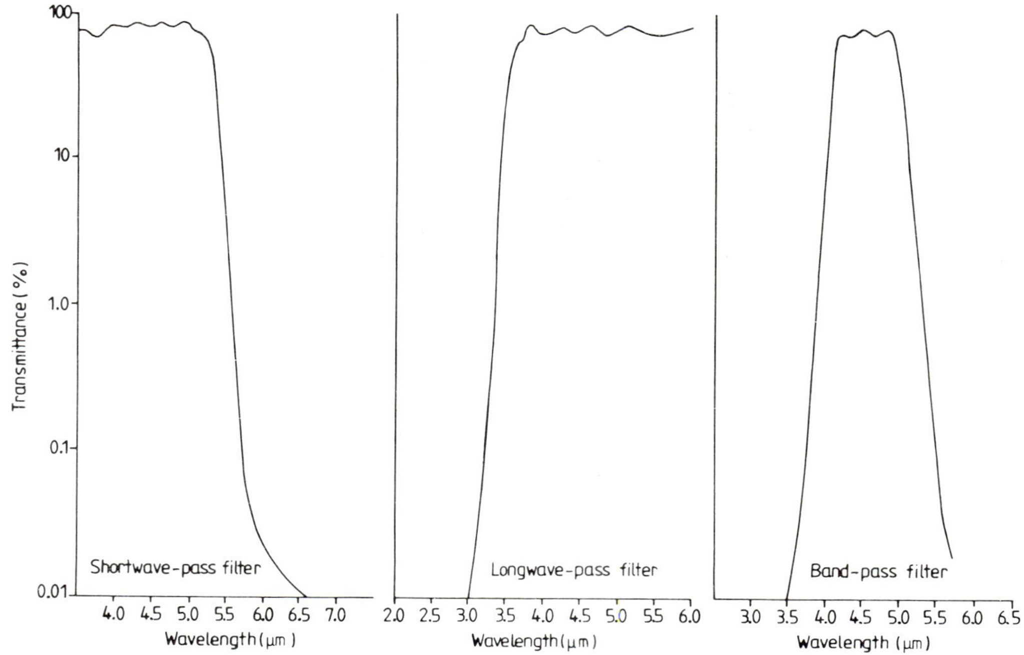

Figure 7.1 The construction of a band-pass filter by placing two separate edge filters in series. (Courtesy of Standard Telephones and Cables Ltd.)

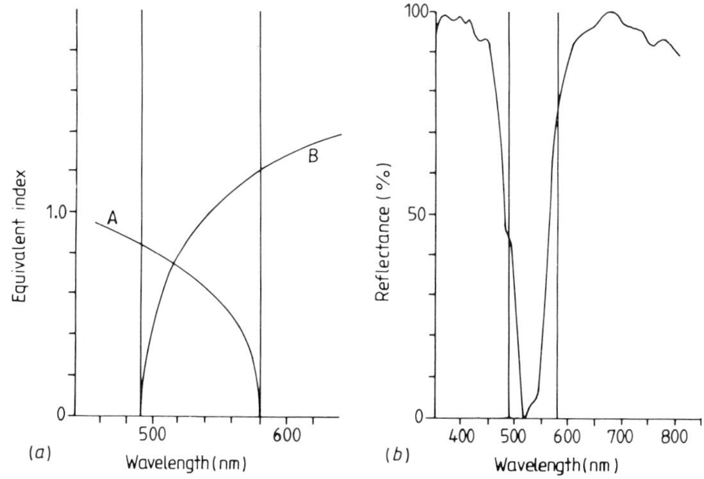

Figure 7.2 (*a*) Equivalent admittances of two stacks made up of symmetrical periods used to form a band-pass filter. A: (0.5*L H* 0.5*L*); B: (0.5*H L* 0.5*H*), where $n_L = 1.38$; $n_H = 2.30$. (*b*) Calculated reflectance curve for a band-pass filter. For the complete design of this filter, made up of two superimposed stacks, one of type A and one of type B, refer to table 7.1. (After Epstein[1].)

Table 7.1†

Index	Phase thickness of each layer measured at 546 nm (degrees)	Index	Phase thickness of each layer measured at 546 nm (degrees)
1.52	Massive	1.38	55.4
1.38	67.3	2.30	33.9
2.30	134.5	1.38	67.9
1.38	122.7	2.30	67.9
2.30	110.8	1.38	67.9
1.38	110.8	2.30	67.9
2.30	110.8	1.38	67.9
1.38	110.8	2.30	67.9
2.38	110.8	1.38	67.9
1.38	110.8	2.30	33.9
2.30	110.8	1.00	Massive

† From Epstein[1].

for maximum transmission is

$$T_{\max} = \frac{T_a T_b}{[1 - (R_a R_b)^{1/2}]^2}$$

but this only holds if the phase conditions are met.

NARROWBAND FILTERS

When we speak of narrowband filters we are referring to filters with bandwidths of perhaps 15 % or less. The main difference between these and the broadband filters described above is that here we are relying on a single assembly to give the pass band and both the adjoining stop bands.

The metal–dielectric Fabry–Perot filter

The simplest type of narrowband thin-film filter is based on the Fabry–Perot interferometer discussed in chapter 5. In its original form, the Fabry–Perot interferometer consists of two identical parallel reflecting surfaces spaced apart a distance *d*. In collimated light, the transmission is low for all wavelengths except for a series of very narrow transmission bands spaced at intervals that are constant in terms of wavenumber. This device can be replaced by a complete thin-film assembly consisting of a dielectric layer bounded by two metallic reflecting layers (figure 7.3). The dielectric layer takes the place of the spacer and is known as the spacer layer. Except that the spacer layer now has an index greater than unity, the analysis of the performance of this thin-film filter is exactly the same as for the conventional etalon, but in other respects there are a few significant differences.

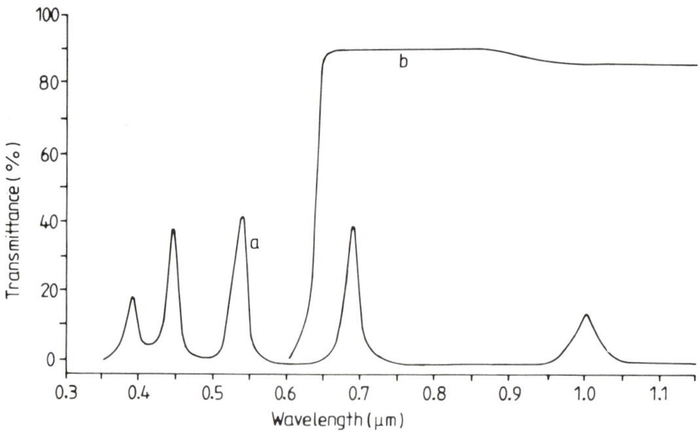

Figure 7.3 Characteristics of a metal–dielectric filter for the visible region (curve a). Curve b is the transmittance of an absorption glass filter that can be used for the suppression of the short wavelength sidebands. (Courtesy of Barr & Stroud Ltd.)

While the surfaces of the substrates should have a high degree of polish, they need not be worked to the exacting tolerances necessary for etalon plates. Provided the vapour stream in the plant is uniform, the films will follow the contours of the substrate without exhibiting thickness variations. This implies that it is possible for the thin-film Fabry–Perot filter to be used in a much lower order than the conventional etalon. Indeed, it turns out in practice that lower orders must be used, because the thin-film spacer layers begin, when thicker than the fourth order or so, to exhibit roughness. This roughness broadens the pass band and reduces the peak transmittance so much that any advantage of the higher order is completely lost. This simple type of filter is known as a metal–dielectric Fabry–Perot to distinguish it from the all-dielectric one to be described later.

It is worthwhile briefly analysing the performance of the Fabry–Perot once again, this time including the effects of phase shift at the reflectors. The starting point for this analysis is equation (2.81):

$$\left.\begin{aligned} T_F &= \frac{T_a T_b}{[1-(R_a R_b)^{1/2}]^2}\,\frac{1}{1+F\sin^2[\frac{1}{2}(\phi_a+\phi_b)-\delta]} \\ F &= \frac{4(R_a R_b)^{1/2}}{[1-(R_a R_b)^{1/2}]^2} \qquad \delta = \frac{2\pi n d\cos\theta}{\lambda} \end{aligned}\right\} \tag{7.1}$$

where the notation is given in figure 2.12. We have adapted equation (2.81) slightly by removing the + and − signs on the reflectances. The analysis which follows is similar to that already performed in chapter 5 except that here we are including the effects of ϕ_a and ϕ_b. The maxima of transmission are given by

$$\frac{2\pi n d\cos\theta}{\lambda} - \frac{\phi_a+\phi_b}{2} = m\pi \qquad m = 0, \pm 1, \pm 2, \pm 3, \ldots \tag{7.2}$$

where we have chosen $-m$ rather than $+m$ because $(\phi_a+\phi_b)/2 < \pi$ by definition. The analysis is marginally simpler if we work in terms of wavenumber instead of wavelength. The positions of the peaks are then given by

$$\frac{1}{\lambda} = \nu = \frac{m\pi + (\phi_a+\phi_b)/2}{2\pi n d\cos\theta} = \frac{1}{2nd\cos\theta}\left(m + \frac{\phi_a+\phi_b}{2\pi}\right). \tag{7.3}$$

Depending on the particular metal, the thickness, and the index of the substrate, and the index of the spacer, the phase shift on reflection ϕ will be either in the first or second quadrant. $(\phi_a+\phi_b)/(2\pi)$ will therefore be positive, between 0 and 1 and roughly in the region of 0.5. The peak wavelength of the filter will therefore be shifted to the shortwave side of the peak which would be expected simply from the optical thickness of the spacer layer.

The resolving power of the thin-film Fabry–Perot filter may be defined in exactly the same way as for the interferometer. As we saw in chapter 5, a convenient definition is

$$\frac{\text{Peak wavelength}}{\text{Halfwidth of pass band}}$$

where the halfwidth is the width of the band measured at half the peak transmission. Now let the pass bands be sufficiently narrow, which is the same as F being sufficiently large, so that near a peak we can replace

$$\frac{\phi_a+\phi_b}{2}-\delta \quad \text{by} \quad -m\pi-\Delta\delta$$

and

$$\sin^2\left(\frac{\phi_a+\phi_b}{2}-\delta\right) \quad \text{by} \quad (\Delta\delta)^2.$$

We are assuming here that ϕ_a and ϕ_b are constant or vary very much more slowly than δ over the pass band.

The half-peak bandwidth, or halfwidth, can be found by noting that at the half-peak transmission points

$$F\sin^2\left(\frac{\phi_a+\phi_b}{2}-\delta\right)=1.$$

Using the approximation given above, this becomes

$$(\Delta\delta_h)^2=\frac{1}{F}$$

i.e. the halfwidth of the pass band

$$2\Delta\delta_h=2/F^{1/2}.$$

The finesse is defined as the ratio of the interval between fringes to the fringe halfwidth, and is written $\mathscr{F}$. The change in δ in moving from one fringe to the next is just π, and the finesse, therefore, is

$$\mathscr{F}=\frac{\pi F^{1/2}}{2}. \tag{7.4}$$

Now $\nu_0/\Delta\nu_h=\delta_0/2\Delta\delta_h$ because $\nu\propto\delta$, where ν_0 and δ_0 are respectively the values of the wavenumber and spacer layer phase thickness associated with the transmission peak, and $\Delta\nu_h$ and $2\Delta\delta_h$ are the corresponding values of halfwidth. The ratio of the peak wavenumber to the halfwidth is then given by

$$\frac{\nu_0}{\Delta\nu_h}=\mathscr{F}\left(m+\frac{\phi_a+\phi_b}{2\pi}\right) \tag{7.5}$$

for a peak of order m, since

$$\delta_0=m\pi+\frac{\phi_a+\phi_b}{2}.$$

The ratio of peak position to halfwidth expressed in terms of wavenumber is exactly the same in terms of wavelength,

$$\frac{\nu_0}{\Delta\nu_h}=\frac{\lambda_0}{\Delta\lambda_h}$$

where λ_0 is given by

$$\lambda_0 = \frac{2nd\cos\theta}{m + (\phi_a + \phi_b)/2\pi} \tag{7.6}$$

and this was discussed in chapter 5. The halfwidth is thus a most useful parameter with which to specify a narrowband Fabry–Perot filter since it can be converted very quickly into a measure of resolution. It has come to be used rather than resolving power for all types of narrowband filter, regardless of whether or not they are Fabry–Perot type. Usually, therefore, $\Delta\lambda_h/\lambda_0$, often expressed as a percentage, is the parameter which is quoted by the manufacturers and users alike. Other measures of bandwidth sometimes quoted along with the halfwidth are the widths measured at 0.9 × peak transmission, at 0.1 × peak transmission, and at 0.01 × peak transmission. For a Fabry–Perot filter, provided the phase shifts on reflection from the reflecting layers are effectively constant over the pass band, these widths are given respectively by one-third of the halfwidth, three times the halfwidth, and ten times the halfwidth. The other measures of bandwidth are used to give some indication of the extent to which, in any given type of filter, the sides of the pass band, compared with those of the Fabry–Perot, can be considered rectangular.

The manufacture of the metal–dielectric filter is straightforward. The main point to watch is that the metallic layers should be evaporated as quickly as possible on to a cold substrate. In the visible and near infrared regions the best results are probably achieved with silver and cryolite, while in the ultraviolet the best combination is aluminium and either magnesium fluoride or cryolite. Wherever possible the layers should be protected by cementing a cover slip over them as soon as possible after deposition. This also serves to balance the assembly by equalising the refractive indices of the media outside the metal layers.

Turner[2] quoted some results for metal–dielectric filters constructed for the visible region which may be taken as typical of the performance to be expected. The filters were constructed from silver reflectors and magnesium fluoride spacers. For a first-order spacer a bandwidth of 13 nm with a peak transmission of 30 % was obtained at a peak wavelength of 531 nm. A similar filter with a second-order spacer gave a bandwidth of 7 nm with peak transmission of 26 % at 535 nm. With metal–dielectric filters the third order is usually the highest used. Because of scattering in the spacer layer, which becomes increasingly apparent in the fourth and higher orders, any benefit which would otherwise arise from using these orders is largely lost.

A typical curve for a metal–dielectric filter for the visible region is shown in figure 7.3. The particular peak to be used is that at 0.69 μm, which is of the third order. The shortwave sidebands due to the higher-order peaks can be suppressed quite easily by the addition of an absorption glass filter, which can be cemented over the metal–dielectric element to act as a cover glass. Such a filter is also shown in the figure and is one of a wide range of absorption glasses which are available for the visible and near infrared and which have longwave-pass characteristics. There are, unfortunately, few absorption filters suitable for the suppression of the longwave sidebands. If the detector which is to be used is not sensitive to these longer wavelengths, then no problem exists and commercial metal–dielectric filters for the visible and near infrared usually possess long-wavelength sidebands beyond the limit of the photocathodes or photographic emulsions, which are the usual detectors for this region. If the longwave-sideband suppression must be included as part of the filter assembly, then there is an advantage in using metal–dielectric filters in the first order, even though the peak transmission for a given bandwidth is much lower, since they do not usually possess long-wavelength sidebands. Theoretically, there will always be a peak corresponding to the zero order at very long wavelengths, but this will not usually appear, partly because the substrate will cut off long before the zero order is reached, and also because the properties of the thin-film materials themselves will change radically. We shall discuss later a special type of metal–dielectric filter, the induced transmission filter, which can be made to have a much higher peak transmission, though with a rather broader halfwidth, without introducing long-wavelength sidebands, and which is often used as a long-wavelength suppression filter.

Silver does not have an acceptable performance for ultraviolet filters and aluminium has been found to be the most suitable metal with magnesium fluoride as the preferred dielectric. In the ultraviolet beyond 300 nm there are few suitable cements (none at all beyond 200 nm) and it is not possible to use cover slips which are cemented over the layers in the way in which filters for the visible region are protected. The normal technique, therefore, is to attempt to protect the filter by the addition of an extra dielectric layer between the final metal layer and the atmosphere. These layers are effective in that they slow down the oxidation of the aluminium which otherwise takes place rapidly and causes a reduction in performance even at quite low pressures. This oxidation has already been referred to in chapter 4. They cannot completely stabilise the filters, however, and slight longwave drifts can occur, as reported by Bates and Bradley[3]. A second function of the final dielectric layer is to act as a reflection-reducing layer at the outermost metal surface and hence to increase the transmittance of the filter. This is not a major effect—the problem of improving metal–dielectric filter performance is dealt with later in this chapter—but any technique which helps to improve performance, even marginally, in the ultraviolet, is very welcome. Some performance curves of first-order metal–dielectric Fabry–Perot filters are shown in figure 7.4.

The formula for transmission of the Fabry–Perot filter can also be used to determine both the peak transmission in the presence of absorption in the reflectors and the tolerance which can be allowed in matching the two reflectors. First of all, let the reflectances be equal and let the absorption be denoted by A, so that

$$R+T+A=1. \tag{7.8}$$

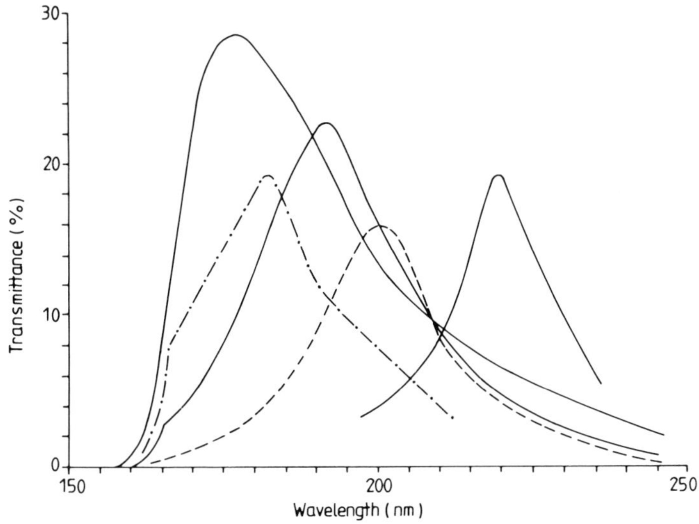

Figure 7.4 Experimental transmittance curves of first-order metal–dielectric filters for the far ultraviolet deposited on Spectrosil B substrates. (After Bates and Bradley[3].)

The peak transmission will then be given by

$$(T_F)_{\text{peak}} = \frac{T^2}{(1-R)^2}$$

and, using equation (7.8),

$$(T_F)_{\text{peak}} = \frac{1}{(1+A/T)^2} \tag{7.9}$$

exactly as for the Fabry–Perot interferometer, which shows that when absorption is present the value of peak transmission is determined by the ratio A/T.

To estimate the accuracy of matching which is required for the two reflectors we assume that the absorption is zero. The peak transmission is given by the expression

$$(T_F)_{\text{peak}} = \frac{T_a T_b}{[1-(R_a R_b)^{1/2}]^2} \tag{7.10}$$

where the subscripts a and b refer to the two reflectors. Let

$$R_b = R_a - \Delta_a \tag{7.11}$$

where Δ_a is the error in matching, so that $T_b = T_a + \Delta_a$. Then we can write

$$(T_F)_{\text{peak}} = \frac{T_a(T_a+\Delta_a)}{\{1-[R_a(R_a-\Delta_a)]^{1/2}\}^2}$$

$$= \frac{T_a(T_a+\Delta_a)}{\{1-R_a[1-\frac{1}{2}(\Delta_a/R_a)+\ \dots\ ]\}^2}. \tag{7.12}$$

Now assume that Δ_a is sufficiently small compared with R_a so that we can take only the first two terms of the expansion in equation (7.12). With some rearrangement the equation becomes

$$(T_F)_{\text{peak}} = \frac{T_a^2}{(1-R_a)^2} \frac{1+(\Delta_a/T_a)}{[1+\frac{1}{2}(\Delta_a/T_a)]^2}. \tag{7.13}$$

The first part of the equation is the expression for peak transmission in the absence of any error in the reflectors, while the second part shows how the peak transmission is affected by errors. The second part of the expression is plotted in figure 7.5 where the abscissa is $T_b/T_a = 1 + \Delta_a/T_a$. Clearly, the Fabry–Perot filter is surprisingly insensitive to errors. Even with reflector transmittance unbalanced by a factor of 3, it is still possible to achieve 75% peak transmission.

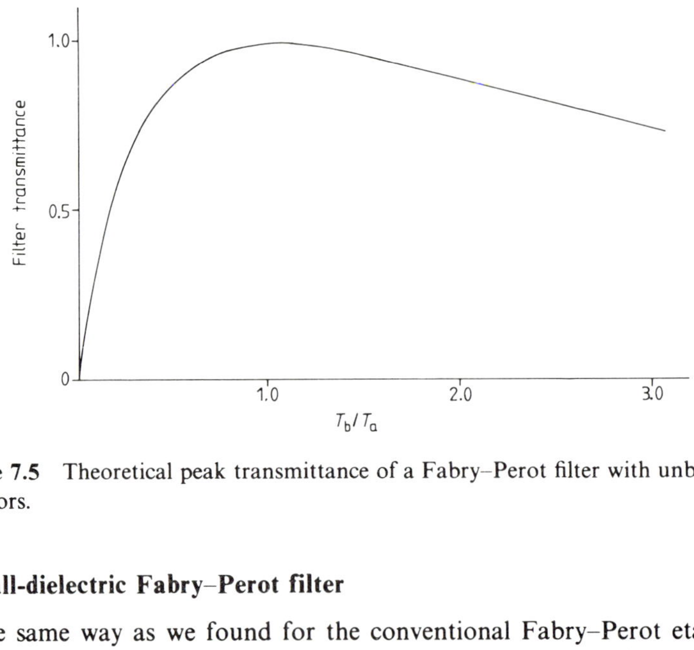

Figure 7.5 Theoretical peak transmittance of a Fabry–Perot filter with unbalanced reflectors.

The all-dielectric Fabry–Perot filter

In the same way as we found for the conventional Fabry–Perot etalon, if improved performance is to be obtained, then the metallic reflecting layers should be replaced by all-dielectric multilayers.

An all-dielectric filter is shown in diagrammatic form in figure 7.6. Basically, this is the same as the conventional etalon with dielectric coatings and with a solid thin-film spacer, and the observations made for the metal–dielectric filter are also valid. Again, the substrate need not be worked to a high degree of flatness although the polish must be good, because, provided the plant geometry is adequate, the films will follow any contours without showing changes in thickness.

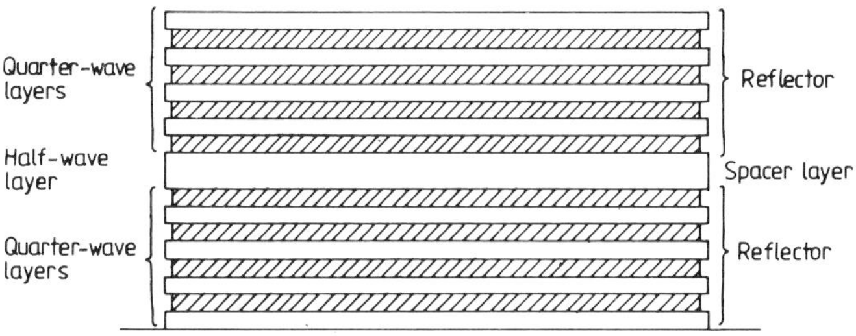

Figure 7.6 The structure of an all-dielectric Fabry–Perot filter.

The bandwidth of the all-dielectric filter can be calculated as follows. If the reflectance of each of the multilayers is sufficiently high, then

$$F = \frac{4R}{(1-R)^2} \simeq \frac{4}{T^2}$$

and

$$\frac{\lambda_0}{\Delta\lambda_h} = m\mathscr{F} = \frac{m\pi F^{1/2}}{2} \simeq \frac{m\pi}{T}. \tag{7.14}$$

Since the maximum reflectance for a given number of layers will be obtained with a high-index layer outermost, there are really only two cases which need be considered and these are shown in figure 7.7. If x is the number of high-index layers in each stack, not counting the spacer layer, then in the case of the high-index spacer, the transmission of the stack will be given by

$$T = \frac{4n_L^{2x} \cdot n_s}{n_H^{2x+1}}$$

and in the case of the low-index spacer by

$$T = \frac{4n_L^{2x-1} n_s}{n_H^{2x}}.$$

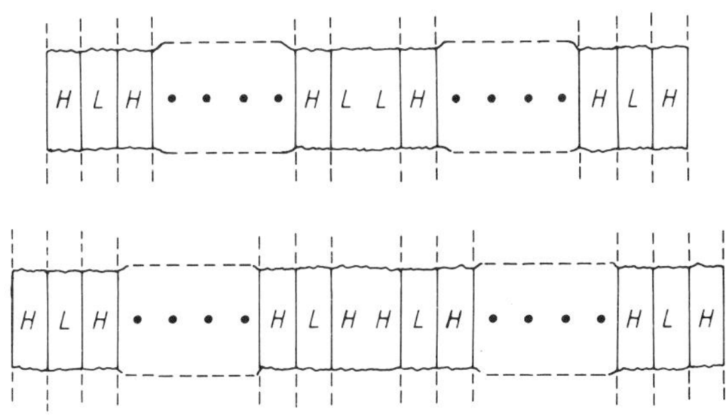

Figure 7.7 The structure of the two basic types of all-dielectric Fabry–Perot filter.

Substituting these results into the expression for bandwidth we find, for the high-index spacer,

$$\frac{\Delta\lambda_h}{\lambda_0} = \frac{4n_L^{2x}n_s}{m\pi n_H^{2x+1}} \tag{7.15}$$

and, for the low-index spacer,

$$\frac{\Delta\lambda_h}{\lambda_0} = \frac{4n_L^{2x-1}n_s}{m\pi n_H^{2x}} \tag{7.16}$$

where we are adopting the fractional halfwidth $\Delta\lambda_h/\lambda_0$ rather than the resolving power $\lambda_0/\Delta\lambda_h$ as the important parameter. This is customary practice.

In these formulae we have completely neglected any effect due to the dispersion of phase change on reflection from a multilayer. As we have already noted in chapter 5, the phase change is not constant. The sense of the variation is such that it increases the rate of variation of $[(\phi_a+\phi_b)/2]-\delta$ with wavelength in the formula for transmission of the Fabry–Perot filter and, hence, reduces the bandwidth and increases the resolving power in equations (7.15) and (7.16). Seeley[4] has studied the all-dielectric filter in detail and by making some approximations in the basic expressions for the filter transmittance, has arrived at formulae for the first-order halfwidths, which, with a little adjustment, become equal to the expressions in (7.15) and (7.16) multiplied by a factor $(n_H-n_L)/n_H$. We can readily extend Seeley's analysis to all-dielectric filters of order m.

We recall that the half-peak points are given by

$$F\sin^2[(2\pi D/\lambda)-\phi] = 1 \tag{7.17}$$

where, since the filter is quite symmetrical, we have replaced $(\phi_1+\phi_2)/2$ by ϕ. It is simpler to carry out the analysis in terms of $g = \lambda_0/\lambda = \nu/\nu_0$. At the peak of the filter we have $g = 1.0$. We can assume for small changes Δg in g that

$$2\pi D/\lambda = m\pi(1+\Delta g)$$

and

$$\phi = \phi_0 + \frac{d\phi}{dg}\Delta g$$

so that equation (7.17) becomes

$$F\sin^2\left(m\pi(1+\Delta g)-\phi_0-\frac{d\phi}{dg}\Delta g\right) = 1.$$

ϕ_0, we know, is 0 or π, and so, using the same approximation as before,

$$F\left(m\pi\Delta g-\frac{d\phi}{dg}\Delta g\right)^2 = 1$$

or

$$\Delta g = F^{-1/2}\left(m\pi-\frac{d\phi}{dg}\right)^{-1}.$$

The halfwidth is $2\Delta g$ so that

$$2\Delta g = \frac{\Delta \nu_h}{\nu_0} = \frac{\Delta \lambda_h}{\lambda_0} = 2F^{-1/2}\left(m\pi - \frac{d\phi}{dg}\right)^{-1}$$

$$= \frac{2}{m\pi F^{1/2}}\left(1 - \frac{1}{m\pi}\frac{d\phi}{dg}\right)^{-1}. \tag{7.18}$$

We now need the quantity $d\phi/dg$. We use Seeley's technique, but, rather than follow him exactly, we choose a slightly more general approach because we shall require the results later. The matrix for a dielectric quarter-wave layer is

$$\begin{bmatrix} \cos\delta & (i\sin\delta)/n \\ in\sin\delta & \cos\delta \end{bmatrix}$$

where, as usual, we are writing n for the optical admittance, which is in free space units. Now, for layers which are almost a quarter-wave we can write

$$\delta = \pi/2 + \varepsilon$$

where ε is small. Then

$$\cos\delta \simeq -\varepsilon \qquad \sin\delta \simeq 1$$

so that the matrix can be written

$$\begin{bmatrix} -\varepsilon & i/n \\ in & -\varepsilon \end{bmatrix}.$$

We limit our analysis to quarter-wave multilayer stacks having high index next to the substrate. There are two cases, even and odd number of layers.

Case I: even number **$(2x)$** ***of layers***

The resultant multilayer matrix is given by

$$\begin{bmatrix} B \\ C \end{bmatrix} = [L][H][L]\ldots[L][H]\begin{bmatrix} 1 \\ n_m \end{bmatrix}$$

where

$$[L] = \begin{bmatrix} -\varepsilon_L & i/n_L \\ in_L & -\varepsilon_L \end{bmatrix}$$

$$[H] = \begin{bmatrix} -\varepsilon_H & i/n_H \\ in_H & -\varepsilon_H \end{bmatrix}.$$

Then

$$\begin{bmatrix} B \\ C \end{bmatrix} = \{[L][H]\}^x\begin{bmatrix} 1 \\ n_m \end{bmatrix}$$

$$= \begin{bmatrix} -\left(\frac{n_4}{n_L}\right) & -i\left(\frac{\varepsilon_L}{n_H} + \frac{\varepsilon_H}{n_L}\right) \\ -i(n_L\varepsilon_H + n_H\varepsilon_L) & -\left(\frac{n_L}{n_H}\right) \end{bmatrix}^x \begin{bmatrix} 1 \\ n_m \end{bmatrix}$$

$$= \begin{bmatrix} M_{11} & iM_{12} \\ iM_{21} & M_{22} \end{bmatrix} \begin{bmatrix} 1 \\ n_m \end{bmatrix}.$$

Our problem is to find expressions for M_{11}, M_{12}, M_{21} and M_{22}. In the evaluation we neglect all terms of second and higher order in ε. Terms in ε appearing in M_{11} and M_{22} are of second and higher order and therefore

$$M_{11} = (-1)^x \left(\frac{n_H}{n_L}\right)^x$$

$$M_{22} = (-1)^x \left(\frac{n_L}{n_H}\right)^x.$$

M_{12} and M_{21} contain terms of first, third and higher orders in ε. The first-order terms are

$$M_{12} = -\left(\frac{\varepsilon_L}{n_H} + \frac{\varepsilon_H}{n_L}\right)\left(-\frac{n_L}{n_H}\right)^{x-1} + \left(-\frac{n_H}{n_L}\right)\left[-\left(\frac{\varepsilon_L}{n_H} + \frac{\varepsilon_H}{n_L}\right)\right]\left(-\frac{n_L}{n_H}\right)^{x-2} + \ldots$$

$$+ \left(-\frac{n_H}{n_L}\right)^p \left[-\left(\frac{\varepsilon_L}{n_H} + \frac{\varepsilon_H}{n_L}\right)\right]\left(-\frac{n_L}{n_H}\right)^{x-p-1} + \ldots$$

$$+ \left(-\frac{n_H}{n_L}\right)^{x-1} \left[-\left(\frac{\varepsilon_L}{n_H} + \frac{\varepsilon_H}{n_L}\right)\right]$$

$$= (-1)^x \left(\frac{\varepsilon_L}{n_H} + \frac{\varepsilon_H}{n_L}\right)\left[\left(\frac{n_L}{n_H}\right)^{x-1} + \left(\frac{n_L}{n_H}\right)^{x-3} + \ldots + \left(\frac{n_H}{n_L}\right)^{x-1}\right]$$

$$= (-1)^x \left(\frac{\varepsilon_L}{n_H} + \frac{\varepsilon_H}{n_L}\right)\left(\frac{n_H}{n_L}\right)^{x-1}\left[\left(\frac{n_L}{n_H}\right)^{2x-2} + \left(\frac{n_L}{n_H}\right)^{2x-4} + \ldots + \left(\frac{n_L}{n_H}\right)^2 + 1\right]$$

$$= (-1)^x \left(\frac{\varepsilon_L}{n_H} + \frac{\varepsilon_H}{n_L}\right)\left(\frac{n_H}{n_L}\right)^{x-1}\left[1 - \left(\frac{n_L}{n_H}\right)^{2x}\right]\left[1 - \left(\frac{n_L}{n_H}\right)^2\right]^{-1}$$

since $(n_L/n_H) < 1$.

Now, provided x is large enough and (n_L/n_H) small enough, we can neglect $(n_L/n_H)^{2x}$ in comparison with 1, and after some adjustment, the expression becomes

$$M_{12} = \frac{(-1)^x n_H n_L (n_H/n_L)^x (\varepsilon_L/n_H + \varepsilon_H/n_L)}{(n_H^2 - n_L^2)}.$$

A similar procedure yields

$$M_{21} = \frac{(-1)^x n_H n_L (n_H/n_L)^x (n_L\varepsilon_H + n_H\varepsilon_L)}{(n_H^2 - n_L^2)}.$$

Case II: odd number $(2x+1)$ *of layers*

The resultant matrix is given by

$$\begin{bmatrix} B \\ C \end{bmatrix} = [H][L][H]\ldots[L][H]\begin{bmatrix} 1 \\ n_m \end{bmatrix}$$

$$= [H]\{[L][H]\}^x\begin{bmatrix} 1 \\ n_m \end{bmatrix}$$

which we can denote by

$$\begin{bmatrix} N_{11} & \mathrm{i}N_{12} \\ \mathrm{i}N_{21} & N_{22} \end{bmatrix}\begin{bmatrix} 1 \\ n_m \end{bmatrix}$$

and which is simply the previous result multiplied by

$$\begin{bmatrix} -\varepsilon_H & \mathrm{i}/n_H \\ \mathrm{i}n_H & -\varepsilon_H \end{bmatrix}.$$

Then

$$N_{11} = -\varepsilon_H M_{11} - M_{21}/n_H = (-1)^{x+1}\left(\frac{n_H}{n_L}\right)^x \frac{(\varepsilon_L n_H n_L + \varepsilon_H n_H^2)}{(n_H^2 - n_L^2)}$$

$$N_{12} = -\varepsilon_H M_{12} + M_{22}/n_H = (-1)^x\left(\frac{n_L}{n_H}\right)^x \frac{1}{n_H}$$

$$N_{21} = n_H M_{11} - \varepsilon_H M_{21} = (-1)^x\left(\frac{n_H}{n_L}\right)^x n_H$$

$$N_{22} = -\varepsilon_H M_{22} - n_H M_{12} = (-1)^{x+1}\left(\frac{n_H}{n_L}\right)^x \frac{n_H^2 n_L(\varepsilon_L/n_H + \varepsilon_H/n_L)}{(n_H^2 - n_L^2)}$$

where terms in $(n_L/n_H)^x$ are neglected in comparison with $(n_H/n_L)^x$.

Phase shift: case I

We are now able to compute the phase shift on reflection. We take, initially, the index of the incident medium to be n_0. Then

$$\begin{bmatrix} B \\ C \end{bmatrix} = \begin{bmatrix} M_{11} & \mathrm{i}M_{12} \\ \mathrm{i}M_{21} & M_{22} \end{bmatrix}\begin{bmatrix} 1 \\ n_m \end{bmatrix}$$

$$= \begin{bmatrix} M_{11} + \mathrm{i}n_m M_{12} \\ n_m M_{22} + \mathrm{i}M_{21} \end{bmatrix}$$

$$\rho = \frac{n_0 B - C}{n_0 B + C} = \frac{n_0(M_{11} + \mathrm{i}n_m M_{12}) - n_m M_{22} - \mathrm{i}M_{21}}{n_0(M_{11} + \mathrm{i}n_m M_{12}) + n_m M_{22} + \mathrm{i}M_{21}}$$

$$= \frac{(n_0 M_{11} - n_m M_{22}) + \mathrm{i}(n_0 n_m M_{12} - M_{21})}{(n_0 M_{11} + n_m M_{22}) + \mathrm{i}(n_0 n_m M_{12} + M_{21})} \qquad (7.19)$$

$$\tan\phi = \frac{2n_0 n_m^2 M_{12} M_{22} - 2n_0 M_{11} M_{21}}{n_0^2 M_{11}^2 - n_m^2 M_{22}^2 + n_0^2 n_m^2 M_{12}^2 - M_{21}^2}.$$

Inserting the appropriate expressions and once again neglecting terms of second and higher order in ε and terms in $(n_L/n_H)^x$, we obtain for ϕ

$$\tan\phi = \frac{-2n_H n_L(n_L\varepsilon_H + n_H\varepsilon_L)}{n_0(n_H^2 - n_L^2)} \tag{7.20}$$

(for $LH \ldots LHLH \,|\, n_m$).

Phase shift: case II

ρ is given by an expression similar to (7.19), in which M is replaced by N. Then, following the same procedure as for case I we arrive at

$$\tan\phi = \frac{-2n_0(\varepsilon_L n_L + \varepsilon_H n_H)}{(n_H^2 - n_L^2)} \tag{7.21}$$

(for $HLH \ldots LHLH \,|\, n_m$).

Equations (7.20) and (7.21) are in a general form which we will make use of later. For our present purposes we can introduce some slight simplification.

$$\delta = \frac{2\pi nd}{\lambda} = 2\pi nd\nu = 2\pi nd\nu_0(\nu/\nu_0) = (\pi/2)g$$

so that

$$\varepsilon_H = \varepsilon_L = (\pi/2)g - \pi/2 = (\pi/2)(g-1).$$

Also, when we consider the construction of the Fabry–Perot filters we see that the incident medium in case I will be a high-index spacer layer and in case II a low index spacer. Thus, for Fabry–Perot filters,

$$\tan\phi = \frac{-\pi n_L}{(n_H - n_L)}(g-1)$$

for both case I and case II.

Now, ϕ is nearly π or 0. Then

$$\frac{d\phi}{dg} = \frac{-\pi n_L}{(n_H - n_L)}$$

which is the result obtained by Seeley. This can then be inserted in equation (7.18) to give

$$\frac{\Delta\nu_h}{\nu_0} = \frac{\Delta\lambda_h}{\lambda_0} = \frac{2}{m\pi F^{1/2}}\left(\frac{n_H - n_L}{n_H - n_L + n_L/m}\right).$$

Then the expressions for the halfwidth of all-dielectric Fabry–Perot filters of mth order become

High-index spacer:

$$\left(\frac{\Delta\lambda_h}{\lambda_0}\right)_H = \frac{4n_m n_L^{2x}}{m\pi n_H^{2x+1}} \frac{(n_H - n_L)}{(n_H - n_L + n_L/m)} \tag{7.22}$$

Low-index spacer:

$$\left(\frac{\Delta\lambda_h}{\lambda_0}\right)_L = \frac{4n_m n_L^{2x-1}}{m\pi n_H^{2x}} \frac{(n_H - n_L)}{(n_H - n_L + n_L/m)} \tag{7.23}$$

which are simply the earlier results multiplied by the factor $(n_H - n_L)/(n_H - n_L + n_L/m)$. It should be noted that these results are for first-order reflecting stacks and mth-order spacer. Clearly the effect of the phase is much greater the closer the two indices are in value and the lower the spacer order m. For the common visible and near infrared materials, zinc sulphide and cryolite, the factor for first-order spacers is equal to 0.43, while for infrared materials such as zinc sulphide and lead telluride it is greater, around 0.57. Figures 7.8 and 7.9 show the characteristics of typical all-dielectric narrowband Fabry–Perot filters.

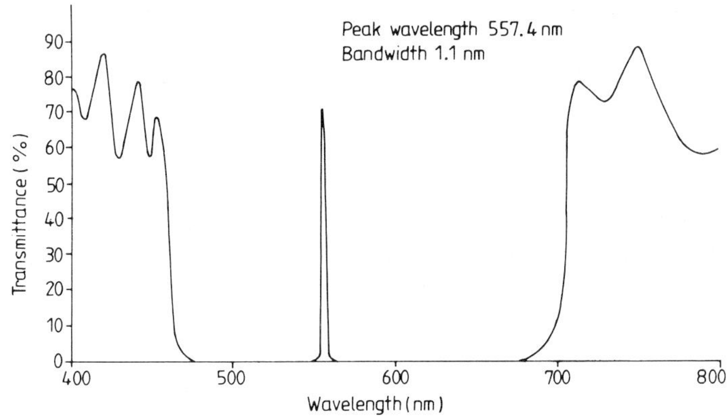

Figure 7.8 Measured transmittance of a narrowband all-dielectric filter with unsuppressed sidebands. Zinc sulphide and cryolite were the thin-film materials used. (Courtesy of Sir Howard Grubb, Parsons & Co Ltd.)

Since the all-dielectric multilayer reflector is effective over a limited range only, sidebands of transmission appear on either side of the peak and in most applications must be suppressed. The shortwave sidebands can be removed very easily by adding to the filter a longwave-pass absorption filter, readily available in the form of polished glass disks from a large number of manufacturers. Unfortunately, it is not nearly as easy to obtain shortwave-pass absorption filters and the rather shallow edges of those which are available tend considerably to reduce the peak transmission of the filter if the sidebands are

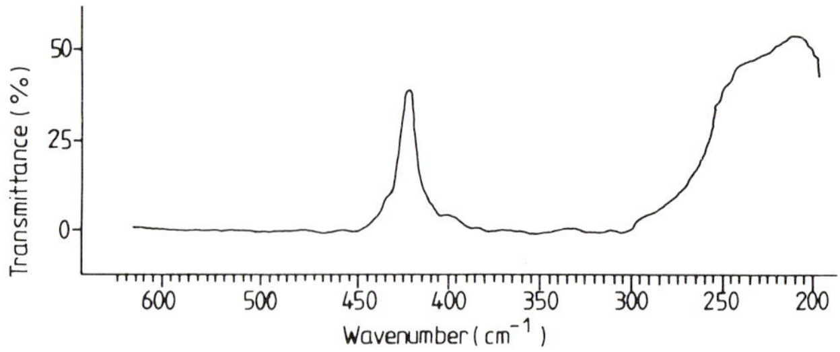

Figure 7.9 Measured transmittance of a Fabry–Perot filter for the far infrared. Design: Air | *LHLHHLH* | Ge with *H* indicating a quarter-wave of germanium and *L* of caesium iodide. The rear surface of the substrate is unbloomed so that the effective transmission of the filter is 50%. (Courtesy of Sir Howard Grubb, Parsons & Co Ltd.)

effectively suppressed. The best solution to this problem is not to use an absorption type of filter at all, but to employ as a blocking filter a metal–dielectric filter of the type already discussed or of the multiple cavity type to be considered shortly. Because metal–dielectric filters used in the first order do not have longwave sidebands, they are very successful in this application. The metal–dielectric blocking filter can, in fact, be deposited over the all-dielectric filter in the same evaporation run provided that the layers are monitored using the narrowband filter itself as the test glass—this is known as direct monitoring—but more frequently a completely separate metal–dielectric filter is used. The various components which go to make up the final filter are cemented together in one assembly.

Before we leave the Fabry–Perot filters we can examine the effects of absorption losses in the layers in a manner similar to that already employed in chapter 5, where we were concerned with quarter-wave stacks. The problem has been investigated by many workers. The account which follows relies heavily on the work of Hemingway and Lissberger[5], but with slight differences.

We apply the method of chapter 5 directly. There, we recall, we showed that the loss in a weakly absorbing multilayer was given by

$$A = (1 - R)\sum \mathscr{A}$$

where, for *quarter-waves*,

$$\mathscr{A} = \beta\left(\frac{n}{y_e} + \frac{y_e}{n}\right)$$

$$\beta = \frac{2\pi kd}{\lambda} = \frac{2\pi nd}{\lambda}\frac{k}{n} = \frac{\pi}{2}\frac{k}{n}.$$

y_e is the admittance of the structure on the emergent side of the layer, in free space units, $n - ik$ is the refractive index of the layer and d is the geometrical thickness. For quarter-waves, $nd = \pi/4$.

The scheme is shown in table 7.2 where the admittance y_e is given at each

Table 7.2

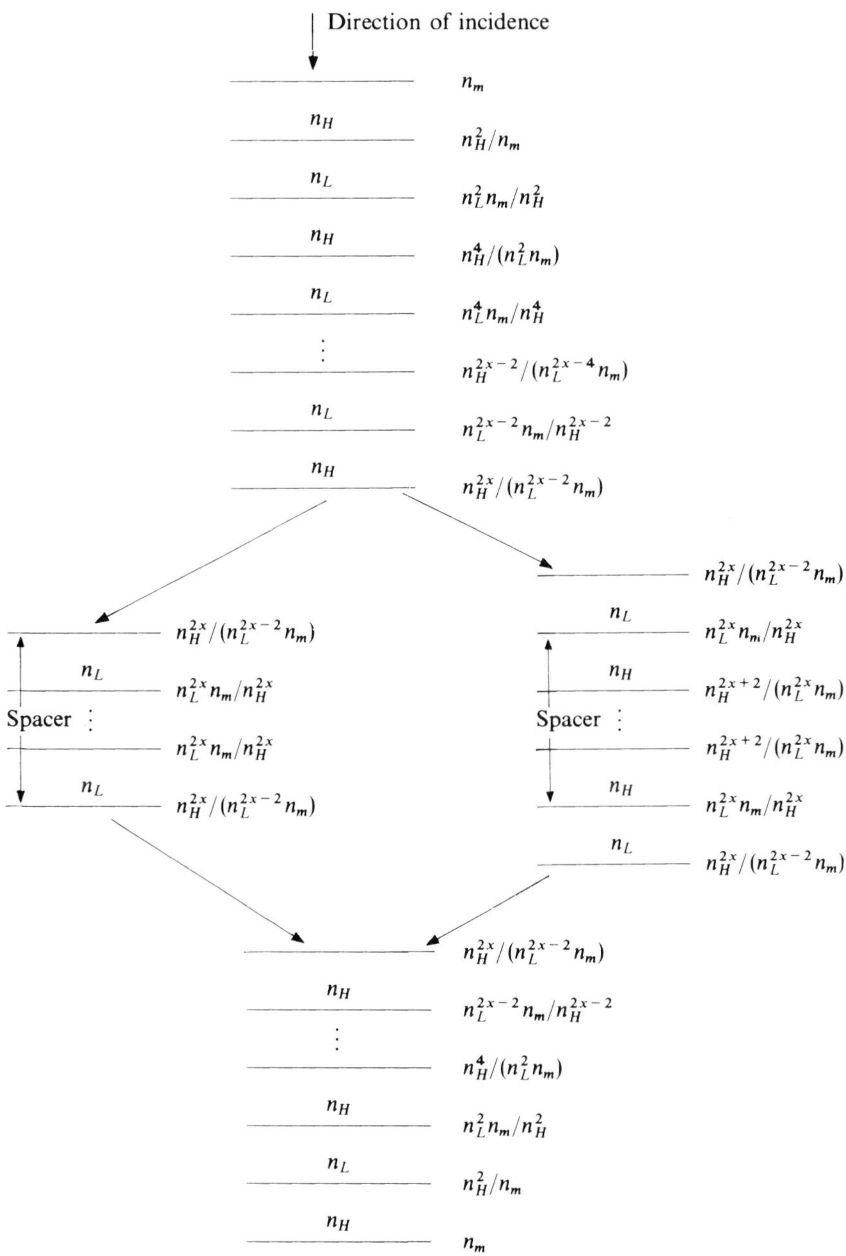

Direction of incidence
n_m
n_H
n_H^2/n_m
n_L
$n_L^2 n_m/n_H^2$
n_H
$n_H^4/(n_L^2 n_m)$
n_L
$n_L^4 n_m/n_H^4$
$n_H^{2x-2}/(n_L^{2x-4} n_m)$
n_L
$n_L^{2x-2} n_m/n_H^{2x-2}$
n_H
$n_H^{2x}/(n_L^{2x-2} n_m)$
$n_H^{2x}/(n_L^{2x-2} n_m)$
n_L
$n_L^{2x} n_m/n_H^{2x}$
Spacer
$n_L^{2x} n_m/n_H^{2x}$
n_L
$n_H^{2x}/(n_L^{2x-2} n_m)$
$n_H^{2x}/(n_L^{2x-2} n_m)$
n_L
$n_L^{2x} n_m/n_H^{2x}$
n_H
$n_H^{2x+2}/(n_L^{2x} n_m)$
Spacer
$n_H^{2x+2}/(n_L^{2x} n_m)$
n_H
$n_L^{2x} n_m/n_H^{2x}$
n_L
$n_H^{2x}/(n_L^{2x-2} n_m)$
$n_H^{2x}/(n_L^{2x-2} n_m)$
n_H
$n_L^{2x-2} n_m/n_H^{2x-2}$
$n_H^4/(n_L^2 n_m)$
n_H
$n_L^2 n_m/n_H^2$
n_L
n_H^2/n_m
n_H
n_m

interface and where alternative schemes for either high- or low-index spacers are included. The reflecting stacks are assumed to begin with high-index layers of which there are x per reflector, not counting the spacer.

We consider the case of low-index spacers first.

$$
\begin{aligned}
\sum \mathscr{A} = {} & \beta_H\left(\frac{n_m}{n_H}+\frac{n_H}{n_m}\right)+\beta_L\left(\frac{n_H^2}{n_L n_m}+\frac{n_L n_m}{n_H^2}\right) \\
& + \beta_H\left(\frac{n_L^2 n_m}{n_H^3}+\frac{n_H^3}{n_L^2 n_m}\right)+\beta_L\left(\frac{n_H^4}{n_L^3 n_m}+\frac{n_L^3 n_m}{n_H^4}\right)+\ldots \\
& + \beta_L\left(\frac{n_H^{2x-2}}{n_L^{2x-3} n_m}+\frac{n_L^{2x-3} n_m}{n_H^{2x-2}}\right)+\beta_H\left(\frac{n_H^{2x-1}}{n_L^{2x-2} n_m}+\frac{n_L^{2x-2} n_m}{n_H^{2x-1}}\right) \\
& + m\left[\beta_L\left(\frac{n_H^{2x}}{n_L^{2x-1} n_m}+\frac{n_L^{2x-1} n_m}{n_H^{2x}}\right)+\beta_L\left(\frac{n_L^{2x-1} n_m}{n_H^{2x}}+\frac{n_H^{2x}}{n_L^{2x-1} n_m}\right)\right] \\
& + \beta_H\left(\frac{n_H^{2x-1}}{n_L^{2x-2} n_m}+\frac{n_L^{2x-2} n_m}{n_H^{2x-1}}\right)+\ldots+\beta_H\left(\frac{n_H}{n_m}+\frac{n_m}{n_H}\right)
\end{aligned}
$$

where the final set of terms is a repeat of the first and where the spacer consists of $2m$ quarter-waves. Rearranging, we find

$$
\begin{aligned}
\sum \mathscr{A} = {} & 2\beta_H\left(\frac{n_m}{n_H}+\frac{n_L^2 n_m}{n_H^3}+\frac{n_L^4 n_m}{n_H^5}+\ldots+\frac{n_L^{2x-2} n_m}{n_H^{2x-1}}\right) \\
& + 2\beta_H\left(\frac{n_H}{n_m}+\frac{n_H^3}{n_L^2 n_m}+\frac{n_H^5}{n_L^4 n_m}+\ldots+\frac{n_H^{2x-1}}{n_L^{2x-2} n_m}\right) \\
& + 2\beta_L\left(\frac{n_L n_m}{n_H^2}+\frac{n_L^3 n_m}{n_H^4}+\ldots+\frac{n_L^{2x-3} n_m}{n_H^{2x-2}}\right) \\
& + 2\beta_L\left(\frac{n_H^2}{n_L n_m}+\frac{n_H^4}{n_L^3 n_m}+\ldots+\frac{n_H^{2x-2}}{n_L^{2x-3} n_m}\right) \\
& + 2m\beta_L\left(\frac{n_H^{2x}}{n_L^{2x-1} n_m}\quad\frac{n_L^{2x-1} n_m}{n_H^{2x}}\right)
\end{aligned}
$$

where we have combined similar terms due to the two mirrors and where the final term is due to the spacer. The first four terms are geometric series and therefore, since $(n_L/n_H) < 1$,

$$
\begin{aligned}
\sum \mathscr{A} = {} & 2\beta_H \frac{n_m}{n_H}\frac{[1-(n_L/n_H)^{2x}]}{[1-(n_L/n_H)^2]} \\
& + 2\beta_H \frac{n_H^{2x-1}}{n_L^{2x-2} n_m}\frac{[1-(n_L/n_H)^{2x-2}]}{[1-(n_L/n_H)^2]}
\end{aligned}
$$

$$+ 2\beta_L \frac{n_L n_m}{n_H^2} \frac{[1-(n_L/n_H)^{2x-2}]}{[1-(n_L/n_H)^2]}$$

$$+ 2\beta_L \frac{n_H^{2x-2}}{n_L^{2x-3} n_m} \frac{[1-(n_L/n_H)^{2x-2}]}{[1-(n_L/n_H)^2]}$$

$$+ 2m\beta_L \left[\frac{n_H^{2x}}{n_L^{2x-1} n_m} + \frac{n_L^{2x-1} n_m}{n_H^{2x}} \right].$$

(n_L/n_H) will usually be rather less than unity and x will normally be large and so we can make the usual approximation and neglect terms such as $(n_L/n_H)^{2x}$ in the numerators and also those terms which have (n_m/n_H) as a factor compared with $(n_L/n_m)\,(n_H/n_L)^{2x-1}$ etc. Then the expression simplifies to

$$\sum \mathscr{A} = 2\beta_H \frac{n_H^{2x-1}}{n_L^{2x-2} n_m} \frac{1}{[1+(n_L/n_H)^2]}$$

$$+ 2\beta_L \frac{n_H^{2x-2}}{n_L^{2x-3} n_m} \frac{1}{[1+(n_L/n_H)^2]}$$

$$+ 2m\beta_L \frac{n_H^{2x}}{n_L^{2x-1} n_m}.$$

But

$$\beta_H = \frac{2\pi n_H d}{\lambda} \frac{k_H}{n_H} = \frac{\pi}{2} \frac{k_H}{n_H}$$

$$\beta_L = \frac{\pi}{2} \frac{k_L}{n_L}.$$

Thus

$$\sum \mathscr{A} = \frac{\pi k_H (n_H^{2x}/n_m n_L^{2x-2}) + \pi k_L (n_H^{2x}/n_m n_L^{2x-2})}{(n_H^2 - n_L^2)} + \frac{\pi m k_L n_H^{2x}}{n_L^{2x} n_m}$$

$$= \frac{\pi n_H^{2x}}{n_m n_L^{2x}} \left(\frac{n_L^2 k_H + n_L^2 k_L}{(n_H^2 - n_L^2)} + m k_L \right).$$

The absorption is then given by $A = (1-R)\sum \mathscr{A}$. If the incident medium has index n_0, then, since the terminating admittance in table 7.2 is n_m,

$$R = \left(\frac{n_0 - n_m}{n_0 + n_m} \right)^2$$

and therefore

$$(1-R) = \frac{4 n_0 n_m}{(n_0 + n_m)^2}.$$

The above expression for $\sum \mathscr{A}$ should, therefore, be multiplied by the factor $4n_0 n_m/(n_0 + n_m)^2$ to yield the absorption. However, the filters should be designed so that they are reasonably well matched into the incident medium and therefore this factor will be unity, or sufficiently near unity. The absorption is then given by $\sum \mathscr{A}$. That is:

$$A = \frac{\pi n_H^{2x}}{n_m n_L^{2x}} \left(\frac{n_L^2 k_H + n_L^2 k_L}{(n_H^2 - n_L^2)} + m k_L \right) \tag{7.24}$$

for low-index spacers.

For high-index spacers we work through a similar scheme and, with the same approximations, we arrive at

$$A = \frac{\pi n_H^{2x}}{n_m n_L^{2x}} \left(\frac{n_L^2 k_H + n_H^2 k_L}{(n_H^2 - n_L^2)} + m k_H \right) \tag{7.25}$$

for high-index spacers.

It should be noted that, since x is the number of high-index layers, the filter represented by equation (7.25) will be narrower than that represented by equation (7.24) for equal x.

A useful set of alternative expressions can be obtained if we substitute equations (7.22) and (7.23) into equations (7.24) and (7.25) to give:

High-index spacer

$$A = 4 \frac{\lambda_0}{\Delta\lambda_\mathrm{h}} \frac{\{k_L + k_H [m + (1-m)(n_L/n_H)^2]\}}{(n_H + n_L)[m + (1-m)(n_L/n_H)]} \tag{7.26}$$

Low-index spacer

$$A = 4 \frac{\lambda_0}{\Delta\lambda_\mathrm{h}} \frac{\{k_L (n_H/n_L)[m + (1-m)(n_L/n_H)^2] + (n_L/n_H) k_H\}}{(n_H + n_L)[m + (1-m)(n_L/n_H)]}. \tag{7.27}$$

Figure 7.10 shows the value of A plotted for Fabry–Perot filters with $n_H = 2.35$ and $n_L = 1.35$, typical of zinc sulphide and cryolite. $(\lambda_0/\Delta\lambda_\mathrm{h})$ is taken as 100 and k_H and k_L as either zero or 0.0001. The effect of other values of $(\lambda_0/\Delta\lambda_\mathrm{h})$ or k can be estimated by multiplying by an appropriate factor. The approximations are reasonable for $k(\lambda_0/\Delta\lambda_\mathrm{h})$ less than around 0.1.

It is difficult to draw any general conclusions from figure 7.10 because the results depend on the relative magnitudes of k_H and k_L. However, except in the case of very low k_L, the high-index spacer is to be preferred. There are very good reasons connected with performance when tilted, with energy grasp and with the manufacture of filters, for choosing high- rather than low-index spacers.

In the visible and near infrared regions of the spectrum, materials such as zinc sulphide and cryolite are capable of halfwidths of less than 0.1 nm with useful peak transmittance. Uniformity is, however, a major difficulty for filters of such narrow bandwidths. At the 90%-of-peak points, the Fabry–Perot filter has a width which is one third of the halfwidth. It is a good guide that the uniformity of the filter should be such that the peak wavelength does not vary

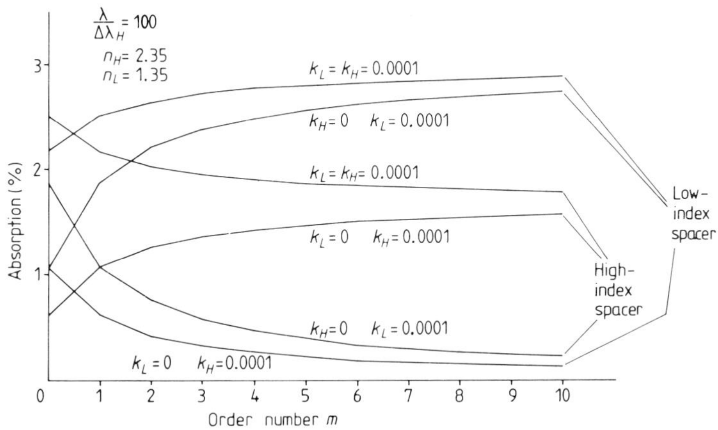

Figure 7.10 The value (expressed as a percentage) of the absorptance, as a function of the order number m, of Fabry–Perot filters with $\lambda_0/\Delta\lambda_h$ of 100 and values of extinction coefficients k_H and k_L of 0.0001 or zero. Other values can be accommodated by multiplying by an appropriate factor. n_H is taken as 2.35 and n_L as 1.35. The results are derived from equations (7.26) and (7.27).

by more than one third of the halfwidth over the entire surface of the filter. This means that the effective increase in halfwidth due to the lack of uniformity is kept within some $4\frac{1}{2}\%$ of the halfwidth and the reduction in peak transmittance to less than 3% (these figures can be calculated using the expressions derived later for assessing the performance of filters in uncollimated incident light). For filters of less than 0.1 nm halfwidth this rule implies a variation of not more than 0.03 nm or 0.006% in terms of layer thickness, a very severe requirement even for quite small filters. Halfwidths of 0.3–0.5 nm are less demanding and can be produced more readily provided considerable care is taken. For narrower filters use is often made of the solid etalon filters now to be described.

The solid etalon filter

A solid etalon filter, or, as it is sometimes called, a solid spacer filter, is a very high-order Fabry–Perot filter in which the spacer consists of an optically worked plate or a cleaved crystal. Thin-film reflectors are deposited on either side of the spacer in the normal way, so that the spacer also acts as the substrate. The problems of uniformity which exist with all-thin-film narrowband filters are avoided and the thick spacer does not suffer from the increased scattering losses which always seem to accompany the higher-order thin-film spacers. The solid etalon filter is very much more robust and stable than the conventional air-spaced Fabry–Perot etalon, while the manufacturing difficulties are comparable. The high order of the spacer implies a small interval between orders and a conventional thin-film narrowband filter must be used in series with it to eliminate the unwanted orders.

An early account of the use of mica for the construction of filters of this type is that of Dobrowolski[6] who credits Billings with being the first to use mica in this way, achieving halfwidths of 0.3 nm. Dobrowolski obtained rather narrower pass bands and his is the first complete account of the technique. Mica can be cleaved readily to form thin sheets with flat parallel surfaces, but there is a complication due to the natural birefringence of mica which means that the position of the pass band depends on the plane of polarisation. This splitting of the pass band can be avoided by arranging the thickness of the mica such that it is a half-wave plate, or multiple half-wave, at the required wavelength. If the two refractive indices are n_0 and n_e, this implies

$$\frac{2\pi(n_0 - n_e)d}{\lambda} = p\pi \qquad p = 0, \pm 1, \pm 2, \ldots$$

The order of the spacer will then be given by

$$m = \frac{n_0 p}{(n_0 - n_e)} \qquad \text{or} \qquad \frac{n_e p}{(n_0 - n_e)}$$

depending on the plane of polarisation. The difference between these two values is p, but, since p is small, the bandwidth will be virtually identical. The separation of orders for large m is given approximately by λ/m. Dobrowolski found that the maximum order separation, corresponding to $p = 1$, was given by 1.64 nm at 546.1 nm. With such spacers, around 60 μm thick, filters with halfwidths around 0.1 nm, the narrowest 0.085 nm, were constructed. Peak transmission ranged up to 50% for the narrower filters and up to 80% for slightly broader ones with around 0.3 nm halfwidth.

More recent work on solid etalon filters has concentrated on the use of optically worked materials as spacers. These must be ground and polished so that the faces have the necessary flatness and parallelism. The most complete account so far of the production of such filters is by Austin[7]. Fused silica spacers as thin as 50 μm have been produced with the necessary parallelism for halfwidths as narrow as 0.1 nm in the visible region, while thicker discs can give bandwidths as narrow as 0.005 nm. A 50 μm fused silica spacer gives an interval between orders of around 1.4 nm in the visible region which allows the suppression of unwanted orders to be fairly readily achievable by conventional thin-film narrowband filters.

The process of optical working tends to produce an error in parallelism over the surface of the spacer which is ultimately independent of the thickness of the spacer. Let us denote the total range of spacer thickness due to this lack of parallelism and to any deviation from flatness by Δd. This variation in spacer thickness causes the peak wavelength of the filter to vary. We can take an absolute limit for these variations as half the bandwidth of the filter. Then the resultant halfwidth will be increased by just over 10% and the peak transmittance reduced by just over 7% (once again using the expressions which we will shortly establish for filter performance in uncollimated light). We can write

$$\Delta\lambda_0/\lambda_0 = \Delta D/D = \Delta d/d \leqslant 0.5\Delta\lambda_h/\lambda_0$$

where D is the optical thickness nd of the spacer, $\Delta\lambda_0$ is the error in peak wavelength and $\Delta\lambda_h$ is the halfwidth. But

$$\text{Resolving power} = \lambda_0/\Delta\lambda_h = m\mathscr{F}$$

and hence, since

$$D = m\lambda_0/2$$

$$\mathscr{F} \leqslant \frac{0.25\lambda_0}{\Delta D}.$$

Now the attainable ΔD in the visible region is of the order of $\lambda/100$ and this means that the limiting finesse is around 25, independent of the spacer thickness. High resolving power then has to be achieved by the order number m which determines both the spacer thickness $D = m\lambda_0/2$ and the interval between orders λ_0/m. For a halfwidth of 0.01 nm at, say, 500 nm the resolving power is 50 000. The finesse of 25 implies an order number of 2000, a spacer optical thickness of 500 μm and an interval between orders of 0.25 nm. This very restricted range between orders means that it is very difficult to carry out sideband blocking by a thin-film filter directly. Instead, a broader solid etalon filter can be used with its correspondingly greater interval between orders. It, in its turn, can be suppressed by a thin-film filter. For a halfwidth of 0.1 nm, a spacer optical thickness of 50 μm is required which gives an interval between orders of 2.5 nm.

The temperature coefficient of peak wavelength change of solid etalon filters with fused silica spacers is 0.005 nm $^\circ C^{-1}$ and the filters may be finely tuned by altering this temperature.

Candille and Saurel[8] have used Mylar foil as the spacer. Their filters were strictly of the multiple cavity type described later in this chapter. The Mylar acted as substrate and a high-order spacer. One of the reflectors included a low-order Fabry–Perot filter which served both as blocking filter to eliminate the additional unwanted orders of the Mylar section and as an additional cavity to steepen the sides of the pass band. The position of the pass band could be altered by varying the tension in the Mylar. The filters were not as narrow as the other solid etalon filters which have been mentioned, halfwidths of 0.8–1.0 nm being obtained.

Solid etalon filters have also been constructed for the infrared. Smith and Pidgeon[9] used a polished slab of germanium some 780 μm thick working at around 700 cm^{-1} in the 400th order. Both faces were coated with a quarter-wave of zinc sulphide followed by a quarter-wave of lead telluride to give a reflectance of 62%, a fringe halfwidth of 0.1 cm^{-1} and an interval between orders of 1.6 cm^{-1}. This particular arrangement was designed so that the lines in the R-branch of the CO_2 spectrum, which are spaced $1.6\ \mathrm{cm}^{-1}$ apart at around 14.5 μm, should be exactly matched by a number of adjacent orders. Order sorting was not, therefore, a problem.

Roche and Title[10] have reported a range of solid etalon filters for the infrared. These filters are some 13 mm in diameter, have resolving powers in the region of 3×10^4 and the techniques used for their construction are as reported by Austin[7]. For wavelengths equal to or shorter than 3.5 μm, fused silica spacers are quite satisfactory. For longer wavelengths Yttralox, a combination of yttrium and thorium oxides, was found most satisfactory. With this material, solid etalon filters were produced which at 3.334 μm had halfwidths as low as 0.2 nm and at 4.62 μm, 0.8 nm. At these wavelengths, the attainable finesse was 30–40 and the current limit to the halfwidth which can be achieved is the permissible interval between orders which determines the arrangement of subsidiary blocking filters.

The effect of varying the angle of incidence

As we have seen with other types of thin-film assembly the performance of the all-dielectric Fabry–Perot varies with angle of incidence, and this effect is particularly important when considering, for instance, the allowable focal ratio of the pencil being passed by the filter or the maximum tilt angle in any application. The variation with angle of incidence is not altogether a bad thing because the effect can be used to tune filters which would otherwise be off the desired wavelength—very important from the manufacturer's point of view because it enables him to ease a little the otherwise almost impossibly tight production tolerances.

The effect of tilting has been studied by a number of workers, particularly by Dufour and Herpin[11], Lissberger[12], Lissberger and Wilcock[13] and Pidgeon and Smith[14]. For our present purposes we follow Pidgeon and Smith since their results are in a slightly more suitable form.

Simple tilts in collimated light

The phase thickness of a thin film at oblique incidence is

$$\delta = 2\pi nd \cos\theta/\lambda$$

which can be interpreted as an apparent optical thickness of $nd\cos\theta$ which varies with angle of incidence so that layers seem thinner when tilted. Although the optical admittance changes with tilts, in narrowband filters the predominant effect is the apparent change in thickness which moves the filter pass band to shorter wavelengths.

For an ideal Fabry–Perot filter with spacer layer index n^*, where the reflectors have constant phase shift of zero or π regardless of the angle of incidence or wavelength, we can write for the position of peak wavelength in the mth order

$$2\pi n^* d\cos\theta/\lambda = m\pi$$

i.e.

$$(2\pi n^* d/\lambda_0)\ \ g\cos\theta = m\pi$$

i.e.

$$g\cos\theta = 1$$

$$\Delta g = \left(\frac{1}{\cos\theta} - 1\right).$$

If the angle of incidence is θ_i in air then

$$\theta = \sin^{-1}(\sin\theta_i/n^*)$$

and Δg is given in terms of θ_i and n^*. The effect of tilting, then, in this ideal filter can be estimated simply from a knowledge of the index of the spacer and the angle of incidence. For small angles of incidence, the shift is given by

$$\Delta g = \Delta\nu/\nu_0 = \Delta\lambda/\lambda_0 = \theta_i^2/2n^{*2}. \tag{7.28}$$

The index of the spacer n^* determines its sensitivity to tilt: the higher the index, the less the filter is affected.

In the case of a real filter, the reflectors are also affected by the tilting and so the calculation of the shift in peak wavelength is more involved. It has, however, been shown by Pidgeon and Smith that the shift is similar to that which would have been obtained from an ideal filter with spacer index n^*, intermediate between the high and low indices of the layers of the filter. n^* is known as the effective index. This concept of the effective index holds good for quite high angles of incidence, up to 20° or 30° or even higher, depending on the indices of the layers making up the filter.

We can estimate the effective index for the filter by a technique similar to that already used for metal–dielectrics (equation (7.3)). We retain our assumption of small angle of incidence and small changes in g around the value which corresponds to the peak at normal incidence.

The peak position is given, as before, by

$$\sin^2[(2\pi nd\cos\theta/\lambda) - \phi] = 0 \tag{7.29}$$

with, at normal incidence,

$$\sin^2[(2\pi nd/\lambda_0) - \phi_0] = 0. \tag{7.30}$$

Now ϕ_0 is 0 or π and so equation (7.30) is satisfied by

$$2\pi nd/\lambda_0 = m\pi \qquad m = 0,\ 1,\ 2\ \ldots$$

The analysis is once again easier in terms of g $(= \lambda_0/\lambda = \nu/\nu_0)$. Equation (7.29) becomes

$$\sin^2[(2\pi nd/\lambda_0)g\cos\theta - \phi_0 - \Delta\phi] = 0.$$

We write

$$g = 1 + \Delta g \qquad \text{and} \qquad \cos\theta \simeq 1 - \theta^2/2.$$

However, we should work in terms of θ_i, the external angle of incidence, which we assume is referred to free space (if not, then we make the appropriate correction). Then

$$n \sin\theta = n_i \sin\theta_i = \sin\theta_i$$

and, using equation (7.31),

$$\sin^2[(2\pi nd/\lambda_0) - \phi_0 + m\pi\Delta g - (m\pi\theta_i^2/2n^2) - \Delta\phi] = 0$$

is the condition for the new peak position. This requires

$$m\pi\Delta g - (m\pi\theta_i^2/2n^2) - \Delta\phi = 0. \tag{7.32}$$

Now $\Delta\phi$ is a function of θ and Δg and to evaluate it we return to equations (7.20) and (7.21). The layers in the reflectors are all quarter-waves and so ε is given by

$$\pi/2 + \varepsilon = (2\pi nd/\lambda_0) g \cos\theta = (\pi/2)(1+\Delta g)(1-\theta^2/2)$$

but

$$\theta = \theta_i/n$$

so that

$$\varepsilon = (\pi/2)\Delta g - \pi\theta_i^2/4n^2$$

with n being either n_L or n_H for ε_L or ε_H respectively.

At this stage we are forced to consider high-index and low-index spacers separately.

Case I: high-index spacers

From equation (7.20) we have, inserting n_H for n_0,

$$\begin{aligned}
\Delta\phi &= -\frac{2n_L^2}{(n_H^2 - n_L^2)}\varepsilon_H - \frac{2n_H n_L}{(n_H^2 - n_L^2)}\varepsilon_L \\
&= \frac{2n_L^2}{(n_H^2 - n_L^2)}\left(\frac{\pi}{2}\Delta g - \frac{\pi\theta_i^2}{4n_H^2}\right) - \frac{2n_H n_L}{(n_H^2 - n_L^2)}\left(\frac{\pi}{2}\Delta g - \frac{\pi\theta_i^2}{4n_L^2}\right) \\
&= -\frac{\pi n_L}{(n_H - n_L)}\Delta g + \frac{\pi}{2}\frac{(n_L^2 - n_L n_H + n_H^2)}{n_H^2 n_L (n_H - n_L)}\theta_i^2
\end{aligned}$$

and equation (7.32) becomes

$$m\pi\Delta g - \frac{m\pi\theta_i^2}{2n_H^2} + \frac{\pi n_L \Delta g}{(n_H - n_L)} - \frac{\pi}{2}\frac{(n_L^2 - n_L n_H + n_H^2)}{n_H^2 n_L (n_H - n_L)}\theta_i^2 = 0$$

giving, after some manipulation and simplification

$$\Delta g = \frac{1}{n_H^2}\frac{[(m-1) - (m-1)(n_L/n_H) + (n_H/n_L)]}{[m - (m-1)(n_L/n_H)]}\left(\frac{\theta_i^2}{2}\right).$$

But, comparing the expression with equation (7.28) we find

$$n^{*2} = \frac{n_H^2[m - (m-1)(n_L/n_H)]}{[(m-1) - (m-1)(n_L/n_H) + (n_H/n_L)]}$$

or

$$n^* = n_H \left(\frac{m - (m-1)(n_L/n_H)}{(m-1) - (m-1)(n_L/n_H) + (n_H/n_L)} \right)^{1/2}. \tag{7.33}$$

For first-order filters

$$n^* = (n_H n_L)^{1/2} \tag{7.34}$$

which is the result obtained by Pidgeon and Smith. As $m \to \infty$ then $n^* \to n_H$, as we would expect.

Case II: low-index spacer

The analysis is exactly as for case I except that equation (7.21) is used and the n in equation (7.32) becomes n_L.

$$n^* = n_L \left(\frac{m - (m-1)(n_L/n_H)}{m - m(n_L/n_H) + (n_L/n_H)^2} \right)^{1/2}. \tag{7.35}$$

For first-order filters

$$n^* = \frac{n_L}{[1 - (n_L/n_H) + (n_L/n_H)^2]^{1/2}} \tag{7.36}$$

which is, again, the expression given by Pidgeon and Smith and we note again that as $m \to \infty$ then $n^* \to n_L$.

Typical curves showing how the effective index n^* varies with order number for both low- and high-index spacers are given in figure 7.11.

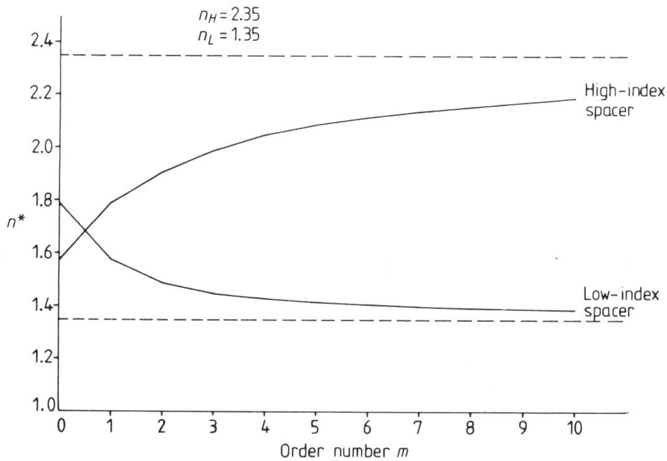

Figure 7.11 The effective index n^* plotted against order number m for Fabry–Perot filters constructed of materials such as zinc sulphide, $n = 2.35$, and cryolite, $n = 1.35$. The results were calculated from expressions (7.35) and (7.36).

Pidgeon and Smith made experimental measurements on narrowband filters for the infrared. The designs in question were

(*a*) $L \mid \mathrm{Ge} \mid LHLH\ LL\ HLH \mid \mathrm{Air}$

(*b*) $L \mid \mathrm{Ge} \mid LHLHL\ HH\ LHLH \mid \mathrm{Air}$

where *H* represents a quarter-wave thickness of lead telluride and *L* of zinc sulphide, and where the peak wavelength was in the vicinity of 15 μm. Calculations of shift were carried out by the approximate method using n^* and by the full matrix method without approximations. The results using n^* matched the accurate calculations up to angles of incidence of 40° to an accuracy representing ±2% change in n^*. The experimental points showed good agreement with the theoretical estimates. Some of the results are shown in figures 7.12 and 7.13.

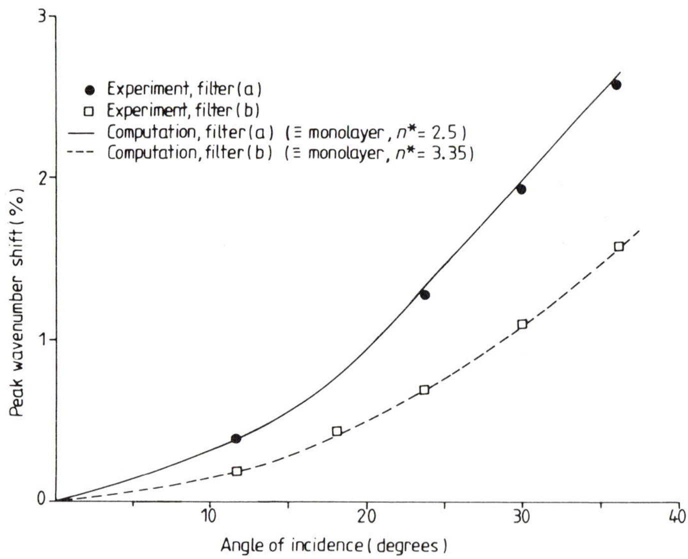

Figure 7.12 The shift of peak wavenumber with scanning angle for two Fabry–Perot filters in collimated light. In both cases the monolayer curves fit the computed curves to ±2% in *n*. (After Pidgeon and Smith[14].)

The angle of incidence may be in a medium other than free space, in which case equation (7.28) becomes

$$\Delta g = \Delta\lambda_0/\lambda = \Delta\nu_0/\nu = \tfrac{1}{2}(n_i\theta_i/n^*)^2 \qquad (7.37)$$

where θ_i is measured in radians.

If θ_i is measured in degrees, then

$$\Delta g = \Delta\lambda_0/\lambda_0 = \Delta\nu_0/\nu_0 = 1.5 \times 10^{-4}(n_i/n^*)^2\theta_i^2. \qquad (7.38)$$

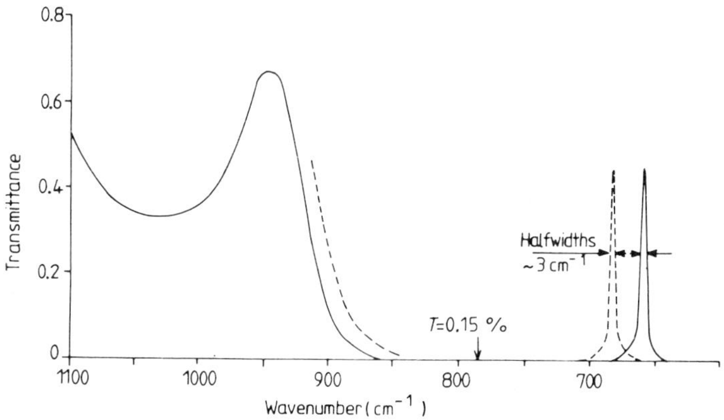

Figure 7.13 Measured transmittance of two filters of type (*b*) (see p 264). Design: Air|*HLHL HH LHLHL*|Ge substrate |*L*|Air (*H* = PbTe, *L* = ZnS). (After Pidgeon and Smith[14].)

Effect of an incident cone of light

The analysis can be taken a stage further to arrive at expressions for the degradations of peak transmission and bandwidth which become apparent when the incident illumination is not perfectly collimated. Essentially the same results have been obtained by Lissberger and Wilcock[13] and by Pidgeon and Smith[14].

It is assumed first of all that, in collimated light, the sole effect of tilting a filter is a shift of the characteristic towards shorter wavelengths or greater wavenumbers, leaving the peak transmittance and bandwidth virtually unchanged. The performance in convergent or divergent light is then given by integrating the transmission curve over a range of angles of incidence. The analysis is simpler in terms of wavenumber or of g, rather than wavelength. If ν_0 is the wavenumber corresponding to the peak at normal incidence and ν_Θ to the peak at angle of incidence Θ, then it is plausible that the resultant peak, when all angles of incidence in the cone from 0 to Θ are included, should appear at a wavenumber given by the mean of the above extremes. We shall show, shortly, that this is indeed the case. The new peak is given by

$$\nu_m = \nu_0 + \tfrac{1}{2}\Delta\nu' \tag{7.39}$$

where

$$\Delta\nu' = \nu_\Theta - \nu_0 = \nu_0\,\Theta^2/2n^{*2}.$$

The effective bandwidth of the filter will, of course, appear broader and, since the process is, in effect, a convolution of a function with bandwidth W_0, which is the width of the filter at normal incidence, and another function with bandwidth $\Delta\nu'$, the change in peak position produced by altering the angle of incidence from 0 to Θ, it seems likely that the resultant bandwidth might be given by the square root of the sum of their squares. This too is indeed the case, as we shall also show.

$$W_\Theta^2 = W_0^2 + (\Delta\nu')^2. \tag{7.40}$$

The peak transmission falls and is given by

$$\hat{T}_\Theta = \left(\frac{W_0}{\Delta\nu'}\right)\tan^{-1}\left(\frac{\Delta\nu'}{W_0}\right). \tag{7.41}$$

The analysis is as follows.

We consider incident light in the form of a cone with semiangle Θ, that is a cone of focal ratio $1/(2\tan\Theta)$. We assume that in collimated light the effect of tilting the filter is simply to move the characteristic towards shorter wavelengths, leaving the bandwidth and peak transmittance unchanged.

For small values of θ, the flux incident on the filter is proportional to $\theta\,\mathrm{d}\theta$. The resultant transmittance of the filter is then given by the total flux transmitted divided by the total flux incident.

The total flux incident is proportional to

$$\int_0^\Theta \theta\,\mathrm{d}\theta = \tfrac{1}{2}\Theta^2.$$

The total flux transmitted is proportional to

$$\int_0^\Theta \theta T\,\mathrm{d}\theta.$$

We can, for small values of θ and Δg, set

$$T = \frac{1}{1-\{(2/\Delta g_\mathrm{h})[\Delta g-(\theta_\mathrm{i}^2/2n^{*2})]\}^2}$$

where Δg_h is the halfwidth at normal incidence of the filter in units of g. This expression follows directly from the concept of n^*. The transmittance of the filter is then given by

$$\begin{aligned}
T &= \frac{2}{\Theta^2}\int_0^\Theta \frac{\theta\,\mathrm{d}\theta}{1+\{(2/\Delta g_\mathrm{h})(\Delta g-(\theta_\mathrm{i}/2n^{*2})]\}^2} \\
&= -\frac{2}{\Theta^2}\frac{n^{*2}\Delta g_\mathrm{h}}{2}\left[\tan^{-1}\left\{\frac{2}{\Delta g_\mathrm{h}}\left(\Delta g-\frac{\theta_\mathrm{i}}{2n^{*2}}\right)\right\}\right]_0^\Theta \\
&= \frac{1}{2}\frac{\Delta g_\mathrm{h}}{(\Theta^2/2n^{*2})}\left\{\tan^{-1}\left(2\frac{\Delta g}{\Delta g_\mathrm{h}}\right)-\tan^{-1}\left[2\left(\frac{\Delta g}{\Delta g_\mathrm{h}}-\frac{\Theta^2}{2n^{*2}}\frac{1}{\Delta g_\mathrm{h}}\right)\right]\right\}
\end{aligned} \tag{7.42}$$

$$= \frac{1}{2}\frac{\Delta g_\mathrm{h}}{(\Theta^2/2n^{*2})}\left[\tan^{-1}\left(\frac{(2/\Delta g_\mathrm{h})(\Theta^2/2n^{*2})}{1+(2/\Delta g_\mathrm{h})^2\{\Delta g[\Delta g-(\Theta^2/2n^{*2})]\}}\right)\right]. \tag{7.43}$$

This is a maximum when

$$\Delta g = \frac{1}{2}\frac{\Theta^2}{2n^{*2}}.$$

But $\Theta^2/(2n^{*2})$ is the shift in the position of the peak at angle of incidence Θ. Thus in a cone of light of semiangle Θ, the peak wavelength of the filter is given by the mean of the value at normal incidence and that at the angle Θ corresponding to equation (7.39). The value of the peak transmittance is then, from equation (7.42),

$$\frac{\Delta g_h}{(\Theta^2/2n^{*2})}\tan^{-1}\left(\frac{\Theta^2/2n^{*2}}{\Delta g_h}\right)$$

which corresponds to equation (7.41).

The half-peak points are given by

$$(7.43) = \tfrac{1}{2}\,(\text{peak } T)$$

i.e.

$$\frac{1}{2}\cdot\frac{\Delta g_h}{(\Theta^2/2n^{*2})}\tan^{-1}\left(\frac{(2/\Delta g_h)(\Theta^2/2n^{*2})}{1+(2/\Delta g_h)^2\{\Delta g[\Delta g-(\Theta^2/2n^{*2})]\}}\right)$$

$$=\frac{1}{2}\frac{\Delta g_h}{(\Theta^2/2n^{*2})}\tan^{-1}\left(\frac{\Theta^2/2n^{*2}}{\Delta g_h}\right)$$

which is satisfied by

$$1+\left(\frac{2}{\Delta g_h}\right)\left[\Delta g\left(\Delta g-\frac{\Theta^2}{2n^{*2}}\right)\right]=2$$

i.e.

$$\Delta g\left(\Delta g-\frac{\Theta^2}{2n^{*2}}\right)-\left(\frac{\Delta g_h}{2}\right)^2=0.$$

We are interested in the difference between the roots of the equation which is the width of the characteristic

$$(\Delta g_1-\Delta g_2)=\left[\left(\frac{\Theta^2}{2n^{*2}}\right)^2+(\Delta g_h)^2\right]^{1/2}$$

which corresponds exactly to equation (7.40).

Since

$$\tan^{-1}x = x-\frac{x^3}{3}+\frac{x^5}{5}-\frac{x^7}{7}+\ldots \qquad \text{for } |x|\leqslant 1$$

for small values of $(\Delta\nu'/W_0)$ we can write

$$\hat{T}_\Theta = 1-\frac{1}{3}\left(\frac{\Delta\nu'}{W_0}\right)^2. \tag{7.44}$$

If FR denotes the focal ratio of the incident light, then, for values of around 2 to infinity, it is a reasonably good approximation that

$$\Theta = 1/[2(FR)].$$

Using this, we find another expression for $\Delta\nu'$ which can be useful:

$$\Delta\nu' = \frac{\nu_0}{8n^{*2}(FR)^2}.$$

We can extend this analysis still further to the case of a cone of semiangle Θ incident at an angle other than normal, provided we make some simplifying assumptions. If the angle of incidence of the cone is χ then the range of angles of incidence will be $\chi \pm \Theta$.

If $\chi < \Theta$ then we can assume that the result is simply that for a normally incident cone of semiangle $\chi + \Theta$.

If $\chi > \Theta$ then we have three frequencies, ν_0 corresponding to normal incidence, ν_1 to angle of incidence $\chi - \Theta$, and ν_2 to angle of incidence $\chi + \Theta$. The new filter peak can be assumed to be

$$\frac{1}{2}(\nu_1 + \nu_2) = \frac{\chi^2 + \Theta^2}{2n^{*2}}\nu_0 \qquad (\chi \text{ and } \Theta \text{ in radians})$$

$$= \frac{1.52 \times 10^{-4}(\chi^2 + \Theta^2)}{n^{*2}}\nu_0 \qquad (\chi \text{ and } \Theta \text{ in degrees}). \tag{7.45}$$

The halfwidth is

$$[W_0^2 + (\nu_2 - \nu_1)^2]^{1/2}$$

where

$$(\nu_2 - \nu_1) = \frac{2\chi\Theta}{n^{*2}}\nu_0 \qquad (\chi \text{ and } \Theta \text{ in radians})$$

$$= \frac{6.09 \times 10^{-4}\chi\Theta}{n^{*2}}\nu_0 \qquad (\chi \text{ and } \Theta \text{ in degrees}) \tag{7.46}$$

and the peak transmittance is

$$\frac{W_0}{(\nu_2 - \nu_1)}\tan^{-1}\left(\frac{(\nu_2 - \nu_1)}{W_0}\right) \simeq 1 - \frac{1}{3}\left(\frac{(\nu_2 - \nu_1)}{W_0}\right)^2. \tag{7.47}$$

$(\nu_2 - \nu_1)$ is proportional to $\Theta\chi$ and Hernandez[15] has found excellent agreement between measurements made on real filters and calculations from these expressions for values of $\Theta\chi$ up to 100 degrees^2.

We can illustrate the use of these expressions in calculating the performance of a zinc sulphide and cryolite filter for the visible region. We assume that this is a low-index first-order filter with a bandwidth of 1%.

For this filter we calculate that $n^* = 1.55$. We take 10% reduction in peak transmittance as the limit of what is acceptable. Then, from equation (7.47)

$$(\nu_2 - \nu_1)/W_0 = 0.55$$

and the increased halfwidth which corresponds to this reduction in peak transmittance is

$$(1+0.55^2)^{1/2}\,W_0 = 1.14\,W_0$$

or an increase of 14% over the basic width.

At normal incidence, the cone semiangle which can be tolerated is given by

$$1.5\times10^4(\Theta^2/n^{*2}) = \Delta\nu = 0.55\,W_0 = 0.55\times0.01 \qquad (\Theta \text{ in degrees})$$

i.e.

$$\Theta = [1.55^2\times0.55\times0.01/(1.5\times10^{-4})]^{1/2} = 9.4^\circ.$$

Such a cone at normal incidence will cause a shift in the position of the peak towards shorter wavelengths or higher frequencies of

$$\tfrac{1}{2}(\Delta\nu'/\nu_0) = (\tfrac{1}{2}\times0.55\times0.01) = 0.275\%.$$

Used at oblique incidence in a cone of illumination we have

$$(6.09\times10^{-4}\chi\Theta/n^{*2})\nu_0 = \nu_2-\nu_1 = 0.55\times0.01$$

i.e.

$$\chi\Theta = \frac{1.55^2\times0.55\times0.01}{6.09\times10^{-4}} = 21.7\text{ degrees}^2$$

which means that the filter can be used in a cone of semiangle 2° up to an angle of incidence of 21.7/2 = 10.9° or of semiangle 3° up to an angle of incidence of 7° and so on.

One very important result is the shift in peak wavelength in a cone at normal incidence which indicates that if a filter is to be used at maximum efficiency in such an arrangement, its peak wavelength at normal incidence in collimated light should be slightly longer to compensate for this shift.

Sideband blocking

There is a disadvantage in the all-dielectric filter that the high-reflectance zone of the reflecting coating is limited in extent and hence the rejection zone of the filter is also limited. In the near ultraviolet, visible and near infrared regions, the transmission sidebands on the shortwave side of the peak can usually be suppressed, or blocked, by an absorption filter with a longwave-pass characteristic in the same way as for metal–dielectric filters. The longwave sidebands are more of a problem. These may be outside the range of sensitivity of the detector and therefore may not require elimination, but if they are troublesome then the usual technique for removing them is the addition of a metal–dielectric first-order filter with no longwave sidebands. It is usually very much broader than the narrowband component in order that the peak transmittance may be high. The metal–dielectric component is usually added as a separate component, but it can be deposited over the basic Fabry–Perot. Rather than a simple Fabry–Perot filter, a double cavity metal–dielectric is commonly used. Multiple cavity filters are the next topic of discussion.

MULTIPLE CAVITY FILTERS

The transmission curve of the basic all-dielectric Fabry–Perot filter is not of ideal shape. It can be shown that one half of the energy transmitted in any order lies outside the halfwidth (assuming an even distribution of energy with frequency in the incident beam). A more nearly rectangular curve would be a great improvement. Further, the maximum rejection of the Fabry–Perot is completely determined by the halfwidth and the order. The broader filters, therefore, tend to have poor rejection as well as a somewhat unsatisfactory shape.

When tuned electric circuits are coupled together, the resultant response curve is rather more rectangular and the rejection outside the pass band rather greater than a single tuned circuit, and a similar result is found for the Fabry–Perot filter. If two or more of these filters are placed in series, much the same sort of double peaked curve is obtained; it has, however, a much more promising shape than the single filter. The filters may be either metal–dielectric or all-dielectric and the basic form is

| reflector | half-wave | reflector | half-wave | reflector |

known as a double half-wave or DHW filter or as a double cavity or *two-cavity* filter. Some typical examples of all-dielectric DHW or two-cavity filters are shown in figure 7.14.

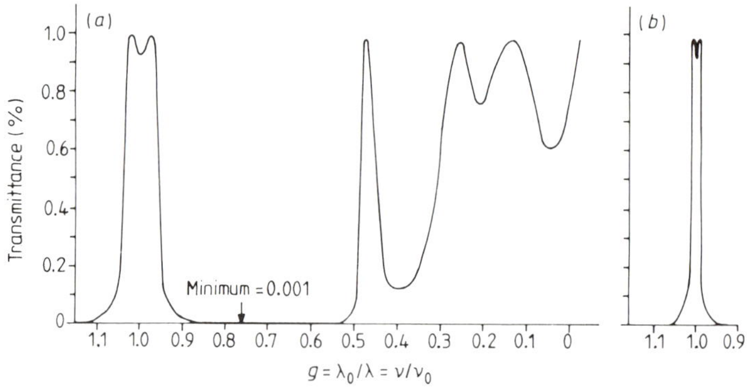

Figure 7.14 (*a*) Computed transmittance of *HLLHLHLLH*. (*b*) Computed transmittance of *HLHHLHLHHLH*. In both cases $n_H = 4.0$ and $n_L = 1.35$. (After Smith[18].)

Such filters were certainly constructed by A F Turner and his co-workers at Bausch and Lomb in the early 1950s but the results were published only as quarterly reports in the Fort Belvoir Contract Series over the period 1950–68[16]. The earliest filters were of the triple half-wave type, known at Bausch and Lomb as WADIS (Wide Band All Dielectric Interference filters)[17]. Double half-wave, or two-cavity, filters came later but were in routine use at

Bausch and Lomb certainly by 1957. They were initially known as TADIS. The Fort Belvoir Contract† Reports make fascinating reading and show just how advanced the work at Bausch and Lomb was at that time. Use was being made of the concept of equivalent admittance for the design both of WADI filters and of the edge filters for blocking the sidebands. Multilayer antireflection coatings were also well understood.

The first complete account of a theory applicable to multiple half-wave filters was published by Smith[18] and it is his method that we follow first here.

The reflecting stacks in the classical Fabry–Perot filter have more or less constant reflectance over the pass band of the filter. A dispersion of phase change on reflection does, as we have seen, help to reduce the bandwidth, but this does so without altering the basic shape of the pass band shape. Smith suggested the idea of using reflectors with much more rapidly varying reflectance to achieve a better shape. The essential expression for the transmission of the complete filter has already been derived on p 52 where we have assumed $\beta = 0$, that is, no absorption in the spacer layer. From Smith's formula, equation (2.80),

$$T = \frac{|\tau_a^+|^2 |\tau_b^+|^2}{(1 - |\rho_a^-||\rho_b^+|)^2} \left[1 + \frac{4|\rho_a^-||\rho_b^+|}{(1 - |\rho_a^-||\rho_b^+|)^2} \sin^2 \frac{\phi_a + \phi_b - 2\delta}{2}\right]^{-1} \tag{7.48}$$

it can be seen that high transmission can be achieved at any wavelength if, and only if, the reflectances on either side of a chosen spacer layer are equal. Of course the phase condition must be met too, but this can be arranged by choosing the correct spacer thickness to make

$$\left|\frac{\phi_a + \phi_b}{2} - \delta\right| = m\pi.$$

In these expressions, the symbols have the same meanings as given in figure 2.12.

Smith pointed out the advantage of having reasonably low reflectance in the region around the peak wavelength, which means that absorption is less effective in limiting the peak transmittance. In the Fabry–Perot filter, low reflectance means wide bandwidth, but Smith limited the bandwidth by arranging for the reflectances to begin to differ appreciably at wavelengths only a little removed from the peak. This is illustrated in figure 7.15. The figure shows what is the simplest type of DHW filter, which has construction *HHLHH*. The *HH* layers are the two half-wave spacers and the *L* layer is a coupling layer. In the discussion which follows, for simplicity we shall ignore any substrate. The behaviour of the filter is described in terms of the reflectances on either side of one of the two spacers. R_1 is the reflectance of the interface between the

† These reports were obtainable from the Engineer Research and Development Laboratories, Fort Belvoir, Virginia 22060, USA, but are now out of print.

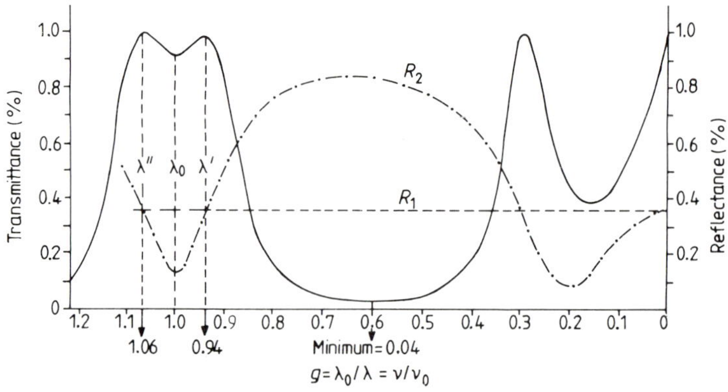

Figure 7.15 Computed transmittance of *HHLHH* and explanatory reflectance curves R_1 and R_2 ($n_H = 4.0$, $n_L = 1.35$). (After Smith[18].)

high index and the surrounding medium, which we take as air with index unity, and is a constant. R_2 is the reflectance of the assembly on the other side of the spacer and is low at the wavelength at which the spacer is a half-wave and rises on either side. At wavelengths λ' and λ'', the reflectances R_1 and R_2 are equal and we would expect to see high transmission if the phase condition is met, which in fact it is. The transmission of the assembly is also shown in the figure and the shape can be seen to consist of a steep-sided pass band with two peaks close together and only a slight dip in transmission between the peaks, much more like the ideal rectangle than the shape of the Fabry–Perot filter.

Smith's formula for the transmittance of a filter can be written:

$$T(\lambda) = T_0(\lambda)\frac{1}{1 + F(\lambda)\sin^2[(\phi_1 + \phi_2)/2 - \delta]} \tag{7.49}$$

where

$$T_0(\lambda) = \frac{(1 - R_1)(1 - R_2)}{[1 - (R_1 R_2)^{1/2}]^2} \tag{7.50}$$

$$F(\lambda) = \frac{4(R_1 R_2)^{1/2}}{[1 - (R_1 R_2)^{1/2}]^2}. \tag{7.51}$$

Both these quantities are now variable since they involve R_2, which is a variable. The form of the functions is also shown in figure 7.16. At wavelengths removed from the peak, $T_0(\lambda)$ is low and $F(\lambda)$ is high, the combined effect being to increase the rejection. In the region of the peak, T_0 is high, and, just as important, F is low, producing high transmittance which is not sensitive to the effects of absorption. As we have shown before, the peak transmittance is dependent on the quantity A/T, where A is the absorptance and T the transmittance of the reflecting stacks. Clearly, the greater T is, the higher A can be for the same overall filter transmittance.

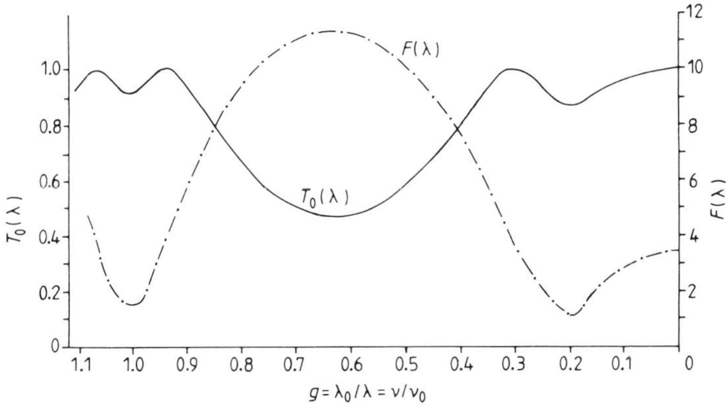

Figure 7.16 $T_0(\lambda)$ and $F(\lambda)$ for *HHLHH*. (After Smith[18].)

The typical double-peaked shape of the double half-wave filter results from the intersection of the R_1 and R_2 curves at two separate points. Two other cases can arise. The curves can intersect at one point only, in which case the system has a single peak whose transmittance is theoretically unity, or the curves may never intersect at all, in which case the system will show a single peak of transmittance rather less than unity, the exact magnitude depending on the relative magnitudes of R_1 and R_2 at their closest approach. This third case is to be avoided in design. For the twin-peaked filter, a requirement is that the trough in the centre between the two peaks should be shallow, which means that R_1 and R_2 should not be very different at λ_0.

Having examined the simplest type of DHW filter, we are in a position to study more complicated ones. What we have to look for is a system of two reflectors, where one of the reflectors remains reasonably constant over the range of interest and where the other should be equal, or nearly equal, to the first over the pass band region, but should increase sharply outside the pass band. The straightforward Fabry–Perot filter has effectively zero reflectance at the peak wavelength, but the reflectance rapidly rises on either side of the peak. If, then, a simple quarter-wave stack is added to the Fabry–Perot, the resultant combination should have the desired property, that is, the reflectance equal to that of the simple stack at the centre wavelength and increasing sharply on either side. We can therefore use a simple stack as one reflector, with more or less constant reflectance, on one side of the spacer, and, on the other side, an exactly similar stack combined with a Fabry–Perot filter. This will result in a single-peaked filter since the reflectances in this way will be exactly matched at λ_0. The double-peaked transmission curve will be obtained if the reflectance of the stack plus the Fabry–Perot filter is arranged to be just a little less than the reflectance of the stack by itself. This is the arrangement that is more often used and it involves the insertion of an extra quarter-wave layer between the stack and the Fabry–Perot. This layer appears as a sort of coupling layer in the filter. Figure 7.17 should make the situation clear.

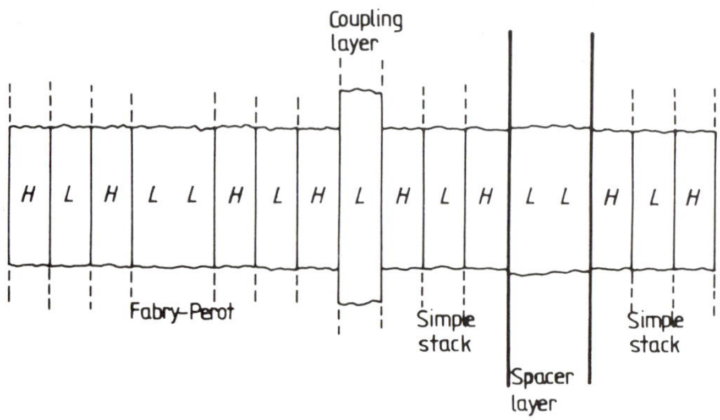

Figure 7.17 The construction of a DHW filter.

So far we have not given any consideration to the substrate of the filter. The substrate will be on one side of the spacer and will alter the reflectance on that side. This change in reflectance can easily be calculated, particularly if the substrate is considered to be on the same side of the spacer as the simple stack. The constant reflectance R_1 of the simple stack will generally be large, and if the substrate index is given by n_s, then the transmittance of the stack on its own, $(1-R_1)$, will become either $(1-R_1)/n_s$ if the index of the layer next to the substrate is low, or $n_s(1-R_1)$ if it is high.

Since this change in reflectance could be considerable, especially if n_s is large, the substrate must be taken into account in the design and this should be done right from the beginning. The substrate can be considered part of the simple stack and R_1 can be adjusted to include it. Provided the reflectances of the two assemblies on either side of the spacer layer are arranged always to be equal at the appropriate wavelengths, the transmittance of the complete filter will be unity.

For example, let us consider the case of a filter deposited on a germanium substrate using zinc sulphide for the low-index layers and germanium for the high ones. Let the spacer be of low index and let the reflecting stack on the germanium substrate be represented by Ge | *LHLL*, where the *LL* layer is the spacer. The transmittance of the stack into the spacer layer will be approximately $T_1 = 4n_L^3/n_H^2 n_{\mathrm{Ge}}$, which, since the substrate is the same material as the high-index layer, becomes $4n_L^3/n_H^3$. On the other side of the spacer layer we make a start with the combination *LLHLH*|air, representing the basic reflecting stack, where *LL* once again is the spacer layer. This has transmission $T_2 = 4n_L^3/n_H^4$, which is $1/n_H$ times T_1. Clearly this is too unbalanced and an adjustment to this second stack must be made. If a low-index layer is added next to the air, then the transmission becomes $T_2 = 4n_L^5/n_H^4$. Since n_L^2 is approximately equal to the index of germanium, the transmittances T_1 and T_2 are now equal and the Fabry–Perot filter can be added to the second stack to give the desired shape to the reflectance curve. The Fabry–Perot can take any form, but it is convenient here to use a combination almost exactly the same as the combination of two stacks and a spacer layer which has already been arrived at. The complete design of filter is then:

$$\text{Ge}\,|\,LH\;LL\;HLH\;L\;HLH\;LL\;HLH\,|\,\text{air}$$

and the performance of the filter is shown in figure 7.18.

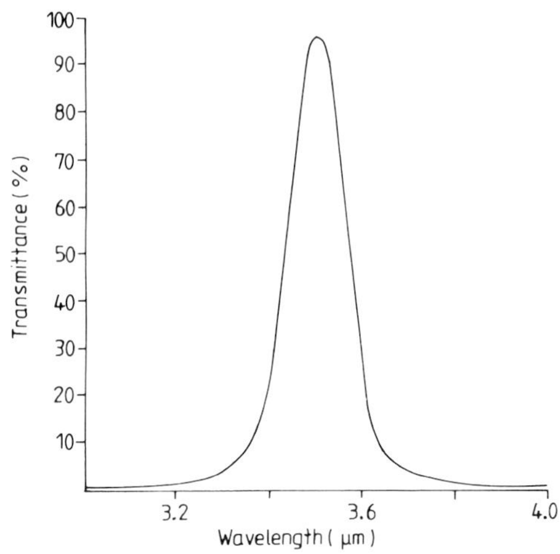

Figure 7.18 Computed transmittance of the double half-wave filter. Design: Air|$LHLLHLHLHLHLLHLH$|Ge. The substrate is germanium ($n = 4.0$); $H = $ germanium ($n = 4.0$), $L = $ zinc sulphide ($n = 2.35$) and the incident medium is air ($n = 1.0$).

An alternative way of checking whether or not the filter is going to have high transmission uses the concept of absentee half-wave layers. The layers in DHW filters are usually either of quarter- or half-wave thickness at the centre of the pass band, as in the above filter, and we can take it as an example to illustrate the method. First we note that the two spacers are both half-wave layers and that they can be eliminated without affecting the transmission. The filter, at the centre wavelength, will have the same transmittance as

$$\text{Ge}\,|\,LH\;HLH\;L\;HLH\;HLH\,|\,\text{air}.$$

In this there are two sets of HH layers which can be eliminated in the same way, leaving two sets of LL layers which can be removed in their turn. Almost all the layers in the filter can be eliminated in this way leaving ultimately

$$\mathrm{Ge}\,|\,L\,|\,\mathrm{air}.$$

As we already know, a single quarter-wave of zinc sulphide is a good antireflection coating for germanium, and so the transmittance of the filter will be high in the centre of the pass band. Any type of DHW filter can be dealt with in this way.

Knittl[19,20] has used an alternative multiple beam approach to study the design of DHW filters. Basically he has applied a multiple beam summation to the first cavity, the results of which are then used in a multiple beam summation for the second cavity. This yields an expression which is not unlike Smith's, although slightly more complicated, but which has the advantage that it is only the phase which varies across the pass band. The magnitude of the reflection and transmission coefficients can be safely assumed constant and this means that the parameters which involve these quantities are also constant. The form of the expression for overall transmittance is then very much easier to manipulate so that the positions and values of maxima and minima in the pass band can be readily determined. We shall not deal further with the method here, because it is already well covered by Knittl[20].

Of course, the possible range of designs does not end with the DHW filter. Other types of filter exist involving even more half-waves. An early type of filter, which we have already mentioned, was the WADI which was devised by Turner and which consisted of a straightforward Fabry–Perot filter, to either side of which was added a half-wave layer together with several quarter-wave layers. The function of these extra layers was to alter the phase characteristics of the reflectors on either side of the primary spacer layer, so that the pass band was broadened and at the same time the sides became steeper. Similarly, it is possible to repeat the basic Fabry–Perot element used in the DHW filter once more to give a triple half-wave or THW filter, which has a similar bandwidth but steeper sides. WADI and THW filters are much the same thing, although the original design philosophy was a little different, and usually the term THW is taken as referring to all types having three half-wave spacers. Even more spacer layers may be used giving multiple half-wave filters. The method which we have been using for the analysis of the filters becomes rather cumbersome when many half waves are involved—even the simple method for checking that the transmittance is high in the pass band breaks down, for reasons which will be made clear in the next section, where we shall consider a very powerful design method which has been devised by Thelen.

Thelen's method of analysis

We have not yet arrived at any ready way of calculating the bandwidth of DHW and THW filters. The design method has merely ensured that the transmittance of the filter is high in the pass band and that the shape of the transmission curve is steep-sided. The bandwidth can be calculated, but to arrive at a prescribed bandwidth in the design has to be achieved by trial and error. It can indeed be calculated using the formula for transmittance

$$T = T_0 \frac{1}{1 + F_0 \sin^2 \delta}$$

but this can be very laborious as the phases of the reflectances have to be included in δ. This expression has been very useful in achieving an insight into the basic properties of the multiple half-wave filter, but, for systematic design, a method based on the concept of equivalent admittance will be found much more useful.

As was shown in chapter 6, any symmetrical assembly of thin films can be replaced by a single layer of equivalent admittance and optical thickness which both vary with wavelength, but which can be calculated. This concept has been used by Thelen[21] in the development of a very powerful systematic design method which predicts all the performance features of the filters including the bandwidth. The basis of the method is the splitting of the multiple half-wave filter into a series of symmetrical periods, the properties of which can be predicted by finding the equivalent admittance. Take for example the design we have already examined

$$\text{Ge}\,|\,LHLLHLHLHLHLLHLH\,|\,\text{air}.$$

This can be split up into the arrangement

$$\text{Ge}|LHL \quad LHLHLHLHL \quad LHLH|\text{air}.$$

The part of the filter which determines its properties is the central section $LHLHLHLHL$ which is a symmetrical assembly. It can therefore be replaced by a single layer having the usual series of high-reflectance zones where the admittance is imaginary, and pass zones where the admittance is real. We are interested in the latter because they represent the pass bands of the final filter. The symmetrical section must then be matched to the substrate and the surrounding air, and matching layers are added for that purpose on either side. This is the function of the remaining layers of the filter. The condition for perfect matching is easily established because the layers are all of quarter-wave optical thicknesses.

A most useful feature of this design approach is that the central section of the filter can be repeated many times, steepening the edges of the pass band and improving the rejection without affecting the bandwidth to any great extent.

In order to make predictions of performance straightforward, Thelen has computed formulae for the bandwidth of the basic sections. We use Thelen's technique here, with some slight modifications, in order to fit in with the pattern of analysis already carried out for the Fabry–Perot. In order to include filters of order higher than the first, we write the basic period as

$$H^m LHLHLH \ldots LH^m \qquad \text{or} \qquad L^m HLHLHL \ldots HL^m$$

where there are $2x+1$ layers, $x+1$ of the outermost index and x of the other, and m is the order number. We have already mentioned how Seeley[4], in the course of developing expressions for the Fabry–Perot filter, arrived at an approximate formula for the product of the characteristic matrices of quarter-wave layers of alternating high and low indices. Using an approach similar to Seeley's, we can put the characteristic matrix of a quarter-wave layer in the form:

$$\begin{bmatrix} -\varepsilon & \mathrm{i}/n \\ \mathrm{i}n & -\varepsilon \end{bmatrix} \tag{7.52}$$

where $\varepsilon = (\pi/2)(g-1)$ and $g = \lambda_0/\lambda$. This expression is valid for wavelengths close to that for which the layer is a quarter-wave. First let us consider m odd, and write m as $2q+1$. Then, to the same degree of approximation, the matrix for H^m or L^m is

$$(-1)^q \begin{bmatrix} -m\varepsilon & \mathrm{i}/n \\ \mathrm{i}n & -m\varepsilon \end{bmatrix}.$$

Neglecting terms of second and higher order in ε, then the product of the $2x-1$ layers making up the symmetrical period is

$$\begin{bmatrix} M_{11} & \mathrm{i}M_{12} \\ \mathrm{i}M_{21} & M_{22} \end{bmatrix} \tag{7.53}$$

where

$$M_{11} = M_{22} = (-1)^{x+2q}(-\varepsilon)\left[m\left(\frac{n_1}{n_2}\right)^x + \left(\frac{n_1}{n_2}\right)^{x-1} + \left(\frac{n_1}{n_2}\right)^{x-2} + \ldots + \left(\frac{n_2}{n_1}\right)^{x-1} + m\left(\frac{n_2}{n_1}\right)^x \right]$$

$$\mathrm{i}M_{12} = \mathrm{i}(-1)^x/[(n_1/n_2)^x n_1]$$

and

$$\mathrm{i}M_{21} = \mathrm{i}(-1)^x[(n_1/n_2)^x n_1].$$

Now it is not easy from this expression to derive the halfwidth of the final filter analytically. Instead of deriving the halfwidth, therefore, Thelen chose to define the edges of the pass band as those wavelengths for which

$$\frac{1}{2}\left| M_{11} + M_{22} \right| = 1$$

or, since $M_{11} = M_{22}$,

$$|M_{11}| = 1.$$

These points will not be too far removed from the half peak transmission points, especially if the sides of the pass band are steep. Applying this to equation (7.53), we obtain

$$|M_{11}| = \varepsilon\left[m\left(\frac{n_1}{n_2}\right)^x + \left(\frac{n_1}{n_2}\right)^{x-1} + \ldots + \left(\frac{n_2}{n_1}\right)^{x-1} + m\left(\frac{n_2}{n_1}\right)^x \right]. \tag{7.54}$$

Now, this expression is quite symmetrical in terms of n_1 and n_2. Then if we replace n_1 and n_2 by n_H and n_L, regardless of which is which, we will obtain the same expression

$$\varepsilon\left[m\left(\frac{n_H}{n_L}\right)^x+\left(\frac{n_H}{n_L}\right)^{x-1}+\left(\frac{n_H}{n_L}\right)^{x-2}+\ldots+\left(\frac{n_L}{n_H}\right)^{x-1}+m\left(\frac{n_L}{n_H}\right)^x\right]=1$$

i.e.

$$\varepsilon\left[(m-1)\left(\frac{n_H}{n_L}\right)^x+(m-1)\left(\frac{n_L}{n_H}\right)^x+\left(\frac{n_H}{n_L}\right)^x\left(\frac{1-(n_L/n_H)^{x+1}}{1-(n_L/n_H)}\right)\right]=1$$

where we have used the formula for the sum of a geometric series just as in the case of the Fabry–Perot. We now neglect terms of power x or higher in (n_L/n_H) to give

$$\varepsilon\left(\frac{n_H}{n_L}\right)^x\left((m-1)+\frac{1}{1-(n_L/n_H)}\right)=1$$

i.e.

$$\varepsilon=\left(\frac{n_L}{n_H}\right)^x\frac{[1-(n_L/n_H)]}{[m-(m-1)(n_L/n_H)]}. \tag{7.55}$$

The bandwidth will be given by

$$\left|\frac{\Delta\lambda_B}{\lambda_0}\right|=\left|\frac{\Delta\nu_B}{\nu_0}\right|=2(g-1)=\frac{4\varepsilon}{\pi}$$

so that, manipulating equation (7.55) slightly,

$$\left|\frac{\Delta\lambda_B}{\lambda_0}\right|=\frac{4}{m\pi}\left(\frac{n_L}{n_H}\right)^x\frac{(n_H-n_L)}{(n_H-n_L+n_L/m)}. \tag{7.56}$$

The equivalent admittance is given by

$$\eta_E=\left(\frac{M_{21}}{M_{12}}\right)^{1/2}=\left(\frac{n_1}{n_2}\right)^x n_1. \tag{7.57}$$

The case of m even, i.e. $m=2q$, is arrived at similarly. Here the matrix of H^m or L^m is

$$(-1)^q\begin{bmatrix}1 & im\varepsilon/n\\ im\varepsilon n & 1\end{bmatrix}$$

and a similar multiplication, neglecting terms higher than first in ε gives

$$\frac{\Delta\lambda_B}{\lambda_0}=\frac{4}{m\pi}\left(\frac{n_L}{n_H}\right)^x\frac{(n_H-n_L)}{(n_H-n_L+n_L/m)}$$

that is, exactly as equation (7.56), but

$$\eta_E=\left(\frac{n_2}{n_1}\right)^{x-1}n_2 \tag{7.58}$$

for equivalent admittance. This is to be expected since the layers L^m or H^m act as absentees because of the even value of m.

Table 7.3

Matching combination for germanium	Symmetrical period	Matching combination for air
Ge|L	LHL	|air (already matched)
Ge|LH	$HLHLH$	H|air
Ge|LHL	$LHLHLHL$	LH|air
Ge|$LHLH$	$HLHLHLHLH$	HLH|air
Ge|$LHLHL$	$LHLHLHLHLHL$	$LHLH$|air

L: ZnS, $n_L = 2.35$ H: Ge, $n_H = 4.0$

The validity of any of these combinations can easily be tested. Take for example the fourth one, with the nine-layer period in the centre. Here the equivalent admittance of the symmetrical period is $E = n_H^5/n_L^4$. The $LHLH$ section between the germanium substrate and the centre section transforms the admittance into

$$\frac{n_L^4}{n_H^4}\frac{n_H^5}{n_L^4} = n_H$$

which is a perfect match for germanium. The matching section at the other end is HLH and this transforms the admittance into

$$\frac{n_H^4}{n_L^2}\frac{n_L^4}{n_H^5} = \frac{n_L^2}{n_H}$$

which, because zinc sulphide is a good antireflection material for germanium, gives a good match for air.

For higher-order filters, the method of designing the matching layers is similar. However, we can choose, if we wish, to add half-wave layers to that part of the matching assembly next to the symmetrical period in order to make the resulting cavity of the same order as the others. For example, the period $HHHLHLHLHLHHH$, based on the fourth example of table 7.3, can be matched either by Ge|$LHLH$ and HLH|air, as shown, or by Ge|$LHLHHH$ and $HHHLH$|air, making all cavities of identical order regardless of the number of periods.

This method, then, gives the information necessary for the design of multiple half-wave filters. The edge steepness and rejection in the stop bands will determine the number of basic symmetrical periods in any particular case. Usually, because of the approximations which have been used in establishing the various formulae, and also because the definition used for bandwidth is not necessarily the halfwidth although it would not be too far removed from it, it is advisable to check the design by accurate computation before actually manufacturing the filter. It may also be advisable to make an estimate of the permissible errors which can be tolerated in the manufacture because it is pointless attempting to achieve a performance beyond the capabilities of the process. The result will just be worse than if a less demanding specification had been attempted. The estimation of manufacturing errors is a subject which has not received much attention in the literature on thin-film filters. A brief discussion of permissible errors is given in chapter 10, pp 434–43, with some examples of calculations applied to multiple half-wave filters. Typical multiple half-wave filters are shown in figure 7.20.

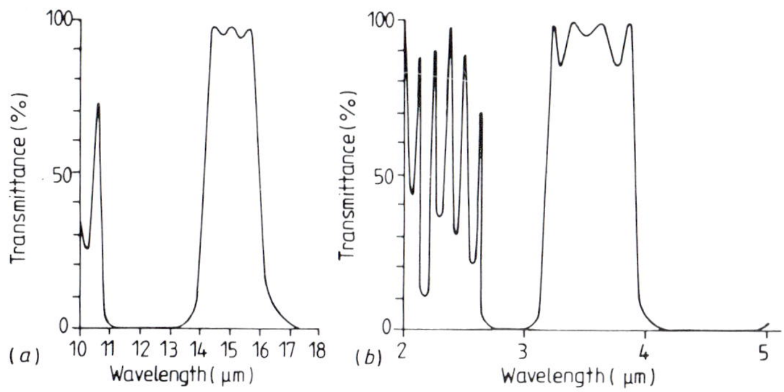

Figure 7.20 (*a*) Transmittance of a multiple half-wave filter. Design: Air|*HHLH LHHL LHHLH*|Ge with H = PbTe ($n = 5.0$); L = ZnS ($n = 2.35$), $\lambda_0 = 15$ μm. (*b*) Transmittance of a multiple half-wave filter. Design: Air|*HHLH LHHLH LHHLH LHHLH LHH*|silica H = Ge ($n = 4.0$); L = ZnS ($n = 2.35$); silica substrate ($n = 1.45$) $\lambda_0 = 3.5$ μm (courtesy of Sir Howard Grubb, Parsons & Co Ltd).

Effect of tilting

A feature of the design not so far mentioned is the sensitivity to changes in angle of incidence. Thelen[21] has examined this aspect and for those types which involve symmetrical periods consisting of quarter-waves of alternating high and low index and where the spacers are of the first order, he arrived at exactly the same expressions as those of Pidgeon and Smith for the Fabry–Perot. As far as angular dependence is concerned, the filter behaves as if it were a single layer with an effective index of

$$n^* = (n_1 n_2)^{1/2}$$

where $n_1 > n_2$ or

$$n^* = \frac{n_1}{[1 - (n_1/n_2) + (n_1/n_2)^2]^{1/2}}$$

where $n_2 > n_1$.

For higher-order filters, therefore, we should be safe in making use of expressions (7.33) and (7.35).

Losses in multiple cavity filters

Losses in multiple cavity filters can be estimated in the same way as for the Fabry–Perot filter. There are so many possible designs that a completely general approach would be very involved. However, we can begin by assuming that the basic symmetrical unit is perfectly matched at either end. The scheme of admittances through the basic unit will then be as shown in table 7.4.

Table 7.4

	n_1^x/n_2^{x-1}
n_1	
	n_2^{x-1}/n_1^{x-2}
n_2	
	n_1^{x-2}/n_2^{x-3}
n_1	
	n_2^{x-3}/n_1^{x-4}
$\vdots$	
	n_1^{x-2}/n_2^{x-3}
n_2	
	n_2^{x-1}/n_1^{x-2}
n_1	
	n_1^x/n_2^{x-1}

x layers of n_1
$(x-1)$ layers of n_2

Then, in the same way as for the Fabry–Perot, we can write

$$\begin{aligned}\sum\mathscr{A} &= \beta_1\left[\left(\frac{n_1}{n_2}\right)^{x-1}+\left(\frac{n_2}{n_1}\right)^{x-1}\right]+\beta_2\left[\left(\frac{n_2}{n_1}\right)^{x-2}+\left(\frac{n_1}{n_2}\right)^{x-2}\right]\\ &\quad+\beta_1\left[\left(\frac{n_1}{n_2}\right)^{x-3}+\left(\frac{n_2}{n_1}\right)^{x-3}\right]+\ldots+\beta_2\left[\left(\frac{n_2}{n_1}\right)^{x-2}+\left(\frac{n_1}{n_2}\right)^{x-2}\right]\\ &\quad+\beta_1\left[\left(\frac{n_1}{n_2}\right)^{x-1}+\left(\frac{n_2}{n_1}\right)^{x-1}\right]\\ &=\beta_1\left\{\left[\left(\frac{n_1}{n_2}\right)^{x-1}+\left(\frac{n_1}{n_2}\right)^{x-3}+\ldots+\left(\frac{n_2}{n_1}\right)^{x-1}\right]\right.\\ &\quad\left.+\left[\left(\frac{n_2}{n_1}\right)^{x-1}+\left(\frac{n_2}{n_1}\right)^{x-2}+\ldots+\left(\frac{n_1}{n_2}\right)^{x-1}\right]\right\}\end{aligned}$$

$$+\beta_2\left\{\left[\left(\frac{n_2}{n_1}\right)^{x-2}+\left(\frac{n_2}{n_1}\right)^{x-4}+\ldots+\left(\frac{n_1}{n_2}\right)^{x-2}\right]\right.$$
$$\left.+\left[\left(\frac{n_1}{n_2}\right)^{x-2}+\left(\frac{n_1}{n_2}\right)^{x-4}+\ldots+\left(\frac{n_2}{n_1}\right)^{x-2}\right]\right\}.$$

We note that the second expression of each pair is the same as the first with inverse order.

The layers are quarter waves and so we can write, as before,

$$\beta_1=\frac{\pi}{2}\frac{k_1}{n_1}\qquad\text{and}\qquad\beta_2=\frac{\pi}{2}\frac{k_2}{n_2}.$$

Once again we divide the cases into high- and low-index spacers.

Case I: high-index spacers

We replace n_1 by n_H, k_1 by k_H, n_2 by n_L and k_2 by k_L. Then, neglecting, as before, terms in $(n_L/n_H)^x$ compared with unity,

$$\sum\mathscr{A}=\frac{\pi(k_H/n_H)(n_H/n_L)^{x-1}}{1-(n_L/n_H)^2}+\frac{\pi(k_L/n_L)(n_H/n_L)^{x-2}}{1-(n_L/n_H)^2}$$
$$=\pi\left(\frac{n_H}{n_L}\right)^x\frac{n_L(k_H+k_L)}{(n_H^2-n_L^2)}$$

or, using

$$\Delta\lambda_\mathrm{B}/\lambda_0=\frac{2}{\pi}\sin^{-1}\left(\frac{2(n_1/n_2-1)}{(n_1/n_2)^x}\right)$$

and replacing $\sin^{-1}\theta$ by θ

$$\frac{\Delta\lambda_\mathrm{B}}{\lambda_0}=\frac{4}{\pi}\frac{(n_H/n_L-1)}{(n_H/n_L)^x}$$

i.e.

$$\sum\mathscr{A}=4\left(\frac{\lambda_0}{\Delta\lambda_\mathrm{B}}\right)\frac{(k_H+k_L)}{(n_H+n_L)}.$$

Now, this is the loss of one basic symmetrical unit. If further basic units are added each will have the same loss. In addition, there are the matching stacks at either end of the filter. We will not be far in error if we assume that they add a further loss equal to one of the basic symmetrical units. We can also assume that $R=0$ so that the absorption loss becomes

$$A=(p+1)\pi\left(\frac{n_H}{n_L}\right)^x\frac{n_L(k_H+k_L)}{(n_H^2-n_L^2)}\qquad(7.61)$$

or

$$A=4(p+1)\left(\frac{\lambda_0}{\Delta\lambda_\mathrm{B}}\right)\frac{(k_H+k_L)}{(n_H+n_L)}\qquad(7.62)$$

where p is the number of basic units, i.e. $(p+1)=2$ for a DHW and $(p+1)=3$ for a THW and so on.

Case II: low-index spacer

In the same way

$$A=(p+1)\pi\left(\frac{n_H}{n_L}\right)^x\frac{(n_H^2k_L+n_L^2k_H)}{n_H(n_H^2-n_L^2)} \tag{7.63}$$

or

$$A=4(p+1)\left(\frac{\lambda_0}{\Delta\lambda_B}\right)\frac{[k_L(n_H/n_L)+k_H(n_L/n_H)]}{(n_H+n_L)}. \tag{7.64}$$

Expressions (7.62) and (7.64) are $(p+1)$ times the absorption of Fabry–Perot filters with the same halfwidth, a not unexpected result.

Further information

The examples of multiple half-wave filters so far described have been for the infrared, but of course they can be designed for any region of the spectrum where suitable thin-film materials exist. An account of filters for the visible and ultraviolet is given by Barr[22]. All-dielectric filters, both of the Fabry–Perot and multiple cavity types for the near ultraviolet are described by Nielson and Ring[23]. They used combinations of cryolite and lead fluoride and of cryolite and antimony trioxide, the former for the region 250–320 nm and the latter for 320–400 nm. Apart from the techniques required for the deposition of these materials, the main difference between such filters and those for the infrared is that the values of the high and low refractive indices are much closer together, requiring more layers for the same rejection. Nielson and Ring's filters contained basic units of seventeen or nineteen layers, in most cases, so that complete DHW filters consisted of 31 or 39 layers respectively. Malherbe[24] has described a lanthanum fluoride and magnesium fluoride filter for 202.5 nm in which the basic unit had 51 layers (high-index first-order spacer), the full design being $(HL)^{12}H\ H(LH)^{25}H(LH)^{12}$ with a total number of 99 layers, giving a measured bandwidth of 2.5 nm.

PHASE DISPERSION FILTER

The phase dispersion filter represents an attempt to find an approach to the design of narrowband filters which would avoid some of the manufacturing difficulties inherent in Fabry–Perot filters. The Fabry–Perot becomes increasingly difficult to manufacture as halfwidths are reduced below 0.3 % of peak wavelength. Attempts to improve the position by using higher-order spacers are not effective when the spacer becomes thicker than the fourth order because of what has been described as increased roughness of the spacer. Much more is now known about the Fabry–Perot filter and the causes of manufacturing difficulties, and those will be dealt with in some detail in a subsequent chapter. Although the phase dispersion filter was not, as it turned out, the solution to the narrowband filter problem, nevertheless it does have very interesting properties and the philosophy behind the design is worth discussing.

The reflecting stack with extended bandwidth which was originally intended for classical Fabry–Perot plates and was described in chapter 5 shows a large dispersion of the phase change on reflection and this suggested to Baumeister and Jenkins[25] that it might form the basis for a new type of filter in which the narrow bandwidth would depend almost entirely on this phase dispersion rather than on the very high reflectances of the reflecting stacks. They called this type of filter a 'phase dispersion filter'. It consists quite simply of a Fabry–Perot all-dielectric filter which has, instead of the conventional dielectric quarter-wave stacks on either side of the spacer layer, reflectors consisting of the staggered multilayers. The rapid change in phase causes the bandwidth of the filter and the position of the peak to be much less sensitive to the errors in thickness of the spacer layer than would otherwise be the case.

The results which they themselves[25] and also with Jeppesen[26] eventually achieved were good, although they never quite succeeded in attaining the performance possible in theory. This prompted a study[27] of the influence of errors in any of the layers of a filter on the position of the peak. The idea behind this study was that random errors in both thickness and uniformity in layers other than the spacer might be responsible for the discrepancy between theory and practice. If, in a practical filter, the errors were causing the peak to vary in position over the surface of the filter, then the integrated response would exhibit a rather wider bandwidth and lower transmittance than those of any very small portion of the filter, which might well be attaining the theoretical performance. It seemed possible that there might be a design of filter which could yield the minimum sensitivity to errors and therefore give the minimum possible bandwidth with a given layer 'roughness'.

Giacomo *et al*'s findings[27] can be summarised as follows (the notation in their paper has been slightly altered to agree with that used throughout this book): the peak of an all-dielectric multilayer filter is given by

$$\frac{\phi_a + \phi_b}{2} - \delta = m\pi \tag{7.65}$$

where

$$\delta = \frac{2\pi n d_s}{\lambda} = 2\pi n d_s \nu$$

the symbols having their usual meanings.

For a change Δd_i in the ith layer, Δd_j in the jth layer and Δd_s in the spacer, the corresponding change in the wavenumber of the peak $\Delta\nu$ is given by

$$\sum_i \frac{\partial\phi_a}{\partial d_i}\Delta d_i + \sum_j \frac{\partial\phi_b}{\partial d_j}\Delta d_j - 2\frac{\partial\delta}{\partial d_s}\Delta d_s + \left(\frac{\partial\phi_a}{\partial\nu} + \frac{\partial\phi_b}{\partial\nu} - 2\frac{\partial\delta}{\partial\nu}\right)\Delta\nu = 0. \tag{7.66}$$

Now

$$\frac{\partial\delta}{\partial d_s} = 2\pi n\nu = \frac{\delta}{d_s} \tag{7.67}$$

and

$$\frac{\partial\delta}{\partial\nu} = 2\pi n d_s = \frac{\delta}{\nu} \tag{7.68}$$

and also, since d_i and ν appear in the individual thin-film matrices only in the value of $\delta_i = 2\pi n_i d_i \nu$, then

$$\sum_i \frac{\partial\phi_a}{\partial d_i}\Delta_0 d_i = \frac{\partial\phi_a}{\partial\nu}\Delta_0\nu$$

and similarly for ϕ_b, where Δ_0 indicates that the changes in d_i are related by

$$\frac{\Delta_0 d_i}{d_i} = \frac{\Delta_0\nu}{\nu}.$$

This gives

$$\frac{\partial\phi_a}{\partial\nu} - \sum_i\left(\frac{\partial\phi_a}{\partial d_i}\frac{d_i}{\nu}\right) \tag{7.69}$$

which is independent of the particular choice of Δ_0 used to arrive at it. A similar expression holds for ϕ_b. Using equations (7.67), (7.68) and (7.69) in equation (7.66):

$$\sum_i \frac{\partial\phi_a}{\partial d_i}\Delta d_i + \sum_j \frac{\partial\phi_b}{\partial d_j}\Delta d_j - 2\delta\frac{\Delta d_s}{d_s} + \left(\sum_i \frac{\partial\phi_a}{\partial d_i}d_i + \sum_j \frac{\partial\phi_b}{\partial d_j}d_j - 2\delta\right)\frac{\Delta\nu}{\nu} = 0$$

i.e.

$$\frac{\Delta\nu}{\nu} = -\left[-2\delta\alpha_s + \sum_i\left(\frac{\partial\phi_a}{\partial d_i}d_i\alpha_i\right) + \sum_j\left(\frac{\partial\phi_b}{\partial d_j}d_j\alpha_j\right)\right]$$
$$\times\left[-2\delta + \sum_i\left(\frac{\partial\phi_a}{\partial d_i}d_i\right) + \sum_j\left(\frac{\partial\phi_b}{\partial d_j}d_j\right)\right]^{-1} \tag{7.70}$$

where

$$\alpha_i = \frac{\Delta d_i}{d_i} \qquad \text{etc.}$$

Now, in a real filter, the fluctuations in thickness, or 'roughness', will be completely random in character, and in order to deal with the performance of any appreciable area of the filter, we must work in terms of the mean square deviations. Each layer in the assembly can be thought of as being a combination of a large number of thin elementary layers of similar mean thicknesses but which fluctuate in a completely random manner quite independently of each other. The RMS variation in thickness of any layer in the filter can then be considered to be proportional to the square root of its thickness. This can be written:

$$\varepsilon_i = k a_i^{1/2}$$

where k can be assumed to be the same for all layers regardless of thickness. If a_i is the RMS fractional variation of the ith layer, then

$$a_i = \frac{\varepsilon_i}{d_i} = \frac{k}{d_i^{1/2}}$$

where

$$a_i^2 = \overline{\alpha_i^2}.$$

We now define β as being

$$\beta^2 = \overline{\left(\frac{\Delta\nu}{\nu}\right)^2}.$$

Then

$$\beta^2 = \left\{4\delta^2 a_s^2 + \sum_i \left[\left(\frac{\partial\phi_a}{\partial d_i} d_i\right)^2 a_i^2\right] + \sum_j \left[\left(\frac{\partial\phi_b}{\partial d_j} d_j\right)^2 a_j^2\right]\right\} \times \left[-2\delta + \sum_i \left(\frac{\partial\phi_a}{\partial d_i} d_i\right) + \sum_j \left(\frac{\partial\phi_b}{\partial d_j} d_j\right)\right]^{-2}$$

which gives

$$\beta^2 = \left(k^2 \sum_{k=1}^{q} \frac{1}{d_k} A_k^2\right)\left(\sum_{k=1}^{q} A_k\right)^{-2} \tag{7.71}$$

where

$$A_k = \frac{\partial\phi_a}{\partial d_k} d_k \qquad \text{or} \qquad \frac{\partial\phi_b}{\partial d_k} d_k \qquad \text{or} \qquad -2\delta$$

whichever is appropriate. q is the number of layers in the filter. This expression will be a minimum when

$$A_k/d_k = A_l/d_l = \ldots \tag{7.72}$$

Then

$$\beta^2 = k^2/T \tag{7.73}$$

where T is the total thickness of the filter.

In the general case,

$$\beta \geqslant k/T^{1/2}$$

and one might hope to attain a limiting resolution of

$$R = T^{1/2}/k. \tag{7.74}$$

The condition written in equation (7.62) can be developed with the aid of equation (7.59) into

$$\frac{\partial\phi_a}{\partial d_k} = \frac{\partial\phi_b}{\partial d_l} = -4\pi n\nu$$

so that

$$\nu\left(\frac{\partial\phi_a}{\partial\nu}\right) = \sum_i \frac{\partial\phi_a}{\partial d_i} d_i = -4\pi n\nu d_m$$

and likewise for reflector b, where d_m = total thickness of the appropriate reflector and a is the index of the spacer. This gives

$$\frac{\partial \phi_a}{\partial \nu} = -4\pi n d_m. \tag{7.75}$$

This condition is necessary but not sufficient for the resolution to be a maximum and it can be used as a preliminary test of the suitability of any particular multilayer reflector which may be employed.

The classical quarter-wave stack is very far from satisfying it but the staggered multilayer is much more promising. In their paper, Giacomo *et al* compare a staggered multilayer reflector with a conventional quarter-wave stack. Both reflectors have fifteen layers, and the results are quoted for the broadband reflector at 17000 cm^{-1} and for the conventional reflector at 20000 cm^{-1}.

Equation (7.75) can be written

$$\sum_i \frac{\partial \phi_a}{\partial d_i} d_i = \sum_i \frac{\partial \phi_a}{\partial \alpha_i} = -4\pi n \nu d_m.$$

Now, from table 7.5,

$$-\sum_i \frac{\partial \phi_a}{\partial \alpha_i} = 30.662$$

and

$$4\pi n \nu d_m = 34.5$$

so that on the preliminary basis of equation (7.75) the prospects look extremely good. However this is not a sufficient condition. We must calculate the actual relationship between β and k and compare it with the theoretical condition given by equation (7.73). Now

$$A_i = d_i \frac{\partial \phi}{\partial d_i} = \frac{\partial \phi}{\partial \alpha_i}$$

which is the last column given for each reflector. This can be used in equation (7.71) giving for a filter using the broadband reflector

$$\beta = 1.023k$$

which can be compared with the value obtained in the same way for the conventional quarter-wave stack of table 7.5:

$$\beta = 1.289k.$$

For a total filter thickness of 2.35 μm the theoretical minimum value of β is given by (7.73) as

$$\beta = 0.652k$$

(k having units of $\mu m^{1/2}$).

Thus, although the phase dispersion filter using the reflectors shown in table 7.5 appears to be promising on the basis of the criterion (7.75), in the event its

Table 7.5

Layer number	Broadband film Thickness d_i (μm)	Index n	$\partial\phi/\partial d_i$ (μm^{-1})	$\partial\phi/\partial\alpha_i$	Classical film Thickness d_i (μm)	$\partial\phi/\partial d_i$ (μm^{-1})	$\partial\phi/\partial\alpha_i$
Substrate	—	1.52	—	—	—	—	—
1	0.0751	2.30	0.32	0.024	0.0543	0.01	0.001
2	0.1279	1.35	0.60	0.076	0.0926	0.02	0.002
3	0.0751	2.30	1.97	0.148	0.0543	0.05	0.003
4	0.1235	1.35	1.85	0.229	0.0926	0.06	0.005
5	0.0626	2.30	4.75	0.298	0.0543	0.16	0.009
6	0.1299	1.35	4.60	0.597	0.0926	0.16	0.015
7	0.0681	2.30	11.68	0.795	0.0543	0.48	0.026
8	0.0957	1.35	10.63	1.018	0.0926	0.48	0.044
9	0.0566	2.30	30.85	1.746	0.0543	1.39	0.075
10	0.0859	1.35	30.37	2.608	0.0926	1.39	0.128
11	0.0504	2.30	78.33	3.948	0.0543	4.03	0.219
12	0.0805	1.35	62.33	5.019	0.0926	4.03	0.373
13	0.0450	2.30	121.58	5.471	0.0543	11.69	0.635
14	0.0767	1.35	65.41	5.015	0.0926	11.69	1.082
15	0.0450	2.30	81.59	3.672	0.0543	33.92	1.843
Medium of incidence	—	1.35	—	—	—	—	—
Σ	1.1978	—	506.8	30.662	1.0829	69.53	4.460

After Giacomo *et al*[27].

performance is somewhat disappointing. It is, however, certainly better than the straightforward classical filter. So far no design which better meets the condition of equation (7.72) has been proposed.

Some otherwise unpublished results obtained by Ritchie[28] are shown in figure 7.21. This filter used zinc sulphide and cryolite as the materials on glass as

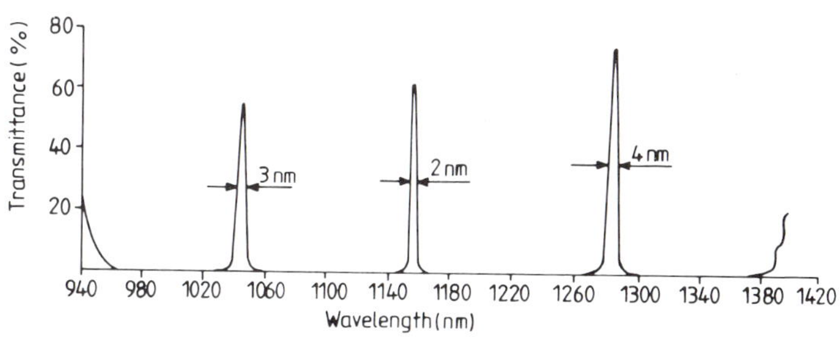

Figure 7.21 The measured transmittance of a 35-layer phase-dispersion filter. The design is given in table 7.5. (After Ritchie[28].)

Table 7.6

Layer number	Material	Optical thickness as fraction of monitoring wavelength
1	ZnS	0.2375
2	Na_3AlF_6	0.2257
3	ZnS	0.2143
4	Na_3AlF_6	0.2036
5	ZnS	0.1934
6	Na_3AlF_6	0.1838
7	ZnS	0.1746
8	Na_3AlF_6	0.1649
9	ZnS	0.1576
10	Na_3AlF_6	0.1498
11	ZnS	0.1423
12	Na_3AlF_6	0.1352
13	ZnS	0.1285
14	Na_3AlF_6	0.1220
15	ZnS	0.1159
16	Na_3AlF_6	0.1101
17	ZnS	0.1046
Spacer	Na_3AlF_6	0.5000

These seventeen layers are followed by another seventeen layers which are a mirror image of the first seventeen.

substrate. Its design is given in table 7.6. An experimental filter monitored at 1.348 μm gave peaks with corresponding bandwidths of

1.047 μm, bandwidth 3.0 nm

1.159 μm, bandwidth 2.5 nm

1.282 μm, bandwidth 4.0 nm.

Theoretically, the bandwidths should have been 0.8 nm, 1.7 nm and 4.6 nm respectively.

MULTIPLE CAVITY METAL–DIELECTRIC FILTERS

Metal–dielectric filters are indispensable in suppressing the longwave sidebands of narrowband all-dielectric filters, and as filters in their own right, especially in the extreme shortwave region of the spectrum. Unlike all-dielectric filters, however, they possess the disadvantage of high intrinsic absorption. In single Fabry–Perot filters this means that the pass bands must be wide in order to achieve reasonable peak transmission and the shape is far from ideal. It is possible to combine metal–dielectric elements into multiple cavity filters which, because of their more rectangular shape, are more satisfactory but, again, losses can be high.

The accurate design procedure for such metal–dielectric filters can be lengthy and tedious and frequently they are simply designed by trial and error as they are manufactured. We have already mentioned the metal–dielectric Fabry–Perot filter. These filters may be coupled together simply by depositing them one on top of the other with no coupling layer in between.

We can illustrate this by choosing silver as our metal, which we can give an index of 0.055 – i3.32 at 550 nm[29]. The thickness of the spacer layer in the Fabry–Perot filter, as we have already noted, should be rather thinner than a half-wave at the peak wavelength to allow for the phase changes in reflection at the silver/dielectric interfaces. This phase change varies only slowly with silver thickness when it is thick enough to be useful as a reflector and we can assume, as a reasonable approximation, that it is equal to the limiting value for infinitely thick material. We can then use equation (4.5) to calculate the thickness of the spacer layer. Equation (4.5) calculates for us exactly one half of the filter because it gives the thickness of the dielectric material to yield real admittance with zero phase change at the outer surface of the metal–dielectric combination. Adding a second exactly similar structure with the two dielectric layers facing each other, so that they join to form a single spacer, yields a Fabry–Perot filter in which the phase condition, equation (7.2), is satisfied.

Let us choose a spacer of index 1.35, similar to that of cryolite. Then half the spacer thickness is given by

$$D_{\mathrm{f}} = \frac{1}{4\pi}\tan^{-1}\left(\frac{2\beta n_{\mathrm{f}}}{n_{\mathrm{f}}^2 - \alpha^2 - \beta^2}\right) \tag{7.76}$$

where $\alpha - \mathrm{i}\beta$ is the index of the metal and n_{f} that of the cryolite and the angle is in the first or second quadrant.

With $\alpha - \mathrm{i}\beta = 0.055 - \mathrm{i}3.32$ and $n_{\mathrm{f}} = 1.35$ we find

$$D_{\mathrm{f}} = 0.188\,55$$

so that the spacer thickness should be 0.3771 full waves.

We can choose a metal layer thickness of 35 nm, quite arbitrarily, simply for the sake of illustration. Our Fabry–Perot filter is then

Glass	Ag 35 nm	Cryolite 0.3771 full waves	Ag 35 nm	Glass

(the geometrical thickness being quoted for the silver and the optical thickness for the cryolite) and the DHW filter is exactly double this structure:

Glass	Ag 35 nm	Cryolite $D = 0.3771$	Ag 70 nm	Cryolite $D = 0.3771$	Ag 35 nm	Glass.

Curves of these filters are shown in figure 7.22. The peaks are slightly displaced from 550 nm because of the approximations inherent in the design procedure.

The Fabry–Perot has reasonably good peak transmission but its typical triangular shape means that its rejection is quite poor even at wavelengths far from the peak. The DHW filter has better shape but rather poorer peak transmittance. The rejection can be improved by increasing the metal thickness, but at the expense of peak transmission.

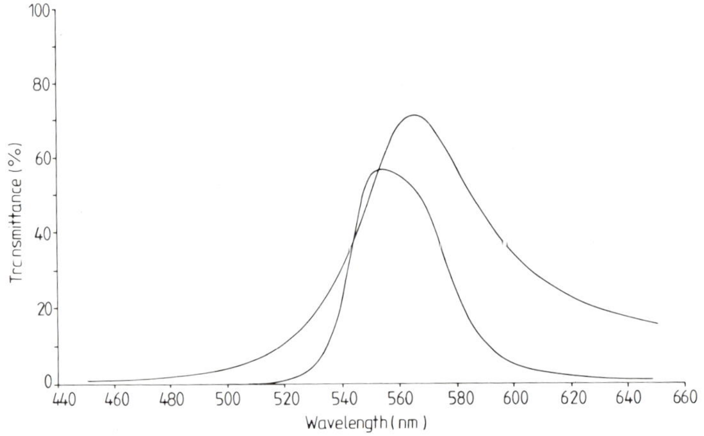

Figure 7.22 The transmittance as a function of wavelength of filters of design:

Glass	Silver 35 nm	Cryolite $0.3771\lambda_0$	Silver 35 nm	Glass

and

Glass	Silver 35 nm	Cryolite $0.3771\lambda_0$	Silver 70 nm	Cryolite $0.3771\lambda_0$	Silver 35 nm	Glass

where $\lambda_0 = 550$ nm, $n - ik = 0.055 - i3.32$ and $n_{cryolite} = 1.35$. Dispersion in the materials has been neglected.

The design approach we have described is quite crude and simply concentrates on ensuring that the peak of the filter is centred near the desired wavelength. Peak transmittance and bandwidth are either accepted as they are or a new metal thickness is tried. Performance is in no way optimised.

The unsatisfactory nature of this design procedure led Berning and Turner[30] to develop a new technique for the design of metal–dielectric filters in which the emphasis is on ensuring that maximum transmittance is achieved in the filter pass band. For this purpose they devised the concept of potential transmittance and created a new type of metal–dielectric filter known as the induced-transmission filter.

The induced-transmission filter

Given a certain thickness of metal in a filter, what is the maximum possible peak transmission, and how can the filter be designed to realise this transmission? This is the basic problem tackled and solved by Berning and Turner[30]. The development of the technique as given here is based on their approach, but it has been adjusted and adapted to conform more nearly to the general pattern of this book.

The concept of potential transmittance has already been touched on in chapter 2 and used in the analysis of losses in dielectric multilayers. We recall that the potential transmittance ψ of a layer or assembly of layers is defined as the ratio of the intensity leaving the rear surface to that actually entering at the front surface, and it represents the transmittance which the layer or assembly of layers would have if the reflectance of the front surface were reduced to zero. Thus, once the parameters of the metal layer are fixed, the potential transmittance is determined entirely by the admittance of the structure at the exit face of the layer. Furthermore, it is possible to determine the particular admittance which gives maximum potential transmittance. To achieve this transmittance it is sufficient to add a coating to the front surface to reduce the reflectance to zero. The maximum potential transmittance is a function of the thickness of the metal layer.

The design procedure is then as follows: the optical constants of the metal layer at the peak wavelength are given. Then the metal layer thickness is chosen and the maximum potential transmittance together with the matching admittance at the exit face of the layer which is required to produce that level of potential transmittance is found. Often a minimum acceptable figure for the maximum potential transmittance will exist and that will put an upper limit on the metal layer thickness. A dielectric assembly which will give the correct matching admittance when deposited on the substrate must then be designed. The filter is then completed by the addition of a dielectric system to match the front surface of the resulting metal–dielectric assembly to the incident medium. Techniques for each of these steps will be developed. The matching admittances for the metal layer are such that the dielectric stacks are efficient in matching over a limited region only, outside which their performance falls off rapidly. It is this rapid fall in performance that defines the limits of the pass band of the filter.

Before we can proceed further, we require some analytical expressions for the potential transmittance and for the matching admittance. This leads to some lengthy and involved analysis, which is not difficult but rather time-consuming.

(*a*) *Potential transmittance*

We limit the analysis to an assembly in which there is only one absorbing layer, the metal. The potential transmittance is then related to the matrix for the assembly, as shown in chapter 2

$$\begin{bmatrix} B_i' \\ C_i' \end{bmatrix} = [M] \begin{bmatrix} 1 \\ Y_e \end{bmatrix}$$

where $[M]$ is the characteristic matrix of the metal layer and Y_e is the admittance of the terminating structure. Then the potential transmittance ψ is given by

$$\psi = \frac{T}{(1-R)} = \frac{Re(Y_e)}{Re(B_i' C_i'^*)}. \tag{7.77}$$

Let

$$Y_e = X + iZ.$$

Then

$$\begin{bmatrix} B_i' \\ C_i' \end{bmatrix} = \begin{bmatrix} \cos\delta & (i\sin\delta)/y \\ iy\sin\delta & \cos\delta \end{bmatrix} \begin{bmatrix} 1 \\ X+iZ \end{bmatrix}$$

where

$$\begin{aligned} \delta &= 2\pi(n-ik)d/\lambda = 2\pi nd/\lambda - i2\pi kd/\lambda \\ &= \alpha - i\beta \\ \alpha &= 2\pi nd/\lambda \\ \beta &= 2\pi kd/\lambda. \end{aligned}$$

If free space units are used, then

$$y = n - ik.$$

Now,

$$\begin{aligned} (B_i' C_i'^*) &= [\cos\delta + i(\sin\delta/y)(X+iZ)][iy\sin\delta + \cos\delta(X+iZ)]^* \\ &= [\cos\delta + i(\sin\delta/y)(X+iZ)][-iy^*\sin\delta^* + \cos\delta^*(X-iZ)] \\ &= -iy^*\cos\delta\sin\delta^* + \frac{\sin\delta\sin\delta^* y^{*2}(X+iZ)}{yy^*} \\ &\quad + \cos\delta\cos\delta^*(X-iZ) + \frac{i\sin\delta\cos\delta^* y^*(X-iZ)(X+iZ)}{yy^*}. \end{aligned}$$

We require the real part of this and we take each term in turn.

$$\begin{aligned} -iy^*\cos\delta\sin\delta^* = -i(n+ik)(\cos\alpha\cosh\beta + i\sin\alpha\sinh\beta)(\sin\alpha\cosh\beta \\ + i\cos\alpha\sinh\beta) \end{aligned}$$

and the real part of this, after a little manipulation, is

$$Re(-iy^*\cos\delta\sin\delta^*) = n\sinh\beta\cosh\beta + k\cos\alpha\sin\alpha.$$

Similarly

$$\begin{aligned} Re\left(\frac{\sin\delta\sin\delta^* y^{*2}(X+iZ)}{yy^*}\right) = \frac{X(n^2-k^2)-2nkZ}{(n^2+k^2)}(\sin^2\alpha\cosh^2\beta \\ + \cos^2\alpha\sinh^2\beta) \end{aligned}$$

$$Re[\cos\delta\cos\delta^*(X-iZ)] = X(\cos^2\alpha\cosh^2\beta + \sin^2\alpha\sinh^2\beta)$$

$$\mathrm{Re}\left(\frac{\mathrm{i}\sin\delta\cos\delta^* y^*(X-\mathrm{i}Z)(X+\mathrm{i}Z)}{yy^*}\right)$$

$$=\frac{X^2+Z^2}{(n^2+k^2)}(n\sinh\beta\cosh\beta-k\sin\alpha\cos\alpha).$$

The potential transmittance is then

$$\psi=\left(\frac{(n^2-k^2)-2nk(Z/X)}{(n^2+k^2)}(\sin^2\alpha\cosh^2\beta+\cos^2\alpha\sinh^2\beta)\right.$$
$$+(\cos^2\alpha\cosh^2\beta+\sin^2\alpha\sinh^2\beta)$$
$$+\frac{1}{X}(n\sinh\beta\cosh\beta+k\cos\alpha\sin\alpha)$$
$$\left.+\frac{X^2+Z^2}{X(n^2+k^2)}(n\sinh\beta\cosh\beta-k\cos\alpha\sin\alpha)\right)^{-1}. \tag{7.78}$$

(b) Optimum exit admittance

Next we find the optimum values of X and Z. From equation (7.78)

$$\frac{1}{\psi}=\left(\frac{q[n^2-k^2-2nk(Z/X)]}{[n^2+k^2]}+r+\frac{p}{X}+\frac{s(X^2+Z^2)}{X(n^2+k^2)}\right) \tag{7.79}$$

where p, q, r and s are shorthand for the corresponding expressions in equation (7.78). For an extremum in ψ, we have an extremum in $1/\psi$ and hence

$$\frac{\partial}{\partial X}\left(\frac{1}{\psi}\right)=0 \qquad \text{and} \qquad \frac{\partial}{\partial Z}\left(\frac{1}{\psi}\right)=0$$

i.e.

$$\frac{q2nkZ}{X^2(n^2+k^2)}-\frac{p}{X^2}+\frac{s}{(n^2+k^2)}-\frac{sZ^2}{X^2(n^2+k^2)}=0 \tag{7.80}$$

and

$$\frac{q(-2nk)}{X(n^2+k^2)}+\frac{2sZ}{X(n^2+k^2)}=0. \tag{7.81}$$

From equation (7.81):

$$Z=nkq/s$$

and, substituting in equation (7.80),

$$X^2=p(n^2+k^2)/s-n^2k^2q^2/s^2.$$

Then, inserting the appropriate expressions for p, q and s, from equation (7.79)

$$X=\left(\frac{(n^2+k^2)(n\sinh\beta\cosh\beta+k\sin\alpha\cos\alpha)}{(n\sinh\beta\cosh\beta-k\sin\alpha\cos\alpha)}\right.$$
$$\left.-\frac{n^2k^2(\sin^2\alpha\cosh^2\beta+\cos^2\alpha\sinh^2\beta)^2}{(n\sinh\beta\cosh\beta-k\sin\alpha\cos\alpha)^2}\right)^{1/2} \tag{7.82}$$

$$Z = \frac{nk(\sin^2 \alpha \cosh^2 \beta + \cos^2 \alpha \sinh^2 \beta)}{(n \sinh \beta \cosh \beta - k \sin \alpha \cos \alpha)}. \tag{7.83}$$

We note that for β large $X \to n$ and $Z \to k$, that is:

$$Y_e \to (n + ik) = (n - ik)^*.$$

(c) *Maximum potential transmittance*

The maximum potential transmittance can then be found by substituting the values of X and Z, calculated by equations (7.82) and (7.83), into equation (7.78). All these calculations are best performed by computer or calculator and so there is little advantage in developing a separate analytical solution for maximum potential transmittance.

(d) *Matching stack*

We have to devise an assembly of dielectric layers which, when deposited on the substrate, will have an equivalent admittance of

$$Y = X + iZ.$$

This is illustrated diagramatically in figure 7.23 where a substrate of admittance $(n_s - ik_s)$ has an assembly of dielectric layers terminating such that the final equivalent admittance is $(X + iZ)$. Now, the dielectric layer circles are executed in a clockwise direction always. If we therefore reflect the diagram in the x-axis and then reverse the direction of the arrows, we get exactly the same set of circles—that is, the layer thicknesses are exactly the same—but the order is reversed (it was ABC and is now CBA) and they match a starting admittance of $X - iZ$, i.e. the complex conjugate of $(X + iZ)$, into a terminal admittance of $(n_s + ik_s)$, i.e. the complex conjugate of the substrate index. In our filters the substrate will have real admittance, i.e. $k_s = 0$, and it is a more straightforward problem to match $(X - iZ)$ into n_s than n_s into $(X + iZ)$.

There is an infinite number of possible solutions, but the simplest involves adding a dielectric layer to change the admittance $(X - iZ)$ into a real value and then to add a series of quarter-waves to match the resultant real admittance into the substrate. We will illustrate the technique shortly with several examples. At the moment we recall that the necessary analysis was carried out in chapter 4. There we showed that a film of optical thickness D given by

$$D = \frac{1}{4\pi} \tan^{-1}\left(\frac{2Zn_f}{(n_f^2 - X^2 - Z^2)}\right) \tag{7.84}$$

(where the tangent is taken in the first or second quadrant) will convert an admittance $(X - iZ)$ into a real admittance of value

$$\mu = \frac{2Xn_f^2}{(X^2 + Z^2 + n_f^2) + [(X^2 + Z^2 + n_f^2)^2 - 4X^2 n_f^2]^{1/2}}. \tag{7.85}$$

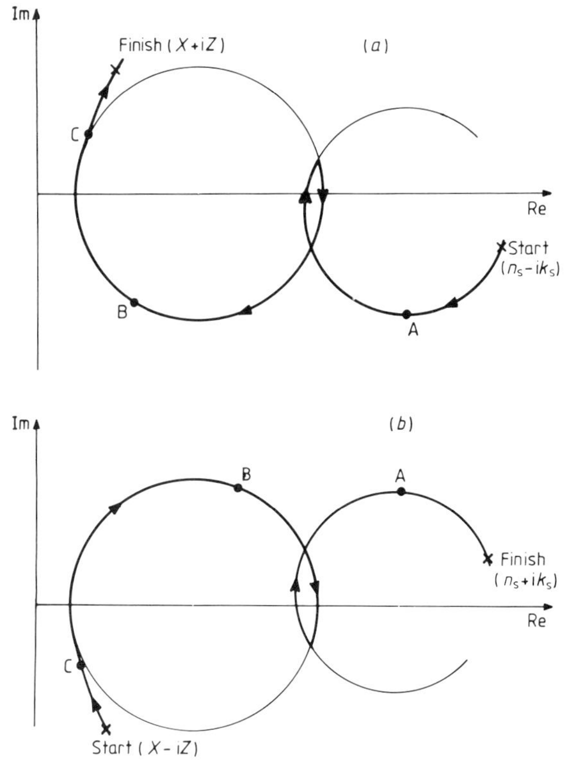

Figure 7.23 (a) A sketch of the admittance diagram of an arbitrary dielectric assembly of layers matching a starting admittance of $(n_s - ik_s)$ to the final admittance of $(X + iZ)$. (b) The curves of figure 7.23(a) reflected in the real axis and with the directions of the arrows reversed. This is now a multilayer identical to (a) but in the opposite order and connecting an admittance of $(X - iZ)$ (i.e. $(X + iZ)^*$) to one of $(n_s + ik_s)$ (i.e. $(n_s - ik_s)^*$).

n_f can be of high or low index, but μ will always be lower than the index of the substrate (except in very unlikely cases) because it is the first intersection of the locus of n_f with the real axis which is given by equations (7.84) and (7.85). Since the substrate will always have an index greater than unity, then the quarter-wave stack to match μ to n_s must start with a quarter-wave of low index. Alternate high- and low-index layers follow, the precise number being found by trial and error.

In order to complete the design, we need to know the equivalent admittance at the front surface of the metal layer and then we construct a matching stack to match it to the incident medium.

(e) *Front surface equivalent admittance*

If the admittance of the structure at the exit surface of the metal layer is the optimum value $(X + iZ)$ given by equations (7.82) and (7.83), then it can be shown that the equivalent admittance which is presented by the front surface of the metal layer is simply the complex conjugate $(X - iZ)$. The analytical proof of this requires a great deal of patience, although it is not particularly difficult. Instead, let us use a logical justification.

Consider a filter consisting of a single metal layer matched on either side to the surrounding media by dielectric stacks. Let the transmittance of the assembly be equal to the maximum potential transmittance and let the admittance of the structure at the rear of the metal layer be the optimum admittance $(X + iZ)$ given by equations (7.82) and (7.83). Let the equivalent admittance at the front surface be $(\xi + i\eta)$ and let this be matched perfectly to the incident medium. Now we know that the transmittance is the same regardless of the direction of incidence. Let us turn the filter around, therefore, so that the transmitted light proceeds in the opposite direction. The transmittance of the assembly must be the maximum potential transmittance once again. The admittance of the structure at what was earlier the input, but is now the new exit face of the metal layer, must therefore be $(X + iZ)$. But, since the layers are dielectric and the medium is of real admittance, this must also be the complex conjugate of $(\xi + i\eta)$, that is, $(\xi - i\eta)$. $(\xi + i\eta)$ must therefore be $(X - iZ)$, which is what we set out to prove.

The procedure for matching the front surface to the incident medium is therefore exactly the same as that for the rear surface and, indeed, if the incident medium is identical to the rear exit medium, as in a cemented filter assembly, then the front dielectric section can be an exact repetition of the rear.

Examples of filter designs

We can now attempt some filter designs. We choose the same material, silver, as we did for the Fabry–Perot and the DHW filters earlier. Once again, arbitrarily, we select a thickness of 70 nm. The wavelength we retain as 550 nm, at which the optical constants of silver are 0.055 − i3.32.

The filter is to use dielectric materials of indices 1.35 and 2.35 corresponding to cryolite and zinc sulphide respectively. The substrate is glass, $n = 1.52$, and the filter will be protected by a cemented cover slip so that we can also use $n = 1.52$ for the incident medium.

$$\alpha = 2\pi nd/\lambda \quad = 0.043\,98$$

$$\beta = 2\pi kd/\lambda \quad = 2.6549$$

and from equations (7.82) and (7.83) we find the optical admittance

$$X + iZ = 0.4572 + i\,3.4693.$$

Substituting this in equation (7.78) gives

$$\psi = 80.50\,\%.$$

We can choose to have either a high- or a low-index spacer. Let us choose first a low index and from equation (7.84) we obtain an optical thickness for the 1.35 index layer of 0.19174 full waves. Equation (7.85) yields a value of 0.05934 for μ which must be matched to the substrate index of 1.52. We start with a low-index quarter-wave and simply work through the sequence of possible admittances:

$$\frac{n_L^2}{\mu}, \quad \frac{n_H^2\mu}{n_L^2}, \quad \frac{n_L^4}{n_H^2\mu}, \quad \frac{n_H^4\mu}{n_L^4} \quad \text{etc}$$

until we find one sufficiently close to 1.52. The best arrangement in this case involves three layers of each type.

$$\frac{n_H^6\mu}{n_L^6} = 1.6511$$

equivalent to a loss of 0.2% at the interface with the substrate.

The structure so far is then

$$|\text{Ag}|\, L''\ LHLHLH\,|\,\text{Glass} \qquad (7.86)$$

with $L'' = 0.19174$ full waves. This can be combined with the following L layer into a single layer $L' = 0.25 + 0.19174 = 0.44174$ full waves, i.e.

$$|\text{Ag}|\, L'\ HLHLH\,|\text{Glass}.$$

Since the medium is identical to the substrate then the matching assembly at the front will be exactly the same as that at the rear so that the complete design is

$$\text{Glass}|\,HLHLH\ L'\,\text{Ag}\ L'\ HLHLH\,|\,\text{Glass}$$

with

Ag	70 nm (geometrical thickness)
L'	0.44174 full waves (optical thickness)
H, L	0.25 full waves
λ_0	550 nm.

The performance of this design is shown in figure 7.24(*a*). Dispersion of the silver has not been taken into account to give a clearer idea of the intrinsic characteristics. The peak is indeed centred at 550 nm with transmittance virtually that predicted.

A high-index matching layer can be handled in exactly the same way. For an index of 2.35, equation (7.84) yields an optical thickness of 0.1561 and equation (7.85) gives a value of 0.1426 for μ. Again, the matching quarter-wave stack should start with a low-index layer. There are two possible arrangements, H' representing 0.1561 full waves.

$$\text{Ag}\ H'\ LHLH\,|\,\text{Glass} \qquad (a)$$

with $n_H^4\mu/n_L^4 = 1.310$, i.e. a loss of 0.6% at the glass interface, or

$$\text{Ag } H' \; LHLH \,|\, \text{Glass} \qquad (b)$$

with $n_L^6/n_H^4\mu = 1.392$ representing a loss of 0.2% at the glass interface.

We choose alternative (*b*) and the full design can then be written

$$\text{Glass} \,|\, HLHL \; H' \; \text{Ag} \; H' \; LHLH \,|\, \text{Glass}$$

with

Ag	70 nm (geometrical thickness)
H'	0.1561 full waves (optical thickness)
H, L	0.25 full waves.

The performance of this design is shown in figure 7.24(*b*), where, again, the dispersion of silver has not been taken into account. Peak transmission is virtually as predicted.

When, however, we plot the performance of any of these designs, including

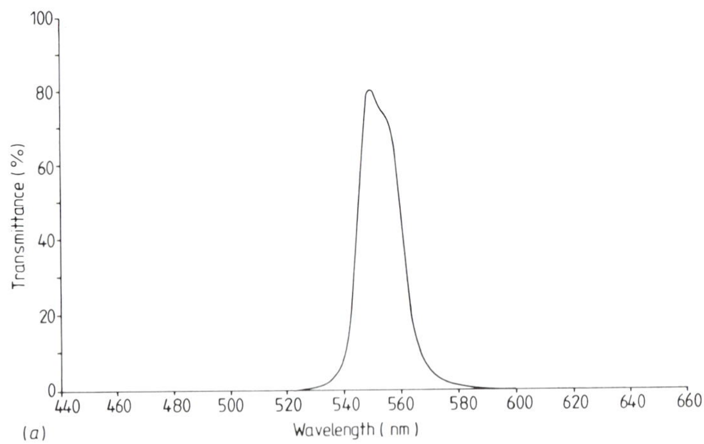

Figure 7.24 (*a*) Calculated performance of the design:

$$\text{Glass} \,|\, HLHLHL' \; \text{Ag} \; L' \; HLHLH \,|\, \text{Glass}$$

where

$n_{\text{Glass}} = 1.52$		
Ag = 70 nm	(geometrical thickness) of index	0.055-i3.32
$H = 0.25\,\lambda_0$	(optical thickness) of index	2.35
$L = 0.25\,\lambda_0$	(optical thickness) of index	1.35
$L' = 0.4417\,\lambda_0$	(optical thickness) of index	1.35
$\lambda_0 = 550$ nm.		

Dispersion has been neglected.

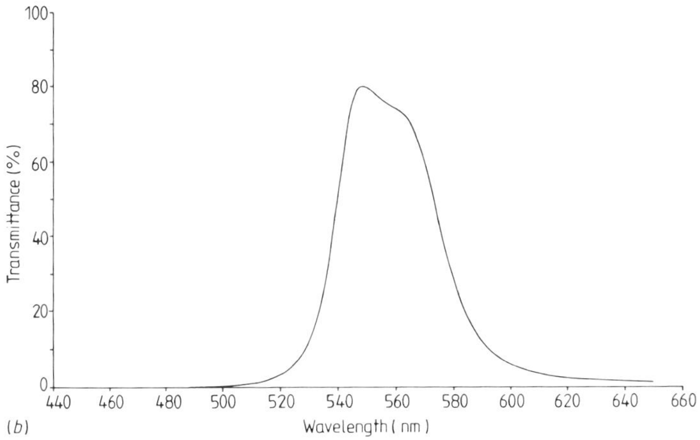

Figure 7.24 (*b*) Calculated performance of the design:

$$\text{Glass}|HLHL\ H'\ \text{Ag}\ H'\ LHLH|\text{Glass}$$

where

$n_{Glass} = 1.52$		
$\text{Ag} = 70$ nm	(geometrical thickness) of index	0.055-i3.32
$H = 0.25\,\lambda_0$	(optical thickness) of index	2.35
$L = 0.25\lambda_0$	(optical thickness) of index	1.35
$H' = 0.1561\lambda_0$	(optical thickness) of index	2.35
$\lambda_0 = 550$ nm.		

Dispersion has been neglected.

the metal–dielectric Fabry–Perot and DHW filters over an extended wavelength region, we find that the performance at longer wavelengths appears very disappointing. One example, the low-index matched induced-transmission filter, is shown in figure 7.25(*a*). In the case of the Fabry–Perot and the DHW, the rise is smoother, but is of a similar order of magnitude. The reason for the rise is, in fact, our assumption of zero dispersion. This means that β is reduced as λ increases. α is always quite small and the performance of the metal layers is determined principally by β. Silver, however, over the visible and near infrared, shows an increase in k which corresponds roughly to the increase in λ so that k/λ is roughly constant (to within around $\pm 20\%$) over the region 400 nm–2.0 μm. This completely alters the picture and is the reason why the first-order metal–dielectric filters do not show longwave sidebands.

Taking dispersion into account, the performance of the induced transmission filter improves considerably and is shown in figure 7.25(*b*). The rejection is, however, not particularly high, being between 0.01 and 0.1 % transmittance over most of the range with an increase to 0.15 % in the vicinity of 860 nm. This level of rejection can be acceptable in some applications and the induced-transmission filter represents a very useful, inexpensive general purpose filter. The dispersion which improves the performance on the longwave side of the

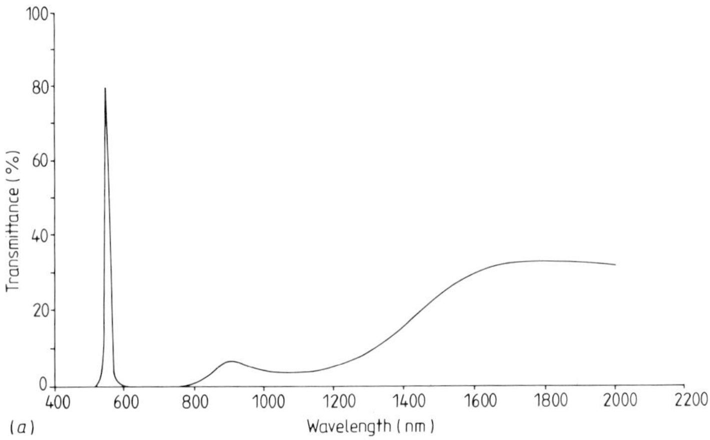

Figure 7.25 (*a*) The design of figure 7.24(*a*) computed over a wider spectral region neglecting dispersion.

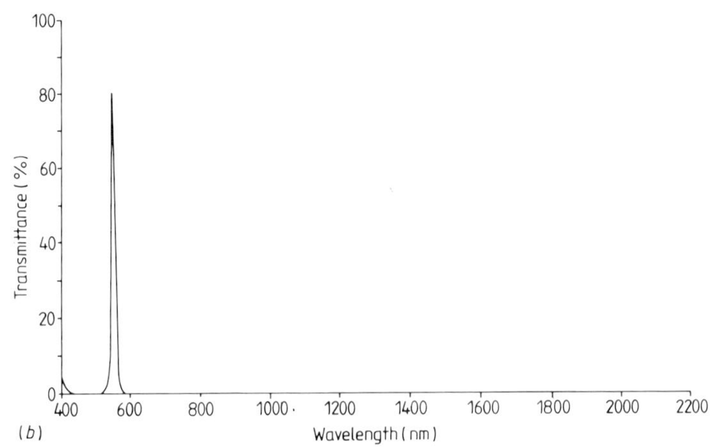

Figure 7.25 (*b*) The design of figure 7.24(*a*) computed this time including dispersion. The rise in transmittance at longer wavelengths has vanished but there is now obvious transmittance at 400 nm.

peak degrades it on the shortwave side, and to complete the filter it is normal to add a longwave-pass absorption glass filter which is cemented to the induced transmission component.

To improve the rejection of the basic filter it is necessary to add further metal layers. The simplest arrangement is to have these extra metal layers of exactly the same thickness as the first. The potential transmittance of the complete filter will then be the product of the potential transmittances of the individual layers. The terminal admittances for all the metal layers can be arranged to be optimum quite simply, giving optimum performance for the filter. All that is required is a dielectric layer in between the metal layers which is twice the thickness given by equation (7.84) for the first matching layer. We can see why this is by imagining a matching stack on the substrate overcoated with the first metal layer. Since its terminal admittance will be optimum, the input admittance will be the complex conjugate, as we have discussed already. Addition of the thickness given by equation (7.84) renders the admittance real, that is, the admittance locus has reached the real axis. Addition of a further identical thickness must give an equivalent input admittance which is the complex conjugate of the metal input admittance and hence is equal to the optimum admittance. This can be repeated as often as desired.

Returning to our example, a two-metal layer induced-transmission filter will have peak transmittance, if perfectly matched, of $\psi = (0.80501)^2$, that is, 64.8 %, a three-metal layer should have $\psi = (0.805\,01)^3$ that is, 52.17 %, and so on.

The designs, based on the low-index matching layer version, are then, from equation (7.86)

$$\begin{aligned}&\text{Glass}\,|\,HLHLHL\ L''\ \text{Ag}\ L''L''\ \text{Ag}\ L''\ LHLHLH\,|\,\text{Glass}\\ &= \text{Glass}\,|\,HLHLHL'\ \text{Ag}\ L'''\ \text{Ag}\ L'\ HLHLH\,|\,\text{Glass}\end{aligned} \tag{7.87}$$

where

$$\begin{aligned} L' &= 0.25 + 0.191\,74 = 0.441\,74 \text{ full waves}\\ L'' &= 0.191\,74 \text{ full waves}\\ L''' &= 2\times 0.191\,74 = 0.383\,48 \text{ full waves}\\ \text{Ag} &= 70 \text{ nm}\end{aligned}$$

and

$$\text{Glass}\,|\,HLHLH\ L'\ \text{Ag}\ L'''\ \text{Ag}\ L'''\ \text{Ag}\ L'\ HLHLH\,|\,\text{Glass}. \tag{7.88}$$

Unfortunately, these designs, although they do have the peak transmittance predicted, possess a poor pass band shape, in that it has a hump on the longwave side. To eliminate this hump, it is necessary to add an extra half-wave layer to each of the layers marked L''', i.e.

$$\text{Glass}\,|\,HLHLHL'\ \text{Ag}\ L''''\ \text{Ag}\ L'\ HLHLH\,|\,\text{Glass} \tag{7.89}$$

and

$$\text{Glass}\,|\,HLHLH\ L'\ \text{Ag}\ L''''\ \text{Ag}\ L''''\ \text{Ag}\ L'\ HLHLH\,|\,\text{Glass} \tag{7.90}$$

where

$$L'''' = 0.5 + 0.383\,48 = 0.883\,48 \text{ full waves.}$$

Figure 7.26 shows the form of designs (7.87) and (7.88) and the hump can clearly be seen together with the improved shape of designs (7.89) and (7.90).

Dispersion was not included in the computation of figure 7.26. To examine

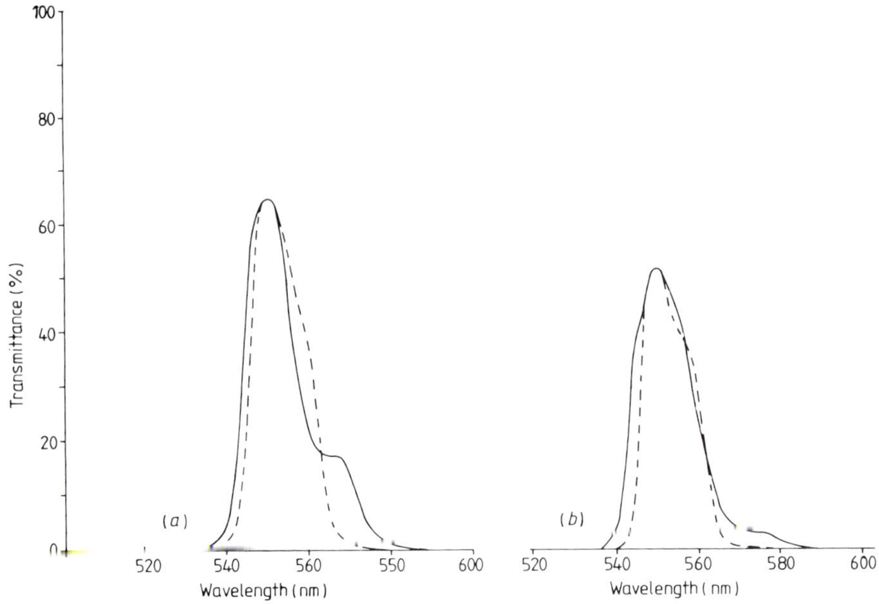

Figure 7.26 Performance, neglecting dispersion, of (*a*) two-metal-layer designs and (*b*) three-metal-layer designs of induced-transmission filter. The full curves denote (7.87) and (7.88) and there is a spurious shoulder on the longwave side of the peak in each case. This can be eliminated by the addition of half-wave decoupling layers as the broken lines show. They are derived from (7.89) and (7.90) respectively.

the rejection over an extended region, we must include the effects of dispersion. Unfortunately, the modified designs (7.89) and (7.90) act as metal–dielectric–metal (M–D–M is a frequently used shorthand notation for such a filter) and metal–dielectric–metal–dielectric–metal (M–D–M–D–M) filters at approximately 1100 nm which gives a very narrow leak, rising to around 0.15% in the former and 0.05% in the latter. Elsewhere, the rejection is excellent, of the order of 0.0001% at 900 nm and 0.000015% at 1.05 μm for the former and 0.0000001% at 900 nm and 3×10^{-9}% at 1.05 μm for the latter.

If the leak is unimportant, then the filter can be used as it is with the addition of a longwave-pass filter of the absorption type as before. For the suppression of all-dielectric filter sidebands, it is better to use filters of type (7.87) and (7.88) since the shape of the sides of the pass band is relatively unimportant. The rejection of these filters is slightly better than that of (7.89) and (7.90) and, of course, the leak is missing (figure 7.27).

The bandwidth of the filters is not an easy quantity to predict analytically and the most straightforward approach is simply to compute the filter profile.

Berning and Turner[30] show that a figure of merit indicating the potential

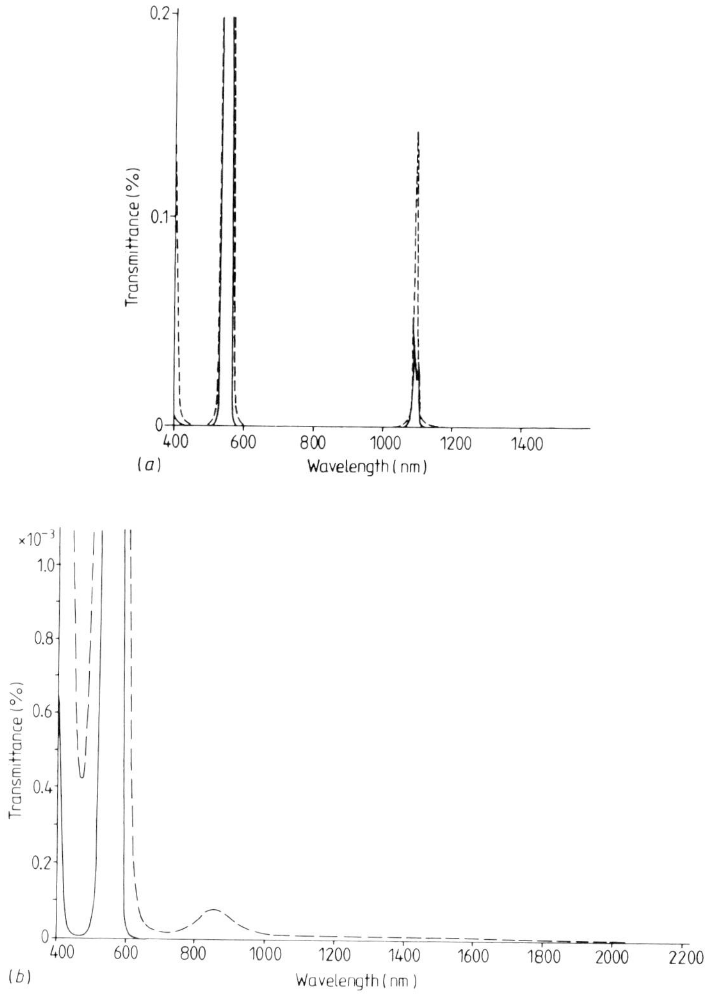

Figure 7.27 (*a*) Calculation, including dispersion, of the performance of the designs of (7.89) (broken curve) and (7.90) over an extended spectral range. These designs include the half-wave decoupling layers and the penalty for the improved pass band shape is the narrow transmission spike near 1.05 μm. (*b*) Calculation, including dispersion, of the original designs (7.87) (broken curve) and (7.88). The transmission spike is no longer there but the pass band shape includes the shoulder (off scale).

usefulness of a metal is the ratio k/n. The higher this ratio, the better is the performance of the completed filter.

Induced-transmission filters for the visible region having only one single metal layer are relatively straightforward to manufacture. The thickness of the metal layer can be arrived at by trial and error. If the metal layer is less than optimum in thickness, the effect will be a broadening of the pass band and a rise in peak transmission at the expense of an increase in background transmission remote from the peak. A splitting of the pass band will also become noticeable with the appearance eventually, if the thickness is further reduced, of two separate peaks. If, on the other hand, the silver layer is made too thick, the effect will be a narrowing of the peak with a reduction of peak transmission. The best results are usually obtained with a compromise thickness where the peak is still single in shape but where any further reduction in silver thickness would cause the splitting to appear. A good approximation in practice, which can be used as a first attempt at a filter, is to deposit the first dielectric stack and to measure the transmission. The silver layer can then be deposited using a fresh monitor glass so that the optical density is twice that of the dielectric stack. The second spacer and stack can then be added on yet another fresh monitor. A measurement of the transmission of the complete filter will quickly indicate which way the thickness of the silver layer should be altered in order to optimise the design. Usually, one or two tests are sufficient to establish the best parameters. If, after this optimising, the background rejection remote from the peak is found to be unsatisfactory, then not enough silver is being used. As the thickness was chosen to be optimum for the two dielectric sections, a pair of quarter-wave layers should be added to each in the design and the trial and error optimisation repeated. This will also narrow the bandwidth, but this is usually preferable to high background transmission.

In the ultraviolet the available metals do not have as high a performance as, for instance, silver in the visible, and it is very important, therefore, to ensure that the design of a filter is optimised as far as possible; otherwise a very inferior performance will result. An important paper in this field is that by Baumeister *et al*[31]. Aluminium is the metal commonly used for this region and measured and computed results obtained by these workers for filters with aluminium layers are shown in figure 7.28. The performance which has been achieved is most satisfactory and the agreement between practical and theoretical curves is good.

Induced-transmission filters have been the subject of considerable study by many workers. Metal–dielectric multilayers are reviewed by MacDonald[32]. A useful, recent account of induced-transmission filters is given by Lissberger[33]. Multiple cavity induced-transmission filters have been described by Maier[34]. An alternative design technique for metal–dielectric filters involving symmetrical periods has been published by Macleod[35]. Symmetrical periods for metal–dielectric filter design have also been used by McKenney[36] and by Landau and Lissberger[37].

MEASURED FILTER PERFORMANCE

Not a great deal has been published on the measured performance of actual filters and the main source of information for a prospective user is always the

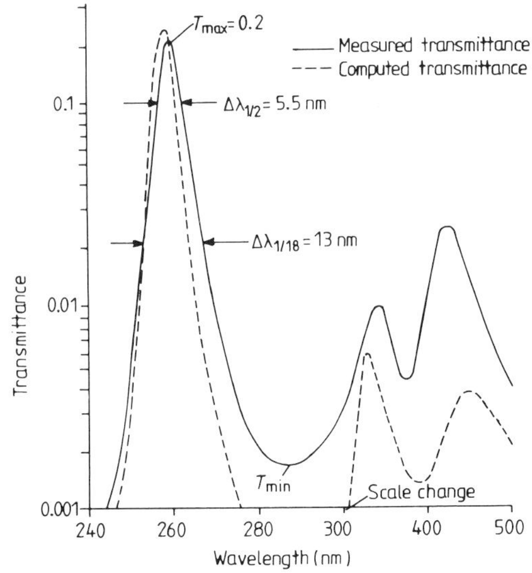

Figure 7.28 Computed and measured transmittance of an induced transmission filter for the ultraviolet. Design:

$$\text{Air}|HLHLHLH\ 1.76L\ \text{Al}\ 1.76L\ HLHLHLH|\text{Quartz}$$

where $H = PbF_2$ ($n_H = 2.0$) and $L = Na_3AlF_6$ ($n_L = 1.36$). The physical thickness of the aluminium layer is 40 nm and $\lambda_0 = 253.6$ nm. (After Baumeister, Costich and Pieper[31].)

literature issued by manufacturers. Performance of current production filters tends to improve all the time so that inevitably such information does not remain up to date for long. Two papers[38,39] quote the results of a number of tests on commercial filters, and, although they were written some time ago, they will still be found useful sources of information.

Blifford examined the performance of the products of four different manufacturers, covering the region 300–1000 nm. The variation of peak wavelength with angle of incidence was found to be similar to the relationship already established (see p 260). Unfortunately, information on the design and materials is lacking, so that the expression for the effective index cannot be checked. The sensitivities to tilt varied from $P = 0.22$ to $P = 0.51$, where P corresponds to the quantity $1/n^{*2}$ in equation (7.39). Blifford suggests that an average value of 0.35 for P would probably be the best value to assume in any case where no other data were available. Changes in peak transmittance with angle of incidence were found, but were not constant from one filter to another and apparently must always be measured for each individual filter. Possibly, the effect is due to the absorption filters which are used for sideband suppression and which, because they do not show any shift in edge wavelength with angle of incidence, may cut into the pass band of the interference section at large angles of incidence. In most cases examined, the change in peak transmission was less than 10% for angles of 5°–10°.

The variation in peak transmittance over the surface of the filter was also measured in a few cases. For a typical filter with a peak wavelength of 500 nm and a bandwidth not explicitly mentioned, but probably 2.1 nm (from information given elsewhere in the paper), the extremes of peak transmission were 54% and 60%. This is, in fact, one aspect of a variation of peak wavelength, bandwidth and peak transmittance which frequently occurs, although the magnitude can range from very small to very large. The cause is principally the adsorption of water vapour from the atmosphere before a cover slip can be cemented over the layers and it is dealt with in greater detail in chapter 9. Infrared filters appear to suffer less from this defect than visible and near infrared filters.

Another parameter measured by Blifford was the variation of peak wavelength with temperature. Variation of the temperature from -60°C to $+60$°C resulted in changes of peak wavelength of from $+0.01$ nm $°C^{-1}$ to $+0.03$ nm $°C^{-1}$. The relationship was found to be linear over the whole of this temperature range with little, if any, change in the pass band shape and peak transmittance. In most cases, the temperature coefficients of bandwidth and peak transmittance were found to be less than 0.001 nm $°C^{-1}$. Filters for the visible region have also been the subject of a detailed study by Pelletier and his colleagues[40]. The shift with temperature for any filter is a function of the coefficients of optical thickness change with temperature, depending on the design of the filter and especially on the material used for the spacers. Measurements made on different filter designs yielded the following coefficients of optical thickness for the individual layer materials:

zinc sulphide	$(4.8 \pm 1.0) \times 10^{-5}\ °C^{-1}$
cryolite	$(3.1 \pm 0.7) \times 10^{-5}\ °C^{-1}$.

Hysteresis is frequently found when temperature cycling narrowband filters over an extended temperature range. The hysteresis is particularly pronounced when the filters are uncemented and when they are heated towards 100°C. It is usually confined to the first cycle of temperature, takes the form of a shift of peak wavelength towards shorter wavelengths and is caused by the desorption of water which is discussed again in chapter 9.

An effect of a different kind, although related, is the subject of a contribution by Title and his colleagues[41, 42]. A permanent shift of a filter characteristic towards shorter wavelengths amounting to a few tenths of nanometres accompanied by a distortion of pass band shape was produced by a high level of illumination. The filters were for the H_α wavelength, 656.3 nm, and the changes were interpreted as due to a shift in the properties of the zinc sulphide material, the fundamental nature of the shift being unknown. Zinc sulphide can be transformed into zinc oxide by the action of ultraviolet light, especially in the presence of moisture, and the shifts that were observed could probably have been caused by such a mechanism.

The possibility of variations in filter properties both over the surface of the filter and as a function of time, temperature and illumination level should clearly be borne in mind in the designing of apparatus incorporating filters.

A useful survey which compares the performance achievable from different types of narrowband filters was the subject of a report by Baumeister[43].

A study was carried out by Baker and Yen on infrared filters. The effects studied were those of variation in angle of incidence and temperature, and both theoretical and experimental results were quoted.

Accurate calculation of the effects of changes in the angle of incidence yielded a variation of peak wavelength of the expected form, but no significant variation of bandwidth for angles of incidence up to 50°. They also calculated that the peak transmittance and the shape of the pass band should remain unchanged for angles up to 45°. For angles above 50°, both the shape and the peak transmittance gradually deteriorated. The calculations were confirmed by measurements on real filters.

The effects of varying temperatures were also investigated both theoretically and practically. As in the case of the shorter wavelength filters examined by Blifford, they measured a shift towards longer wavelengths with increasing temperature. For temperatures down to liquid helium the filters show little loss of peak transmittance or variation of characteristic pass band shape. However, serious losses in transmittance occurred above 50°C. Although not mentioned in the paper, this is probably due to the use of germanium, either as substrate or one of the layer materials, which always exhibits a marked fall in transmittance at elevated temperatures above 50°C. Baker and Yen make the point that filters designed to be least sensitive to variations in the angle of incidence are usually most sensitive to temperature and vice versa. The temperature coefficients of peak wavelength which they quote vary from $+0.0035\,\%\,{}^\circ C^{-1}$ to $+0.0125\,\%\,{}^\circ C^{-1}$. Unfortunately, neither the materials used in the filters nor the designs are quoted in the paper, but it is likely that the figures will apply to most interference filters for the infrared.

Similar measurements of the temperature shift of infrared filters were made at Grubb Parsons. The materials used were zinc sulphide and lead telluride, and the filters which had first-order high-index spacers gave temperature coefficients of peak wavelength of $-0.0135\,\%\,{}^\circ C^{-1}$. These filters were of the type used in the selective chopper radiometer described in chapter 12. The negative temperature coefficient is usual with filters having lead telluride as one of the layer materials. This negative coefficient in lead telluride is especially useful as it tends to compensate for the positive coefficient in zinc sulphide, and Seeley *et al*[44] have succeeded in designing and constructing filters using lead telluride which have zero temperature coefficient.

REFERENCES

1 Epstein L I 1952 The design of optical filters *J. Opt. Soc. Am.* **42** 806–10
2 Turner A F 1950 Some current developments in multilayer optical films *J. Phys. Radium* **11** 443–60
3 Bates B and Bradley D J 1966 Interference filters for the far ultraviolet (1700 to 2400 Å) *Appl. Opt.* **5** 971–5
4 Seeley J S 1964 Resolving power of multilayer filters *J. Opt. Soc. Am.* **54** 342–6
5 Hemingway D J and Lissberger P H 1973 Properties of weakly absorbing multilayer systems in terms of the concept of potential transmittance *Opt. Acta* **20** 85–96
6 Dobrowolski J A 1959 Mica interference filters with transmission bands of very narrow half-widths *J. Opt. Soc. Am.* **49** 794–806 and 1963 Further developments in mica interference filters *J. Opt. Soc. Am.* **53** 1332 (summary only)
7 Austin R R 1972 The use of solid etalon devices as narrowband interference filters *Opt. Eng.* **11** 65–9
8 Candille M and Saurel J M 1974 Réalisation de filtres "double onde" a bandes passantes très étroites sur supports en matière plastique (mylar) *Opt. Acta.* **21** 947–62
9 Smith S D and Pidgeon C R 1963 Application of multiple beam interferometric methods to the study of CO_2 emission at 15 μm *Mém. Soc. R. Sci. Liège 5ième série* **9** 336–49
10 Roche A E and Title A M 1974 Tilt tunable ultra narrow-band filters for high resolution photometry *Appl. Opt.* **14** 765–70
11 Dufour C and Herpin A 1954 Applications des méthodes matricielles au calcul d'ensembles complexes de couches minces alternées *Opt. Acta* **1** 1–8
12 Lissberger P H 1959 Properties of all-dielectric filters. I—A new method of calculation *J. Opt. Soc. Am.* **49** 121–5
13 Lissberger P H and Wilcock W L 1959 Properties of all-dielectric filters. II—Filters in parallel beams of light incident obliquely and in convergent beams *J. Opt. Soc. Am.* **49** 126–38
14 Pidgeon C R and Smith S D 1964 Resolving power of multilayer filters in non-parallel light *J. Opt. Soc. Am.* **54** 1459–66
15 Hernandez G 1974 Analytical description of a Fabry–Perot spectrometer, 3. Off-axis behaviour and interference filters *Appl. Opt.* **13** 2654–61
16 For example, Reports 4, 5 and 6 of Contract DA-44-009-eng-1113 covering the period January–October 1953
17 Turner A F 1952 Wide pass band multilayer filters *J. Opt. Soc. Am.* **42** 878(*a*)
18 Smith S D 1958 Design of multilayer filters by considering two effective interfaces *J. Opt. Soc. Am.* **48** 43–50
19 Knittl Z 1965 Dielektrische Interferenzfilter mit recheckigen Maximum *Proc. Coll. Thin Films, Budapest* 153–61 (The method is described in detail in reference 20 also)
20 Knittl Z 1976 *Optics of Thin Films* (London: Wiley)
21 Thelen A 1966 Equivalent layers in multilayer filters *J. Opt. Soc. Am.* **56** 1533–8
22 Barr E E 1974 Visible and ultraviolet bandpass filters, in *Optical Coatings, Applications and Utilization* ed G W DeBell and D H Harrison (*Proc. SPIE* **50** 87–118)

23 Neilson R G T and Ring J 1967 Interference filters for the near ultra-violet *J. Physique* **28** C2-270–C2-275 (supplement to no 3–4 March–April)

24 Malherbe A 1974 Interference filters for the far ultraviolet *Appl. Opt.* **13** 1275–6

25 Baumeister P W and Jenkins F A 1957 Dispersion of the phase change for dielectric multilayers. Application to the interference filter *J. Opt. Soc. Am.* **47** 57–61

26 Baumeister P W, Jenkins F A and Jeppesen M A 1959 Characteristics of the phase-dispersion interference filter *J. Opt. Soc. Am.* **49** 1188–90

27 Giacomo P, Baumeister P W and Jenkins F A 1959 On the limiting band width of interference filters *Proc. Phys. Soc.* **73** 480–9

28 Ritchie F S Unpublished work on Ministry of Technology Contract KX/LSO/C.B.70(a)

29 Hass G and Hadley L 1972 Optical constants of metals *American Institute of Physics Hand Book* ed D E Gray (New York: McGraw-Hill) pp 6-124–6-156

30 Berning P H and Turner A F 1957 Induced transmission in absorbing films applied to band pass filter design *J. Opt. Soc. Am.* **47** 230–9

31 Baumeister P W, Costich V R and Pieper S C 1965 Bandpass filters for the ultraviolet *Appl. Opt.* **4** 911–13

32 MacDonald J 1971 Metal–Dielectric Multilayers (London: Adam Hilger)

33 Lissberger P H 1981 Coatings with induced transmission *Appl. Opt.* **20** 95–104

34 Maier R L 1967 2M interference filters for the ultraviolet *Thin Solid Films* **1** 31–7

35 Macleod H A 1978 A new approach to the design of metal–dielectric thin-film optical coatings *Opt. Acta.* **25** 93–106

36 McKenney D B 1969 Ultraviolet interference filters with metal–dielectric stacks *PhD Dissertation* Optical Sciences Center, University of Arizona

37 Landau B V and Lissberger P H 1972 Theory of induced transmission filters in terms of the concept of equivalent layers *J. Opt. Soc. Am.* **62** 1258–64

38 Blifford I H Jr 1966 Factors affecting the performance of commercial interference filters *Appl. Opt.* **5** 105–11

39 Baker M L and Yen V L 1967 The effect of the variation of angle of incidence and temperature on infrared filter characteristics *Appl. Opt.* **6** 1343–51

40 Pelletier F, Roche P and Bertrand L 1974 On the limiting bandwidth of interference filters: influence of temperature during production *Opt. Acta* **21** 927–46

41 Title A M, Pope T P and Andelin J P 1974 Drift in interference filters. 1 *Appl. Opt.* **13** 2675–9

42 Title A M 1974 Drift in interference filters. 2: radiation effects *Appl. Opt.* **13** 2680–4

43 Baumeister P W 1973 Thin films and interferometry *Appl. Opt.* **12** 1993–4

44 Seeley J S, Evans C S, Hunneman R and Whatley A 1976 Filters for the v2 band of CO_2: monitoring and control of layer deposition *Appl. Opt.* **15** 2736–45

8 Tilted coatings

INTRODUCTION

We have already seen in chapter 2 that the characteristics of coatings change when they are tilted with respect to the incident illumination, and the particular way in which they change depends on the angle of incidence. We have studied the shifts that are induced in narrowband filters. Narrowband filters are a simple case because the tilt angle is usually small and we can assume that the major effect is in the phase thickness of the layers, which is affected equally for each plane of polarisation. For larger tilts, however, the admittances are also affected and then the performance for each plane of polarisation differs. Some important applications involve the difference in performance between one plane and the other, which can be controlled to some extent, making possible the construction of phase retarders and polarisers. On the other hand, the differences in performance can create problems, and although it is impossible to cancel the effects completely, there are ways of modifying it so that a more acceptable performance may be achieved. Then there are some, at first sight, strange effects which occur with dielectric-coated reflectors. Under certain conditions and at reasonably high angles of incidence, sharp absorption bands can exist for one plane of polarisation. This can create difficulties with dielectric-overcoated reflectors such as protected silver. The chapter begins with the addition of tilting effects to the admittance diagram, which allows us to explain qualitatively the behaviour of many different types of tilted coatings including overcoated reflectors and which involves a slight modification to the traditional form of the tilted admittances. Next there is a description of polarisers followed by an account of phase retarders. Some coatings where the polarisation splitting is undesirable, such as dichroic filters, are described with ways of reducing this splitting. Finally some antireflection coatings at high angles of incidence are described.

Some of the material in this chapter has already been mentioned and discussed in earlier chapters but here we attempt to introduce a consistent and connected account and so there are some advantages in repeating what has been said before in the present context.

MODIFIED ADMITTANCES AND THE TILTED ADMITTANCE DIAGRAM

The form of the admittances and the phase thickness of a film which is illuminated at oblique incidence are given in chapter 2 and have already been used in considering the performance of some coatings including narrowband filters. They are:

$$\delta = 2\pi d(n^2 - k^2 - n_0^2 \sin^2\theta_0 - 2ink)^{1/2}/\lambda \tag{8.1}$$

where the fourth quadrant solution is correct, and then

$$\eta_s = (n^2 - k^2 - n_0^2 \sin^2\theta_0 - 2ink)^{1/2}\mathscr{Y} \tag{8.2}$$

again in the fourth quadrant, and

$$\eta_p = y^2/\eta_s \tag{8.3}$$

where n, k refer to the film and n_0, θ_0 etc to the incident medium. When the layers are purely dielectric then this is in the simpler form

$$\delta = (2\pi nd\cos\theta)/\lambda \tag{8.4}$$

$$\eta_s = y\cos\theta \tag{8.5}$$

and

$$\eta_p = y/\cos\theta \tag{8.6}$$

where $n\sin\theta = n_0\sin\theta_0$. Expressions (8.4)–(8.6) can be used instead of expressions (8.1)–(8.3) if the $\cos\theta$ term is permitted to become complex.

The calculation of multilayer properties at angles of incidence other than normal simply involves the use of the above expressions instead of those for normal incidence. It should be emphasised that the appropriate tilted values are to be adopted for incident medium and substrate as well as for the films. The use of the admittance diagram is rendered much more complicated because of the change in the incident admittance. The isoreflectance and isophase contours depend on the admittance of the incident medium and we therefore need one set for s-polarisation and one quite different set for p-polarisation, as well as completely new sets each time the angle of incidence is changed. Fortunately, there is a way round this problem, which carries some other advantages as well.

It has been shown by Thelen[1] that the properties of a multilayer are unaffected if all the admittances are multiplied or divided by a constant factor, and indeed it is usual to divide the admittances by $\mathscr{Y}$, the admittance of free space, so that the normal incidence admittance is numerically equal to the refractive index. We now propose an additional correction to the admittances, the dividing of the s-polarised admittances, and the multiplying of the p-polarised admittances, by $\cos\theta_0$. This has the effect of preserving, for both s- and p-polarisation, the admittance of the incident medium at its normal incidence value, regardless of the angle of incidence, and means that the isoreflectance and isophase contours of the admittance diagram retain their normal incidence values whatever the angle of incidence or plane of polarisation. We can call these admittances simply the modified admittances, and the expressions for them become

$$\eta_s = (n^2 - k^2 - n_0^2 \sin^2\theta_0 - 2ink)^{1/2}/\cos\theta_0 \tag{8.7}$$

again in the fourth quadrant, and

$$\eta_p = y^2/\eta_s. \tag{8.8}$$

Or, when the layers are dielectric, the simpler forms are

$$\eta_s = (y\cos\theta)/\cos\theta_0 \tag{8.9}$$

and

$$\eta_p = (y\cos\theta_0)/\cos\theta. \tag{8.10}$$

The values of reflectance, transmittance, absorptance and phase changes on either transmission or reflection are completely unchanged by the adoption of these values for the admittances. Since the expressions involve $\cos\theta_0$ and $\cos\theta$, which are connected by the admittance of the incident medium, then the dependence of the modified admittances on the index of the incident medium will be somewhat different from the unmodified, traditional ones. Nevertheless, we shall see that this does carry some advantages.

We consider first of all purely dielectric materials. In this case, provided that $n_0 \sin\theta_0$ is less than n, the film index, then the two values for the modified admittances are real and positive. If, however, n_0 is greater than n, then there is a real value of θ_0 at which $n_0 \sin\theta_0$ is equal to n. This angle is known as the critical angle, and, for angles of incidence greater than this value, the admittances are imaginary. We will consider what happens for angles of incidence beyond critical later. First we will limit ourselves to angles less than critical where the admittances are real.

First of all, let us consider air of index unity as the incident medium. We recall that all transparent thin-film materials have refractive index greater than unity. In figure 8.1 the modified admittance is shown for a number of thin-film materials as a function of angle of incidence. The p-admittances of all materials cross the line $n = 1$ at the value known as the Brewster angle for which the single-surface p-reflectance is zero. The s-admittances all increase away from the line $n = 1$, so that the single-surface s-reflectance simply increases with angle of incidence. Since all these materials are dielectric, their modified optical thickness is real and therefore, although a correction has to be made for the effect of angle of incidence, quarter- and half-wave layers can be produced at

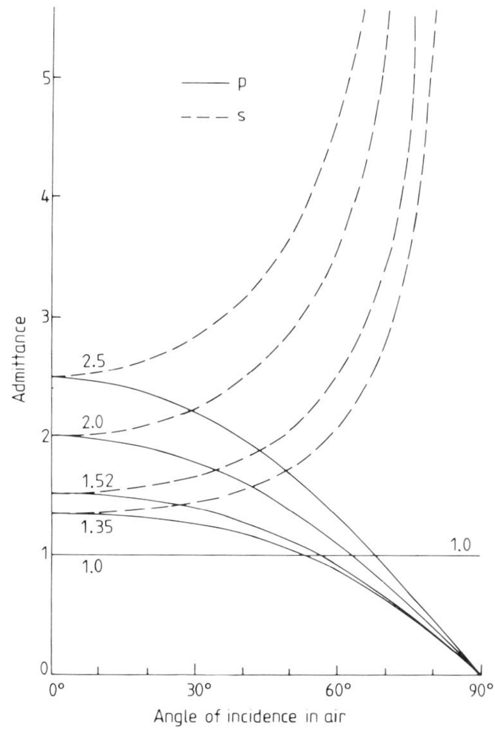

Figure 8.1 Modified p- and s-admittances (i.e. including the extra factor of $\cos\theta_0$) of materials of indices 1.0, 1.35, 1.52, 2.0 and 2.5 for an incident medium of index 1.0.

non-normal incidence just as readily as at normal and it cannot be too greatly emphasised that although the optical thickness changes with angle of incidence, it does not vary with the plane of polarisation.

It is possible to make several deductions directly from figure 8.1. The first is that, for any given pair of indices, the ratio of the s-admittances increases with angle of incidence, while that for p-admittances reduces. Since the width of the high-reflectance zone of a quarter-wave stack decreases with decreasing ratio of these admittances, the width will be less for p-polarised light than for s-polarised. As we shall shortly see, this effect is used in a useful type of polariser. The splitting of the admittance of dielectric layers means also that there is a relative phase shift between p- and s-polarised light reflected from a high-reflectance coating when the layers depart from quarter-waves. This effect can be used in the design of phase retarders and we will give a brief account of this. The diagram also helps us to consider the implications of antireflection coatings for high angles of incidence. A frequent requirement is an antireflection coating for a crown glass of index around 1.52. For a perfect single-layer coating we should have a quarter-wave of material of optical admittance equal to the square root of the product of the admittances of the glass and the incident medium. At normal incidence in air there is, of course, no sufficiently robust material with index as low as 1.23. For greater angles of incidence, the s-polarised reflectance increases still further from its normal incidence value and the admittance required for a perfect single-layer antireflection coating remains outside the range of practical materials, corresponding to still lower indices of refraction. The p-polarised behaviour is, however, completely different, and in the range from approximately 50°–70° the admittance required for the antireflection coating is within the range of what is possible. No coating is required, of course, at the Brewster angle. For angles greater than the Brewster angle, the index required is *greater* than that of the glass. Antireflection coatings for high angles of incidence will be also be discussed shortly.

The behaviour of dielectric materials when the incident medium is of a higher index (one that is within the range of available thin-film materials) is somewhat more complicated. Figure 8.2 shows the way in which the admittances vary when the incident medium is glass of index 1.52. There is the

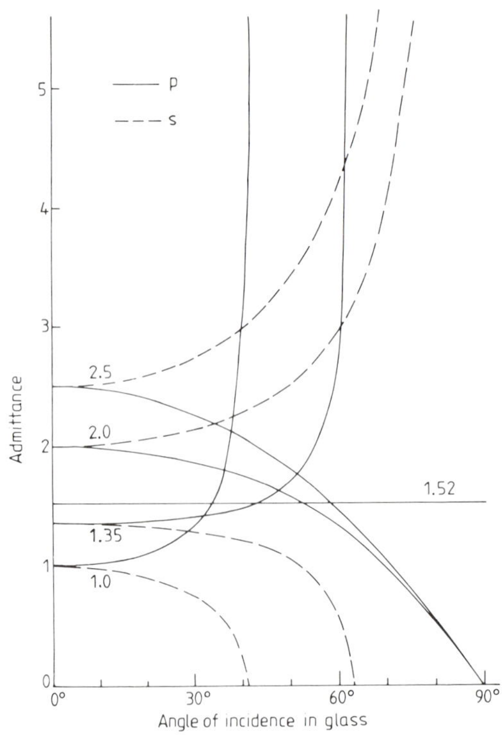

Figure 8.2 Modified p- and s-admittances (i.e. including the extra factor of $\cos\theta_0$) of materials of indices 1.0, 1.35, 1.52, 2.0 and 2.5 for an incident medium of index 1.52.

familiar splitting of the s- and p-polarised admittances which, as before, increases with angle of incidence. For indices which are lower than that of the glass it is possible to reach the critical angle, and at that point the admittances reach either zero or infinity and disappear from the diagram. Their behaviour beyond the critical angle will be discussed shortly. A further very important feature is that, while for indices higher than that of the incident medium the p-polarised admittance falls with angle of incidence, for indices lower than the incident medium the p-polarised admittance rises. All cut the incident medium admittance at the Brewster angle, but now a new phenomenon is apparent. The p-admittance curves for materials of index lower than that of the incident medium intersect the curves corresponding to higher indices. An immediate deduction is that a quarter-wave stack, composed of such pairs of materials, will simply behave, at the angle of incidence corresponding to the point of intersection, as a thick slab of material. Provided the admittances of substrate, thin films and incident medium are not too greatly different, the p-reflectance will be low. The ratio of the s-admittances is large, because their splitting increases with angle of incidence, and so the corresponding s-reflectance is high and the width of the high-reflectance zone is large. This is the basic principle of the MacNeille polarising beam splitter that we will return to in a later section. The range of useful angles of incidence will depend partly on the rate at which the curves of p-polarised admittance diverge on either side of the intersection, and this can be estimated from the diagram.

Apart from the polarisation-splitting of the admittance, the behaviour of dielectric layers at angles of incidence less than critical is reasonably straightforward and does not involve difficulties of a more severe order than exist at normal incidence. When metal films are introduced, however, the difficulties increase and the behaviour becomes still stranger when combined with dielectric materials, especially when used beyond the critical angle. The aim in the remainder of this section is to discuss, in a qualitative fashion, such behaviour and to suggest techniques which can be used for visualisation and prediction. The use of admittance loci will be emphasised.

We know already that the admittance locus of a dielectric layer at normal incidence is a circle centred on the real axis. Tilted dielectric layers at angles of incidence less than critical still have circular loci which can be calculated from the tilted admittances in exactly the same way. Provided the modified admittances are used in constructing the loci then the isoreflectance and isophase circles on the admittance diagram will remain exactly the same as at normal incidence for both p- and s-polarisation.

The admittance of a metal layer is a little more complicated than a dielectric. For a lossless metal in which the refractive index, and hence the optical admittance, is purely imaginary, and given by $-ik$, the loci are a set of circles with centres on the real axis and passing through the points ik and $-ik$, which are on the imaginary axis. Figure 8.3 shows the typical form. The circles are like the dielectric ones, traced out clockwise so that they start on ik and end on

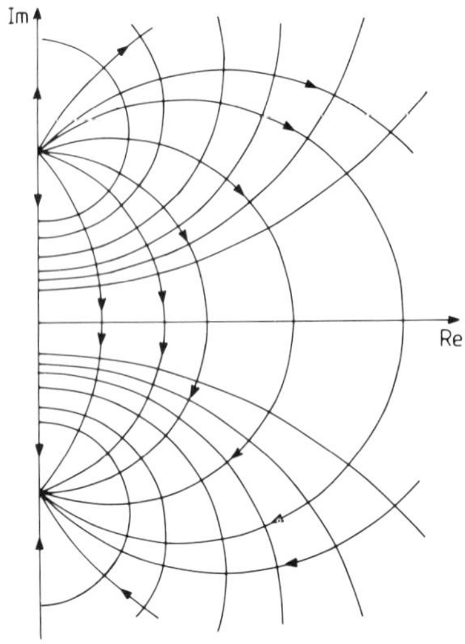

Figure 8.3 Admittance loci for an ideal metal with admittance $-ik$. The loci begin at the point ik and terminate on $-ik$. Equi-thickness contours are also shown at no fixed intervals. Similar loci are obtained for s-polarised FTR layers. For p-polarised FTR layers, the shape of the loci is similar but they are traced in the opposite direction.

$-ik$. Real metallic layers depart somewhat from this ideal model but if the metal is of high performance, i.e. if the ratio k/n is high, then the loci are similar to the perfect case. It is as if the diagram were rotated slightly about the origin so that the points where all circles intersect are $(n, -k)$ and $(-n, k)$ respectively, although the circles can never reach the point $(-n, k)$ since admittance loci are constrained to the first and second quadrants of the Argand diagram. Figure 8.4 shows a set of optical admittance loci calculated for silver, $n - \mathrm{i}k = 0.075 - \mathrm{i}3.41$[2] demonstrating this typical behaviour. The direction of the loci is now better described as terminating on $(n, -k)$, although most are still described in a clockwise direction. We will omit from the discussion in this chapter metals which are not of high optical quality and for which the loci resemble a set of spirals terminating at $(n, -k)$. What happens at oblique incidence?

The optical phase factor at normal incidence is

$$2\pi(n - \mathrm{i}k)d/\lambda \tag{8.11}$$

dominated by the imaginary part. At oblique incidence, it becomes

$$2\pi(n^2 - k^2 - n_0^2 \sin^2\theta_0 - 2\mathrm{i}nk)^{1/2} d/\lambda \tag{8.12}$$

still in the fourth quadrant. Since $n_0 \sin\theta_0$ is normally small compared with k, it has little effect on the phase factor. It reduces the real part slightly and increases the imaginary part, but the effect is small, and the behaviour is

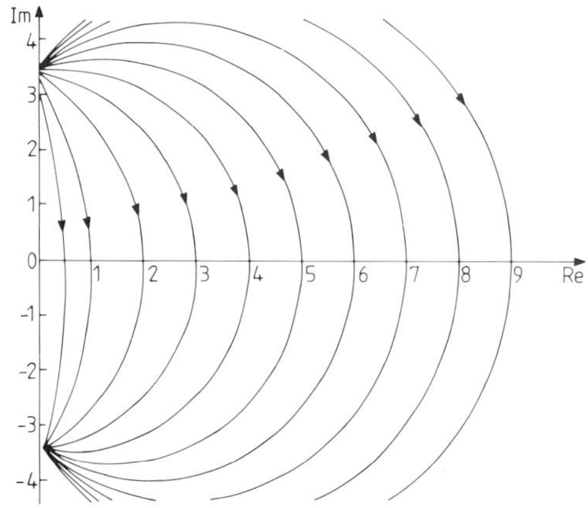

Figure 8.4 Admittance loci for silver at normal incidence in the visible region. The value assumed for the optical constants is 0.075 − i3.41.[2]

essentially similar to that at normal incidence. At an angle of incidence of 80° in air, for example, the phase factor of silver changes from $2\pi(0.075 - i3.41)d/\lambda$ to $2\pi(0.00721 - i3.549)d/\lambda$. The change in the modified admittance, therefore, is mainly due to the $\cos\theta_0$ term. The ratio of real to imaginary parts remains virtually the same, and the p-admittance simply moves towards the origin (both real and imaginary parts reduced) and the s-admittance away from the origin. Thus the principal effect for high-performance metal layers with tilt is an expansion of the circular loci for s-polarisation and a contraction for p-polarisation. The basic form remains the same.

The shift in the modified optical admittance does mean that the phase shift on reflection from a massive metal will vary. For silver at normal incidence, the phase shift will be in the second quadrant. As the angle of incidence increases, the movement of the p-polarised admittance towards the origin implies that the p-polarised phase shift moves towards the first quadrant, entering it at an angle of incidence of just above 70° (i.e. roughly $\cos^{-1}\frac{1}{3}$) while the s-polarised phase shift moves further towards 180°. The reflectance for s-polarised light increases, while for p-polarised light it shows a very slight drop initially to a shallow minimum, but rising thereafter.

Now we examine what happens when a metal layer is overcoated with a dielectric layer. The arrangement is sketched schematically in figure 8.5. Provided the admittance η_f of the dielectric layer is less than $(\eta_m\eta_m^*)^{1/2}$, where η_m is the admittance of the metal layer, the admittance locus will loop outside the line joining the origin to the starting point, as in the diagram. For dielectric layers having admittance greater than that of the incident medium, the reflectance falls while the locus is in the fourth quadrant of the Argand diagram. As the thickness of the dielectric layer increases, the reflectance is reduced until the intersection with the real axis. It then begins to rise, but, at the

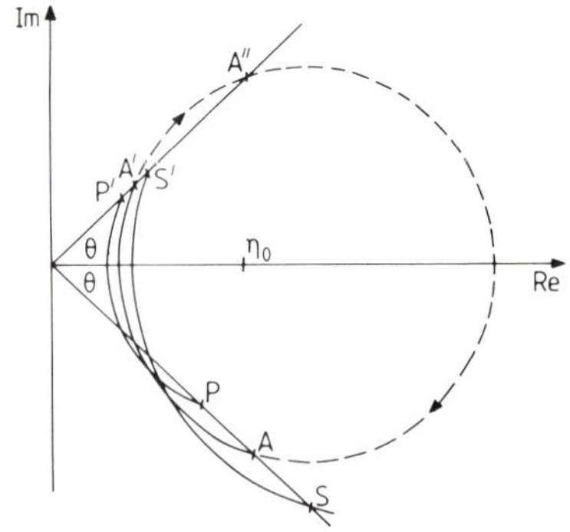

Figure 8.5 Schematic diagram of a dielectric overcoat on a metal surface. At normal incidence the metal admittance is at point A. A′ represents a quarter-wave thickness of material, while A″ represents the point at which the reflectance returns to the starting value. The lowest reflectance is given by the intersection with the real axis between the points A and A′. When tilted, the p-locus is given by PP′ and the s-locus by SS′.

quarter-wave point A′ given by η_f^2/η_m, it is still below the reflectance of the bare metal. Only at point A″ does the reflectance return to its initial level. The drop in reflectance for silver is slight, but for aluminium it is catastrophic. Silver is therefore usually overcoated with a quarter-wave, but aluminium with a half-wave that limits its useful spectral range somewhat.

As the metal–dielectric combination is tilted, the p-admittance of the metal slides towards the origin, the reflectance dropping, while the s-admittance moves away from the origin with a rise in reflectance. The dielectric layer shows a drop in admittance for p-polarised light and an increase for s-polarised. For dielectric coatings that are a quarter-wave or less these changes tend to compensate, and indeed, in silver, slightly overcompensate, the changes in reflectance of the bare metal. The p-reflectance of the overcoated metal tends to be slightly higher than the s-reflectance.

Eventually, for very high angles of incidence, the p-polarised admittance of the dielectric layer falls below the admittance of the incident medium, and now the fourth quadrant portion of the locus represents increasing reflectance. This means that the dielectric overcoating, when thin, instead of reducing the reflectance of the metal, actually enhances it. Thus, depending on the final thickness of the dielectric layer, the reflectance will tend to be high. For s-polarised light, the admittance of the dielectric layer tends to infinity as the angle of incidence tends to 90°. The locus of the dielectric overcoat, therefore, tends more and more towards a vertical line. As the admittance of the metal moves away from the origin, its projection in the real axis moves further to the right, eventually crossing the incident medium admittance and continuing towards infinity. There must, therefore, be an angle of incidence, very high, where the locus of the dielectric overcoat will intersect the real axis at the admittance of the incident medium. If the thickness is chosen so that the locus terminates at this point, then the reflectance of the metal–dielectric combination will be zero. This will occur for one particular value of angle of incidence and for a precise value of the dielectric layer thickness, and the dip in reflectance will show a rapid variation with angle of incidence. Such behaviour, for s-polarised light, of a metal overcoated with a thin dielectric layer was predicted by Nevière and Vincent[3] from a quite different analysis based on a Brewster absorption phenomenon in a lossy waveguide used just under its cutoff thickness. Since the modified admittance for s-polarised light increases with angle of incidence only in the case where its refractive index is greater than that of the incident medium, this is a necessary condition for the observation of the effect. The increased flexibility given by two dielectric layers deposited on a metal has been used to advantage in the design of reflection polarisers[4].

A different phenomenon was observed by Cox *et al*[5] in connection with an infrared mirror of aluminium with a protective overcoat of silicon dioxide. The silicon dioxide is heavily absorbing in the region beyond 8 μm. At a wavelength of just over 8 μm, n and k have values around 0.4 and 0.3 respectively. At normal incidence, the admittance loci of the silicon dioxide are spirals which end on the admittance of the silicon dioxide and are described in a clockwise manner in much the same way as the silver loci already discussed. At non-normal incidence, the s-polarised admittance and the phase factor for the layer remain in the fourth quadrant, and so the behaviour of the silicon oxide is similar to that at normal incidence. The p-polarised admittance, however, moves towards the first quadrant, and enters it at an angle of incidence around 40°. The behaviour of such a material, where the phase thickness is in the fourth quadrant but the optical admittance is in the first, is different from normal materials in that the spirals are now traced out anticlockwise, rather than clockwise. The admittance of aluminium at 8.1 μm is around $18.35 - i55.75$ and, for p-polarised light at an angle of incidence of 60°, the modified admittance becomes $9.176 - i27.87$. The dielectric locus sweeps down towards the real axis, as in figure 8.6, and, in a thickness of 150 nm, terminates in the vicinity of the point (1, 0), so that the reflectance is near zero.

This behaviour is quite unlike the normal behaviour to be expected with lossless dielectric overcoats which have refractive index greater than that of the incident medium. However, we shall see that it does have a certain similarity with one of the techniques for generating surface electromagnetic waves, which we shall be dealing with shortly, where the coupling medium is a dielectric layer of index lower than that of the incident medium, and where the angle of incidence is beyond the critical angle.

We now turn back to dielectric materials and investigate what happens when angles of incidence exceed the critical angle. Equations (8.7), (8.8) and (8.12) are the relevant equations and we have $k = 0$ and $n_0 \sin\theta_0 > n$. The phase thickness at normal incidence, $2\pi nd/\lambda$, becomes, from equation (8.12),

$$2\pi(n^2 - n_0^2 \sin^2\theta_0)^{1/2} d/\lambda$$

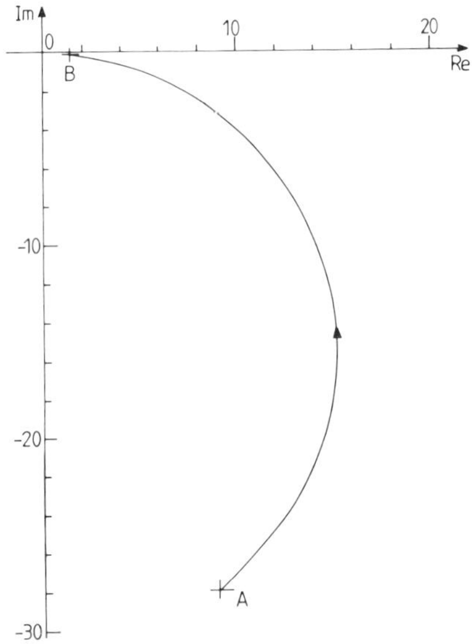

Figure 8.6 p-polarised admittance locus for 150 nm thickness of SiO_2, 0.39 – i0.29, on aluminium, 18.35 – i55.75, at an angle of incidence of 60°. A is the point corresponding to the modified admittance of aluminium and the anticlockwise curvature of the spiral locus carries it into the region of low reflectance.

i.e.

$$-\mathrm{i}2\pi(n_0^2\sin^2\theta_0-n^2)^{1/2}d/\lambda \tag{8.13}$$

at oblique incidence, where, again, the fourth rather than second quadrant solution is correct. The modified admittances are then

$$\begin{aligned}\eta_s &= -\mathrm{i}(n_0^2\sin^2\theta_0)^{1/2}/\cos\theta_0 \qquad \text{(fourth quadrant)}\\ \eta_p &= n^2/\eta_s.\end{aligned} \tag{8.14}$$

Since η_s is negative imaginary, η_p must be positive imaginary. The behaviour of the modified admittance is shown diagrammatically in figure 8.7. For a thin film of material used beyond the critical angle, then, the s-polarised behaviour is indistinguishable from that of an ideal metal. We have a set of circles centred on the real axis, described clockwise and ending on the point η_s which is on the negative imaginary axis. For p-polarised light, the behaviour is, in one important respect, different. Here, the combination of negative imaginary phase thickness and positive imaginary admittance inverts the way in which the circles are described, so that although they are still centred on the origin, they are clockwise and terminate at η_p on the positive imaginary axis. This behaviour plays a significant part in what follows. We assume a beam of light incident on the hypotenuse of a prism beyond the critical angle. Simply for

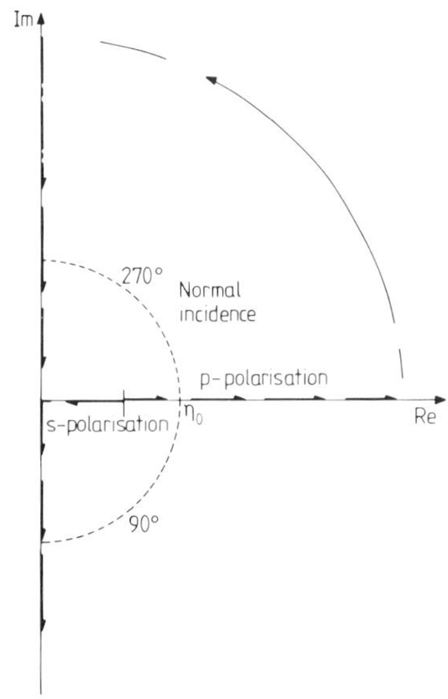

Figure 8.7 The variation of the s-polarised and p-polarised modified admittances of free space with respect to an incident medium of higher index. η_0 is the incident admittance. The s-admittance falls along the real axis until zero at the critical angle and then it turns along the negative direction of the imaginary axis tending to negative imaginary infinity as the angle of incidence tends to 90°. The p-admittance rises along the real axis, passing the point η_0 at the Brewster angle, becoming infinity at the critical angle, switching over to positive imaginary infinity and then sliding down the imaginary axis tending to zero as the angle of incidence tends to 90°.

plotting some of the following figures, we assume a value for the index of the incident medium of 1.52.

For an uncoated hypotenuse, the second medium is air of refractive index unity. The modified admittance for p-polarised light is positive imaginary and, as θ_0 increases, falls down the imaginary axis towards the origin. The reflectance is unity and figure 8.7 shows that the phase shift varies from 180° through the third and fourth quadrants towards 0°. The s-polarised reflectance is likewise unity, but the admittance is negative imaginary, and falls from zero to infinity along the imaginary axis so that the s-polarised phase shift increases with θ_0 from zero, through the first and second quadrants towards 180°. Since the incident medium has admittance 1.52, the circle separating the first and second quadrants and the third and fourth quadrants, which has centre the origin, has radius 1.52.

Now let a thin film be added to the hypotenuse. Since we are treating our glass prism as the incident medium, we should treat the surrounding air as the substrate. Thus the starting admittance for the film is on the imaginary axis. Provided the thin film has no losses, then the admittance of the film–substrate combination must remain on the imaginary axis. If the film admittance is imaginary, the combination admittance will simply move towards the film admittance. If, however, the film admittance is real, the admittance of the combination will move along the imaginary axis in a positive direction, returning to the starting point every half-wave. The lower the modified admittance, the slower the locus moves in the vicinity of the origin and the faster at points far removed from the origin. The variation of phase change between the fourth quadrant and the start of the first quadrant is, therefore, slower, while that between the third and second quadrants is faster than for a higher admittance. Thus there is a wide range of possibilities for varying the relative phase shifts for p- and s-polarisations by choosing an overcoat of higher or lower index and varying the thickness[6, 7].

Given that the starting point is on the axis, then the only way in which the admittance can be made to leave it is by an absorbing layer. We turn to the set of metal loci (figure 8.4) and we can see that for a range of values of starting admittance on the imaginary axis, the metal loci loop around, away from the axis, to cut the real axis. Although figure 8.4 shows the behaviour of metal layers for an incident medium of unity at normal incidence, the tilted behaviour for an incident admittance of 1.52 is quite similar. Figure 8.8 shows the illuminating arrangement and the loci. For a very narrow range of starting values, the metal locus cuts the real axis in the vicinity of the incident admittance, and, if the metal thickness is such that the locus terminates there,

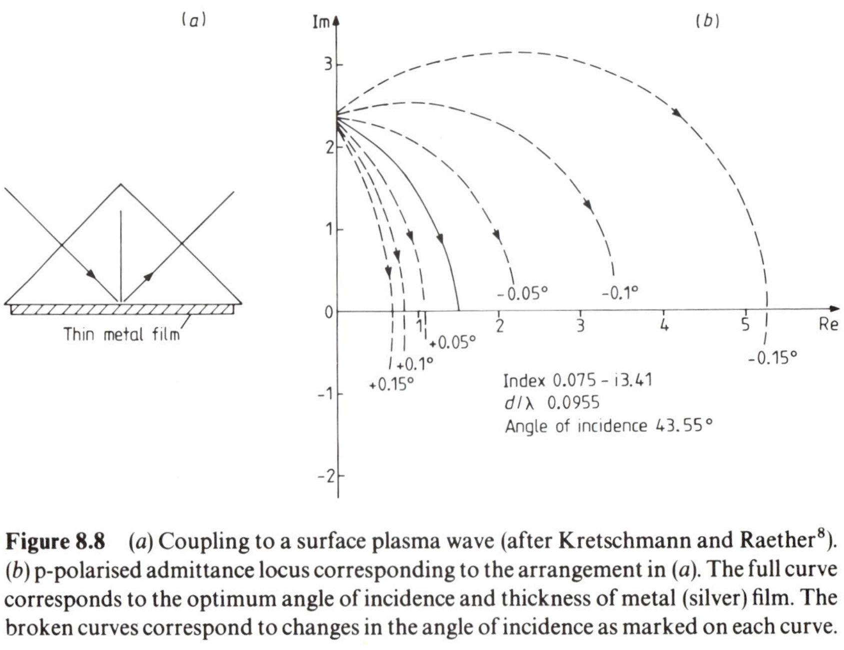

Figure 8.8 (*a*) Coupling to a surface plasma wave (after Kretschmann and Raether[8]). (*b*) p-polarised admittance locus corresponding to the arrangement in (*a*). The full curve corresponds to the optimum angle of incidence and thickness of metal (silver) film. The broken curves correspond to changes in the angle of incidence as marked on each curve.

then the reflectance of the combination will be low. For one particular angle of incidence and metal thickness the reflectance will be zero. It should not be too much of a surprise to find that the condition is very sensitive to angle of incidence. Since the admittance of the metal varies much more slowly than the air substrate, the zero reflectance condition will no longer hold, even for quite small tilts. This very narrow drop in reflectance to a very low value, which has all the hallmarks of a sharp resonance, can be interpreted as the generation of a surface plasma wave on the metal film. This coupling arrangement, devised by Kretschmann and Raether[8], cannot operate for s-polarised light without modification. The admittance of the substrate for s-polarisation is now on the negative part of the real axis and, therefore, any metal which is deposited will simply move the admittance of the combination towards the admittance of the bulk metal.

An alternative coupling arrangement, devised by Otto[9], involves the excitation of surface waves through an evanescent wave in an FTR layer (frustrated total reflectance). We recall that the admittance locus for p-polarisation of a layer used beyond the critical angle is a circle which is described in an anticlockwise direction. This means that such a layer can be used to couple into a massive metal. Here the metal acts as the substrate, with a starting admittance in the fourth quadrant of the Argand diagram. For p-polarised light, the dielectric FTR layer has a circular locus which cuts the real axis. Clearly, then, for the correct angle of incidence and dielectric layer thickness, the reflectance can be made zero. Surface plasma oscillations and their applications are extensively reviewed by Raether[10]. Abelès[11] includes an account of the optical features of such effects in his review of the optical properties of very thin films.

Now let us return to the first case of coupling and let us examine what happens when a thin layer is deposited over the metal next to the surrounding air. The starting admittance is, as before, on the imaginary axis, but now the dielectric layer modifies that position, so that the starting point for the metal locus is changed. Because the metal loci at the imaginary axis are clustered closely together, almost intersecting, a small change in starting point produces an enormous change in the locus, and hence in the point at which it cuts the real axis, leading to a substantial change in reflectance (figure 8.9). This very large change which a thin external dielectric film makes to the internal reflectance of the metal film has been used in the study of contaminant films adsorbed on metal surfaces. Film thicknesses of a few ångstroms have been detected in this way. Provided that the film is very thin, then an additional tilt of the system will be sufficient to pull the intersection of the metal locus with the real axis back to the incident admittance, and so the effect can be interpreted as a shift in the resonance rather than a damping.

This result helps us to devise a method for exciting a similar resonance with s-polarised light. The essential problem is the starting point on the negative imaginary axis, which means that the subsequent metal locus remains within

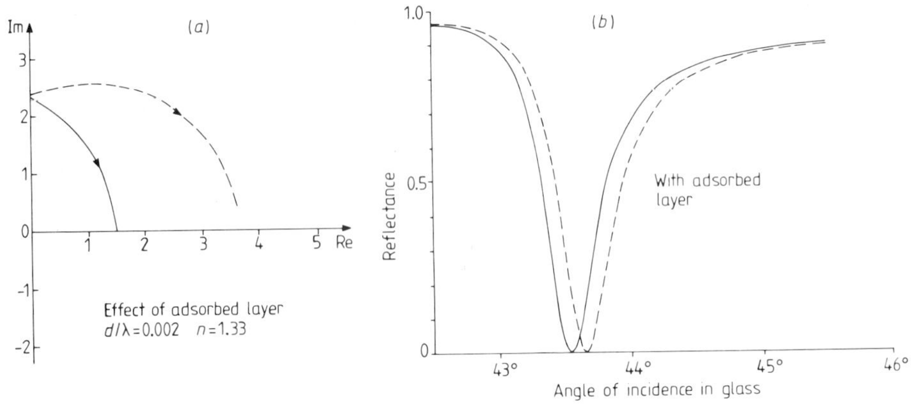

Figure 8.9 (*a*) The effect of a thin adsorbed layer on the surface of the silver in figure 8.8. The full line is the optimum while the broken line is the change in the metal locus due to the adsorbed layer. (*b*) Calculated reflectance as a function of angle of incidence with and without the adsorbed layer.

the fourth quadrant, never crossing the real axis to make it possible to have zero reflectance. The addition of a dielectric layer between the metal surface and the surrounding air can move the starting point for the metal on to the positive part of the imaginary axis so that the coated metal locus can cut the real axis for s-polarised light in just the same way as the uncoated metal in p-polarised light. Moreover, for both p- and s-polarised light, the low reflectance will be repeated for each additional half-wave dielectric layer which is added. This behaviour was used by Greenland and Billington[12] for the monitoring of optical layers intended as spacer layers for metal–dielectric interference filters. The operation of the cavities for inducing absorption devised by Harrick and Turner[13], although designed on the basis of a different approach, can also be explained in this way.

POLARISERS

The Brewster angle polarising beam splitter

This type of beam splitter was first constructed by Mary Banning[14] at the request of S M MacNeille, the inventor of the device[15] which is frequently known as a MacNeille polariser.

The principle of the device is that it is always possible to find an angle of incidence so that the Brewster condition for an interface between two materials of differing refractive index is satisfied. When this is so, the reflectance for the p-plane of polarisation vanishes. The s-polarised light is partially reflected and transmitted. To increase the s-reflectance, retaining the p-transmittance at or very near unity, the two materials may then be made into a multilayer stack. The layer thicknesses should be quarter-wave optical thicknesses at the appropriate angle of incidence.

When the Brewster angle for normal thin-film materials is calculated, it is found to be greater than 90° referred to air as the incident medium. In other words, it is beyond the critical angle for the materials. This presents a problem which is solved by building the multilayer filter into a glass prism so that the light can be incident on the multilayer at an angle greater than critical. The type of arrangement is shown in figure 8.10.

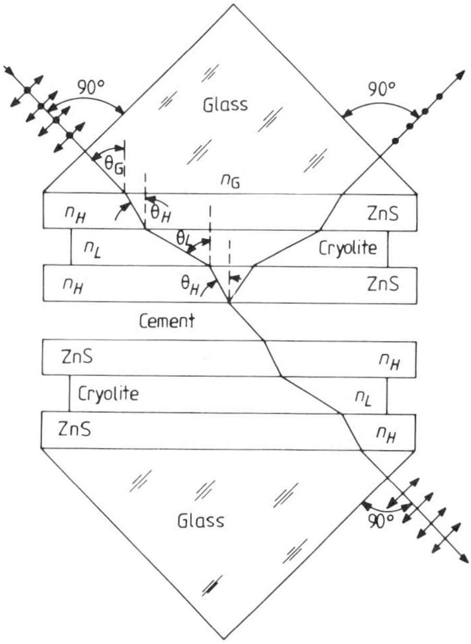

Figure 8.10 Schematic diagram of a polarising beam splitter. (After Banning[14].)

The calculation of the design is quite straightforward. Consider two materials with refractive indices n_H and n_L (where H and L refer to high and low relative indices respectively). The Brewster condition is satisfied when the angle of incidence is such that

$$n_H/\cos\theta_H = n_L/\cos\theta_L \tag{8.15}$$

where

$$n_H \sin\theta_H = n_L \sin\theta_L = n_G \sin\theta_G. \tag{8.16}$$

G refers to the glass of the prism. These equations can be solved easily for θ_H

$$\sin^2\theta_H = \frac{n_L^2}{n_H^2 + n_L^2} \tag{8.17}$$

the form in which we shall use the result. (A more familiar form is $\tan^2\theta_H = n_L^2/n_H^2$.)

Given the layer indices there are two possible approaches to the design. Either we can decide on the refractive index of the glass and then calculate the angle at which the prism must be set, or we can decide on the prism angle, 45° being a convenient figure, and calculate the necessary refractive index of the glass. The approach which was used by Banning was the latter.

First suppose that the condition $\theta_G = 45°$ must be met. Using equations (8.16) and (8.17) we obtain

$$\sin^2\theta_H = \frac{n_G^2\sin^2\theta_G}{n_H^2} = \frac{1}{2}\frac{n_G^2}{n_H^2} \qquad \text{for} \qquad \theta_G = 45°$$

i.e.

$$n_G^2 = \frac{2n_H^2 n_L^2}{n_H^2 + n_L^2} \tag{8.18}$$

the condition obtained by Banning.

If, however, n_G is fixed, then equations (8.16) and (8.17) give

$$\frac{n_G^2\sin^2\theta_G}{n_H^2} = \sin^2\theta_H = \frac{n_L^2}{n_H^2 + n_L^2}$$

i.e.

$$\sin^2\theta_G = \frac{n_H^2 n_L^2}{n_G^2(n_H^2 + n_L^2)}. \tag{8.19}$$

Banning used zinc sulphide with an index of 2.30 and cryolite evaporated at a pressure of 10^{-3} Torr to give a porous layer of index around 1.25. With these indices it is necessary to have an index of 1.55 for the glass if the prism angle is to be 45°. For an index of 1.35, a more usual figure for cryolite, together with zinc sulphide with an index of 2.35, the glass index should be 1.66. Alternatively, for glass of index 1.52, the angle of incidence using the second pair of materials should be 50.5°.

The degree of polarisation at the centre wavelength can also be calculated.

$$R = \left(\frac{\eta_G - (\eta_H^2/\eta_G)(\eta_H/\eta_L)^{n-1}}{\eta_G + (\eta_H^2/\eta_G)(\eta_H/\eta_L)^{n-1}}\right)^2 \tag{8.20}$$

where n is the number of layers and we are assuming n to be odd.

For s-waves:	For p-waves:
$\eta_G = n_G\cos\theta_G$	$\eta_G = n_G/\cos\theta_G$
$\eta_H = n_H\cos\theta_H$	$\eta_H = n_H/\cos\theta_H$
$\eta_L = n_L\cos\theta_L$	$\eta_L = n_L/\cos\theta_L$.

Now, for p-waves, by the conditions we have imposed, $\eta_H = \eta_L$ and

$$R_p = \left(\frac{\eta_G - (\eta_H^2/\eta_G)}{\eta_G + (\eta_H^2/\eta_G)}\right)^2$$

$$= \left[\left(\frac{n_G^2 \cos^2 \theta_H}{n_H^2 \cos^2 \theta_G} - 1 \right) \left(\frac{n_G^2 \cos^2 \theta_H}{n_H^2 \cos^2 \theta_G} + 1 \right)^{-1} \right]^2. \quad (8.21)$$

Similarly,

$$R_s = \left(\frac{n_G^2 \cos^2 \theta_G - n_H^2 \cos^2 \theta_H (n_H \cos \theta_H / n_L \cos \theta_L)^{n-1}}{n_G^2 \cos^2 \theta_G + n_H^2 \cos^2 \theta_H (n_H \cos \theta_H / n_L \cos \theta_L)^{n-1}} \right)^2. \quad (8.22)$$

Now

$$\frac{n_H \cos \theta_L}{n_L \cos \theta_H} = 1$$

so that

$$\frac{n_H \cos \theta_H}{n_L \cos \theta_L} = \frac{n_H^2}{n_L^2}$$

and

$$R_s = \left(\frac{n_G^2 \cos^2 \theta_G - n_H^2 \cos^2 \theta_H (n_H/n_L)^{2(n-1)}}{n_G^2 \cos^2 \theta_G + n_H^2 \cos^2 \theta_H (n_H/n_L)^{2(n-1)}} \right)^2. \quad (8.23)$$

The degree of polarisation in transmission is given by

$$P_T = \frac{T_p - T_s}{T_p + T_s} = \frac{1 - R_p - 1 + R_s}{1 - R_p + 1 - R_s} = \frac{R_s - R_p}{2 - R_p - R_s} \quad (8.24)$$

and in reflection by

$$P_R = \frac{R_s - R_p}{R_s + R_p}. \quad (8.25)$$

It can be seen that in general, for a small number of layers, the polarisation in reflection is better than the polarisation in transmission, but for a large number of layers it is inferior to that in transmission.

The construction of the beam splitter is similar to the cube beam splitter which was considered in chapter 4. Any number of layers can be used in the stack. Banning's original stack consisted of three layers, probably because of practical difficulties at that time. Two stacks were therefore prepared, one on the hypotenuse of each prism making up the cube, as shown in figure 8.10. The two prisms were then cemented together. Nowadays there is little difficulty in depositing 21 layers or more if need be and this can be conveniently deposited on just one prism and the other untreated prism simply cemented to it.

The very great advantage which this type of polarising beam splitter has over the other polarisers such as the pile-of-plates is its wide spectral range coupled with a large physical aperture. Unfortunately, it does suffer from a limited angular field, particularly at the centre of its range, simply because the Brewster condition is met exactly only at the design angle. As the angle of incidence moves away from this condition, a residual reflectance peak for p-polarisation gradually appears in the centre of the range. The performance well away from the centre remains high even for quite large tilts away from optimum. An an example, we can consider a seven-layer ZnS and cryolite beam splitter in glass of index 1.52 designed so that a wavelength of 510 nm corresponds to the centre of the range. At the design angle of 50.4° and at

510 nm the residual p-reflectance is 1.6 %, due to the mismatch between the materials of the stack and the glass prism. (The Brewster angle condition cannot be met for both film materials and the substrate simultaneously—see figure 8.2.) A tilt in the plane of incidence to 55° in glass (that is a tilt to 7° in air) raises the reflectance to 25 % at 510 nm, and over 30 % at 440 nm, since the band centre moves to shorter wavelengths. The reflectance at 650 nm, on the other hand, shows little change. Skew rays present a further difficulty. Polarisation performance is measured with reference to the s- and p-directions associated with the principal plane of incidence containing the axial ray. A skew ray possesses a plane of incidence that is rotated with respect to the principal plane. Thus the s- and p-planes for skew rays are not quite those of the axial ray and although the s-polarised transmittance can be very low there can be a component of the p-polarised light, which is parallel to the axial s-direction and which can represent an appreciable leakage.

A detailed study of the polarising prism has been carried out by Clapham[16].

Plate polariser

The width of the high-reflectance zone of a quarter-wave stack is a function of the ratio of the admittances of the two materials involved. This ratio varies with the angle of incidence and is different for s- and p-polarisations. We recall that

$$\eta_s = n\cos\theta \qquad \text{while} \qquad \eta_p = n/\cos\theta$$

so that

$$\eta_{Hs}/\eta_{Ls} = \cos\theta_H/\cos\theta_L$$

and

$$\eta_{Hp}/\eta_{Lp} = \cos\theta_L/\cos\theta_H$$

whence

$$\frac{(\eta_H/\eta_L)_s}{(\eta_H/\eta_L)_p} = \frac{(\cos\theta_H)^2}{(\cos\theta_L)^2}. \tag{8.26}$$

The factor $(\cos\theta_H)^2/(\cos\theta_L)^2$ is always less than unity so that the width of the high-reflectance zone for p-polarised light is always less than that for s-polarised light. Within the region outside the p-polarised but inside the s-polarised high-reflectance zone, the transmittance is low for s-polarised light but high for p-polarised so that the component acts as a polariser. The region is quite narrow, so that such a polariser will not operate over a wide wavelength range; but for single wavelengths, such as a laser line, it can be very effective. To complete the design of the component it is necessary to reduce the ripple in transmission for p-polarised light and this can be performed using any of the techniques of chapter 6, probably the most useful being Thelen's shifted-period method because it is the performance right at the edge of the pass region which is important. It is normal to use the component as a longwave-pass filter because this involves thinner layers and less material than would a shortwave-pass filter. The rear surface of the component requires an antireflection coating for p-polarised light. We can omit this altogether if the component is used at the Brewster angle. The design of such a polariser is described by Songer[17] who gives the design shown in figure 8.11. Plate polarisers are used in preference to the prism or MacNeille type when high powers are concerned.

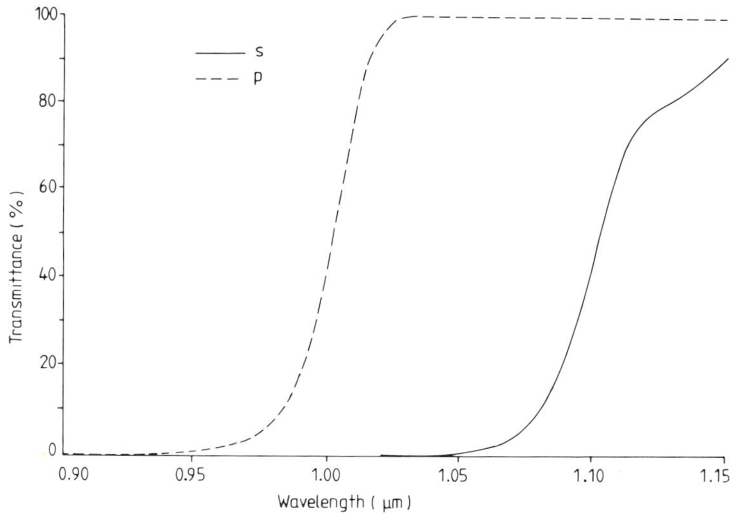

Figure 8.11 Characteristic curve of a plate polariser for 1.06 µm. Design: Air|$(0.5H'L'0.5H')^3$ $(0.5H''L''0.5H'')^8$ $(0.5H'L'0.5H')^3$|Glass where $H' = 1.010H$, $L' = 1.146L$, $H'' = 1.076H$, $L'' = 1.220L$ and with $n_H = 2.25$, $n_L = 1.45$, $\lambda_0 = 0.9$ µm and $\theta_0 = 56.5°$. The solid line indicates s-polarisation and the broken line p-polarisation. (After Songer[17].)

Virtually any coating which possesses a sharp edge between transmission and reflection can potentially be used as a polariser. It has been suggested that narrowband filters have advantages over simple quarter-wave stacks as the basis of plate polariser coatings, because the monitoring of the component during deposition is a more straightforward procedure[18].

Cube polarisers

An advantage of the polariser immersed in a prism is that the effective angle of incidence can be very high—much higher than if the incident medium were air. This enhances the polarisation splitting and gives broader regions of high degree of polarisation than could be the case with air as the incident medium. Even if the Brewster angle condition cannot be reached, there is an advantage in using an immersed design, provided the incident power is not too high. Netterfield[19] has considered the design of such polarisers in some detail and his paper should be consulted for further information.

NON-POLARISING COATINGS

The design of coatings which avoid polarisation problems is a much more difficult task than that of polariser design and there is no completely effective method. The changes in the phase thicknesses of the layers and in the optical admittances are fundamental and cannot be avoided. The best we can hope to do, therefore, is to arrange the sequence of layers so that they give the same performance for p- as for s-polarisation. Clearly, the wider the range of either angle of incidence or of wavelength, the more difficult the task. The techniques which are currently available operate only over very restricted ranges of wavelength and angle of incidence (effectively over a very narrow range of angles). There is a small body of published work but the principal techniques we shall use here rely heavily on techniques devised by Thelen[20, 21].

Edge filters at intermediate angles of incidence

This section is based entirely on an important paper by Thelen[20]. However, the expressions found in the original paper have been altered in order to make the notation consistent with the remainder of this book. Care should be taken, therefore, in reading the original paper. In particular, the x found in the original is defined in a slightly different way.

At angles of incidence which are not so severe that the p-reflectance suffers, the principal effect of operating edge filters at oblique incidence is the splitting between the two planes of polarisation. This limits the edge steepness which can be achieved for light which is unpolarised. Edge filters which have pass regions which are quite limited can be constructed from band-pass filters, but, because band-pass filters are also affected in much the same way, the bandwidth for s-polarised light shrinking and for p-polarised light expanding, they still suffer from the same problem. However, there is a technique which can be used for displacing the pass bands of a band-pass filter to make one pair of edges coincide, resulting in an edge filter of rather limited extent, which for a given angle of incidence has no polarisation splitting. The position of the peak of a band-pass filter can be considered to be a function of both the spacer thickness and the phase shift of the reflecting stacks on either side. At oblique incidence, the relative phase shift between s- and p-polarised light from the reflecting stacks can be adjusted by adding or removing material. This alters the relative positions of the peaks of the pass bands for the two planes of polarisation and, if the adjustment is correctly made, it can make a pair of edges coincide. This, of course, is for one angle of incidence only. As the angle of incidence moves away from the design value, the splitting will reappear.

Rather than apply this technique exactly as we have just described it, we instead adapt the techniques for the design of multiple cavity filters based on symmetrical periods. Let us take a typical multiple cavity filter design:

Incident medium|matching (symmetrical stack)^q matching|substrate.

The symmetrical stack which forms the basis of this filter can be represented as a single matrix which has the same form as that of a single film, as we have already seen in chapter 7. The limits of the pass band are given by those wavelengths for which the diagonal terms of the matrix are unity and the off-diagonal terms are zero. That is, if the matrix is given by

$$\begin{bmatrix} N_{11} & \mathrm{i}N_{12} \\ \mathrm{i}N_{21} & N_{22} \end{bmatrix}$$

then the edges of the pass band are given by

$$N_{11} = N_{22} = \pm 1.$$

The design procedure simply ensures that this condition is satisfied for the appropriate angle of incidence.

We can consider the symmetrical period as a quarter-wave stack of $2x+1$ layers which has two additional layers added, one on either side:

$$fB\ ABAB\ \ldots\ A\ fB$$

where A and B indicate quarter-wave layers and f is a correction factor which is to be applied to the quarter-wave thickness to yield the thicknesses of the detuned outer layers. We can write the overall matrix as $fB\ M\,fB$ where $M = ABAB \ldots A$, giving for the product:

$$\begin{bmatrix} \cos\alpha & \mathrm{i}\sin\alpha/\eta_B \\ \mathrm{i}\eta_B\sin\alpha & \cos\alpha \end{bmatrix}\begin{bmatrix} M_{11} & \mathrm{i}M_{11} \\ \mathrm{i}M_{21} & M_{11} \end{bmatrix}\begin{bmatrix} \cos\alpha & \mathrm{i}\sin\alpha/\eta_B \\ \mathrm{i}\eta_B\sin\alpha & \cos\alpha \end{bmatrix}.$$

Then N_{11} is given by

$$N_{11} = N_{22} = M_{11}\cos 2\alpha - 0.5(M_{12}\eta_B - M_{21}/\eta_B)\sin 2\alpha = \pm 1 \qquad (8.27)$$

for the edge of the zone for each plane of polarisation. This must be satisfied for both planes of polarisation simultaneously for the edges of the pass bands to coincide. In fact, symmetrical periods which are made up of thicknesses other than quarter waves can be used, when some trial and error will be required to satisfy equation (8.27). A computer can be of considerable help. For quarter-wave stacks we seek assistance in the expressions derived in chapter 7 for narrowband filter design. We use expression (7.53), with $m = 1$ and $q = 0$, giving

$$\begin{aligned} M_{11} &= M_{22} = (-1)^x(-\varepsilon)[(\eta_A/\eta_B)^x + \ \ldots\ + (\eta_B/\eta_A)^x] \\ \mathrm{i}M_{12} &= \mathrm{i}(-1)^x/[(\eta_A/\eta_B)^x\eta_A] \qquad (8.28) \\ \mathrm{i}M_{21} &= \mathrm{i}(-1)^x[(\eta_A/\eta_B)^x\eta_A]. \end{aligned}$$

Note that $2x+1$ is now the number of layers in the inner stack. The total number of layers, including the detuned ones, is $2x+3$. Now, using exactly the same procedure as in chapter 7, we can write expressions for the coefficients in equation (8.27) as

$$M_{11} = \frac{(-1)^x(-\varepsilon)\,(n_H/n_L)^x}{(1 - n_L/n_H)} = (-1)^x(-\varepsilon)P$$

and

$$0.5(M_{12}\eta_B + M_{21}/\eta_B) = 0.5(-1)^x[(\eta_B/\eta_A)^{x+1} + (\eta_A/\eta_B)^{x+1}] = (-1)^x Q$$

where

$$P = (\eta_H/\eta_L)^x/(1 - \eta_L/\eta_H) \qquad \text{and} \qquad Q = 0.5(\eta_H/\eta_L)^{x+1}.$$

Then the two equations become

$$\begin{aligned} \pm 1 &= \varepsilon P_\mathrm{p}\cos 2\alpha + Q_\mathrm{p}\sin 2\alpha \\ \pm 1 &= \varepsilon P_\mathrm{s}\cos 2\alpha + Q_\mathrm{s}\sin 2\alpha \end{aligned} \tag{8.29}$$

which give for α and ε:

$$\sin 2\alpha = \pm\frac{P_\mathrm{s} - P_\mathrm{p}}{(P_\mathrm{s}Q_\mathrm{p} - P_\mathrm{p}Q_\mathrm{s})} \tag{8.30}$$

$$\varepsilon = \frac{\pm 1 - Q_\mathrm{p}\sin 2\alpha}{P_\mathrm{p}\cos 2\alpha}. \tag{8.31}$$

Now,

$$\begin{aligned} \varepsilon &= (\pi/2)(1 - g) \qquad \text{where} \qquad g = \lambda_0/\lambda \\ \alpha &= (\pi/2)(\lambda_\mathrm{R}/\lambda) = (\pi/2)(\lambda_\mathrm{R}/\lambda_0)g = (\pi/2)fg \end{aligned}$$

so that

$$f = \alpha/(\pi g/2) = \alpha/(\pi/2 - \varepsilon). \tag{8.32}$$

Two values for f will be obtained. Usually, the larger corresponds to a shortwave-pass and the smaller to a longwave-pass filter.

There are some important points about the particular values of α and ε, which are best discussed within the framework of a numerical example. Let us attempt the design of a longwave-pass filter at 45° in air having a symmetrical period of

$$fL\ HLHLHLH\ fL$$

where H represents an index of 2.35 and L of 1.35. The inner stack has 7 layers, which corresponds to $2x + 1$, so that x in this example is 3. We will use the modified admittances that for this combination are (the subscripts S and A referring to the substrate and to air, respectively):

$$\begin{aligned} \eta_{H\mathrm{s}} &= 3.1694 \qquad & \eta_{L\mathrm{s}} &= 1.6264 \\ \eta_{\mathrm{Ss}} &= 1.9028 & \eta_{\mathrm{As}} &= 1.000 \\ \eta_{H\mathrm{p}} &= 1.7425 & \eta_{L\mathrm{p}} &= 1.1206 \\ \eta_{\mathrm{Sp}} &= 1.2142 & \eta_{\mathrm{Ap}} &= 1.000. \end{aligned}$$

Then

$$P_s = 15.201 \qquad P_p = 10.535$$
$$Q_s = 7.211 \qquad Q_p = 2.923$$

giving $\sin\alpha = \pm 0.1480$.

Now, the outer tuning layers in their unperturbed state will be quarter-waves and so the two solutions we look for will be near $2\alpha = \pi$, that is, in the second and third quadrants. We continue to keep the results in the correct order and find

$$2\alpha = \pi \pm 0.1485 = 2.9931 \qquad \text{or} \qquad 3.2901.$$

Then, in both cases, $\cos 2\alpha = -0.9890$ and so

$$\varepsilon = \pm(1 + 2.923 \times 0.148)/(-10.535 \times 0.9890) = \pm(-0.1375)$$

whence

$$f = (2.9931/2)/[(\pi/2) - 0.1375] = 1.148$$

with

$$g = 1 - 2 \times 0.1375/\pi = 0.9125$$

and

$$f = (2.9931/2)/[(\pi/2) + 0.1375] = 0.876$$

with

$$g = 1 + 2 \times 0.1375/\pi = 1.088.$$

We take the second of these which will correspond to a longwave-pass filter. We now need to consider the matching requirements. Since we are attempting to obtain coincident edges for both planes of polarisation in an edge filter of limited pass band extent, we will interest ourselves in having good performance right at the edge of the pass band with little regard for performance further away. We use the symmetrical period method. The basic period is

$$0.876L\ HLHLHLH\ 0.876L$$

with H and L quarter-waves of indices 2.35 and 1.35 respectively, and tuned for 45°. Calculation of the equivalent admittances for the symmetrical period gives the values for s- and p-polarisation shown in table 8.1. (Again they are modified admittances.) We will arrange matching at $g = 1.08$. Adding a $HLHL$ combination to the period with the L layer next to it yields admittances of 0.9625 for p-polarisation and 1.416 for s. The media we have to match have modified admittances of 1.0 for air and 1.214 for glass for p-polarisation and 1.0 and 1.903 respectively for s. As an initial attempt, therefore, this matching is probably adequate. Since the matching is to be at $g = 1.08$, the thicknesses of the four layers in the matching assemblies must be corrected by the factor 1.0/1.08. To complete the design we need to make sure all layers are tuned for 45° which means multiplying their effective thicknesses for 45° by the factor $1/\cos\theta$. The final design with all thicknesses quoted as their normal incidence values is then

$$\text{Air}|(0.971\,H\ 1.087\,L)^2(1.028\,L(1.049\,H\ 1.174\,L)^3 1.049\,H\ 1.028\,L)^q(1.087\,L\ 0.971\,H)^2|\text{Glass}.$$

Table 8.1 Equivalent admittances and phase thicknesses of the symmetrical period (0.876*L HLHLHLH* 0.876*L*) where *L* and *H* indicate quarter-waves at 45° angle of incidence of index 1.35 and 2.35 respectively.

	s-polarisation		p-polarisation	
g	E (modified)	γ/π	E (modified)	γ/π
1.04	Imaginary values		0.1946	4.4372
1.05	0.0949	4.2955	0.2018	4.4372
1.06	0.1190	4.4454	0.1993	4.5884
1.07	0.1202	4.5786	0.1861	4.6652
1.08	0.0982	4.7211	0.1588	4.7486
1.09	Imaginary values		0.1049	4.8530
1.10	Imaginary values		Imaginary values	

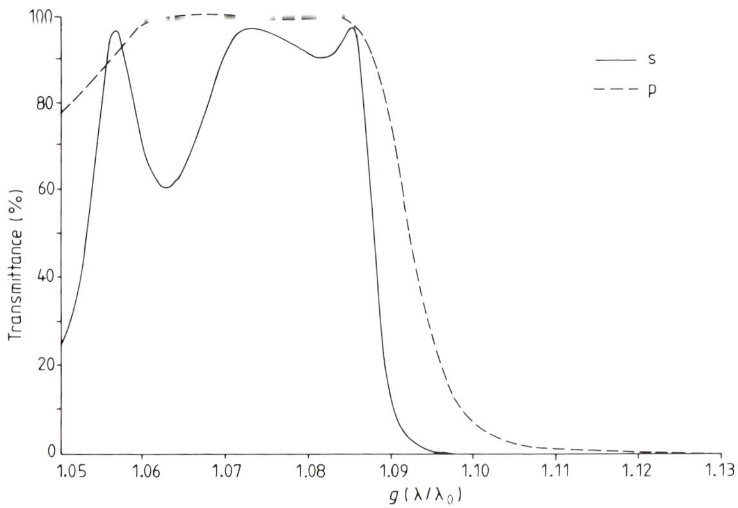

Figure 8.12 Calculated performance of a polarisation-free edge filter designed for use at 45° in air using the method of Thelen[20]. The multilayer structure is given in the text. The full curve indicates s-polarisation and the broken curve p-polarisation.

The performance with $q = 4$ is shown in figure 8.12 along with the performance of a band-pass filter of similar design using unaltered quarter-waves to demonstrate the difference. Since the p-admittances are less effective than the s in achieving high reflectance, the steepness of the edge for s-polarisation is somewhat greater and so the two edges coincide at their upper ends. Adjustment of the factor f can move this point of coincidence up and down the edges. Thelen gives many examples of designs including some which are based on symmetrical periods containing thicknesses other than quarter-waves.

Reflecting coatings at very high angles of incidence

Reflecting coatings at very high angles of incidence suffer catastrophic reductions in reflectance for p-polarisation. This is especially true for coatings that are embedded in glass such as cube beam splitters and we have already seen how they can make good polarisers. The admittances for p-polarised light are not favourable for high reflectance and so to increase the p-reflectance we must use a large number of layers—many more than is usual at normal incidence. The s-reflectance must also at the same time be considerably reduced, otherwise it will vastly exceed what is possible for p-polarisation. The technique we use here is based on yet another method originated by Thelen[21]. A number of authors have studied the problem. For a detailed account of the use of symmetrical periods in the design of reflecting coatings for oblique incidence, the paper by Knittl and Houserkova[22] should be consulted.

We consider a quarter-wave stack. The admittance of such a stack is given at normal incidence by

$$Y = \frac{y_1^2 y_3^2 y_5^2 \ldots y_{sub}}{y_2^2 y_4^2 y_6^2 \ldots} \tag{8.33}$$

with y_{sub} in the numerator, as shown, if the number of layers is even or in the denominator if odd. The reflectance is

$$R = \left(\frac{y_0 - Y}{y_0 + Y}\right)^2$$

in the normal way. Now, if the stack of quarter-waves is considered to be tilted, with the thicknesses tuned to the particular angle of incidence, the expression for reflectance will be similar except that the appropriate tilted admittances must be used. Here we will use the modified admittances so that y_0 will remain the same. Then Y becomes

$$Y = \frac{\eta_1^2 \eta_3^2 \eta_5^2 \ldots \eta_{sub}}{\eta_2^2 \eta_4^2 \eta_6^2 \ldots} \tag{8.34}$$

and in order for the reflectances for p- and s-polarisations to be equal, the modified admittances for p- and s-polarisation must be equal. If we write Δ_1 for (η_{1p}/η_{1s}) and so on, then this condition is

$$\frac{\Delta_1^2 \Delta_3^2 \Delta_5^2 \ldots \Delta_{sub}}{\Delta_2^2 \Delta_4^2 \Delta_6^2 \ldots} = 1. \tag{8.35}$$

(Note that Thelen's paper does not use modified admittances and so includes the incident medium in the formula.) The procedure then is to attempt to find a combination of materials such that condition (8.35) is satisfied and the value of admittance is such that the required reflectance is achieved. This is a matter of trial and error.

An example will help to make the method clear. Table 8.2 gives some figures for modified admittances in glass ($n = 1.52$) and at an angle of incidence of 45°. There is a number of possible arrangements but the most straightforward is to find three materials H, L and M, M being of intermediate index, such that

$$\Delta_H \Delta_L = \Delta_M^2. \tag{8.36}$$

Table 8.2

n_f	$1/\cos\theta$	η_p	η_s	$\Delta(=\eta_p/\eta_s)$
1.35	1.6526	1.5776	1.1553	1.3656
1.38	1.5943	1.5558	1.2241	1.2710
1.45	1.4898	1.5275	1.3765	1.1097
1.52	1.4142	1.5200	1.5200	1.0000
1.57	1.3719	1.5230	1.6185	0.9410
1.65	1.3180	1.5377	1.7705	0.8685
1.70	1.2907	1.5515	1.8627	0.8330
1.75	1.2672	1.5680	1.9531	0.8028
1.80	1.2466	1.5867	2.0419	0.7771
1.85	1.2286	1.6072	2.1295	0.7548
1.90	1.2127	1.6292	2.2158	0.7353
1.95	1.1985	1.6525	2.3010	0.7182
2.00	1.1858	1.6770	2.3853	0.7030
2.05	1.1744	1.7023	2.4687	0.6895
2.10	1.1640	1.7285	2.5514	0.6775
2.15	1.1546	1.7554	2.6334	0.6666
2.20	1.1461	1.7829	2.7147	0.6568
2.25	1.1383	1.8110	2.7955	0.6478
2.30	1.1311	1.8396	2.8757	0.6397
2.35	1.1245	1.8686	2.9554	0.6323
2.40	1.1184	1.8980	3.0347	0.6254

Modified admittances
Incident medium index = 1.52
Angle of incidence = 45°

Then the multilayer structure can be . . . $HMLMHMLMHMLM$ . . . so that the form of the admittance is

$$Y = \frac{\eta_H^2 \eta_L^2 \eta_H^2 \cdots}{\eta_M^2 \eta_M^2 \eta_M^2 \cdots} \tag{8.37}$$

and the number of layers chosen so that adequate reflectance is achieved. The substrate does not appear in (8.37) because it is assumed to be of the same material as the incident medium and so Δ_{sub} is unity. Where the substrate is of a different material there may be a slight residual mismatch but practical difficulties will usually make achievement of an exact match difficult. A set of layers giving an approximate match at 45° has indices 1.35, 2.25 and 1.57. For this combination

$$\frac{\Delta_H \Delta_L}{\Delta_M^2} = \frac{1.3656 \times 0.6478}{0.941^2} = 0.999.$$

The p-admittance increase due to one 4-layer period of that type is

$$\frac{\eta_{Hp}^2 \eta_{Lp}^2}{\eta_{Mp}^4} = \frac{1.811^2 \times 1.578^2}{1.523^4} = 1.518.$$

Eight periods give a value of 28.2, that is a reflectance of 87% for 32 layers. The particular arrangement of *H*, *L* and *M* layers is flexible as long as *H* or *L* are odd and *M* is even. The performance of a coating to this design is shown in figure 8.13. The basic period is four quarter-waves thick. High reflectance zones exist wherever the basic period is an integral number of half-waves thick. Since in this case we have four quarter-waves we expect extra high reflectance zones at $g = 0.5$ and $g = 1.5$. The peak at $g = 0.5$ (i.e. $\lambda = 2 \times 510 = 1020$ nm) is visible at the long wavelength end of the diagram.

Examination of the modified admittances for the materials shows how the

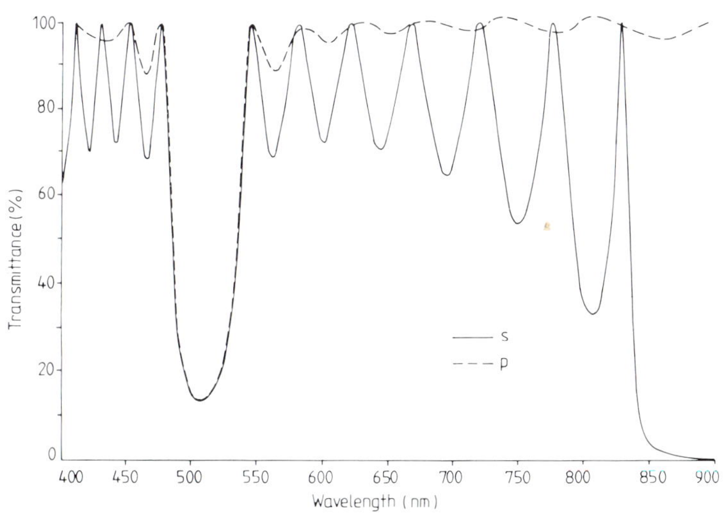

Figure 8.13 Calculated performance of a polarisation-free reflector at an angle of incidence of 45° in glass. The coating was designed using the method of Thelen[21]. Design: Glass |(1.38*H* 1.372*M* 1.653*L* 1.372*M*)^8| Glass with $n_H = 2.25$, $n_M = 1.57$, $n_L = 1.35$, $n_{Glass} = 1.52$ and $\lambda_0 = 510$ nm. The solid line indicates s-polarisation and the broken line p-polarisation.

coating does yield the desired performance. Each second pair of layers tends to reduce the s-reflectance of the preceding pair but slightly to increase the p-reflectance. To achieve high reflectance large numbers of layers are needed. Angular sensitivity is quite high and there is little that can be done to improve it.

Edge filters at very high angles of incidence

It is possible to adapt the treatment of the previous section to design edge filters for use at high angles of incidence. Let us illustrate the method by using the example we have just calculated. Figure 8.13 shows the performance. We wish to use this component as a longwave-pass filter and hence to eliminate the ripple on the longwave side of the peak. The ripple is principally confined to s-polarisation and so we concentrate our efforts there. We will use a symmetrical period approach.

The basic symmetrical period can be either

$$(0.5H\ MLM\ 0.5H) \qquad \text{or} \qquad (0.5L\ MHM\ 0.5L).$$

We use the modified s-admittances that we have already calculated in the previous section and we compute the equivalent admittances as shown in table 8.3. The surrounding material has admittance 1.52 and it appears as though a simple match would be obtained with the $(0.5L\ MHM\ 0.5L)$ combination. We match at $g = 0.88$ where the equivalent admittance is 0.802. To match to 1.52, a quarter-wave of admittance $(0.802 \times 1.52)^{1/2}$ is required. This is 1.104 and corresponds fairly well with the 1.155 admittance of the 1.35 low index material. A quarter-wave at $g = 0.88$ and 45° has a normal incidence thickness of $(1.0/0.88) \times 1.653 \times 0.25$ full waves, that is 1.877 quarter-waves or 0.470 full waves. The full design is then

$$\text{Glass}\,|\,1.877L\ (0.826L\ 1.372M\ 1.138H\ 1.372M\ 0.826L)^q\ 1.877L\,|\,\text{Glass}.$$

The performance of a coating with $q = 10$ is shown in figure 8.14. Shortwave-pass filters or filters with different materials can be designed in the same way. The design is fairly sensitive to materials and to angle of incidence.

ANTIREFLECTION COATINGS

Antireflection coatings at high angles of incidence are a stage more difficult than the design of coatings for normal incidence. Some simplification occurs when only one plane of polarisation has to be considered. Then it is a case of taking the tables for modified optical admittance at the appropriate angle of incidence and designing coatings in much the same way as for normal incidence. The complication is that the range of admittances available is different from the range at normal incidence and also different for the two planes of polarisation. We therefore consider briefly the problem of antireflec-

Table 8.3 Equivalent admittances and phase thicknesses of the symmetrical periods $(0.5LMHM0.5L)$ and $(0.5HMLM0.5H)$ calculated for 45° angle of incidence in glass of index 1.52. $n_H = 2.35$, $n_L = 1.35$ and $n_M = 1.57$.

g	$E_{\text{mod,s}}$ $(0.5HMLM0.5H)$	$E_{\text{mod,s}}$ $(0.5LMHM0.5L)$	γ/π
0.58	Imaginary values		
0.60	18.8985	0.1442	1.0473
0.62	6.8181	0.3965	1.1438
0.64	5.1698	0.5184	1.2061
0.68	4.4178	0.6007	1.2600
0.70	3.9680	0.6613	1.3100
0.72	3.6599	0.7443	1.3577
0.74	3.4300	0.7728	1.4040
0.76	3.2471	0.7949	1.4494
0.78	2.9594	0.8114	4.5382
0.80	2.8362	0.8225	1.5820
0.82	2.7180	0.8281	1.6256
0.84	2.5994	0.8276	1.6691
0.86	2.4741	0.8199	1.7126
0.88	2.3340	0.8024	1.7564
0.90	2.1662	0.7705	1.8005
0.92	1.9467	0.7151	1.8456
0.94	1.6195	0.6135	1.8930
0.96	0.9761	0.3808	1.9489
0.98	Imaginary values		

tion coatings for one plane of polarisation first. In order to simplify the discussion of design we assume an angle of incidence of 60° in air with a substrate of index 1.5 and possible film indices of 1.3, 1.4, 1.5, . . . , 2.5. Real designs will be based on available indices and will therefore be more constrained and may require more layers. The modified admittances with values of $\Delta(=\eta_p/\eta_s)$ are given in table 8.4.

p-polarisation only

At 60° the modified p-admittance of the substrate is only 0.9186 giving a single-surface reflectance for p-polarised light of less than 0.2 %, acceptable for most purposes. The angle of incidence of 60° is only just greater than the Brewster angle. If still lower reflectance is required then a single quarter-wave of admittance given by $(0.9186 \times 1.0000)^{1/2}$, that is 0.9584, is required. This corresponds from the table to an index of 1.6 that is *greater* than the index of the substrate. As the angle of incidence increases still further from 60° the

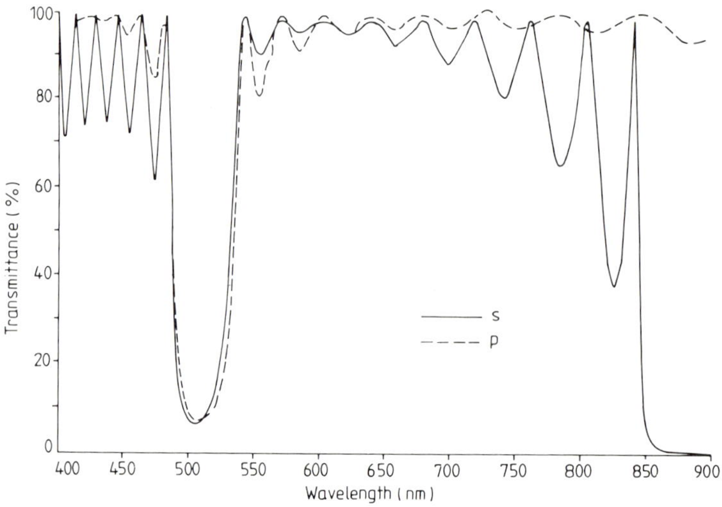

Figure 8.14 Calculated performance of a polarisation-free edge filter at an angle of incidence of 45° in glass. Design: Glass|1.877L (0.826L 1.372M 1.138H 1.372M 0.826L)^{10} 1.877L|Glass with $n_H = 2.25$, $n_M = 1.57$, $n_L = 1.35$, $n_{\text{Glass}} = 1.52$ and $\lambda_0 = 510$ nm. The solid line indicates s-polarisation and the broken line p-polarisation.

Table 8.4

n_f	$1/\cos\theta$	η_p	η_s	$\Delta(=\eta_p/\eta_s)$
1.00	2.0000	1.0000	1.0000	1.0000
1.30	1.3409	0.8716	1.9391	0.4495
1.40	1.2727	0.8909	2.2000	0.4050
1.50	1.2247	0.9186	2.4495	0.3750
1.60	1.1893	0.9514	2.6907	0.3536
1.70	1.1621	0.9878	2.9258	0.3376
1.80	1.1407	1.0266	3.1560	0.3253
1.90	1.1235	1.0673	3.3823	0.3156
2.00	1.1094	1.1094	3.6056	0.3077
2.10	1.0977	1.1526	3.8262	0.3012
2.20	1.0878	1.1966	4.0448	0.2958
2.30	1.0794	1.2414	4.2615	0.2913
2.40	1.0722	1.2867	4.4766	0.2874
2.50	1.0660	1.3325	4.6904	0.2841

Modified admittances
Incident medium index = 1.00
Angle of incidence = 60°

required index will become still greater. Eventually, at very high angles of incidence indeed, the required single layer index will be greater than the highest index available and at that stage designs based on combinations such as Air | HL | Glass will be required with quarter-wave thicknesses at the appropriate angle of incidence. Such coatings operate over a very small range of angles of incidence only and are very difficult to produce with any reasonable degree of success. If at all possible it is better to avoid such designs altogether by redesigning the optical system.

s-polarisation only

The modified s-admittance for the substrate is 2.449 and the required single-layer admittance for perfect antireflection is $(2.4495 \times 1.0000)^{1/2}$ or 1.5650, well below the available range. The problem is akin to that at normal incidence where we do not have materials of sufficiently low index and the solution is similar. We begin by raising the admittance of the substrate to an acceptable level by adding a quarter-wave of higher admittance. In this case a layer of index 1.9 or admittance 3.3823 is convenient and gives a resultant admittance of $3.3823^2/2.449$ or 4.3991 that requires a quarter-wave of admittance $(4.3991 \times 1.0000)^{1/2}$ or 2.0974 to complete the design. This corresponds most nearly to an index of 1.4, admittance 2.2000, and the residual reflectance with such a combination is 0.23 %, a considerable improvement over the 17.7 % reflectance of the uncoated substrate. We cannot expect that such a coating will have a broad characteristic and figure 8.15 confirms it. A small improvement can be

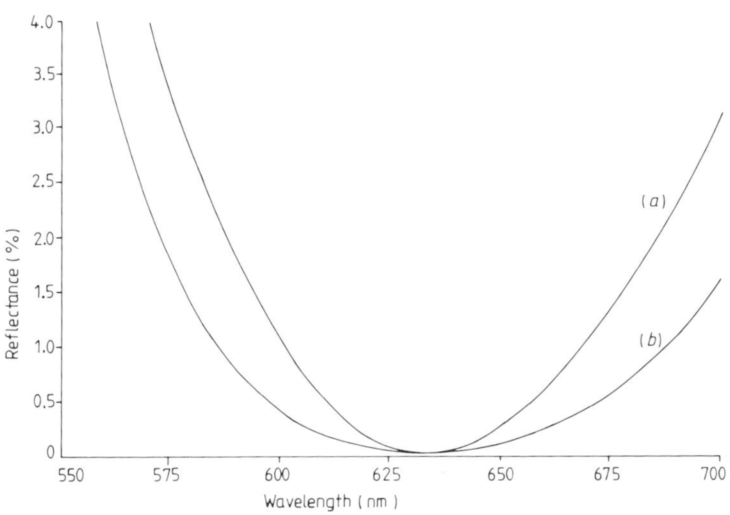

Figure 8.15 Antireflection coatings for s-polarised light at an angle of incidence of 60° in air. (a) Air | $1.273L$ $1.123H$ | Glass, (b) Air | $1.273L$ $1.123H$ $2.682A$ | Glass with $n_L = 1.4$, $n_H = 1.9$, $n_A = 1.3$, $n_{Glass} = 1.5$ and $\lambda_0 = 632.8$ nm.

made by adding a high-admittance half-wave layer between the two quarter-waves or a low-admittance half-wave next to the substrate. The latter is also shown in the figure. In terms of normal incidence thicknesses the two designs are:

$$\text{Air}\,|\,1.273L\ 1.123H\,|\,\text{Glass}$$

and

$$\text{Air}\,|\,1.273L\ 1.123H\ 2.682A\,|\,\text{Glass}$$

where L, H and A indicate quarter-waves at normal incidence of films of index 1.4, 1.9 and 1.3 respectively. The p-reflectance of these designs is very high and they are definitely suitable for s-polarisation only.

Again it is better wherever possible to avoid the necessity for such antireflection coatings by rearranging the optical design of the instrument so that s-polarised light is reflected and p-polarised light is transmitted.

s- and p-polarisation together

The task of assuring low reflectance for both s- and p-polarised light is almost impossible and should only be attempted as a last and very expensive resort. It is possible to arrive at designs that are effective over a narrow wavelength region and one such technique is included here. Again we use the range of indices given in table 8.4 and design a coating to give low s- and p-reflectance on a substrate of index 1.5 in air.

We use quarter-wave layer thicknesses only and a design technique similar to the procedure we have already used for high-reflectance coatings but with an additional condition that the admittance of both substrate and coating for both p- and s-polarisations should be unity to match the incident medium. This implies

$$\frac{\Delta_1^2\Delta_3^2\Delta_5^2\ldots\Delta_{\text{sub}}}{\Delta_2\Delta_4^2\Delta_6^2\ldots}=1 \tag{8.38}$$

and

$$Y=\frac{\eta_{1\text{s}}^2\eta_{3\text{s}}^2\eta_{5\text{s}}^2\ldots\eta_{\text{sub,s}}}{\eta_{2\text{s}}^2\eta_{4\text{s}}^2\eta_{6\text{s}}^2\ldots}=1. \tag{8.39}$$

Equation (8.39) ensures that the reflectance for s-polarised light is zero and equation (8.38) that the p-reflectance equals the s-reflectance. From table 8.4, the starting values are $\Delta_{\text{sub}}=0.3750$ and $\eta_{\text{sub}}=2.4495$. Trial and error shows that with the addition of one single quarter-wave layer, the best result corresponds to an index of 1.3 for which $\Delta_1^2/\Delta_{\text{sub}}=0.4495^2/0.3750=0.5387$ and $\eta_{1\text{s}}^2/\eta_{\text{sub}}=1.9391^2/2.4495=1.5350$. Other combinations give values that are further from unity in each case. Adopting a quarter-wave of index 1.3 as the first layer of the coating we need a further combination of layers that will provide a correction factor of 1.3624 in Δ and of 0.8071 in η_s. An additional single layer will not do, but two-layer combinations of a high- followed by a low-index layer can be found that will correct Δ but which are inadequate in terms of η_s. The two-layer combination that comes nearest to satisfying the requirements is a layer of index 1.8 followed by one of index 1.3 making the design so far:

$$|n = 1.3|n = 1.8|n = 1.3|\text{Glass}.$$

This has an overall Δ of $(0.4495^2 \times 0.4495^2)/(0.3253^2 \times 0.375) = 1.0288$ and a η_s of $(1.9391^2 \times 0.9391^2)/(3.1560^2 \times 2.4495) = 0.5795$. But the combination of index 2.5 followed by 1.4 gives approximately the same correction for Δ but a different correction for η_s. This gives the opportunity of using both combinations in a four-layer arrangement to adjust the value of η_s without altering Δ. The correction factor for Δ is given by $(0.4495^2 \times 0.2841^2)/(0.4050^2 \times 0.3253^2) = 0.9396$ and for η_s by $(1.9391^2 \times 4.6904^2)/(2.2000^2 \times 3.1560^2) = 1.7159$. This then yields an overall value for Δ of $0.9396 \times 1.0288 = 0.9667$ and for η_s of $1.7159 \times 0.5795 = 0.9944$. The seven layers can be put in various orders without altering the reflectance at the reference wavelength. All that is required is that the 1.3 and 2.5 indices should be odd and the 1.4 and 1.8 indices even. Here we put them in descending value of index from the substrate so that the final design is:

$$\text{Air}|1.3409L\ 1.2727A\ 1.3409L\ 1.1407B\ 1.3409L\ 1.1407B\ 1.066H|\text{Glass}$$

with $n_L = 1.30$, $n_A = 1.40$, $n_B = 1.80$ and $n_H = 2.50$.

The calculated performance of this coating for a reference wavelength of 632.8 nm is shown in figure 8.16. As we might have suspected, the width of the

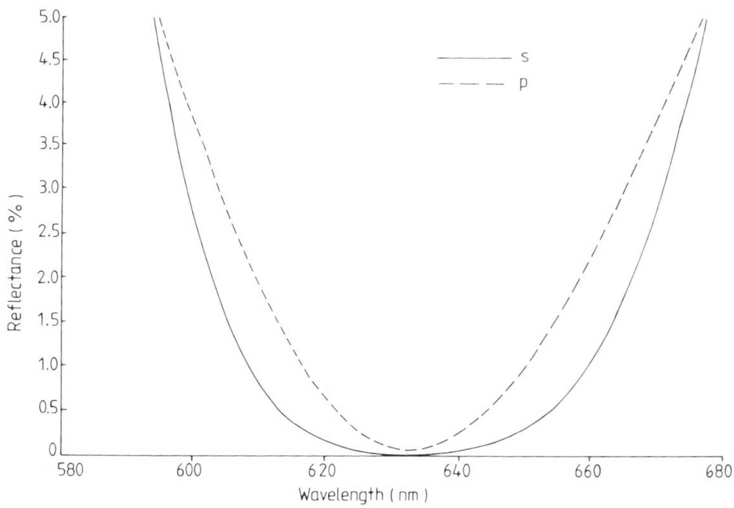

Figure 8.16 Calculated performance of an antireflection coating for glass to have low reflectance for both p- and s-polarisation at an angle of incidence of 60° in air. The solid line indicates s-polarisation and the broken line p-polarisation. $\lambda_0 = 632.8$ nm and the design is given in the text.

zone of low reflectance is narrow. An alternative design arrived at in the same way but for a substrate of index 1.52 and a range of film indices from 1.35 to 2.40 uses ten layers:

$$\text{Air}\,|\,1.3036L\ 1.1748A\ 1.3036L\ 1.1748A\ 1.3036L\ 1.1407B$$
$$1.0722H\ 1.1235C\ 1.0722H\ 1.1235C\,|\,\text{Glass}$$

with $n_L = 1.35$, $n_A = 1.65$, $n_B = 1.80$, $n_C = 1.90$, $n_H = 2.40$, $n_{\text{Glass}} = 1.52$ and $n_{\text{air}} = 1.00$. The performance is similar to that of figure 8.16.

RETARDERS

Achromatic quarter- and half-wave plates

As well as being used in the construction of polarisers, optical thin films can find application in the production of achromatic quarter- and half-wave plates. A quarter-wave plate by definition produces between the two principal planes of polarisation a phase shift of 90°, which corresponds to an optical path difference of a quarter of a wavelength, while a half-wave plate produces a phase shift of 180° corresponding to a half wavelength. These components are generally made from mica, or some other similar birefringent material, cut to such a thickness that the difference in optical pathlength for each plane of polarisation is either a quarter or a half wavelength. A considerable disadvantage of such retardation plates is the rapid variation of the performance of the device with wavelength.

The case of the half-wave plate has been considered by Lostis,[7] who has used a thin film to alter the phase shift on total internal reflection to make it exactly 180°. The arrangement is shown in figure 8.17. The notation for the various refractive indices and thicknesses is shown also in the figure. Let Y indicate the optical admittance with regard to the s-plane of polarisation and Z with respect to the p-plane. Then $Y_r = n_r \cos\phi_r$, $Z_r = n_r/\cos\phi_r$. Once the notation is established the calculation of the reflectances for the two planes of polarisation is an easy matter. The reflectance will be total for both but their phase shifts will depend on the parameters of the thin film. The condition that the relative phase difference between the two planes of polarisation should be 180° can then be asserted and the necessary condition derived for this to be so. Lostis found this condition to be

$$A \tan\beta + B \tan\beta + C = 0 \tag{8.40}$$

where

$$\beta = \frac{2\pi}{\lambda} n_1 d \cos\phi_1$$

$$A = n_1^2 - \left(\frac{n_0 n_2}{n_1}\right)^2$$

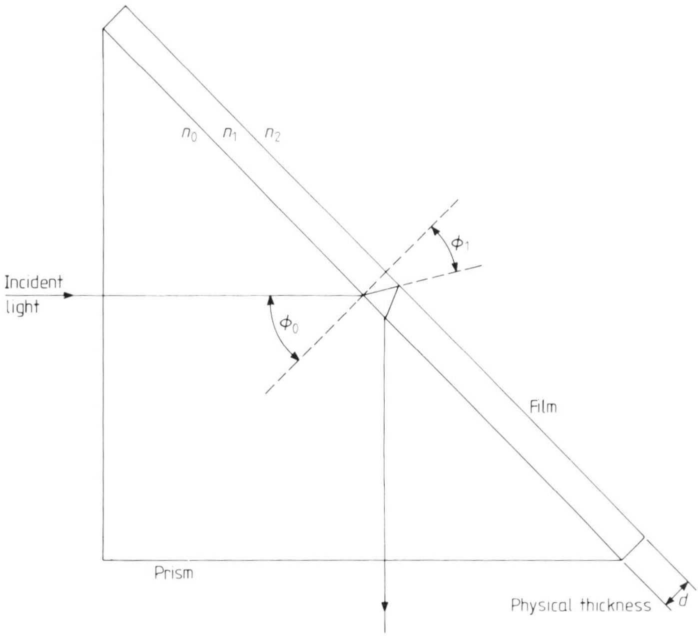

Figure 8.17 A half-wave retardation prism. (After Lostis[7].)

$$B = \frac{\gamma}{n_1 \cos\phi_1}(n_1^2 - n_0^2) + \frac{n_1 \cos\phi_1}{\gamma}\left[\left(\frac{n_0 n_2}{n_1}\right)^2 - n_2^2\right]$$

$$C = n_0 - n_2^2$$

and

$$Y_2 = n_2 \cos\phi_2 = \mathrm{i}\,(n_0^2 \sin^2\phi_0 - n_2^2)^{1/2} = \mathrm{i}\gamma.$$

In the case where the surrounding medium is air, of index 1.0, the necessary condition for the above equation to have a real root is

$$n_0 \leqslant 1.46 \qquad \text{and} \qquad n_1 \geqslant 2.6.$$

When the limiting values are inserted in equation (8.40), the optical thickness of the film is found to be $\lambda/11$. Having arrived at this value the retardation can be calculated for the rest of the visible spectrum and it is found that the retardation does not vary by more than $\pm\lambda/50$ from 400–700 nm. Lostis constructed such a system using a prism of fused silica and a layer of titanium dioxide as the thin film.

The quarter-wave plate made from mica suffers from the same disability as the half-wave plate. It is correct for only one wavelength. Results derived in chapter 2 show that the phase change on total internal reflection varies with the angle of incidence and the plane of polarisation, and the difference in phase between the two principal planes also varies as the angle of incidence varies. With the materials available in the visible region it is not possible with a single reflection to obtain a retardation of 90°, but, with glass of refractive index 1.51 a retardation of 45° is obtained with an angle of incidence of either 48° 37′ or 54° 37′, and with two successive internal reflections the value of 90° can be obtained[23]. This is achieved in a device known as a Fresnel rhomb, shown in figure 8.18. The Fresnel rhomb is almost achromatic in performance, but the dispersion of the glass causes the retardation to increase gradually with decrease in wavelength. A further disadvantage of the Fresnel rhomb is its sensitivity to angle of incidence changes. The performance of the Fresnel rhomb can be considerably improved in both these directions by the addition of a thin-film coating to both surfaces of the rhomb. King[24] has manufactured Fresnel rhombs which show a phase retardation which varies by less than 0.4° over the wavelength range 330–600 nm. These were made from hard crown glass with one surface coated with magnesium fluoride 20 nm thick.

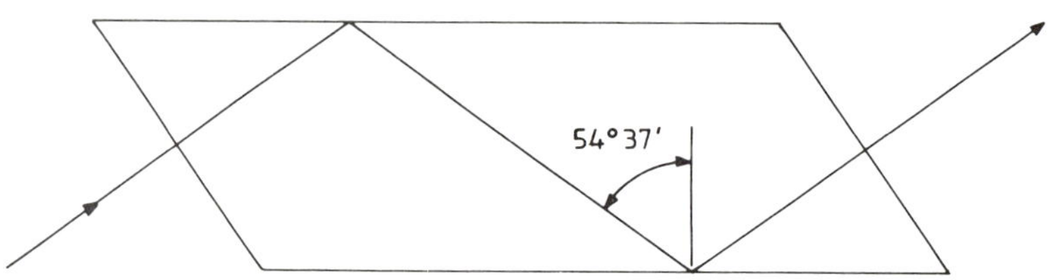

Figure 8.18 A Fresnel rhomb.

Multilayer phase retarders

In recent years there has been a number of applications where reflecting coatings have been required which introduced specified phase retardances between s- and p-polarisation. In particular there is a need in certain types of high-power laser resonators for coatings that introduce a 90° phase shift between s- and p-polarisation at an angle of incidence of 45°. The coatings that have been designed and manufactured for this purpose have been tuned for wavelengths in the infrared and have taken the form of silver films with a multilayer dielectric overcoat. The first published designs were due to Southwell[25,26] who used a computer synthesis technique. Then Apfel[27] devised an analytical approach that we follow here. The principle of operation of the coatings is that an added dielectric layer will not affect the reflectance of a system that has already a reflectance of unity. It will simply alter the phase change on reflection. When the component is used at oblique incidence, the alteration in phase will be different for each plane of polarisation. By adding layers in the correct sequence, eventually any desired phase difference between p- and s-polarisation for a single specified angle of incidence and wavelength can be achieved. In practice a silver layer is used as the basic reflecting coating and, although this has reflectance slightly less than unity, in the infrared it is high enough for it to be possible to neglect any error that might otherwise be introduced. It is of course not necessary to use a metal layer as starting reflector. A dielectric stack would be equally effective but would simply have more layers.

The basis of Apfel's method is a plot of phase retardance, denoted by Apfel as D, against the average phase shift A as a function of thickness of added layer of a given index. For simplicity, we retain this notation but in the rest of what follows we alter both notation and derivation to agree with the remainder of the book.

The starting point of the treatment is a reflector with a reflectance of unity, that is, a surface with imaginary admittance. Let this imaginary admittance be $\mathrm{i}\beta$. Then

$$\rho e^{\mathrm{i}\phi} = e^{\mathrm{i}\phi} = (\eta_0 - \mathrm{i}\beta)/(\eta_0 + \mathrm{i}\beta) \tag{8.41}$$

i.e.

$$\tan(\phi_{\mathrm{sub}}/2) = -\beta/\eta_0. \tag{8.42}$$

Should the incident medium be changed to η_1 then the phase shift becomes

$$\tan(\phi_1/2) = (-\beta/\eta_1) = (\eta_0/\eta_1)\tan(\phi_{\mathrm{sub}}/2). \tag{8.43}$$

Now we add a film of admittance η_1 and phase thickness $\delta_1 = (2\pi/\lambda)n_1 d_1$ to the substrate.

$$\begin{bmatrix} B \\ C \end{bmatrix} = \begin{bmatrix} \cos\delta_1 & \mathrm{i}(\sin\delta_1)/\eta_1 \\ \mathrm{i}\eta_1\sin\delta_1 & \cos\delta_1 \end{bmatrix} \begin{bmatrix} 1 \\ \mathrm{i}\beta \end{bmatrix}$$

$$= \begin{bmatrix} \cos\delta_1 - (\beta/\eta_1)\sin\delta_1 \\ \mathrm{i}(\eta_1\sin\delta_1 + \beta\cos\delta_1) \end{bmatrix}. \tag{8.44}$$

The phase shift is now given, from equation (8.44), as

$$\tan(\phi_0/2) = \frac{-(\eta_1\sin\delta_1 + \beta\cos\delta_1)}{\eta_0[\cos\delta_1 - (\beta/\eta_1)\sin\delta_1]} = (\eta_1/\eta_0)\frac{[(-\beta/\eta_1) - \tan\delta_1]}{[1 + (-\beta/\eta_1)\tan\delta_1]}.$$

The second factor has the form of the tangent of the difference of two angles. Using this and expression (8.43) we have

$$\tan(\phi_0/2) = (\eta_1/\eta_0)\tan(\phi_1/2 - \delta_1). \tag{8.45}$$

This expression is valid for either plane of polarisation simply by inserting the appropriate values of η and δ.

To draw a D–A curve, we choose a starting point given by $D = 2\psi$ and $A = 0$, equivalent to $\phi_{\mathrm{sub,p}} = \psi$ and $\phi_{\mathrm{sub,s}} = -\psi$, and plot the difference in phase against the average phase, all calculated from (8.45). Different values of ψ yield a family of curves. This family of curves can have a scale of thickness marked along them, in the manner of figure 8.19. Note that as curves disappear off the

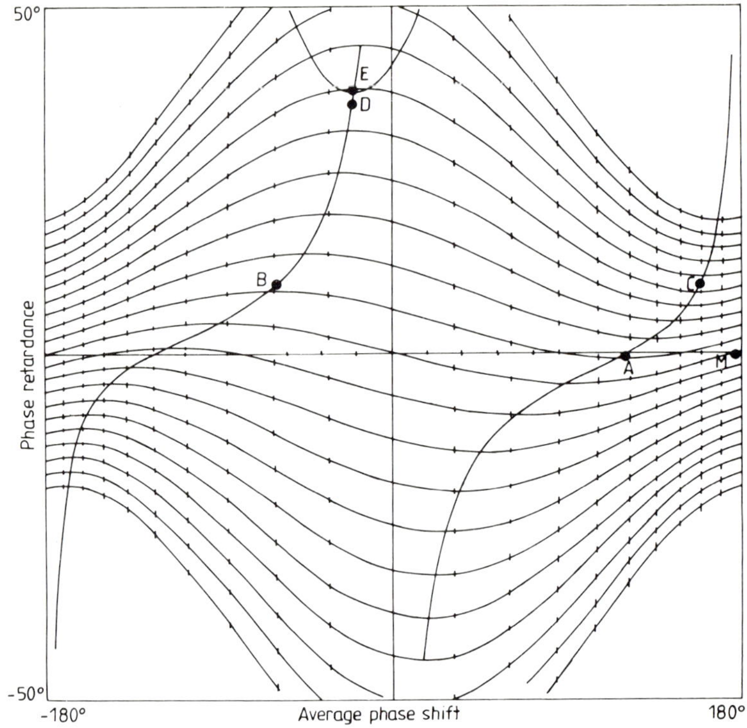

Figure 8.19 Immersed *D–A* plot for a film of index 4.0 in an incident medium of index 2.2 at an angle of incidence of 45° in air. The two *S*-shaped vertical curves mark the extrema of the *D–A* curves. The target retardation of 90° in air is denoted by the *U*-shaped curve at the top of the figure. The letters M, A, B, C, D and E are explained in the text. (After Apfel[27].)

left-hand side of the diagram they reappear at the right-hand side. The relationships for the various quantities may be written

p-polarisation:

$$\begin{aligned} \tan(\phi_{0,p}/2) &= [(y_1\cos\theta_0)/(y_0\cos\theta_1)]\tan[(\phi_{1,p}/2)-\delta_1] \\ \tan(\phi_{1,p}/2) &= [(y_0\cos\theta_1)/(y_1\cos\theta_0)]\tan(\psi/2) \end{aligned} \tag{8.46}$$

s-polarisation:

$$\begin{aligned} \tan(\phi_{0,s}/2) &= [(y_1\cos\theta_1)/(y_0\cos\theta_0)]\tan[(\phi_{1,s}/2)-\delta_1] \\ \tan(\phi_{1,s}/2) &= [(y_0\cos\theta_0)/(y_1\cos\theta_1)]\tan(-\psi/2) \end{aligned} \tag{8.47}$$

where δ_1 is calculated for the appropriate angle of incidence. Then

$$D = \phi_{0,p} - \phi_{0,s} \qquad A = (\phi_{0,p} + \phi_{0,s})/2.$$

The curves now make it possible to determine the phase retardation produced by any thickness of the dielectric material added to any substrate of unity reflectance. To complete the design we need to construct similar diagrams for each dielectric material that is to be used. Since these sets of curves will not coincide, it is possible to reach any point of the diagram simply by moving from one set of curves to the other in succession. Only two dielectric materials are necessary and in that case Apfel shows that a technique of immersion simplifies the diagram. If we imagine that the structure is immersed in a medium of admittance equal to y_1 then

$$n_0 = n_1 \qquad y_0 = y_1$$

and

$$\tan(\phi_0/2) = \tan[(\phi_1/2) - \delta_1]$$

for both planes of polarisation. Then D is a constant and $A = -2\delta_1$, since $\phi_{1,s} = -\phi_{1,p}$.

This result implies that the curves corresponding to the addition of material of index equal to that of the incident medium are horizontal lines on the diagram and can easily be visualised. The only problem we have now is that the target retardation is specified in a medium that will, in general, be different from that of the layer material. We therefore must add to the diagram the specification for retardation in the dummy immersion medium that will give the correct retardation when the dummy medium is removed and replaced by the correct medium. Let the phase retardation required in the correct incident medium be D_f. Then we can write

$$D_f = \phi_{fp} - \phi_{fs} \qquad 2A_f = \phi_{fp} + \phi_{fs}$$

i.e.

$$\phi_{fp} = [(D_f/2) + A_f] \qquad \phi_{fs} = [-(D_f/2) + A_f].$$

Converting ϕ_{fp} and ϕ_{fs} to ϕ_{0p} and ϕ_{0s} the immersed values are

$$\tan(\phi_{0p}/2) = \frac{n_0 \cos\theta_1}{n_1 \cos\theta_0} \tan(\phi_{fp}/2) = \frac{n_0 \cos\theta_1}{n_1 \cos\theta_0} \tan[(D_f/4) + (A_f/2)]$$

$$\tan(\phi_{0s}/2) = \frac{n_0 \cos\theta_1}{n_1 \cos\theta_0} \tan(\phi_{fs}/2) = \frac{n_0 \cos\theta_1}{n_1 \cos\theta_0} \tan[(D_f/4) + (A_f/2)]. \qquad (8.48)$$

Then varying A_f gives the curves. Note that equations (8.48) are similar to (8.46) and (8.47) but with n_0 and n_1 interchanged.

The method is illustrated by figure 8.19, taken from Apfel[27] and showing the design curves for a retarder constructed of films of germanium, index 4.0, and zinc sulphide, index 2.2, to have a retardance of 90° in air at an angle of incidence of 45°. The curves of figure 8.19 are *D–A* curves for germanium immersed in a medium of index 2.2. The *U*-shaped curve in the upper region is the retardation target of 90° in air referred to the dummy medium of 2.2. The *S*-shaped curves running top to bottom mark the maxima of the *D–A* curves while the tick marks are made at intervals of one tenth of a quarter-wave optical thickness. The four layer design: $0.864H$ $0.778L$ $0.674H$ $0.319L$ Ag gives a retardance of 86.8° at the design wavelength and is represented by the trajectory MABCD. Two extra layers would be required to reach exactly 90°. The diagram could be made into a design aid for any desired retardance by adding a family of target curves.

OPTICAL TUNNEL FILTERS

At an earlier stage in the development of narrowband filters a main barrier to their construction was the fabrication of reflecting stacks of sufficiently low loss, and it appeared that the phenomenon of frustrated total internal reflection might offer some hope as a possible solution. This phenomenon has been known for some time. If light is incident on a boundary beyond the critical angle, it will normally be completely reflected. However, the incident light does in fact penetrate a short distance into the second medium, where it decays exponentially. Provided the second medium is somewhat thicker than a wavelength or so, the decay will be more or less complete and the reflectance unity. If, on the other hand, the second medium is made extremely thin, then the decay may not be complete when the wave meets the boundary with the third medium and, if the angle of propagation is then no longer greater than critical, a proportion of the incident light will appear in the third medium and the reflectance at the first boundary will be something short of total. This, as Baumeister[28] has pointed out, is very similar to the behaviour of fundamental particles in tunnelling through a potential barrier, and he has used the term 'optical tunnelling' to describe the phenomenon. The most important feature of the effect, as far as the thin-film filter is concerned, is that the frustrated total reflection can be adjusted to any desired value, simply by varying the thickness of the frustrating layer between the first and third media.

The method of constructing a filter using this effect is very similar to the polarising beam splitter (p 328). The hypotenuse of a prism is first coated with a frustrating layer of lower index so that the light will be incident at an angle greater than critical. This is a function of the prism angle, refractive index, and the refractive index of the frustrating layer. Next follows the spacer layer which must necessarily be of higher index so that a real angle of propagation will exist. This in turn is followed by yet another frustrating layer. The whole is then cemented into a prism block by adding a second prism. The angle at which light is incident on the diagonal face must be greater than the angle ψ given by

$$\sin\psi = n_F/n_G$$

where n_F is the index of the frustrating layer and n_G is the index of the glass of the prism. For $n_F = 1.35$ and $n_G = 1.52$, we find $\psi = 63°$, which is quite an appreciable angle. Usually glass of rather higher index, nearer 1.7, is used to reduce the angle as far as possible.

Although at first sight the optical tunnel or frustrated total reflectance (FTR) filter appears most attractive and simple, there are some tremendous theoretical disadvantages. First there is an enormous shift in peak wavelength between the two planes of polarisation. Typical figures quoted are of the order of 100 nm in the visible region, the peak corresponding to the p-plane of polarisation being at a shorter wavelength. This large polarisation splitting is due to the large angle of incidence at which the device must be used. Another effect of this large angle is that the angle sensitivity of the filter is extremely large. Shifts of 5 nm/degree of arc have been calculated[28].

Added to these disadvantages is the fact that the attempts which have been made to produce FTR filters have been very disappointing in their results, the performance appearing to fall far short of what was expected theoretically. It seems that the difficulties inherent in the construction of the FTR filter are at least as great as those involved in the conventional Fabry–Perot filter. Because of this, interest in the FTR filter has been mainly theoretical and the filter does not appear to be in commercial production.

The theory of the FTR filter has been written up in great detail by Baumeister[28]. Not only has he covered the FTR filter but he has also pointed out that, as far as the theory is concerned, the frustrating layer or, as he has renamed it, the tunnel layer, behaves exactly as a loss-free metal layer. This means that all sorts of filters including induced-transmission filters are possible using tunnel layers. Designs for a number of these are included in the paper. One conclusion which Baumeister reaches is that there appears to be no practical application for the tunnel-layer filter of the induced-transmission and FTR Fabry–Perot types. However he does mention the possibility of a longwave-pass filter constructed from an assembly of many tunnel layers separated by spacer layers and which has the advantage of a limitless rejection zone on the shortwave side of the edge. Even with this type of filter there are some disadvantages which could be serious. The characteristics of the filter near the edge suffer from strong polarisation splitting. This could be overcome by adding a conventional edge filter to the assembly at the front face of the prism. However, the second disadvantage is rather more serious: the appearance of pass bands in the stop region when the filter is tilted in the direction so as to make the angle of incidence more nearly normal. Curves given by Baumeister show a small transmission spike appearing even with a tilt of only 1° internal or 2.7° external with respect to the design value.

REFERENCES

1 Thelen A 1966 Equivalent layers in multilayer filters *J. Opt. Soc. Am.* **50** 1533–8

2 Berning P H and Turner A F 1957 Induced transmission in absorbing films applied to band pass filter design *J. Opt. Soc. Am.* **47** 230–9

3 Nevière M and Vincent P 1980 Brewster phenomena in a lossy waveguide used just under the cut-off thickness *J. Optics* (Paris) **11** 153–9

4 Ruiz-Urbieta M, Sparrow E M and Parikh P D 1975 Two-film reflection polarizers: theory and application *Appl. Opt.* **14** 486–92
5 Cox J T, Hass G and Hunter W R 1975 Infrared reflectance of silicon oxide and magnesium fluoride protected aluminium mirrors at various angles of incidence from 8 μm to 12 μm *Appl. Opt.* **14** 1247–50
6 Clapham P B, Downs M J and King R J 1969 Some applications of thin films to polarization devices *Appl. Opt.* **8** 1965–74
7 Lostis M P 1957 Etude et réalisation d'une lame demi-onde en utilisant les propriétés des couches minces *J. Phys. Rad.* **18** 518–28
8 Kretschmann E and Raether H 1968 Radiative decay of non-radiative surface plasmons excited by light *Z. Naturf.* a **23** 2135–6
9 Otto A 1968 Excitation of non-radiative surface plasma waves in silver by the method of frustrated total reflection *Z. Phys.* **216** 398–410
10 Raether H 1977 Surface plasma oscillations and their applications in *Physics of Thin Films* **9** 145–261 (New York: Academic)
11 Abelès F 1976 Optical properties of very thin films *Thin Solid Films* **34** 291–302
12 Greenland K M and Billington C 1950 The construction of interference filters for the transmission of specified wavelengths *J. Phys. et le Radium* **11** 418–21
13 Harrick N J and Turner A F 1970 A thin film optical cavity to induce absorption of thermal emission *Appl. Opt.* **9** 2111–14
14 Banning M 1947 Practical methods of making and using multilayer filters *J. Opt. Soc. Am.* **37** 792–7
15 MacNeille S M 1946 Beam splitter *US Patent Specification* 2 403 731
16 Clapham P B 1969 The preparation of thin film polarizers *Rep. OP. MET.* 7 National Physical Laboratory, Teddington
17 Songer L 1978 The design and fabrication of a thin film polarizer *Opt. Spectra* **12**(10) 45–50
18 Blanc D, Lissberger P H and Roy A 1979 The design, preparation and optical measurement of thin film polarizers *Thin Solid Films* **57** 191–8
19 Netterfield R P 1977 Practical thin-film polarizing beam splitters *Optica Acta* **24** 69–79
20 Thelen A 1981 Nonpolarizing edge filters *J. Opt. Soc. Am.* **71** 309–14
21 Thelen A 1976 Nonpolarizing interference films inside a glass cube *Appl. Opt.* **15** 2983–5
22 Knittl Z and Houserkova H 1982 Equivalent layers in oblique incidence: the problem of unsplit admittances and depolarization of partial reflectors *Appl. Opt.* **11** 2055–68
23 Born M and Wolf E 1975 *Principles of Optics* 5th edn (Oxford: Pergamon)
24 King R J 1966 Quarter wave retardation systems based on the Fresnel rhomb *J. Sci. Instrum.* **43** 617–22
25 Southwell W H 1979 Multilayer coatings producing 90° phase change *Appl. Opt.* **18** 1875
26 Southwell W H 1980 Multilayer coating design achieving a broadband 90° phase shift *Appl. Opt.* **19** 2688–92
27 Apfel J H 1981 Graphical method to design multilayer phase retarders *Appl. Opt.* **20** 1024–29
28 Baumeister P W 1967 Optical tunnelling and its applications to optical filters *Appl. Opt.* **6** 897–905 (This paper lists 49 references).

9 Production methods and thin-film materials

In this chapter, we shall begin to look at methods employed in the production of thin-film filters. In particular we shall consider the basic process and then go on to examine some of the properties of materials important in thin-film work. First we shall deal briefly with the fundamental process and then look more closely at the plant which is used. Chapter 10 will include a more detailed examination of some of the problems met in production.

Much of this chapter is concerned with the properties of materials, ways of measuring them, and some examples of the results of the measurements of the important parameters. Probably the most important properties from the thin-film point of view are given in the following list, although the order is not that of relative importance, which will vary from one application to another.

1. Optical properties such as refractive index and region of transparency.
2. The method which must be used for the production of the material in thin-film form.
3. Mechanical properties of thin films such as hardness or resistance to abrasion, and the magnitude of any built-in stresses.
4. Chemical properties such as solubility and resistance to attack by the atmosphere, and compatability with other materials.
5. Toxicity.
6. Price and availability.
7. Other properties which may be important in particular applications, for example, electrical conductivity or dielectric constant.

Item 7 is not one on which we can comment further here. On the question of price and availability, item 6, there is also little that can be said. The situation is changing all the time. Many companies are able to offer a wide range of materials completely ready for thin-film production, together with all the necessary information on the techniques which should be used.

THE PRODUCTION OF THIN FILMS

In chapter 1 we saw how the subject could be said to begin with Fraunhofer's preparing thin films by the chemical etching of glass and also by deposition from solution. These and similar methods have been used to some extent in optical thin-film work. Other techniques that, at different stages in the development of the subject, have been, and are still sometimes, employed, include anodic oxidation of aluminium to form a protective coating, and the spraying of material on to a surface either in solution or in the form of a substance that can be chemically converted into the desired material later. Even the substance itself is sometimes sprayed on, possibly after vaporisation in a hot flame. Polymerisation of monomers deposited on surfaces by condensation or from solution is also used occasionally. There are many important processes, used principally in other areas of thin-film technology, which can be classified as chemical vapour deposition, in which materials in vapour form react at the hot substrate surface to form thin layers of the desired material. The chemical reaction can be assisted by energy from an electric discharge, the term *plasma process* being applied. Modern thin-film optical filters are, however, almost entirely manufactured by vacuum deposition processes which can be classified as physical vapour deposition. In these processes the thin films condense from the vapour phase on to the surfaces to be coated which are held at temperatures somewhat lower than the solidification temperature of the films. In order to prevent undesirable reactions taking place in the vapour phase, the process is carried out in an evacuated chamber. The various techniques differ principally in the way in which the material is vaporised. Sputtering consists of bombarding the desired material with ions in a vacuum chamber so that molecules are ejected to collide with and stick to the substrate. It was discovered in the 19th century and is particularly useful for refractory materials. There are also some other ion-assisted processes that are currently under development and may emerge as useful optical thin-film deposition processes. A wide range of such thin-film deposition techniques is fully covered in a useful book by Vossen and Kern[1]. As far as optical coatings are concerned, they tend to be used only in special cases, and the most common and most flexible process for optical coating is that known as vacuum evaporation or thermal evaporation in vacuum.

In thermal evaporation the vapour is produced simply by heating the material that is known as the evaporant. Because of the reduced pressure in the chamber the vapour is given off in an even stream, the molecules appearing to travel in straight lines so that any variation in the thickness of the film that is formed is smooth, and depends principally on the position and orientation of the substrate with respect to the vapour source. The properties of the film are broadly similar to those of the bulk material, although, as we shall see, there are important differences in the detailed microstructure. Precautions that have to be taken to ensure good film quality include scrupulous cleanliness of the substrate surface, near normal incidence of the vapour stream and, sometimes, heating the substrate to temperatures of 200–300°C (or even higher, depending on the material) before commencing deposition. The evaporation is carried out in a sealed chamber that is evacuated to a pressure usually of the order of 10^{-5} mb. The materials to be deposited are melted within the chamber, using one of a number of possible techniques that will be described. The complete plant consists of the chamber together with the necessary pumps, pressure gauges, power supplies for supplying the energy necessary to melt the evaporants, monitoring equipment for the measurement of the thin-film thickness during the process, substrate holding jigs, substrate heaters and the controls. Modern thin-film coating plants are shown in figures 9.1 and 9.2.

In order to evaporate the material, it must be contained in some kind of crucible and it must be heated until molten, unless it sublimes. There is a number of ways of achieving this. The simplest method is to make use of a crucible of refractory metal that acts also as a heater when an electric current is passed through it. The crucibles are elongated in shape with flat contact areas at either end and are commonly referred to as boats. Electrodes within the plant, which are insulated from the structure, act both as terminals and supports. The resistance of the boats is low and high currents, several hundred amps at low voltages, are required to heat them. Because of the high currents and especially to protect the sealing rings, the electrodes are normally water-cooled Figure 9.3 shows a baseplate complete with a set of electrodes and figure 9.4 a

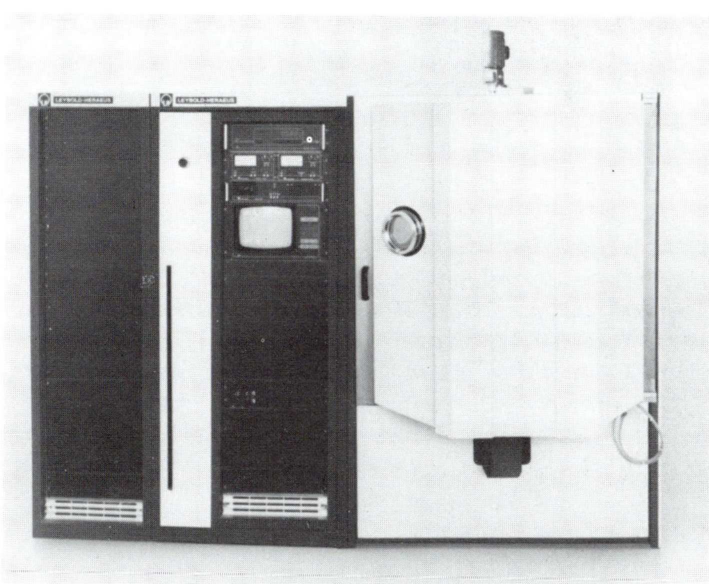

Figure 9.1(*a*) For description, see caption overleaf.

(*b*)

Figure 9.1 Thin-film coating plants. These plants are known as box coaters because the chamber is fabricated in the form of a box with a front door rather than as a bell jar on a baseplate. (*a*) Model A 1100 High Vacuum Deposition System. The latest version of the LEYCOM process control computer is shown. This version displays on the screen the entire vacuum status of the system, the status of the evaporation process and the status of all pre- and post-deposition steps, all of these functions being computer controlled. Part of the OMS 2000 photometer, which is used for real-time *in situ* optical thickness control, can be seen at the lower part of the front door. (Courtesy of Leybold-Heraeus GmbH, Hanau, Germany.) (*b*) BAK 760 High Vacuum Coating System. The microprocessor unit that controls the entire process including pumping system and film thickness is at the top left-hand side of the racks. In this version, film thickness is derived from quartz crystal monitors. Two electron-beam sources are fitted with controllers and power supply in the right-hand rack. (Courtesy of Balzers AG, Balzers, Liechtenstein.) (*c*) Internal view of the chamber of a BAK 760 High Vacuum Coating

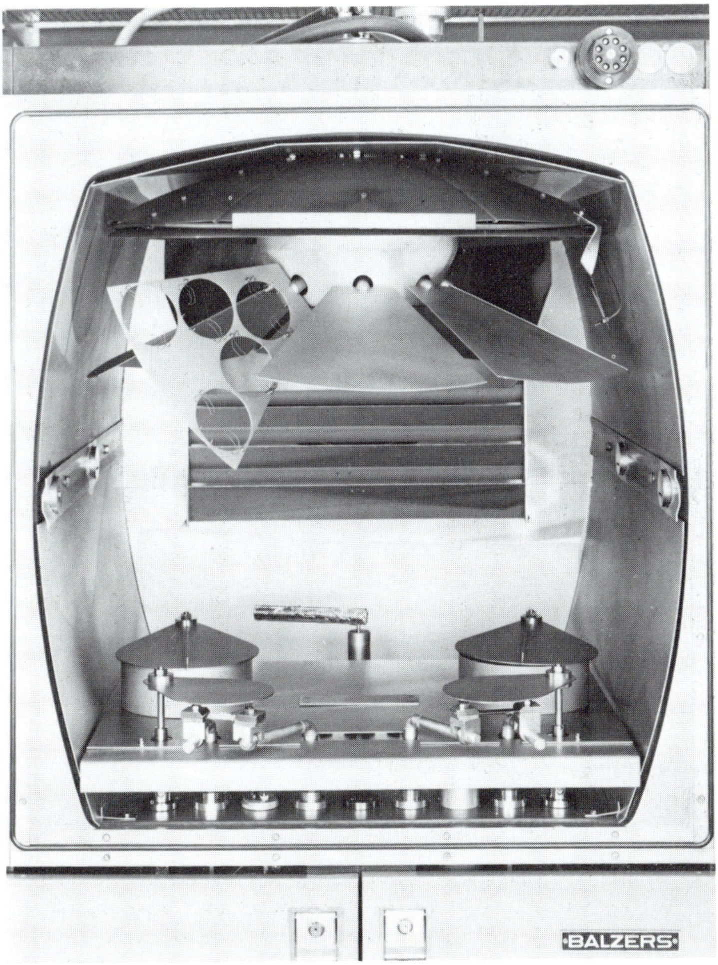

(*c*)

System. The upper part of the chamber is occupied by a reversible calotte so that substrates may be coated on both sides without breaking vacuum. The domed shape at the very top of the chamber, above the calotte, is a radiant heater. In the foreground at the base of the chamber, there are two thermal sources, each with a shutter and one charged with material. Towards the rear of the base, two electron beam sources are surrounded by circular shields and covered with shutters. The glow discharge electrode is a horizontal circular bar at the rear. (Courtesy of Balzers AG, Balzers, Liechtenstein.)

Figure 9.2 Preparing a very large plant for coating a batch of components. The substrate carrier is in the form of a horizontal drum that rotates around the sources and is carried by the chamber door that can be seen on the right. The chamber furniture, as is usual, is covered by aluminium foil for easy subsequent cleaning. (Courtesy of Optical Coating Laboratory Inc, Santa Rosa, California, USA.)

Figure 9.3 The baseplate of a thin-film coating plant showing the electrodes and the shutter used for terminating the layers.

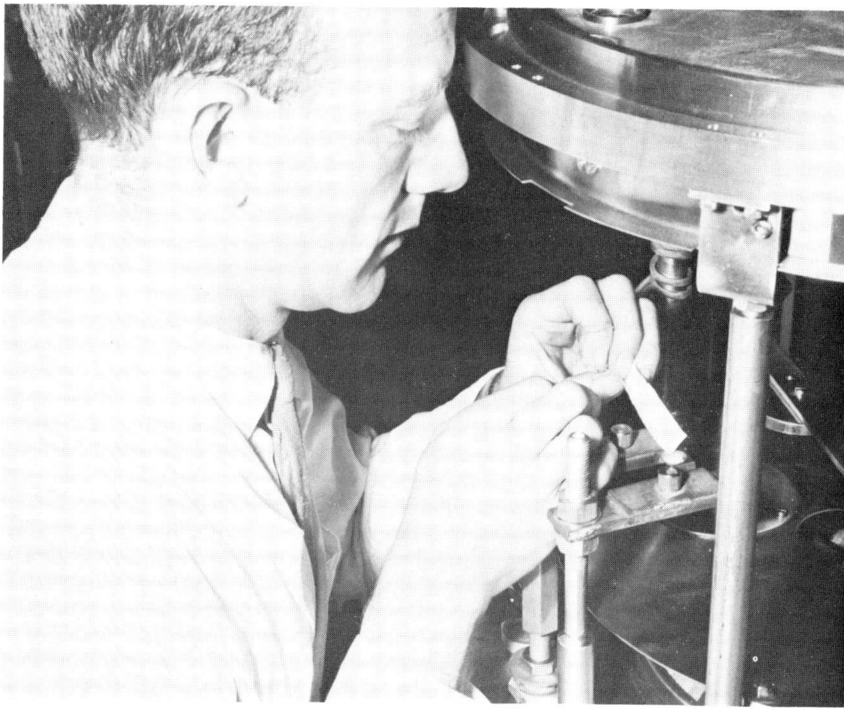

Figure 9.4 A molybdenum boat, mounted between electrodes in an Edwards E19E plant, being charged with material.

molybdenum boat, mounted between electrodes, being charged with material. Tantalum, molybdenum and tungsten are all suitable for the manufacture of boats, tantalum and molybdenum being easily bent and formed, tungsten much less so. A wide range of materials can be evaporated from tantalum, and, of the three, it is the one most frequently used. However, some materials react with it (ceric oxide for example) or with molybdenum, and require the less reactive but rather more difficult tungsten. Considerable skill is required in the manufacture of tungsten boats. To avoid cracking, the tungsten strip should be heated to red heat before bending and only the simplest of shapes can be attempted. Fortunately, a wide range of preformed boats of high quality is available commercially. Certain evaporants react even with tungsten. In some cases a protective liner of alumina can be added, or an alumina crucible surrounded by a tungsten heater can even be used. In other cases, such as aluminium, the reaction is not very fast, and a tungsten wire helix is a satisfactory source. The aluminium, which wets the tungsten, forms droplets along the helix that has its axis horizontal. The area of tungsten in contact with the aluminium for a given evaporation rate is somewhat less, and the thickness of the wire somewhat greater, than for a boat, so that the tungsten is dissolved away more slowly and a greater proportion can be removed before failure. A

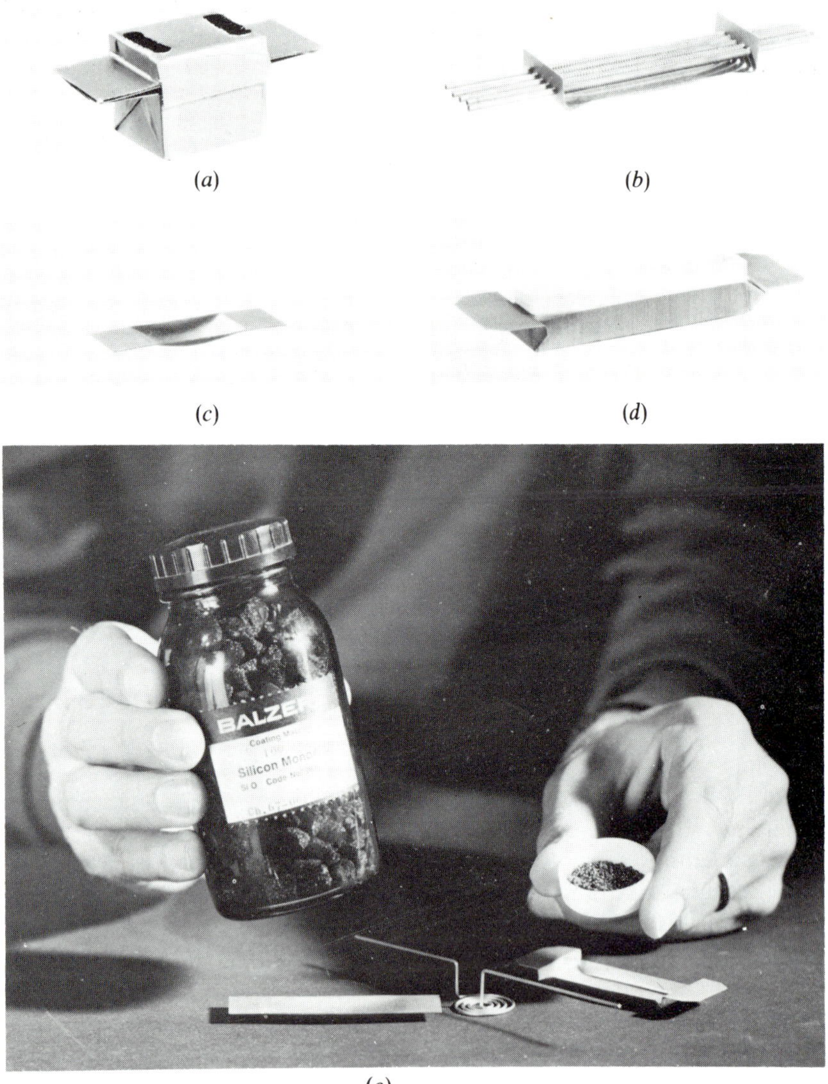

Figure 9.5 Various evaporation sources. (*a*) Tantalum box source (660 A, 1695 W for 1600 °C). (*b*) Tungsten source for large quantities of metals such as aluminium, silver and gold (475 A, 1400 W for 1800 °C). (*c*) Tungsten boat (325 A, 565 W for 1800 °C). (*d*) Aluminium oxide crucible with molybdenum heater. (*e*) Aluminium oxide crucible with tungsten filament. Two tungsten boats can also be seen. (Courtesy of Balzers AG.)

number of different types of boat is shown in figure 9.5.

Materials like zinc sulphide or silicon monoxide, which sublime at not too high a temperature, can be heated in a crucible of alumina, or even fused silica, by radiation from above. A tungsten spiral just above the surface of the material can produce enough heat to vaporise it. This means that the hottest part of the material is the evaporating surface and so the material is much less prone to spitting. One example of such a source is shown in figure 9.5—the crucible is being held in the hand and the spiral is on the table. A development of this type of source is the 'howitzer' source that is shown in figure 9.6, which is particularly useful for zinc sulphide in the infrared as the capacity can be very great[2].

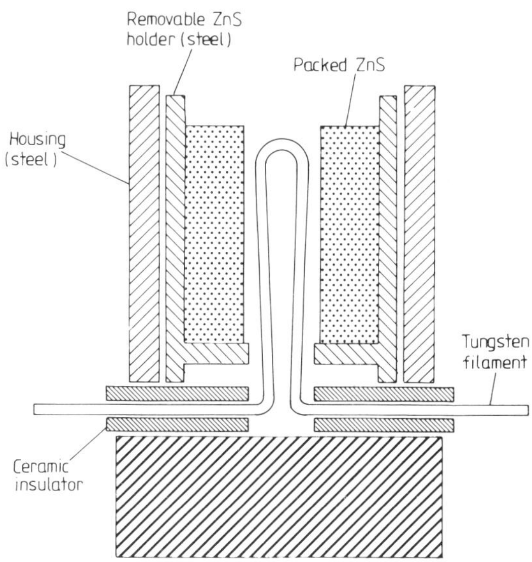

Figure 9.6 The howitzer—a source for evaporating large quantities of ZnS at high deposition rates. The removable ZnS holder shown as steel can also be made of fused silica or alumina and the hairpin filament can be replaced by a tunsten helix. (After Cox and Hass[2].)

Germanium is an example of a material that reacts even with alumina. The reaction is not particularly fast, but the germanium films become contaminated and show higher longwave infrared absorption than is usual. Graphite has been found to be a useful boat material in this case. Supplied in rod form for use as furnace heating elements, it can be easily machined into almost any desired shape or form. Copper, graphite, or one of the refractory metals should be used to make the contacts to the graphite boats. At the high temperatures involved, steel and graphite interact so that the former tends to melt and pit badly and is, therefore, quite unsuitable.

A form of heating which avoids many of the difficulties associated with directly and indirectly heated boats is electron-beam heating, and this is now the preferred technique for most materials, especially the refractory oxides. In this method, the evaporant is contained in a suitable crucible, or hearth, of electrically conducting material, and is bombarded with a beam of electrons to heat and vaporise it. The portion of the evaporant that is heated is in the centre of the exposed surface, and there is a reasonably long thermal conduction path through the material to the hearth that can therefore be held at a rather lower temperature than the melting temperature of the evaporant without prohibitive heat loss. This means that the reaction between the evaporant and the hearth can be inhibited, and the hearth is normally water-cooled to maintain its low temperature. Copper, because of its high thermal conductivity, is the preferred hearth material. The electrons are emitted by a hot filament, normally tungsten, and are attracted to the evaporant by a potential usually between 6 and 10 kV. Various types of electrodes and forms of focusing have been used at different times, but the arrangement that has now been almost universally adopted is what is known as the bent-beam type of gun. The hearth is at the ground potential and the filament is negative with respect to it. The filament and electrodes, usually a plate at filament potential situated close to the filament with a beam-defining slit through which the electrons pass, followed closely by the anode at the same potential as the hearth and incorporating a slightly larger slit so that the beam passes through it, are placed under the hearth, well out of reach of the emitted evaporant. The beam is bent around through rather greater than a semicircle by a magnetic field and focused on the material in the hearth. This avoids the problems of early electron beam systems that had filaments in line of sight of the hearth and hence considerably shortened life due to reactions with the evaporant. Supplementary magnetic fields derived from coils allow the position of the spot to be varied so that the mean can be placed in the centre of the hearth and a raster can be described which increases the area of heated material. This reduces the temperature necessary to maintain the same rate of deposition, improves the efficiency of use of the material in the crucible and makes the electron beam source more stable. A typical electron beam source of this type is shown in figure 9.7.

The electron beam source is particularly useful for materials that react with boats or require very high evaporation temperatures or both. Even in quite small sources, beam currents of up to 1 A at voltages of around 10 kV can be achieved and refractory oxides such as aluminium oxide, zirconium oxide, and hafnium oxide, and reactive semiconductors such as germanium and silicon, can be evaporated readily. Furthermore, materials that can be evaporated quite satisfactorily by directly heated boat, can be evaporated still more easily by electron beam, and so the tendency is to use electron beam sources, once they are installed, for virtually all materials. To improve their flexibility, they can be constructed with multiple pockets in the hearth so that the same source can handle up to four different materials in a single coating cycle. Of course the capacity of each individual pocket in a multiple-pocket version is usually rather less than that of the single-pocket version of the same source. For large-scale production, therefore, or for coatings for the infrared, it is normal to use two or more single-pocket sources.

The temperature of the substrate also plays a part in determining the

Figure 9.7 A four-pocket Supersource™. This is an electron-beam source of the bent-beam type. The water-cooled crucible has four pockets that can be rotated into position at the focus of the electron beam that issues from the slot to the right of the opening in the top of the gun. The sides of the gun are the pole pieces of the focusing and deflecting magnet. (Reproduced by kind permission of Temescal, Berkeley, California, USA, a division of the BOC Group Inc.)

properties of the condensed films. Usually it is the consistency of temperature from one coating run to the next which is of greater importance than the absolute level, although Ritchie[3], working in the far infrared beyond 12 μm, found substrate temperature to be of critical importance and devised ways of controlling it to within 2°C of the experimentally determined optimum. Substrates are often of low thermal conductivity and are mounted on rotating jigs to ensure uniformity of film thickness so that the measurement of the absolute temperature of the substrates is difficult. The heating is usually by means of radiant elements placed a short distance behind the substrates or by tungsten halogen lamps placed so that they illuminate the front surfaces of the substrates, the latter method gaining in popularity. Measurement is most often carried out by placing a thermocouple just in front of the substrate carrier. This will not measure substrate temperature accurately but will give an indication of the constancy of process conditions; frequently this is all that is wanted, anyway. An improvement can be obtained by embedding the thermocouple in a block of material of the same type as the substrates. Thermocouples have been placed on the rotating jig and the signal led out through silver slip rings, but even in this case the temperature of the front surface of the substrates is still not necessarily known to any high degree of accuracy, especially if they are of material of low thermal conductivity such as glass or silica. Rather more accurate results are achievable with substrates of germanium or silicon, frequently used in the infrared. A more consistent technique that is becoming more common is the use of an infrared remote-sensing thermometer that detects infrared radiation from the hot substrates. Usually mounted outside the chamber, this views the substrates through an infrared-transmitting window. The absolute calibration of the device depends on the emittance of the substrate. This varies less for substrates such as glass with dielectric coatings for the visible region than for infrared components. Again, consistency from one run to the next is of prime importance.

Usually metals should be deposited at low substrate temperatures to avoid scatter—particularly important in metal–dielectric filters and in ultraviolet-reflecting coatings, although there is an exception to this rule of thumb in the cases of rhodium and platinum, both of which give substantially better results when deposited hot[4, 5]. There are difficulties in refrigerating substrates, and substrate temperatures below ambient encourage thicker adsorbed gas layers that inhibit the condensation of the films and cause contamination. Thus it is not normal to operate with substrate temperatures below ambient, at which adequate results are obtained. The softer dielectric materials such as zinc sulphide and cryolite can also be deposited at room temperature (except, as we shall see, if zinc sulphide is to be used in the infrared). The harder dielectric materials, however, usually require elevated substrate temperatures, often 200–300°C. These materials include ceric oxide, magnesium fluoride and titanium dioxide. Some of the semiconductors for the infrared must be similarly treated. Frequently, optimum mechanical properties demand deposition at a temperature that is different from that for optimum optical properties and a compromise that depends on the particular application is necessary. Further details will be given when individual materials are discussed.

MEASUREMENT OF THE OPTICAL PROPERTIES

Once a suitable method of producing the particular thin film has been determined, the next step is the measurement of the optical properties. Many methods for this exist and a useful account is given by Heavens[6]. Measurement of the optical constants of thin films is also included in the book by Liddell[7]. Here we shall be concerned with just a few methods that are frequently used.

As we saw in chapter 2, given the optical constants and thicknesses of any series of thin films on a substrate, the calculation of the optical properties is straightforward. The inverse problem, that of calculating the optical constants and thicknesses of even a single thin film, given the measured optical properties, is much more difficult and there is no general analytical solution to the problem of inverting the equations. For an ideal thin film there are three parameters involved, n, k and d, the real and imaginary parts of refractive index and the geometrical thickness, respectively. Both n and k vary with wavelength, which increases the complexity. The traditional methods of measuring optical constants, therefore, rely on special limiting cases that have straightforward solutions.

Perhaps the simplest case of all is represented by a quarter-wave of material on a substrate, both of which are lossless and dispersionless, that is, k is zero and n is constant with wavelength. The reflectance is given by

$$R = \left(\frac{1 - n_f^2/n_s}{1 + n_f^2/n_s}\right)^2 \tag{9.1}$$

where n_f is the index of the film, n_s that of the substrate and the incident medium is assumed to have an index of unity. Then n_f is given by

$$n_f = n_s^{1/2}\left(\frac{1 - R^{1/2}}{1 + R^{1/2}}\right)^{1/2} \tag{9.2}$$

where the refractive index of the substrate n_s must, of course, be known. The measurement of reflectance must be reasonably accurate. If, for instance, the refractive index is around 2.3, with a substrate of glass, then the reflectance should be measured to around one third of a percent (absolute ΔR of 0.003) for a refractive index measurement accurate in the second decimal place. It is sometimes claimed that this method gives a more accurate value for refractive index than the original measure of reflectance since the square root of R is used in the calculation. This may be so, but the value obtained for refractive index will most likely be used in the subsequent calculation of the reflectance of a coating, and therefore the computed figure can be only as good as the original measurement of reflectance. In the absence of dispersion, the curve of reflectance versus wavelength of the film will be similar to that in figure 9.8. The

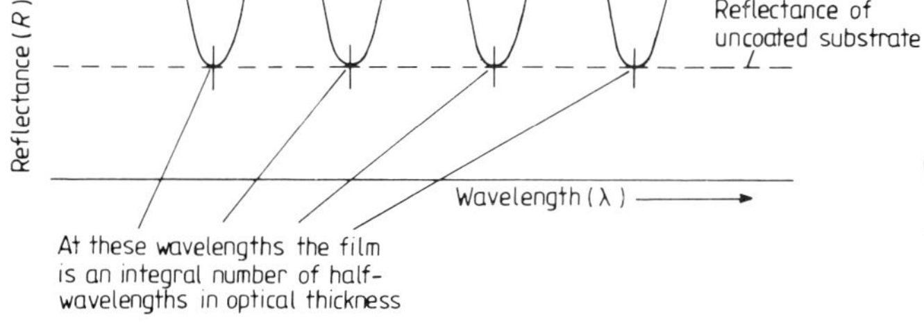

Figure 9.8 The reflectance of a simple thin film.

extrema correspond to integral numbers of quarter-waves, even numbers being half-wave absentees and giving reflectance equal to that of the uncoated substrate, and odd corresponding to the quarter-wave of equations (9.1) and (9.2). Thus it is easy to pick out those values of reflectance which correspond to the quarter-waves.

The technique can be adapted to give results in the presence of slight dispersion. The maxima in figure 9.8 will now no longer be at the same heights but, provided the index of the substrate is known throughout the range, the heights of the maxima can be used to calculate values for film index at the corresponding wavelengths. Interpolation can then be used to construct a graph of refractive index against wavelength. Results obtained by Hall and Ferguson[8] for MgF_2 are shown in figure 9.9.

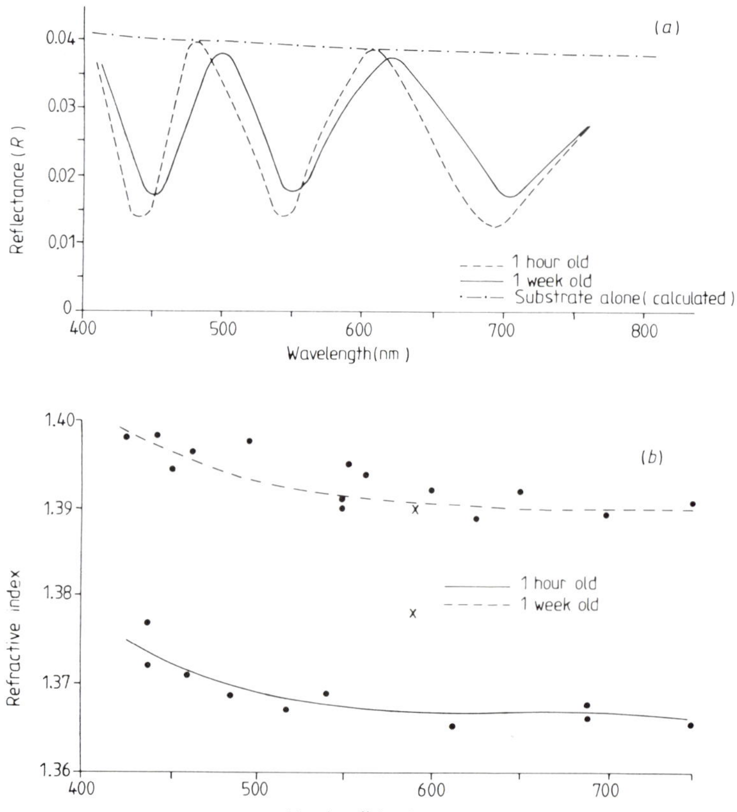

Figure 9.9 The refractive index of magnesium fluoride films. (*a*) The reflectance of a single film. (*b*) The reflectance result transforms into refractive index. The curves are formed by the results from many films. × denotes bulk indices of the crystalline solid. (After Hall and Ferguson[8].)

This simple method yields results which are usually sufficiently accurate for design purposes. If, however, the dispersion is somewhat greater, or if rather more accurate results are required, then the slightly more involved formulae given by Hass, Ramsey and Thun[9] must be applied. It is still assumed that the absorption is negligible. If the curve of reflectance or transmittance of a film possessing dispersion is examined, it will easily be seen that the maxima corresponding to the odd quarter-wave thicknesses are displaced in wavelength from the true quarter-wave points, while the half-wave maxima are unchanged. This shift is due to the dispersion, and measurement of it can yield a more accurate value for the refractive index. In the absence of absorption the turning values of $R, T, 1/R$ and $1/T$ must all coincide. Assuming that the refractive index of the incident medium is unity, that of the substrate n_s, and of the film n_f, then their expression for T becomes

$$T = \frac{4}{n_s + 2 + n_s^{-1} + 0.5n_s^{-1}(n_f^2 - 1 - n_s^2 + n_s^2 n_f^{-2})[1 - \cos(4\pi n_f t_f/\lambda)]}.$$

Since the turning values of T and $1/T$ coincide, the positions of the turning values can be found in terms of t/λ by differentiating the expression for $1/T$ and equating it to zero as follows

$$\frac{1}{T} = \frac{n_s + 2 + n_s^{-1}}{4} + \frac{1}{8n_s}(n_f^2 - 1 - n_s^2 + n_s^2 n_f^{-2})\left(1 - \cos\frac{4\pi n_f t_f}{\lambda}\right)$$

i.e.

$$0 = \frac{d(1/T)}{d(t/\lambda)} = 0.25n_f'(n_s^{-1}n_f - n_s n_f^{-3})\left(1 - \cos\frac{4\pi n_f t_f}{\lambda}\right)$$
$$+ 0.5\pi(n_s^{-1}n_f^2 - n_s^{-1} - n_s + n_s n_f^{-2})\left(n_f + n_f'\frac{t_f}{\lambda}\right)\sin\frac{4\pi n_f t_f}{\lambda}$$

where $n_f' = dn_f/d(t/\lambda)$. That the equation is satisfied exactly at all half-wave positions can easily be seen since both $\sin(4\pi n_f t_f/\lambda)$ and $(1 - \cos 4\pi n_f t_f/\lambda)$ are zero. At wavelengths corresponding to odd quarter-waves a shift does occur and this can be determined by manipulating the above equation into

$$\tan\frac{2\pi n_f t_f}{\lambda} = -2\pi\frac{n_f^5 - (1 + n_s^2)n_f^3 + n_s^2 n_f}{n_f^4 - n_s^2}\left(\frac{n_f}{n_f'} + \frac{t_f}{\lambda}\right). \tag{9.3}$$

Of course it is impossible to solve this equation immediately for n_f because there are too many unknowns. Generally the most useful approach is by successive approximations using the simpler quarter-wave formula (9.1) to obtain a first approximation for the index and the dispersion. It should be remembered that the reflection of the rear surface of the test glass should be taken into account in the derivation of the reflectance curve. It is also important that the test glass should be free from dispersion to a greater degree than the film, otherwise it must also be taken into account with consequent complication of the analysis.

If absorption is present, then formula (9.3) cannot be used. In the case of heavy absorption it can safely be assumed that there is no interference and the value of the extinction coefficient can be calculated from the expression

$$\frac{1-R}{T} = \exp\left(\frac{4\pi k_f d_f}{\lambda}\right)$$

($4\pi k_f d_f/\lambda$, because we are dealing with energies not amplitudes) which gives[9] for k_f

$$k_f = \frac{\lambda}{4\pi d_f \log e} \log\left(\frac{1-R}{T}\right). \tag{9.4}$$

The thin-film designer is not too concerned with very accurate values of heavy absorption. Often it is sufficient merely to know that the absorption is high in a given region and the result given by (9.4) will be more than satisfactory. In regions where the absorption is significant but not great enough to weaken the single-film interference effects, a more accurate method can be used.

Equation (2.69) is valid for any assembly of thin films on a transparent substrate, n_s, and is written

$$\frac{T}{1-R} = \frac{\mathrm{Re}\,(n_s)}{\mathrm{Re}\,(BC^*)}. \tag{9.5}$$

For a single film on a transparent substrate the values of B and C are given by

$$\begin{bmatrix} B \\ C \end{bmatrix} = \begin{bmatrix} \cos\delta_f & (\mathrm{i}\sin\delta_f)/N_f \\ \mathrm{i}N_f\sin\delta_f & \cos\delta_f \end{bmatrix} \begin{bmatrix} 1 \\ n_s \end{bmatrix} = \begin{bmatrix} \cos\delta_f + \mathrm{i}(n_s/N_f)\sin\delta_f \\ n_s\cos\delta_f + \mathrm{i}N_f\sin\delta_f \end{bmatrix}.$$

Now

$$\delta_f = \phi - \mathrm{i}\psi = \frac{2\pi N_f d_f}{\lambda} = \frac{2\pi n_f d_f}{\lambda} - \mathrm{i}\frac{2\pi k_f d_f}{\lambda}. \tag{9.6}$$

We shall assume that k is small compared with n and this implies that ψ will be small compared with ϕ. Now for ϕ sufficiently small

$$\cos\delta = \cos\phi\cosh\psi + \mathrm{i}\sin\phi\sinh\psi \simeq \cos\phi + \mathrm{i}\psi\sin\phi$$

and

$$\sin\delta = \sin\phi\cosh\psi - \mathrm{i}\cos\phi\sinh\psi \simeq \sin\phi - \mathrm{i}\psi\cos\phi$$

which yields the following expression for B and C

$$\begin{bmatrix} B \\ C \end{bmatrix} = \begin{bmatrix} [1-(n_s/n_f)\psi]\cos\phi - (n_s k_f/n_f^2)\sin\phi + \mathrm{i}[\psi + (n_s/n_f)]\sin\phi \\ (n_s + n_f\psi)\cos\phi + k_f\sin\phi + \mathrm{i}(n_f + n_s\psi)\sin\phi \end{bmatrix}. \tag{9.7}$$

At wavelengths where the optical thickness is an integral number of quarter-wavelengths, $\sin\phi$ or $\cos\phi$ is zero, and we can neglect terms in $\cos\phi\sin\phi$. The value of the real part of (BC^*) is then given by

$$\mathrm{Re}\,(BC^*) = \cos^2\phi\left(1+\frac{n_s}{n_f}\psi\right)(n_s+n_f\psi)+\sin^2\phi\left(\psi+\frac{n_s}{n_f}\right)(n_f+n_s\psi)$$

$$= \left[n_s+\left(\frac{n_s^2}{n_f}+n_f\right)\psi\right] \tag{9.8}$$

and when substituted in (9.5) yields, when n_s is inserted for η_m,

$$\frac{1-R}{T} = 1+\left(\frac{n_s}{n_f}+\frac{n_f}{n_s}\right)\psi \tag{9.9}$$

giving for k_f (using the expression (9.6) in (9.9))

$$k_f = \frac{\lambda}{2\pi d_f[(n_s/n_f)+(n_f/n_s)]}\,\frac{1-R-T}{T}. \tag{9.10}$$

This expression is accurate only close to the turning values of the reflectance or transmittance curves.

In the case of low absorption the index should also be corrected. Hall and Ferguson[10] give the following expression

$$n_f = \left(\frac{n_s(1+\sqrt{R})}{1-\sqrt{R}}\right)^{1/2} + \frac{\pi k_f d_f}{\lambda}\left(\frac{1+\sqrt{R}}{1-\sqrt{R}} - n_s\right) \tag{9.11}$$

where R is the value of reflectance of the film at the reflectance maximum.

In the methods discussed so far, we have been assuming that the thickness of the film is unknown, except inasmuch as it can be deduced from the measurements of reflectance and transmittance, and the extrema have been the principal indicator of film thickness. However, it is possible accurately to measure film thickness in other ways, such as multiple beam interferometry, or electron microscopy, or by using a stylus step-measuring instrument. Once there is an independent accurate measure of physical thickness, the problem of calculating the optical constants becomes much simpler. The most frequently used technique of this type was devised by Hadley (see Heavens[6] for a description). Since two optical constants, n_f and k_f, are involved at each wavelength, two parameters must be measured, and these can most conveniently be R and T. In the ideal form of the technique, if now a value of n_f is assumed, then by trial and error one value of k_f can be found, which, together with the known geometrical thickness and the assumed n_f, yields the correct measured value of R, and then a second value of k_f that similarly yields the correct value of T. A different value of n_f will give two further values of k_f, and so on. Proceeding thus, we can plot two curves of k_f against n_f, one corresponding to the T values and the other to the R values, and, where they intersect, we have the correct values of n_f and k_f for the film. The angle of intersection of the curves gives an indication of the precision of the result.

Hadley, at a time when such calculations were exceedingly cumbersome, produced a book of curves giving the reflectance and transmittance of films as a function of the ratio of geometrical thickness to wavelength, with n_f and k_f as parameters, which greatly speeded up the process. Nowadays, the method can be readily programmed and precision estimates incorporated. This method can be applied to any thickness of film, not just at the extrema, although maximum precision is achieved, as we might expect, near optical thicknesses of odd quarter-waves, while, at half-wave optical thicknesses, it is unable to yield any results. As with many other techniques, it suffers from multiple solutions particularly when the films are thick and in practice a range of wavelengths is employed, which adds an element of redundancy and helps to eliminate some of the less probable solutions.

Hadley's method involves simple iteration and does not require any very powerful computing facilities. Even in the absence of Hadley's precalculated curves, it can be accommodated on a programmable calculator of modest capacity. It does, however, involve the additional measurement of film thickness, which is of a different character from the measurements of R and T. A different approach that has been developed by Pelletier and his colleagues in Marseilles[11], and requires the use of powerful computing facilities, retains the measurement of R and T, but, instead of an independent measure of film thickness, adds the measurement of R', the reflectance of the film from the substrate side. Now we have three parameters to calculate at each wavelength and three measurements and it might appear possible that all three could be calculated by a process of iteration, just like the Hadley method; but the precision that it is possible to attain is poor and it breaks down completely when there is no absorption. To overcome this difficulty, the Marseilles method uses the fact that the geometrical thickness of the film does not vary with wavelength, and therefore, if information over a spectral region is used, there will be sufficient redundancy to permit an accurate estimate of geometrical thickness. Then once the thickness has been determined, a computer method akin to refinement finds accurate values of the optical constants n_f and k_f over the whole wavelength region. For dielectric layers of use in optical coatings, k_f will usually be small, and often negligible, over at least part of the region and a preliminary calculation involving an approximate value of n_f is able to yield a value for geometrical thickness, which in most cases is sufficiently accurate for the subsequent determination of the optical constants. Given the thickness, R and T, as we have seen, should in fact be sufficient to determine n_f and k_f. But this would mean discarding the extra information in R', and so the determination of the optical constants uses successive approximations to minimise a figure of merit consisting of a weighted sum of the squares of the differences between measured T, R and R' and the calculated values of the same quantities using the assumed values of n_f and k_f. Although seldom necessary, the new values of the optical constants can then be used in an improved estimate of the geometrical thickness, and the optical constants recalculated. For an estimate of precision, the changes in n_f and k_f to change the values of T, R and R' by a prescribed amount, usually 0.3 %, are calculated. Invariably, there are regions around the wavelengths for which the film is an integral number of half-waves thick, where the errors are greater than can be accepted and results in these regions are rejected. In practice the films are deposited over half of a substrate, slightly wedged to eliminate the effects of multiple reflections, and measurements are made of R and R' and T and T' on both coated and uncoated portions of the substrate. This permits the optical constants of the substrate to be estimated, and the redundancy in the measurements of T and T', the transmittance measured in the opposite direction, gives a check on the stability of the apparatus. A very large number of different dielectric thin-film materials have been measured in this way and a typical result is shown in figure 9.10.

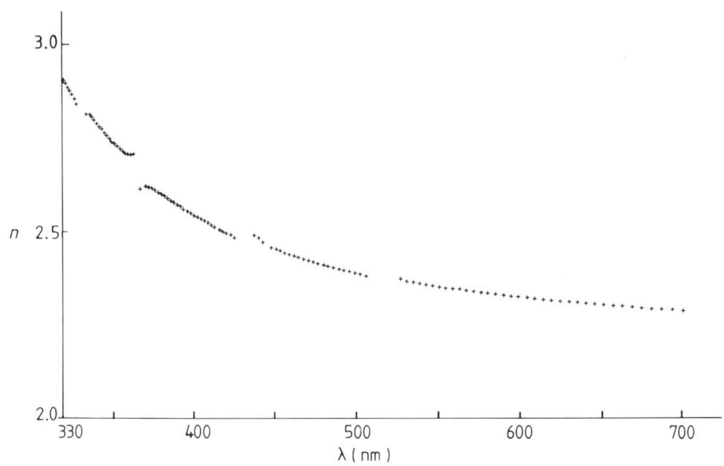

Figure 9.10 The refractive index of a film of zinc sulphide. The slight departure from a smooth curve is due to structural imperfections suggesting that even in this case of a very well behaved optical material there is some very slight residual inhomogeneity. (After Pelletier, Roche and Vidal[11].)

A straightforward technique is due to Manifacier, Gasiot and Fillard[12]. Provided the absorption in a thin film is small then the transmittance at the quarter- and half-wave points is a fairly simple function of n_f, k_f and d_f. Unfortunately, the transmittances at these points for one single film can only be measured for different wavelengths. The optical constants of the film are functions of wavelength and an iterative process involving interpolation is necessary to extract their values. In their method, therefore, Manifacier, Gasiot and Fillard begin by interpolating the actual values of transmittance by drawing two envelope curves around the transmittance characteristic for the film. These envelope curves are then supposed to mark the loci of quarter-wave and half-wave points assuming that the thickness of the film were to vary by a small amount. This gives at each wavelength point two values of transmittance corresponding to the two envelopes and therefore to the transmittances that a film of thickness an integral number of half-waves or an odd number of quarter-waves would have at that particular wavelength. These transmittances are denoted by T_{max} and T_{min} respectively for a film of high index on a substrate of lower index. For such a film we can write

$$\alpha = \frac{C_1[1-(T_{max}/T_{min})^{1/2}]}{C_2[1+(T_{max}/T_{min})^{1/2}]} \tag{9.12}$$

where

$$\alpha = \exp(-4\pi k_f d_f/\lambda) \tag{9.13}$$

$$4\pi n_f d_f/\lambda = m\pi \quad \text{(quarter- or half-wave thickness)}$$

$$\begin{aligned} C_1 &= (n_f + n_0)(n_s + n_f) \\ C_2 &= (n_f - n_0)(n_s - n_f) \\ T_{max} &= 16 n_0 n_s n_f^2 \alpha/(C_1 + C_2\alpha)^2 \\ T_{min} &= 16 n_0 n_s n_f^2 \alpha/(C_1 - C_2\alpha)^2. \end{aligned} \tag{9.14}$$

Then from (9.12) and (9.14), if we define N as

$$N = \frac{n_0^2 + n_s^2}{2} + 2n_0 n_s \frac{T_{max} - T_{min}}{T_{max} T_{min}}$$

n_f is given by

$$n_f = [N + (N^2 - n_0^2 n_s^2)^{1/2}]^{1/2}. \tag{9.15}$$

Once n_f has been determined, equation (9.12) can be used to find a value for α. The thickness d_f can then be found from the wavelengths corresponding to the various extrema and the extinction coefficient k_f from the values of d_f and α. The method has the advantage of explicit expressions for the various quantities, which makes it easily implemented on machines as small as programmable calculators. Unfortunately, as with many of the other techniques, the results can suffer from appreciable errors in the presence of inhomogeneity.

Powerful computers bring the advantage that we no longer need to devise methods of optical constant measurement with the principal objective of ease of calculation. Instead, methods can be chosen simply on the basis of precision of results, regardless of the complexity of the analytical techniques that are required. This is the approach advocated by Hansen[13], who has developed a reflectance attachment making it possible to measure the reflectance of a thin film for virtually any angle of incidence and plane of polarisation, the particular measurements carried out being chosen to suit each individual film.

For rapid, straightforward measurement of refractive index, a method due to Abelès[14] is particularly useful. The method depends on the fact that the reflectance for p-polarisation (TM waves) is the same for substrate and film at an angle of incidence that depends only on the indices of film and incident medium and not at all on either substrate index or film thickness, except, of course, that layers that are a half-wave thick at the appropriate angle of incidence and wavelength will give a reflectance equal to the uncoated substrate regardless of index. This can easily be seen from the following equations.

$$n_f \sin\theta_f = n_0 \sin\theta_0. \tag{9.16}$$

This is just Snell's law, equation (2.27). Then from equation (2.37)

$$n_f/\cos\theta_f = n_0/\cos\theta_0 \tag{9.17}$$

must hold if the reflectances are to be equal. n_f is the index of the film and n_0 that of the medium (that will usually be air of refractive index unity). The equations can be solved for θ_0, giving

$$\tan\theta_0 = n_f/n_0. \tag{9.18}$$

The measurement of index reduces to the measurement of the angle θ_0 at which the reflectances are equal. This is best carried out on a goniometer table using a monochromator feeding a collimator. Heavens[6] shows that the greatest accuracy of measurement is, once again, obtained when the layer is an odd number of quarter-waves thick at the appropriate angle of incidence. This is because there is then the greatest difference in the reflectances of the coated and uncoated substrate for a given angular misalignment from the ideal. It is possible to achieve an accuracy of around 0.002 in refractive index provided the film and substrate indices are within 0.3 of each other, but not equal. Hacskaylo[15] has developed an improved method based on the Abelès technique. It involves incident light that is plane polarised with the plane of polarisation almost but not quite parallel to the plane of incidence. The reflected light is passed through an analyser and the analyser angle, for which the reflected light from the uncoated substrate and from the film-coated substrate are equal, is plotted against the angle of incidence. A very sharp zero at the angle satisfying the Abelès condition is obtained, which permits accuracies of 0.0002–0.0006 in the measurement of indices in the range 1.2–2.3. It is not necessary for the film index to be close to the substrate index.

Unfortunately, the behaviour of real thin films is more complicated than we have been assuming. They are often inhomogeneous, that is, their refractive index varies throughout their thickness. They may also be anisotropic, although little work has been done on this aspect of their behaviour, but the possibility should be borne in mind when considering which methods to use for index determination.

Provided that the variation of index throughout the film is either a smooth increase or a smooth decrease, so that there are no extrema within the thickness of the film, the highest and lowest values being at the film boundaries, then we can use a very simple technique to determine the difference in behaviour at the quarter-wave and half-wave points, which would be obtained with an inhomogeneous film. We assume that the film is absorption-free and that its properties can be calculated by a multiple-beam approach, similar to that in chapter 2, which considers the amplitude reflection and transmission coefficients at the boundary only. The various symbols are defined in figure 9.11 and are similar to those already used in figure 2.10. Now, however, we assume that the index of that part of the film next to the substrate is n_b and that next to the surrounding medium is n_a. The corresponding admittances are y_b and y_a. The only reflections that take place are assumed to be at either of the two interfaces. There is one further complication, also indicated in the figure, before we can sum the multiple beams to arrive at transmittance and reflectance. A beam propagating from the outer surface of the film to the inner is assumed to suffer no loss by reflection and, therefore, the intensity is unaltered Since intensity is proportional to the square of the electric amplitude times admittance, a beam that is of amplitude $\mathscr{E}_a^+$ just inside interface a, will have amplitude $(y_a/y_b)^{1/2}\mathscr{E}_a^+$ at interface b. The correction will be reversed in travelling from b back to a. This is in addition to any phase changes. The inverse correction applies to magnetic amplitudes. Since the correction cancels out for each double pass it does not affect the result for resultant reflectance but it must be taken into account when the multiple beams are being summed for the calculation of transmittance. The derivation of the necessary expressions proceeds as in equations (2.48)–(2.55) in chapter 2 although here we restrict ourselves to normal incidence.

$$E_b = E_{1b}^+ + E_{1b}^- \tag{9.19}$$

$$H_b = y_b E_{1b}^+ - y_b E_{1b}^- \tag{9.20}$$

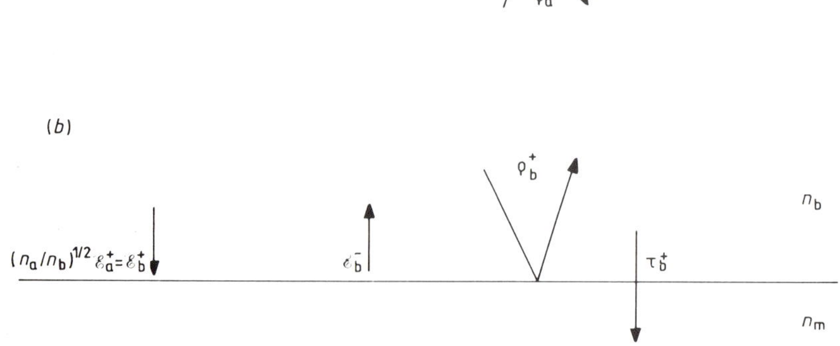

Figure 9.11 Inhomogeneous film quantities used in the development in the text of the matrix expression for an inhomogeneous layer.

giving

$$E_{1b}^{+} = 0.5[(H_b/y_b) + E_b] \qquad H_{1b}^{+} = 0.5(H_b + y_b E_b)$$

$$E_{1b}^{-} = 0.5[-(H_b/y_b) + E_b] \qquad H_{1b}^{-} = 0.5(H_b - y_b E_b).$$

Then the various rays are transferred to interface a

$$\begin{aligned} E_{1a}^{+} &= 0.5[(H_b/y_b) + E_b](y_b/y_a)^{1/2}e^{i\delta} \\ E_{1b}^{-} &= 0.5[-(H_b/y_b) + E_b](y_b/y_a)^{1/2}e^{-i\delta} \\ H_{1b}^{+} &= 0.5(H_b + y_b E_b)(y_a/y_b)^{1/2}e^{i\delta} \\ H_{1b}^{-} &= 0.5(H_b - y_b E_b)(y_a/y_b)^{1/2}e^{-i\delta} \end{aligned}$$

giving

$$\begin{aligned} E_b &= E_{1b}^{+} + E_{1b}^{-} \\ &= (y_b/y_a)^{1/2}(\cos\delta)E_b + \frac{i\sin\delta}{(y_a y_b)^{1/2}} H_b \\ H_b &= y_b E_{1b}^{+} - y_b E_{1b}^{-} \\ &= i(y_a y_b)^{1/2}(\sin\delta)E_b + (y_a/y_b)^{1/2}(\cos\delta)H_b. \end{aligned}$$

The characteristic matrix for the layer is then given by

$$\begin{bmatrix} (y_b/y_a)^{1/2}(\cos\delta) & \dfrac{i\sin\delta}{(y_a y_b)^{1/2}} \\ i(y_a y_b)^{1/2}(\sin\delta) & (y_a/y_b)^{1/2}(\cos\delta). \end{bmatrix} \tag{9.21}$$

Jacobsson[16] considers the calculation of inhomogeneous layer properties in detail and includes this expression.

Now we consider cases where the layer is either an odd number of quarter-waves or an integral number of half-waves. We apply the expression (9.21) in the normal way and find the well known relations

$$R = \left(\frac{y_0 - y_a y_b/y_s}{y_0 + y_a y_b/y_s}\right)^2 \qquad \text{for a quarter-wave} \tag{9.22}$$

and

$$R = \left(\frac{y_0 - y_a y_s/y_b}{y_0 + y_a y_s/y_b}\right)^2 \qquad \text{for a half-wave.} \tag{9.23}$$

The expression for a quarter-wave layer is indistinguishable from that of a homogeneous layer of admittance $(y_a y_b)^{1/2}$ and so it is impossible to detect the presence of inhomogeneity from the quarter-wave result. The half-wave expression is quite different. Here the layer is no longer an absentee layer and cannot therefore be represented by an equivalent homogeneous layer. The shifting of the reflectance of the half-wave points from the level of the uncoated substrate in absorption-free layers is a sure sign of inhomogeneity and can be used to measure it.

The Hadley method of deriving the optical constants takes no account of inhomogeneity. Any inhomogeneity, therefore, introduces errors. The

Marseilles method, however, includes half-wave points and therefore has sufficient information to accommodate inhomogeneity. Expression (9.21) is a good approximation when the inhomogeneity is not too large and when the admittances y_a and y_b are significantly different from those of substrate and incident medium. To avoid any difficulties due to the model, the Marseilles group actually use a model for the layer consisting of ten homogeneous sublayers with linearly varying values of n but identical values of k and thickness d. Measurements are made of R and T over a wide wavelength range, usually 300–800 nm. The half-wave points still give the principal information on the degree of inhomogeneity. They are also affected by the extinction coefficient k and this has also to be taken into account. The quarter-wave points are insensitive to inhomogeneity and, provided R and T are measured, can therefore yield values for k that can be interpolated and used at the half-wave points. One half-wave point within the region of measurement can be used to give a measure of inhomogeneity that is assumed constant over the rest of the region. Several half-wave points can yield values of inhomogeneity that can be fitted to a Cauchy expression, that is, an expression of the form

$$\Delta n/n = A + B/\lambda^2 + C/\lambda^4.$$

Details of the technique are given by Borgogno, Lazarides and Pelletier[17]. Some of their results are shown in figure 9.12.

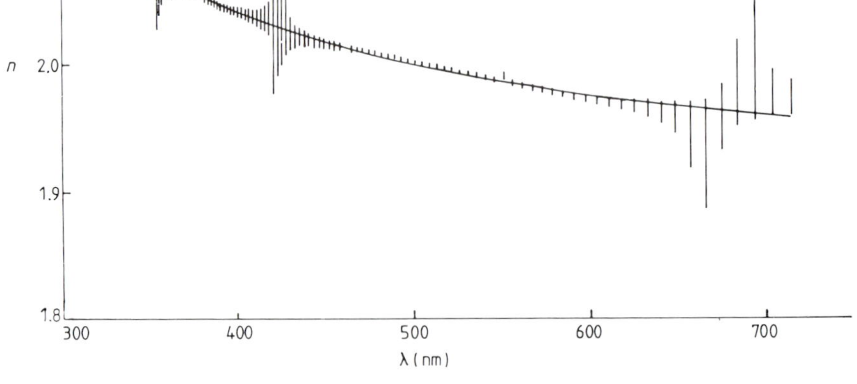

Figure 9.12 Values of mean index and the uncertainty Δn calculated for hafnium oxide using an inhomogeneous film model. The Cauchy coefficients for n are: $A = 1.9165$, $B = 2.198 \times 10^4\ \text{nm}^2$, $C = -3.276 \times 10^8\ \text{nm}^4$ and for $\Delta n/n$ are: $A' = -5.39 \times 10^{-2}$, $B' = -1.77 \times 10^3\ \text{nm}^2$. (After Borgogno, Lazarides and Pelletier[17].)

Netterfield[18] measured the variation in reflectance of a film at a single wavelength as it was deposited. If the assumption is made that the part of the film which is already deposited is unaffected by subsequent material, then the values of reflectance associated with extrema can be used to calculate a profile of the refractive index throughout the thickness of the layer. Some results obtained for cryolite, in this way, are shown in figure 9.13.

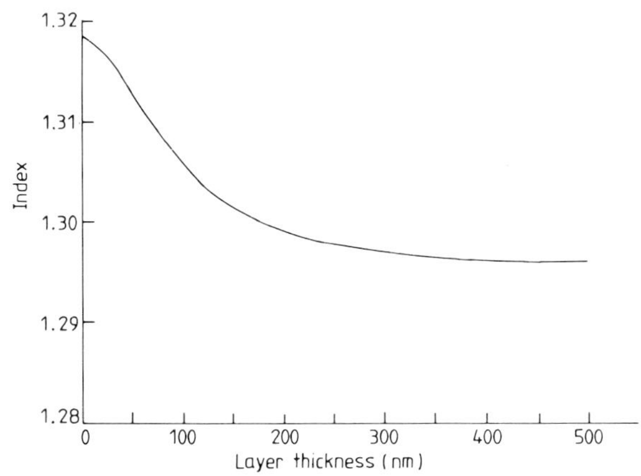

Figure 9.13 Graph of the index profile of cryolite layers at $\lambda = 633$ nm, derived from fitting a formula, $n^2 = A + [B/(t^2 + C)]$, where t is the thickness coordinate, to curves of the variation of reflectance *in vacuo* of a cryolite film deposited over a zinc sulphide film of varying thickness. $A = 1.6773$, $B = 5.0431 \times 10^2$ nm^2, $C = 8.2986 \times 10^3$ nm^2. (After Netterfield[18].)

MEASUREMENT OF THE MECHANICAL PROPERTIES

From the point of view of optical coatings, the importance of the mechanical properties of thin films is primarily in their relation to coating stability, that is, the extent to which coatings will continue to behave as they did when removed from the coating chamber, even when subjected to disturbances of an environmental and/or mechanical nature. There are many factors involved in stability, many of which are neither easy to define nor to measure and there are still great difficulties to be overcome. The approach used in quality assurance in manufacture, discussed further in chapter 10, is entirely empirical. Tests are devised which reproduce, in as controlled a fashion as possible, the disturbances to which the coating will be subjected in practice, and samples are simply subjected to these tests and inspected for signs of damage. Sometimes the tests are deliberately made more severe than those expected in use. Coating performance specifications are normally written in terms of such test levels.

Stress is measured by depositing the material on a thin flexible substrate which becomes deformed. The deformation is then measured and the value of stress necessary to cause it calculated. The substrate may be of any suitable material; glass, mica, silica, metal, for example, have all been used. The form of the substrate is often a thin strip, supported so that part of it can deflect, and either the deflection is measured in some way or a restoring force is applied to restore the strip to its original position. Usually the deflection, or the restoring force, is measured continuously during deposition. Optical microscopes, capacitance gauges, piezoelectric devices and interferometric techniques are some of the successful methods.

A useful survey of the field of stress measurement in thin films in general is given by Hoffman[19,20]. A particularly useful paper which deals solely with dielectric films for optical coatings is that by Ennos[21]. Ennos used a thin strip of fused silica as substrate, simply supported at each end on ball bearings so that the centre of the strip was free to move. An interferometric technique with a helium–neon laser as the light source was used to measure the movement of the strip. The strip was made one mirror of a Michelson interferometer of novel design, shown in figure 9.14(*a*). Since the laser light was plane polarised, the upper surface of the prism was set at the Brewster angle to eliminate losses by reflection of the emergent beam. Apart from the more obvious advantages of large coherence length and high collimation, the laser beam made it possible to line up the interferometer with the bell jar of the plant in the raised position (see figure 9.14(*b*)). No high quality window in the plant was necessary, the glass jar of quite poor optical quality proving adequate. To complete the arrangement, the laser light was also directed on a test flat for the optical monitoring of film thickness. A typical record obtained with the apparatus is also shown in figure 9.14(*c*). The calibration of the fused-silica strip was determined both by calculation and by measurement of deflection under a known applied load. Curves plotted for a wide range of materials showing the variation of stress in the films during the actual growth as a function both of film thickness and evaporation conditions are included in the paper, some examples being shown in figure 9.15. It is of particular interest to note the frequent drop in stress when the films are exposed to the atmosphere. This is principally due to adsorption of water vapour, an effect which will be considered further towards the end of this chapter.

The interferometric technique has been further improved more recently by Roll and Hoffman[22] and applied to optical thin films by Ledger and Bastien[23,24]. The Michelson interferometer of Ennos is replaced by a cat's-eye interferometer, using circular disks as sensitive elements, which is very much less temperature-sensitive, and this has enabled the measurement of stress levels over a wide range of substrate temperatures. Examination of the differences in thermally induced stress for identical films on different substrate materials, when substrate temperature is varied after deposition, has permitted the measurement of the elastic moduli and thermal expansion coefficients of the thin-film materials. Although the measured value of expansion coefficient

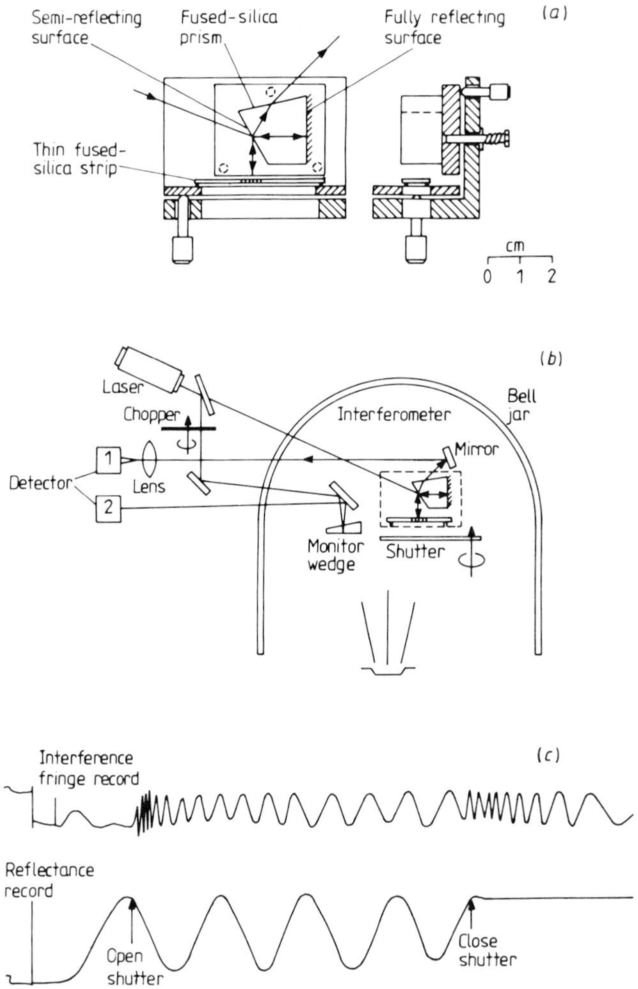

Figure 9.14 (*a*) Film-stress interferometer. (*b*) Experimental arrangement for continuous measurement of film stress during evaporation. (*c*) Recorder trace of fringe displacement and film reflectance. (After Ennos[21].)

for bulk thorium fluoride crystals is small and negative, the values for thorium fluoride thin films were consistently large and positive, varying from 11.1×10^{-6} to 18.1×10^{-6} °C. Young's modulus for the same samples varies from 3.9×10^5 to 6.8×10^5 kg cm^{-2} (that is 3.9×10^{10} to 6.8×10^{10} Pa).

Very recently Pulker has studied the relationship between stress levels and the microstructure of optical thin films, developing further some ideas of Hoffman. The work is surveyed in reference 25. Good agreement between measured levels of stress and those calculated from the model has been

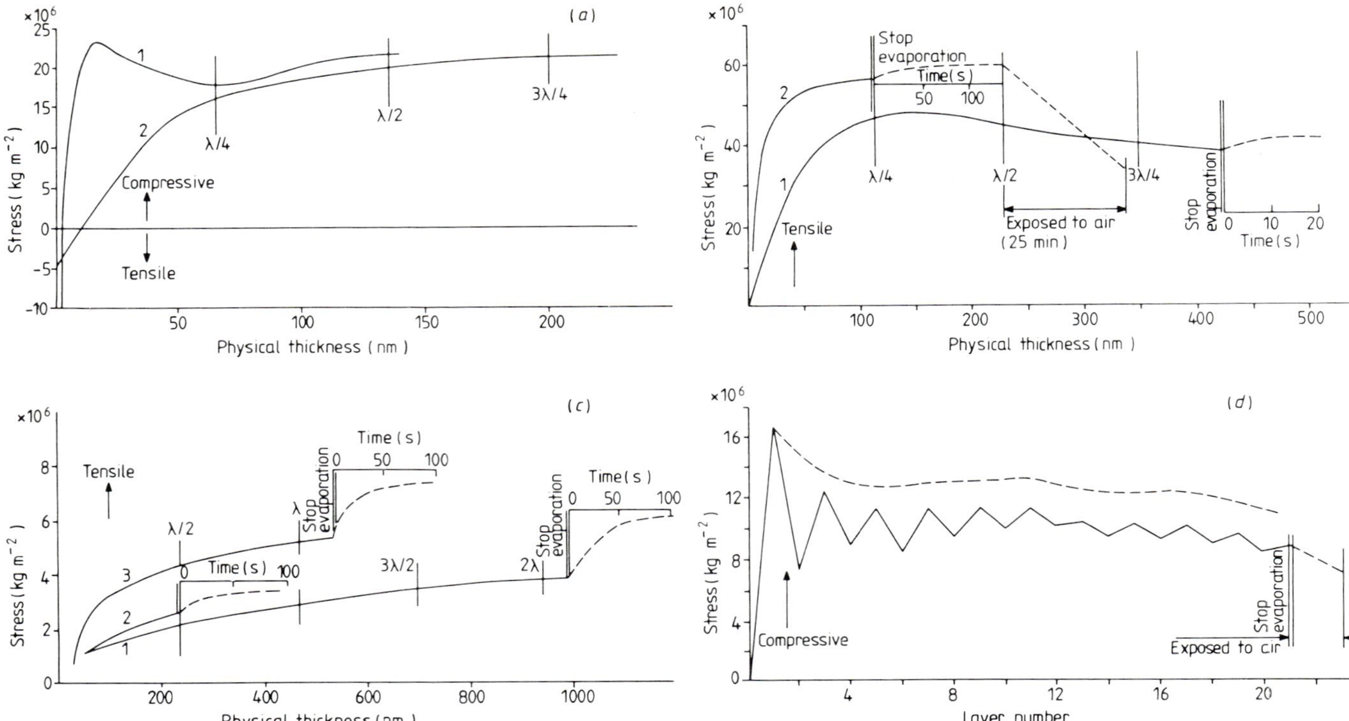

Figure 9.15 (*a*) Film stress in evaporated zinc sulphide on fused silica at ambient temperature. Evaporation rate: 1 : 0.25 nm s^{-1}. 2: 2.2 nm s^{-1}. (*b*) Film stress in magnesium fluoride: 1: Direct evaporation from molybdenum, evaporation rate 4.2 nm s^{-1}. 2: Indirect radiative heating, evaporation rate 1.2 nm s^{-1}. (*c*) Cryolite and chiolite evaporated by indirect radiative heating. 1: Cryolite, evaporation rate 3.5 nm s^{-1}. 2: Cryolite, evaporation rate 2.3 nm s^{-1}. 3: Chiolite, evaporation rate 4 nm s^{-1}. (*d*) Zinc sulphide–cryolite multilayer. Twenty-one layers $(HL)^{10}H$. Resultant average stress after each evaporation plotted. Broken curve shows upper limit of film stress reached during the warm-up period before the evaporation of a layer commenced. (After Ennos[21].)

achieved, but perhaps the most spectacular feature has been the demonstration, in accord with the theory, that small amounts of impurity can have a major effect on stress. The impurities congregate at the boundaries of the columnar grains of the films and reduce the forces of attraction between neighbouring grains, thus reducing stress. Small amounts of calcium fluoride in magnesium fluoride, around 4 mol %, reduce stress by some 50 %.

Abrasion resistance is another mechanical property that is of considerable importance and yet extremely difficult to define in any terms other than empirical. This is probably principally because abrasion resistance is not a single fundamental property but rather a combination of factors such as adhesion, hardness, friction, packing density and so on. Various ways of specifying abrasion resistance exist but all depend on arbitrary empirical standards. In the United Kingdom, reliance is placed on a standard that was originated and is still maintained by Sira Institute (formerly the British Scientific Instrument Research Association). The standard involves a pad,made from rubber loaded with a particular grade of emery, which is drawn over the surface of the film under a controlled load for a given number of strokes. Signs of visible damage show that the coating has failed the test. Similar standards based on the use of cheesecloth or of a standard eraser exist in the United States. Most of the tests suffer from the fact that they do not give a measure of the degree of abrasion resistance but are merely of a go/no-go nature. There is a modification of the Sira test, described in chapter 10, which does permit a measure of abrasion resistance to be derived from the extent of the damage caused by a controlled amount of abrasion. This is probably the best arrangement yet devised, but even here the results vary considerably with film thickness and coating design so that it is far from an absolute measure of a fumdamental thin-film property. Abrasion resistance is, therefore, primarily a quality control tool. It will be considered further in chapter 10.

Adhesion is another important mechanical property that presents difficulties in measurement. What we usually think of as adhesion is the magnitude of the force necessary to detach unit area of the film from the substrate or from a neighbouring film in a multilayer. However, accurate measures of this type are impossible.

Quality control testing is, as for many of the other mechanical properties, of a go/no-go nature. A strip of adhesive tape is stuck to the film and removed. The film fails if it delaminates along with the tape.

Jacobsson and Kruse[26] have studied the application of a direct-pull technique to optical thin films. In principle, the adhesive forces between film and substrate can be measured simply by applying a pull to a portion of the film until it breaks away, and, indeed, this is a technique which is used for other types of coatings, such as paint films. The test technique is straightforward and consists of cementing the flat end of a small cylinder to the film, and then pulling the cylinder, together with the portion of film under it, off the substrate, in as near normal a direction as possible. The force required to accomplish this is the measure of the force of adhesion. Great attention to detail is required. The end of the cylinder must be true, must be cemented to the film so that the thickness of cement is constant and so that the axis of the cylinder is vertical. The pull applied to the cylinder must have its line of action along the cylinder axis, normal to the film surface. The precautions to be taken, and the tolerances which must be held, are considered by Jacobsson and Kruse. Their cylindrical blocks were optically polished at the ends, and, in order more nearly to ensure a pull normal to the surface, the film and substrate were cemented between two cylinders, the axes of which were collinear. The mean value of the force of adhesion between 250 nm thick ZnS films and a glass substrate was found to be 2.3×10^7 Pa, which rose to 4.3×10^7 Pa when the glass substrate was subjected to 20 minutes of ion bombardment before coating. Zinc sulphide films evaporated on to a layer of SiO, some 150 nm thick, gave still higher adhesion figures of 5.4×10^7 Pa. The increases in adhesion due to the ion bombardment and the SiO were consistent, and the scatter in successive measures of adhesion was small, some 30% in the worst case.

An alternative method of measuring the force of adhesion is the scratch test, devised originally by Heavens[27], and improved and studied in detail by Benjamin and Weaver[28,29], who applied it to a range of metal films. Again, in principle, it is a straightforward test that nevertheless is very complex in interpretation. A round-ended stylus is drawn across the film-coated substrate under a series of increasing loads, and the point at which the film under the stylus is removed from the surface is a measure of the adhesion of the film. Benjamin and Weaver were able to show that plastic deformation of the substrate under the stylus subjected the interface between film and substrate to a shear force, directly related to the load on the stylus by the expression[28]:

$$F = [a/(r^2 - a^2)^{1/2}] - P \qquad (9.24)$$

where

$a = [W/(\pi P)]^{1/2}$
P is the indentation hardness of the substrate
r is the radius of the stylus point
a is the radius of the circle of contact
W is the load on the stylus
F is the shear force.

The shear force is roughly proportional to the root of the load on the stylus. For the film just to be removed by drawing the stylus across it, the shear force had just to be great enough to break the adhesive bonds. Using this apparatus, Benjamin and Weaver were able to confirm quantitatively, what had been qualitatively observed before, that the adhesion of aluminium deposited at pressures around 10^{-5} Torr (1.3×10^{-5} mb) on glass was initially poor, of values similar to van der Waals forces, but that after some 200 hours it improved to reach values consistent with chemical bonding. Aluminium deposited at higher pressures, around 10^{-3} Torr (1.3×10^{-3} mb) gave consistently high bonding immediately after deposition. This is attributed to the formation of an oxide bonding layer between aluminium and glass, and a series of experiments demonstrated the importance of such oxide layers in other metal films on glass. On alkali halide crystals, the initial bonding at van der Waals levels showed no subsequent improvement with time[30]. More recently the scratch test has been studied by Laugier[31,32] who has included the effects of friction during the scratching action in the analysis. Zinc sulphide has been shown to exhibit an unusual aging behaviour in that it occurs in two well defined stages. After a period of some 18–24 hours after deposition the adhesion increases by as much as a factor of 4 from an initially low figure. After a period of 3 days the adhesion then begins to increase further, and after a further 7 days reaches a final maximum that can be some 20 times the initial figure. This is attributed to the formation of zinc oxide at the interface between layer and substrate, first free zinc at the interface combining with oxygen that has diffused through the layer from the outer surface and then later zinc that has diffused to the boundary from within the layer.

Unfortunately, none of these adhesion tests is entirely satisfactory. Some of the difficulties are related to consistency of measurement, but the greatest problem is the nature of the adhesion itself. The forces which attach a film to a substrate, or one film to another, are all very large (usually greater than 100 tons in^{-2}) but also of very short range. In fact, they are principally between one atom and the next. The short range of the forces has two major consequences. First, the forces can be blocked by a single atom or molecule of contaminant, and so adhesion is susceptible to even the slightest contamination. A single monomolecular layer of contaminant is sufficient to destroy completely the adhesion between film and substrate. A small fraction of a monomolecular layer is enough to affect it adversely. Second, although the force of adhesion is large, the work required to detach the coating, the product of the force and its range, can be quite small. Coatings usually fail in adhesion in a progressive manner rather than suddenly and simultaneously over a significant area, and in such peel failures, it is the work, rather than the force, required to detach the coating—the work of adhesion, as it is usually called—that is the important parameter. This work can be considered as the supply of the necessary surface energy associated with the fresh surfaces exposed in the adhesion failure together with any work lost in the plastic deformation of film and/or substrate. With some metal films, particularly deposited on plastic, there is evidence that an electrostatic double layer gradually forms, which contributes positively to the adhesion[25].

In the tape test, the adhesive forces are comparatively very weak, but their long range allows them to be applied simultaneously over a relatively large area. Thus the film is unlikely to be detached from the substrate unless it is very weakly bonded, and even then it may not be removed unless there is a stress concentrator that can start the delamination process. Sometimes this is provided by scribing a series of small squares into the coating when the tape will tend to lift out complete squares.

In the case of the direct-pull technique, it is exceedingly difficult to avoid a progressive failure rather than a simultaneous rupturing of the bonds over the entire area of the pin. Unevenness in the thickness of the adhesive, or a pull which is not completely central, can cause a progressive failure with consequent reduction in the force measured. Even when the greatest care is taken it is unlikely that the true force of adhesion will be obtained and the test is useful principally as a quality control vehicle. Poor adhesion will tend to give a very much reduced force.

The scratch test suffers from additional problems. Many of the films used in optical coatings shatter when a sufficiently high load is applied before any delamination from the substrate takes place. Film hardness and brittleness therefore enter into the test results and rarely with dielectric materials does a clean scratch occur. Again the test becomes useful as a comparison between nominally similar coatings rather than an absolute one. Goldstein and DeLong have had some success recently in the assessment of dielectric films using microhardness testers to scratch the films[33]. Clearly, there is still a definite need for a useful quantitative test that would measure the work of adhesion for dielectric layers.

The chemical resistance of the film is also of some importance, particularly in connection with the effects of atmospheric moisture, to be considered later. In this latter respect, the solubility of the bulk material is a useful guide, although it should always be remembered that, in thin film form, the ratio of surface area to volume is extremely large and greatly magnifies any tendency towards solubility present in the bulk material. As in so many other thin-film phenomena, the magnitude of the effect depends very much on the particular thickness of material, on the other materials present in the multilayer, on the particular evaporation conditions, as well as the particular type of test used. However, a broad classification into moisture resistant (materials such as titanium oxide, silicon oxide and zirconium oxide), slightly affected (materials such as zinc sulphide), and badly affected (materials such as sodium fluoride), can be made.

TOXICITY

In thin-film work, as indeed in any other field where much use is made of a variety of chemicals, the possibility that a material may be toxic should always be borne in mind. Fortunately, most of the materials in common use in thin-film work are reasonably innocuous, but there are occasions where distinctly hazardous materials must be used. The thin-film worker would be wise to check this point before using a new material. The technical literature on thin films, being primarily concerned with physical and chemical properties, seldom mentions the toxic nature of the materials. For example, thorium fluoride, oxyfluoride and oxide, are materials that are extensively covered in the literature, but it is only rarely mentioned that these materials are radioactive[34], and that appropriate precautions should be taken. Some of the thallium salts are useful infrared materials, but these compounds are particularly toxic. Fortunately, manufacturers' literature is becoming a useful source of information on toxicity, and , in any cases of doubt, the manufacturer should always be consulted.

As long as toxic material is confined to a bottle there is little danger, but as soon as the bottle is opened, material can escape. A major objective, in the use of toxic materials, is to confine them in a well defined space, in which suitable precautions may be taken. If material is allowed to escape from this space, so that dangerous concentrations can exist outside, then it may be impossible to prevent an accident. It may be necessary to include the whole laboratory in the danger zone and to take special precautions in cleaning up on leaving. Special clothing, extending to respirators, may even be required while in the laboratory. On the other hand, plants may be isolated from the remainder of the production area by special dust-containing cabinets, complete with air circulation and filtration units. Most of the material evaporated in a process ends up as a coating on the inside of the plant and on the jigs and fixtures, where it usually forms a powdery deposit. The greatest danger is in the subsequent cleaning. Some of the solvents and cleaning fluids which can be used in the process give off harmful vapours. A good rule when dealing with potentially hazardous chemicals is to limit the total quantity on the premises to a minimum. This puts an upper bound on the magnitude of any major disaster but also, even if no other precautions are taken, minimises any leakage. It is also good from the psychological point of view. It should also be remembered that many poisons are cumulative in action, and while a slight dose received in the course of a short experiment may not be particularly harmful, the same dose, repeated many times in the course of several years, may do irreparable damage. Thus, the research worker may get away with a particular process which he operates only enough times to prove it, but the production worker will be expected to operate this process day in and day out, possibly for years. The safety standards in the production shop must therefore be of the highest standard. The thin-film worker in industry should make certain that the medical officer of the works is fully aware of the materials currently in use, so that any necessary precautions can be taken before any trouble occurs.

In general, unless positively dangerous materials are involved, the same precautions should be taken as in any chemical laboratory. There is a number of textbooks[35,36] that are useful sources of information on the subject.

SUMMARY OF SOME OF THE PROPERTIES OF THE COMMON MATERIALS

So far, little has been said about the actual properties of the more useful materials employed in thin-film work. The list which follows is far from being exhaustive, but gives the more important properties of some commonly used materials. Often the properties of a particular material appear to vary from plant to plant and sometimes even from operator to operator. This is a symptom of the lack of tight control, which is unfortunately a frequent feature of optical thin-film work, and generally the worker should measure the particular parameters in his own plant and process. Published figures tend to be more of a guide than anything else. The lack of control, of course, could be altered, but only with the expenditure of much time and money, which always poses the question whether the market for thin-film products is sufficiently large to justify the outlay.

The material probably used more than any others in thin-film work is magnesium fluoride. This has an index of approximately 1.39 in the visible (see figure 9.9) and is used extensively in lens blooming. In the simplest case this is generally a single layer. Early workers used fluorite as the blooming material but this was found to be rather soft and vulnerable and was subsequently replaced by magnesium fluoride. Magnesium fluoride can be evaporated from a tantalum or molybdenum boat, and the best results are obtained when the substrate is hot at a temperature of perhaps 200°C or so. When magnesium fluoride is evaporated, trouble can sometimes be experienced through spitting and flying out of material from the boat. This is thought to be caused by thin coatings of magnesium oxide round the grains of magnesium fluoride in the evaporant. Magnesium oxide has a rather higher melting point than magnesium fluoride and the grains tend to explode once they have reached a certain temperature. It is important, therefore, to use a reasonably pure grade of material, preferably one which is specifically intended for blooming.

Probably the easiest materials of all to handle are zinc sulphide and cryolite. They have a good refractive index contrast in the visible, the index of zinc sulphide being around 2.35 and that of cryolite around 1.35. Both materials sublime rather than melt, and can be deposited from a tantalum or molybdenum boat or else from a howitzer (described on p 365). Although these materials are not particularly robust, they are so easy to handle that they are very much used, especially in the construction of multilayer filters for the visible and near infrared which can subsequently be protected by a cemented cover slip. The substrates need not be heated for the deposition of the materials when intended for the visible. Zinc sulphide is also a particularly useful material in the infrared out to about 25 μm. In the infrared, however, the substrates must be heated for the best performance. The conditions are given in the paper by Cox and Hass[2], who state the best conditions to be on substrates which have been heated to around 150°C and cleaned with an effective glow discharge just prior to the evaporation and certainly not more than 5 minutes beforehand. Films produced under these conditions will withstand several hours' boiling in 5% salt water, exposure to humid atmospheres and cleaning with detergent and cotton wool.

A trick, which has sometimes been used with zinc sulphide to improve its durability, is bombardment of the growing film with electrons. This can be achieved by positioning a negatively biased hot filament, somewhere near the substrate carrier, in such a way that the filament is shielded from the arriving evaporant, but is in line of sight of the substrates. This process is still not entirely understood, but a fairly recent paper[37] suggests that an important factor is the modification of the crystal structure of the zinc sulphide layers by electron bombardment. Resistively heated boats produce a mixture of the cubic zincblende and the hexagonal wurtzite structure, while electron-beam sources produce purely the zincblende modification. The hexagonal form is a high temperature modification which, it is suspected, will tend to transform into the lower temperature cubic modification, particularly when water vapour is present, a transformation accompanied by a weakening of adhesion, and even delamination. Deliberate electron bombardment of growing zinc sulphide films from boat sources results in films with entirely cubic structure and with the improved stability expected from that structure.

For more durable films in the visible region, use can be made of cerium dioxide as the high-index layer and magnesium fluoride as the low-index layer. Cerium dioxide is readily evaporated from a tungsten boat (it reacts strongly with molybdenum, producing dense white powdery coatings which completely cover the inside of the system). The procedure to be followed is given by Hass, Ramsey and Thun[38]. Unless the material is one of the types prepared especially for vacuum evaporation, it should first be fired in air at a temperature of around 700–800°C. If this procedure is not followed the films will have a lower refractive index. Even with these precautions cerium dioxide is an awkward material to handle. It tends to form inhomogeneous layers and the index varies throughout the evaporation cycle as the material in the tungsten boat is used up. It is therefore difficult to achieve a very high performance from cerium dioxide layers, in terms of maximum transmission from a filter or from an antireflection coating, and the chief use is in the production of high-reflectance coatings, for high-power lasers for example, where high reflectance coupled with low loss is the primary requirement and transmission in the pass region is not as important.

Titanium dioxide is extremely robust but has a rather high melting point of 1925°C which makes it very difficult to evaporate directly. One of the most successful methods[39] has been the initial evaporation of pure titanium metal which is then subsequently oxidised in air by heating it to temperatures of 400–500°C. To obtain the highest possible index it is important to evaporate the titanium metal as quickly as possible at as low a pressure as possible so that little oxygen is dissolved in the film. On oxidation in air, indices of around 2.65 can be attained. If the deposit is partially oxidised beforehand, the index is usually rather lower, of the order of 2.25. The titanium metal is best evaporated from a tungsten boat. Other methods used involve the reaction between atmospheric moisture and titanium tetrachloride. Titanium dioxide is formed when atmospheric moisture mixes with the vapour of hot titanium tetrachloride and can be made to condense on the surface of a component which is introduced into the vapour. Best results on glass are obtained when the temperature of the glass is maintained at around 200°C. Both of these methods are useful for single layers but are almost impossibly complicated where multilayers are required. A possible alternative method which has been used with success is reactive sputtering. Sputtering is the process of bombardment of the material to be deposited with high-energy positive ions so that molecules are ejected and deposited on the substrate. Reactive sputtering is the same process except that the gas in the chamber is one which can and does react with the material as it is sputtered. Usually this gas is oxygen and in this case it reacts with the titanium to produce titanium dioxide without requiring any subsequent oxidation.

More recently another method known as reactive evaporation[40–42] has been developed (initially by M Auwärter and colleagues). The problem with the direct evaporation of titanium dioxide is that the very high temperatures which are required cause the titanium dioxide to be reduced so that absorption appears in the film. It has been found that the reduced titanium oxide can be reoxidised to titanium dioxide during the deposition by ensuring that there is sufficient oxygen present in the atmosphere within the chamber. The oxidation takes place actually on the surface of the substrate rather than in the vapour stream, and the pressure of the residual atmosphere of oxygen must be arranged to be high enough for the necessary number of oxygen molecules to collide with the substrate surface. If the pressure is too high, then the film becomes porous and soft. There is therefore a range of pressures over which the process works best, usually 5×10^{-5} to 2×10^{-4} Torr. However, it is not possible to give hard and fast figures because they vary from plant to plant and depend on the particular evaporation conditions such as substrate temperature and speed of evaporation. The conditions must therefore be established by trial and error in each plant. Because it melts at a rather lower temperature than either the pure metal or the dioxide, titanium monoxide is the usual form of evaporant. A tungsten boat must be used and the evaporation should proceed slowly enough to ensure that complete oxidisation takes place. This means that several minutes should be allowed for a thickness corresponding to a quarter-wave in the visible region. Provided the rate of evaporation is kept substantially constant then the refractive index of the film is constant at a value of around 2.25 in the visible region. The titanium dioxide remains transparent throughout the visible, the absorption in the ultraviolet becoming intense at around 350 nm.

The most complete account of the properties of titanium dioxide, and the way in which they depend on deposition conditions, is that of Pulker, Paesold and Ritter[43]. The behaviour is exceedingly complicated and the results depend on starting material, oxygen pressure, rate of deposition, and substrate temperature. The evaporation of Ti_3O_5 as starting material gave more consistent results than were obtained with other possible starting materials, probably because the composition of the material in the source did not vary during the course of successive evaporations. With other forms of titanium oxide, the composition varied as the material was used up, tending in each case towards Ti_3O_5.

Apfel[44] has pointed out the slight conflict between high optical properties and durability. Optical absorption falls as the substrate temperature is reduced and the residual gas pressure is raised. At the same time, the durability of the layers is adversely affected, and a compromise, which depends on the actual application, is usually necessary. Substrate temperatures between 200–300°C are usually satisfactory, with gas pressures around 10^{-4} Torr (1.3×10^{-4} mb).

An exactly similar reactive method can be used for silicon oxide[40,42]. Silicon monoxide is a convenient starting material which, in its own right, is a useful material for the infrared. The silicon monoxide can be evaporated readily from a tantalum boat or, as the material sublimes rather than melts, a howitzer source. Provided there is sufficient oxygen present, the silicon monoxide will oxidise to the form Si_2O_3 that has a refractive index of 1.52–1.55 and exhibits excellent transmission from just on the longwave side of 300 nm out to 8 μm[45].

An interesting effect involving the ultraviolet irradiation of films of Si_2O_3 has been reported[46]. With ultraviolet intensity corresponding to a 435 W quartz-envelope Hanovia lamp at a distance of 20 cm, the refractive index of the film, after around 5 hours' exposure, drops to 1.48 (at 540 nm). This change in refractive index appears to be due to an alteration in the structure of the film, rather than in the composition, that remains Si_2O_3. At the same time as the reduction in refractive index, an improvement in the ultraviolet transmission is observed, the films becoming transparent to beyond 200 nm. Longer exposure to ultraviolet, around 150 hours, does eventually alter the composition of the films to SiO_2. These changes appear to be permanent. Si_2O_3 is a particularly useful material for protecting aluminium mirrors, and this method of improvement by ultraviolet irradiation opens the way to greatly improved mirrors for the quartz ultraviolet[47].

Heitmann[48] has made considerable improvements to the reactive process by ionising the oxygen in a small discharge tube through which the gas is admitted to the coating chamber. The degree of ionisation is not high, but the reactivity of the oxygen is improved enormously, and the titanium oxide and silicon oxide films produced in this way have appreciably less absorption than those deposited by the conventional reactive process. The silicon oxide films show infrared absorption bands characteristic of the SiO_2 form rather than the more usual Si_2O_3. The technique has been further improved by Ebert[49] and his colleagues who have developed a more efficient hollow-cathode ion source, and extended the method to materials such as beryllium oxide, with useful transmittance in the ultraviolet.

Other materials which have been found useful in thin films are the oxides and fluorides of a number of the rare earths. Ceric oxide, although possibly strictly not a rare earth, has already been mentioned[38]. Cerium fluoride forms very stable films of index 1.63 at 550 nm when evaporated from a tungsten boat. Similarly, the oxides of lanthanum, praseodymium and yttrium, and their fluorides, form excellent layers when evaporated from tungsten boats. Their properties are summarised in the appendix. A full account of their properties is given by Hass, Ramsay and Thun[9]. The properties of the rare earth oxides have been the subject of a recent study[50] that has shown that better transparency, especially in the ultraviolet, is obtained when electron-beam evaporation is used.

Then there is a number of other hard oxide materials which were extremely difficult to evaporate until the advent of the high-power electron-beam gun, and so were used only relatively infrequently, if at all. Zirconium dioxide[50,51] is a very tough, hard material which has good transparency from around 350 nm to some 7 μm. It tends to give inhomogeneous layers, the degree of inhomogeneity depending principally on the substrate temperature. Hafnium oxide[50,52] has good transparency to around 235 nm, and an index around 2.0 at 300 nm, so that it is a good high-index material for that region. Both yttrium and hafnium oxide have been found to be good protecting layers for aluminium in the 8–12 μm region[53,54], which avoid the drop in reflectance at high angles of incidence associated with SiO_2 and with Al_2O_3.

In the infrared many more possibilities are available. Semiconductors all exhibit a sudden transition from opacity to transparency at a certain wavelength known as the intrinsic edge. This wavelength corresponds to the energy gap between the filled valence band of electrons and the empty conduction band.

At wavelengths shorter than this gap, quanta are absorbed in the material because they are able to transfer their energy to the electrons in the filled valence band by lifting them into the empty conduction band. At wavelengths longer than this value, the quantum energy is not sufficient, and apart from a little free-carrier absorption, there is no mechanism for absorbing the energy and the material appears transparent until the lattice vibration bands at rather long wavelengths are needed. For the commoner semiconductors, silicon and germanium, this wavelength is 1.1 μm and 1.65 μm respectively. Thus both of these materials are potentially useful in the infrared. The great advantage which they possess is their high refractive index, 3.5 for silicon and 4.0 for germanium. Silicon, however, is not at all easy to evaporate because it reacts strongly with any crucible material, and almost the only way of dealing with it is to use an electron gun with a water-cooled crucible so that the cold silicon in contact with the crucible walls acts as its own container. Germanium, on the other hand, is a most useful material and straightforward techniques have been devised to handle it. Tungsten boats can be used provided that the total thickness of material to be deposited is not too great, 2 or 3 μm say, because germanium reacts with tungsten. Molybdenum boats have been used with greater success[55].

The best method is, however, to use a crucible made from graphite and heated directly or indirectly when the germanium films obtained are extremely pure and free from absorption. An alternative method is electron bombardment when the hearth material can be graphite or water-cooled copper. There are other semiconductors of use: Tellurium[56,57] has an index of 5.1 at 5 μm, good transmission from 3.5 μm to at least 12 μm, and can be evaporated easily from a tantalum boat. Lead telluride[3,58–63] has an even higher index of around 5.5 with good transmission from 3.4 μm out to beyond 20 μm. A tantalum boat is the most suitable source. Care must be taken not to overheat the material; the temperature should be just enough to cause the evaporation to proceed, otherwise some alteration in the composition of the film will take place, causing an increase in free-carrier absorption and consequent fall-off in longwave transparency. The substrates should be heated, best results being obtained with temperatures around 250°C, but as this will be too great for the low-index film which is usually zinc sulphide, a compromise temperature which is rather lower, usually around 150°C, is often used for both materials. One difficulty with lead telluride is the ease with which it can be upset by impurities that cause free-carrier absorption. It is extremely important to use pure grades of material and this applies to the accompanying zinc sulphide as well as the lead telluride, especially if the material is to be used at the longwave end of its transparent region. Lead telluride also appears to be incompatible with a number of other materials, particularly some of the halides, presumably because material diffuses into the lead telluride generating free carriers. An annealing process which can in certain circumstances improve the transmission of otherwise absorbing films of lead telluride in the region beyond 12 μm is described by Evans and Seeley[59].

Lead telluride can in some circumstances behave in a curious way immediately after deposition[62,63]. The optical thickness of the material is observed to grow during a period of around 15 minutes while the layer is still under vacuum. Typical gains in optical thickness of a half-wave layer are of the order of 0.007 full waves, although in any particular case it varies considerably and can often be zero. The reasons for this behaviour are not clear but the layers do not exhibit any further instability, once they have ceased growing. It is simply a matter of allowing for this behaviour in the monitoring process.

A wide range of low index materials is used in the infrared. Zinc sulphide[2,10] in comparison with the high-index semiconductors has a relatively low index. If an electron-beam source is not available, then zinc sulphide should be deposited from a tantalum boat, or, better still, a howitzer, on substrates freshly cleaned by a glow discharge and held at temperatures of around 150°C, if the maximum durability is to be obtained. Zinc sulphide films so treated will withstand boiling for several hours in 5% salt solution, cleaning with cotton wool, and exposure to moist air, without damage[2]. Silicon monoxide is another possibility[2,64]. It can also be deposited from a tantalum boat or a howitzer. The deposition rate should be fast and the pressure low, of the order of 10^{-5} Torr (1.3×10^{-5} mb) or less if possible. The refractive index is around 1.85 at 1 μm and falls to 1.6 at 7 μm. A strong absorption band prevents use of the material beyond 8 μm. Thorium fluoride, unfortunately radioactive, is much used, and there are many other materials, such as fluorides of lead, lanthanum, barium, cerium, for example, and oxides such as titanium, yttrium, hafnium and cerium. Some details of these and other materials are given in the appendix.

Mixtures of materials are now receiving attention both in deliberately inhomogeneous films and in homogeneous films where an intermediate index between the two components of the mixture is required to improve the evaporation properties of an otherwise difficult material.

Jacobsson and Martensson[65] used mixtures of cerium oxide and magnesium fluoride, of zinc sulphide and cryolite, and of germanium and magnesium fluoride, with the relative concentration of the two components varying smoothly throughout the films to produce inhomogeneous films with a refractive index variation of a prescribed law. Some of the results they obtained for antireflection coatings were mentioned in chapter 3. To produce the mixture, two separate sources, one for each material, were used; they were evaporated simultaneously but with independent rate controls. Apparently no difficulty in obtaining reasonable films was experienced, the mixing taking place without causing absorption to appear.

Fujiwara[66,67] was interested in the production of homogeneous films for antireflection coatings[68]. The three-layer quarter–half–quarter coating for glass requires a film of intermediate index which is rather difficult to obtain with a simple material, and the solution adopted by Fujiwara was to use a mixture of two materials, one having a refractive index lower than the required value and the other higher. The two combinations which were tried successfully were cerium oxide and cerium fluoride[66,68], and zinc sulphide and cerium fluoride[67,68]. These were simply mixed together in powder form in a certain known proportion by weight and then evaporated from a single source. The mixture evaporated giving an index which was sufficiently reproducible for antireflection coating purposes. The range of indices obtainable with the cerium oxide–cerium fluoride mixture was 1.60–2.13, and with the cerium fluoride–zinc sulphide mixture 1.58–2.40. One interesting feature of the second mixture was that, although zinc sulphide on its own is not particularly robust, in the form of a mixture with more than 20 % by weight of cerium fluoride the robustness was greatly increased, the films withstanding boiling in distilled water for 15 minutes without any deterioration. Curves are given for refractive index against mixing ratio in the papers.

Mixtures of zinc sulphide and magnesium fluoride have also been studied by Yadava, Sharma and Chopra[69]. The refractive index of the mixture varies between the indices of magnesium fluoride and zinc sulphide, depending on the mixing ratio, and the absorption edge varies from that of zinc sulphide to that of magnesium fluoride in a nonlinear fashion. The same authors[69,70] have studied the use of assemblies of large numbers of alternate very thin discrete layers of the components instead of mixtures. For a wide range of material combinations, $ZnS–MgF_2$, $ZnS–MgF_2–SiO$, Ge–ZnS, $ZnS–Na_3AlF_6$ for example, the results were similar to those expected from the evaporation of mixtures of the same materials.

Quartz is a particularly difficult material to evaporate because of its high melting point and also because of its transparency to infrared which makes it difficult to heat. It was found by the Libbey–Owens–Ford Glass Company[71] that quartz could be thermally evaporated readily if some pretreatment was carried out. This consisted of combining the quartz with a metallic oxide, a vast number of different oxides being suitable; the oxide can be mixed intimately with the quartz, coated on the outer surface of quartz chunks or, in some cases where the oxide has a rather lower melting temperature than the quartz, mixed very crudely. Only a small quantity of the oxide is required and the evaporation is carried out in the conventional manner from a tungsten source. The oxides mentioned include aluminium, titanium, iron, manganese, cobalt, copper, cerium and zinc. Working along similar lines it has been discovered by workers at Balzers AG[72,73] that cerium oxide mixed with other oxides improves the oxidation and increases the transparency and ease of evaporation. Materials such as titanium dioxide are difficult to evaporate without absorption, and the most successful method is reactive evaporation in oxygen, which produces absorption-free films, although the process is rather time consuming because the evaporation must proceed slowly. With the addition of a small amount of cerium oxide—the mixture can vary from 1 : 1 to 8 : 1 titanium oxide (the monoxide, the dioxide or even the pure metal) to cerium oxide—hard films free from absorption even when evaporated quickly at pressures of 10^{-5} Torr are readily obtained. Apparently this effect is not limited to titanium oxide, and a vast range of different materials which have been successfully tried is given. Other rare earth oxides and mixtures of rare earth oxides can also take the place of the cerium dioxide.

Stetter and his colleagues[51] have pointed out the advantage of oxygen-depleted materials as source material for electron-beam evaporation, in that composition changes little if at all during evaporation, which leads to more consistent film properties. The extra oxygen is supplied, in the usual way, from the residual atmosphere in the plant. The depleted materials also have higher thermal and electrical conductivity. A mixture of ZrO_2 and $ZrTiO_4$, sintered at high temperature under high vacuum and oxygen-depleted, was developed. This material, designated 'Substance no 1', when evaporated from an electron-beam system in a residual oxygen pressure of $1–2 \times 10^{-4}$ Torr (1.3–2.5 $\times 10^{-4}$ mb) with substrate temperature 270°C, and condensation rate of the order of 10 nm min^{-1}, gives homogeneous layers of refractive index 2.15 (at 500 nm). Such a value of index is ideal for the quarter–half–quarter antireflection coating for the visible region.

Butterfield[74] has produced films of a mixture of germanium and selenium. For composition varying from 35–50 atomic % of germanium, glassy films with refractive index in the range 2.4–3.1, with good transparency from 1.5–15 μm, could be produced. The starting material was an alloy of germanium and selenium in the correct portions, produced by melting the pure substances in an evacuated quartz tube. The evaporation source was a graphite boat.

It is likely that much more work will be carried out on mixtures, because of the apparent ease with which the deposition can be performed to give a side range of refractive indices, many of which are not available by other means. The theory of the optical properties of mixtures is covered in a useful review by Jacobsson[75], who also gives further information on mixtures, and on inhomogeneous layers.

FACTORS AFFECTING LAYER PROPERTIES

One of the most significant features of optical thin films is the way in which their properties and behaviour differ from those of identical materials in bulk form. This is, of course, also true for thin films in areas other than optics. Almost always, the performance of the film is poorer than that of the corresponding bulk material. Refractive index is usually lower, although, very occasionally, for some semiconductor materials it can be slightly higher, losses greater, durability less, and stability inferior. There is also a sensitivity to deposition conditions, especially substrate temperature.

Heitman[76] has studied the influence of parameters, such as the residual gas pressure within the plant and the rate of deposition, on the refractive indices of cryolite and thorium fluoride. Raising the residual gas (nitrogen) pressure from 4×10^{-6} Torr (5.3×10^{-6} mb) in one case, and 2×10^{-6} Torr (2.6×10^{-6} mb) in another, to 2×10^{-5} Torr (2.6×10^{-5} mb) had no measurable effect, within the accuracy of the experiment ($\pm 0.1\%$ for thorium fluoride and $\pm 0.3\%$ for cryolite) while a further increase in residual pressure to 2×10^{-4} Torr (2.6×10^{-4} mb) gave a drop in index of 1.5% for cryolite, and 1.4% for thorium fluoride. At this higher pressure, the mean free path of the nitrogen molecules was less than the distance between boat and substrate, and the decrease in refractive index was probably caused by increased porosity of the layers. This tends to confirm that the mean free path of the residual gas molecules should be kept longer than the source–substrate distance, but that any further increases in mean free path beyond this have little effect. Heitman concluded that the mean free path of the molecules is the important parameter, not the ratio of the numbers of evaporant molecules to residual gas molecules impinging on the substrate in unit time, which appeared to have no effect on refractive index. He also found that changes in the rate of deposition, from a quarter-wave in 0.5 minutes (measured at 632.8 nm) to a quarter-wave in 1.5 minutes, caused a decrease in refractive index of 0.6% in both cases, but that a further decrease to a quarter-wave in 5 minutes produced only slight variations.

Heitman's results are probably best interpreted in terms of slight changes in film structure, induced by the variations in deposition conditions. Layer structure is, in fact, the most significant factor in determining the properties of optical thin films and the way in which they differ from the same material in bulk form. During the past ten years or so, there has been an increasing interest in the structure of, and structural effects in, optical thin films.

A useful technique for the study of thin-film structure is electron microscopy. The examination of thin-film coatings has involved the development of techniques for fracturing multilayers and for replicating the exposed sections. Pearson, Lissberger, Pulker and Guenther[77–80] have all made substantial contributions in this area and their results show that the layers in optical coatings have, almost invariably, a pronounced columnar structure, with the columns running across the films normal to the interfaces. To their investigations, we can add those of Movchan and Demchishin[81] and then Thornton[82], who investigated the effects of substrate temperature and, in Thornton's case, residual gas pressure, on the structure of evaporated and sputtered films. This showed that a critical parameter in vacuum deposition of thin films is the ratio of the temperature of the substrate T_s to the melting temperature T_m of the evaporant. For values of this ratio lower than around 0.5, the structure of the layers is intensely columnar, the columns running along the direction of growth. Increased gas pressure forces the growth into a more pronounced columnar mode even for slightly higher values of substrate temperature.

Because the most useful materials in optical thin films are all of high melting point, substrate temperatures can never be higher than a small fraction of the evaporant melting temperature, and so the structure of thin films is almost invariably a columnar one, with the columns running along the direction of growth, normal to the film interfaces. The columns are several tens of nanometres across and roughly cylindrical in shape. They are packed in an approximately hexagonal fashion with gaps in between the columns, which take the form of pores running completely across the film, and there are large areas of column surface which define the pores and are in this way exposed to the surrounding atmosphere.

Packing density p defined as:

$$p = \frac{\text{Volume of solid part of film (i.e. columns)}}{\text{Total volume of film (i.e. pores plus columns)}}$$

is a very important parameter. It is usually in the range 0.75–1.0 for optical thin films, most often 0.8–0.95, and seldom as great as unity. A packing density that is less than unity reduces the refractive index below that of the solid material of the columns. A useful expression that is reasonably accurate for films of low index[83,84] connects the indices of the film n_f, those of the solid part of the film n_s and the voids n_v, and the packing density p:

$$n_f = pn_s + (1 - p)n_v. \qquad (9.25)$$

For films of higher index, 2.0 and above, the expression is no longer accurate, but nevertheless is still often employed. If the value of packing density has been derived from optical measurements by using equation (9.25), as is frequently the case, then, of course, the expression can, and should be, used. In any event, it gives an indication of the correct trend. For an alternative expression that is more complicated and gives a better fit, although still far from ideal, the paper by Harris and colleagues[84] should be consulted.

Packing density is a function of substrate temperature, usually, but not always, increasing with substrate temperature, and of residual gas pressure, decreasing with rising pressure. Film refractive index, therefore, is also affected by substrate temperature and residual gas pressure. The columns frequently vary in cross-sectional area as they grow outwards from the substrate surface, which is a major cause of film inhomogeneity. Substrate temperature is a difficult parameter to measure and to control so that consistency in technique, heating for the same period each batch, identical rates of deposition, pumping for the same period before commencing deposition and so on, is of major importance in assuring a stable and reproducible process. Changing the substrate dimensions, especially substrate thickness, from one run to the next can cause appreciable changes in film properties. Such changes are even more marked in the case of reactive processes where the residual gas pressure is raised, and where a reaction between evaporant and residual atmosphere takes place at the growing surface of the film. Thus it should not be surprising that a very high proportion of test runs are required in any manufacturing sequence.

The variation in refractive index is not the only feature of film behaviour associated with the columnar structure. The pores between the columns permit the penetration of atmospheric moisture into the film, where, at low relative humidity, it forms an adsorbed layer over the surfaces of the columns and, at medium relative humidity, actually fills the pores with liquid water due to capillary condensation. Moisture adsorption has been the subject of considerable study by Ogura[85,86], who used the variation in adsorption with relative humidity to derive information on the pore structure of the films. The moisture, since it has a different refractive index (around 1.33) from the 1.0 of the air which it displaces from the voids, causes an increase in the refractive index of the films. Since the geometrical thickness of the film does not change, the increase of film index during adsorption is accompanied by a corresponding increase in optical thickness. Exposure of a film to the atmosphere, therefore, usually results in a shift of the film characteristic to a longer wavelength. Such shifts in narrowband filters have been the subject of considerable study. Schildt, Steudel and Walther[87] found that for freshly prepared filters of zinc sulphide and magnesium fluoride, constructed for the region 400–500 nm, the variation in peak wavelength could be expressed as

$$\Delta\lambda = q \log_{10} P$$

where q is a constant varying from around 1.4 for filters which had aged to around 8.3 for freshly prepared filters, and P is the partial pressure of water vapour measured in Torr (P should be replaced by 0.76 P if P is measured in mb) and $\Delta\lambda$ is measured in nm. $\Delta\lambda$ was arbitrarily chosen as zero when the pressure was 1 Torr (1.3 mb). This relationship was found to hold good for the pressure range 1 to approximately 20 Torr (1.3–26 mb). The filters settled down to the new values of peak wavelength some 10–20 minutes after exposure to a new level of humidity began. They found that the shifted values of peak wavelength could be stabilised by cementing cover slips over the layers using an epoxy resin. Koch[88,89] showed that the characteristics of narrowband filters became quite unstable during adsorption until the filters reached an equilibrium state. Macleod and Richmond[90], Richmond[91] and Lee[92] have made detailed studies of the effects of adsorption on the characteristics of narrowband filters. The results are applicable to all types of multilayer coating. The shifts in the characteristics are due, as we have seen, to the filling of the pores of the film with liquid water. In multilayers, the pores of one film are not always directly connected with the pores of the next, and the penetration of atmospheric moisture is frequently a slow and complex process in which a limited number of penetration pores take part from which the moisture spreads across the coating in increasing circular patches. The coating may take several weeks to reach equilibrium and, afterwards, will exhibit some instability should the environmental conditions change. The patches, which can sometimes be seen with the naked eye as a flecked or mottled appearance, can be made more visible if the coating is viewed in monochromatic light, at or near a wavelength for which there is a rapid variation of transmittance. The edge of an edge filter, or the pass band of a narrowband filter, are especially suitable. Wet patches show a shift in wavelength which changes them from high to low transmittance, or vice versa, and they can be readily photographed as was done for figures 9.16 and 9.17.

The drift of the filters towards longer wavelengths, which occurs on exposure to the atmosphere, varies considerably in magnitude with both the materials and the spectral region and there is frequently considerable hysteresis on desorption. In the infrared the layers are thick, and many of the semiconductor materials that are used as high-index materials have high packing density. This means that moisture-induced drift is less of a general problem than it is in the visible and ultraviolet regions of the spectrum, although it is important in some applications. In the visible region, drifts can be as high as 10 nm, and sometimes greater, towards longer wavelengths. The gradual stabilisation of the coating as it reaches equilibrium is frequently referred to as aging or settling.

It is not simply in generating optical shifts that moisture is a problem for coatings. It has major mechanical and sometimes chemical effects as well. The stress in the coating is transmitted across the gaps between the columns, again by short-range forces. These forces can be very easily blocked by water molecules. An alternative explanation of the phenomenon is that the moisture which coats the surfaces of the columns reduces the surface energy to something approaching that of liquid water. Since the surface energy is an important factor in the stress/strain balance in the film, the result of the moisture adsorption is a change in the stress level. The stress is usually tensile and the moisture reduces it, usually significantly. We have already mentioned Pulker's work on impurities in thin films and their reduction of stress levels in a similar way[25]. Adhesion, too, is affected by moisture. The materials used for thin films have usually very high surface energies and then the work of adhesion is correspondingly high. The presence of liquid water in a film can cause a reduction in the surface energy of the exposed surfaces of at least an order of magnitude. If water is present at the site of an adhesion failure and can take part in a process of bond transfer, rather than bond rupture followed by adsorption, then it will reduce the work of adhesion, and it is more likely that the failure will propagate. There is frequently enough strain energy in a film to supply the required work. The penetration sites for the moisture patches are probably associated with defects which may act as stress concentrators where adhesion failures driven by the internal strain energy in the films may originate. All the ingredients for a moisture-assisted adhesion failure are present and it is frequently at such sites that delamination is first observed. Blistering is a similar form of adhesion failure frequently associated with moisture penetration sites.

We have already mentioned in chapter 7 that changes in temperature cause changes in the spectral characteristics of coatings, narrowband filters having characteristics which are probably most sensitive to such alterations. For small temperature changes, the principal effect is a simple shift towards longer wavelengths with increasing temperature. For the materials commonly used in the visible region of the spectrum, the shift is of the order of 0.003 % $^\circ C^{-1}$, while for infrared filters it can be greater, and a useful figure is 0.005 % $^\circ C^{-1}$, although it can be as high as 0.0125 % $^\circ C^{-1}$. It must be emphasised that these figures depend strongly on the particular materials used. Filters of lead telluride and zinc sulphide can actually have negative coefficients greater than 0.01 % $^\circ C^{-1}$ and, using these materials, it is even possible to design a filter that has zero temperature coefficient[61].

With greater positive changes of, say, 60 °C or more, it is usual for the moisture in the filter to desorb partially, causing an abrupt shift towards shorter wavelengths (see figure 9.18). This shift is not recovered immediately on cooling to room temperature, and so considerable hysteresis is apparent in the behaviour[93]. Subsequent temperature cycling, before readsorption of any

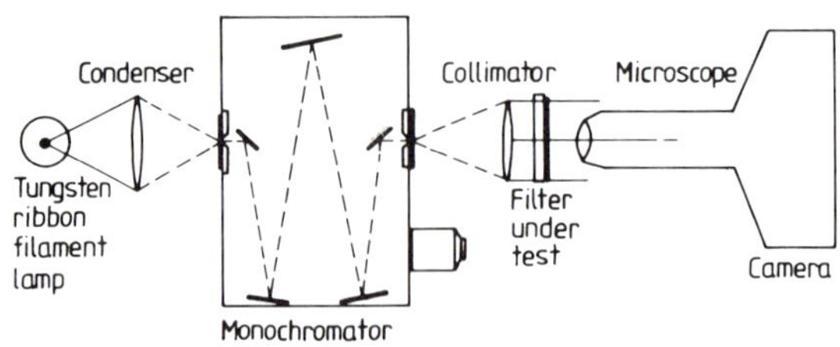

Figure 9.16 Moisture penetration patterns in a multilayer of zinc sulphide and cryolite. (*a*) Sketch of the apparatus for observing the phenomenon. Short slits that are virtually pin holes are used in the monochromator. (After Macleod and Richmond[90].)

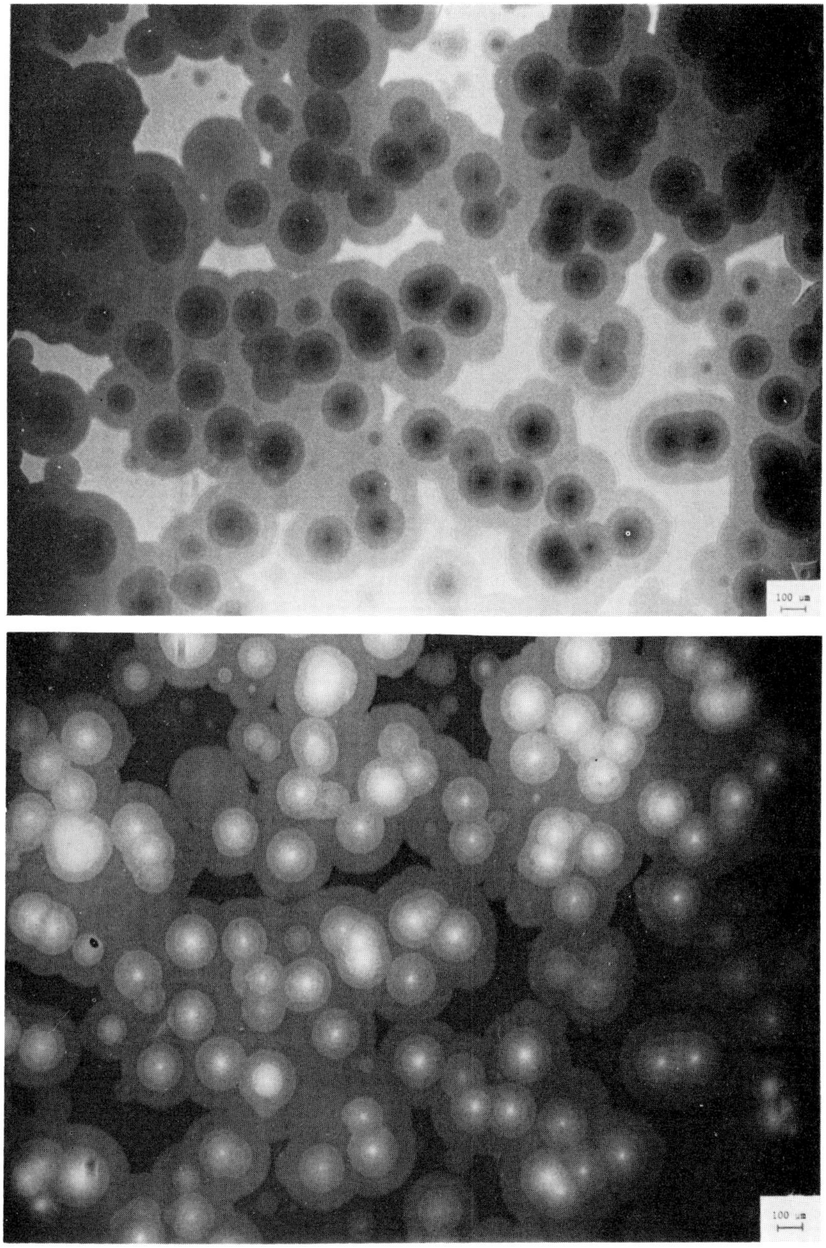

(*b*) Photograph of moisture penetration patterns in a zinc sulphide and cryolite filter some two weeks after coating. The relative humidity was approximately 50 % during this time. The upper photograph was taken at a wavelength of 488.5 nm and the lower at 512.8 nm. The dark patches of the upper photograph correspond to the light patches of the lower showing that a wavelength shift rather than absorption is responsible for the patterns. (After Lee[92].)

Figure 9.17 Mosture penetration patterns in a multilayer of zirconium dioxide and silicon dioxide. The photographs were taken immediately after removal from the coating chamber. The wavelength for the upper photograph was 543 nm, and that for the lower, 553 nm. (After Lee[92].)

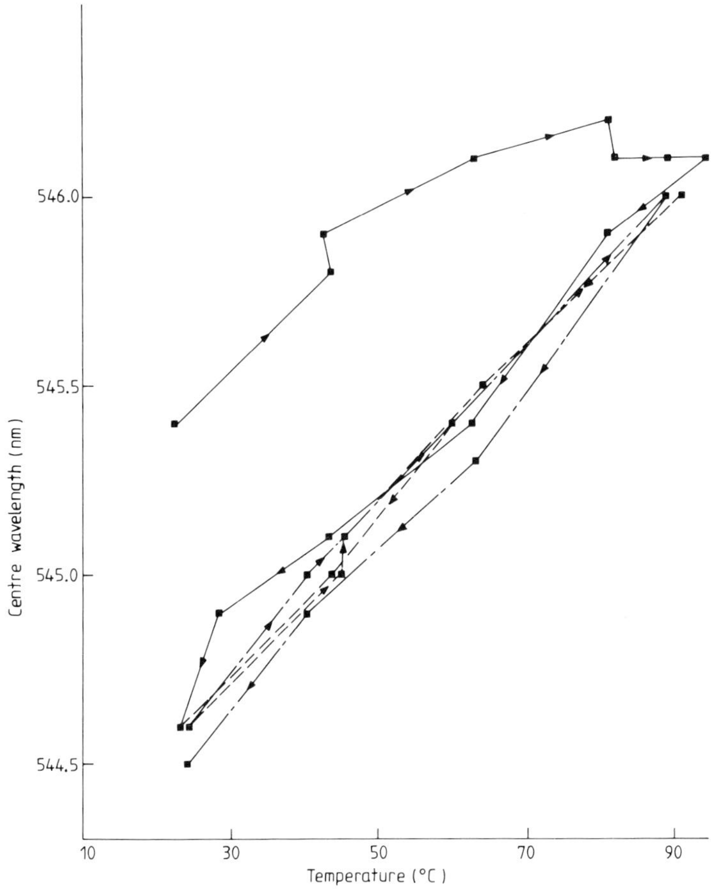

Figure 9.18 Record of the variation of peak wavelength with temperature for a filter of design:

$$\text{Air} \mid (HL)^4 6H\,(LH)^4 \mid \text{Glass}$$

with L = cryolite and H = zinc sulphide. (After Roche *et al*[93].)

moisture, will then exhibit no hysteresis. Eventually, if maintained at room temperature, the filter will readsorb moisture and drift gradually back to its initial wavelength. Exposure to higher temperatures still, over 100°C, can cause permanent changes which appear to be related to minute alterations in the structure of the layers, altering the adsorption behaviour so that some materials become less ready to adsorb moisture while others show more rapid adsorption[90-92].

Coatings which are subjected to very low temperatures usually shift towards shorter wavelengths, consistent with their behaviour at elevated temperatures.

Filters are not usually affected mechanically except for laminated components that run the risk of breaking because of differential contraction and/or expansion.

There are losses associated with all layers which can be divided into scattering and absorption. In absorption, the energy which is lost from the primary beam is dissipated within the coating and usually appears as heat. In scattering, the flux lost is deflected and re-emerges from the coating in a different direction. Absorption is a material property which may be intrinsic or due to impurities. A deficiency of oxygen, for example, can cause absorption in most of the refractory oxide materials. Scattering is usually due to defects in the coating which can be classified into volume or surface defects. Surface defects are simply a departure from the smooth flat surfaces of the ideal film. Such departures can be due to roughness of the substrate surface which tends to be reproduced at each interface in a multilayer, or to the columnar structure of the layers which results in a nodular appearance of the film boundaries. Volume defects are local variations of optical constants and are usually dust particles, pinholes or fissures in the coating.

Losses in thin films are of particular importance in the laser field where they determine the limiting performance of multilayers. A major problem in the production of high quality laser coatings is dust which emanates from the sources and from the powdery deposit which forms on the cold walls of the chamber. If this dust can be eliminated, only possible if the strictest attention is paid to detail and the most involved precautions are taken, then the remaining source of scattering loss is the roughness of the interfaces between the layers and between multilayer and substrate. If great care is exercised, then, in the visible and near infrared regions, the total losses, that is, absorption and scattering, can be reduced below 0.05% (for some very special applications losses towards one tenth of this figure have been achieved) and the power-handling capability of the coatings can be of the order of 4 J cm^{-2} for pulses of 1 ns or less at 1.06 μm. Recent studies of scattering in thin-film systems include references 94–99.

Laser damage is still very much a research topic with no universally accepted detailed models of damage mechanisms. It seems clear that thermal effects associated with absorption, either local or general, are the principal source of damage in continuous-wave applications. Small defects appear less important. In pulsed-laser applications, damage incidence seems to be related much more to the peak electric field to which the coating is subjected and the importance of small, local defects in the coatings is strongly suspected. In those spectral regions where water absorbs strongly considerable importance is now being attached to the presence of liquid water within the films. In other parts of the spectrum its role is less clear, but it may well play a part. Laser damage has been surveyed by Lowdermilk[100] and there is a series of annual conference proceedings on the subject, the most recent at the time of writing being reference 101.

FURTHER INFORMATION

Further information on the subject matter of this chapter will be found in a number of reviews and textbooks. Maissel and Glang[102] is an excellent source of information on film preparation and property measurement. Although optical coatings are not explicitly considered, the information applies equally well to them and there is a section on optical characterisation. Ritter[103,104] gives much useful information in two reviews of optical thin-film materials. The book by Holland[105], although written some thirty years ago, is still very useful. The book by Pulker[106], just published while this book was in proof, is an exceptionally useful and comprehensive source on the subject of this chapter.

REFERENCES

1 Vossen J L and Kern W (eds) 1978 *Thin Film Processes* (New York: Academic)
2 Cox J T and Hass G 1958 Antireflection coatings for germanium and silicon in the infrared *J. Opt. Soc. Am.* **48** 677–80
3 Ritchie F S 1970 Multilayer filters for the infrared region 10–100 microns *PhD Thesis* University of Reading
4 Coulter J K, Hass G and Ramsay J B 1973 Optical constants and reflectance and transmittance of evaporated rhodium films in the visible *J. Opt. Soc. Am.* **63** 1149–53
5 Hass G and Ritter E 1967 Optical film materials and their applications *J. Vac. Sci. Technol.* **4** 71–9
6 Heavens O S 1964 Measurement of optical constants of thin films in *Physics of Thin Films* **2** 193–238
7 Liddell H M 1981 *Computer-aided techniques for the design of multilayer filters* (Bristol: Adam Hilger)
8 Hall J F and Ferguson W F C 1955 Dispersion of zinc sulphide and magnesium fluoride films in the visible spectrum *J. Opt. Soc. Am.* **45** 74–5
9 Hass G, Ramsay J B and Thun R 1959 Optical properties of various evaporated rare earth oxides and fluorides *J. Opt. Soc. Am.* **49** 116–20
10 Hall J F and Ferguson W F C 1955 Optical properties of cadmium sulphide and zinc sulphide from 0.6 micron to 14 micron *J. Opt. Soc. Am.* **45** 714–18
11 Pelletier E, Roche P and Vidal B 1976 Determination automatique des constantes optiques et de l'épaisseur de couches minces: application aux couches diélectriques *Nouv. Rev. Opt.* **7** 353–62
12 Manifacier J C, Gasiot J and Fillard J P 1976 A simple method for the determination of the optical constants *n*, *k*, and the thickness of a weakly absorbing thin film *J. Phys. E: Sci. Instrum.* **9** 1002–1004
13 Hansen W 1973 Optical characterization of thin films: theory *J. Opt. Soc. Am.* **63** 793–802
14 Abelès F 1950 La détermination de l'indice et de l'épaisseur des couches minces transparentes *J. Phys. Rad.* **11** 310–14
15 Hacskaylo M 1964 Determination of the refractive index of thin dielectric films *J. Opt. Soc. Am.* **54** 198–203

16 Jacobsson R 1975 Inhomogeneous and coevaporated homogeneous films for optical applications in *Physics of Thin Films* ed G Hass, M H Francombe and R W Hoffman **8** 51–98
17 Borgogno J P, Lazarides B and Pelletier E 1982 Automatic determination of the optical constants of inhomogeneous thin films *Appl. Opt.* **21** 4020–29
18 Netterfield R P 1976 Refractive indices of zinc sulphide and cryolite in multilayer stacks *Appl. Opt.* **15** 1969–73
19 Hoffman R W 1966 The mechanical properties of thin condensed films in *Physics of Thin Films* ed G Hass and R E Thun **3** 211–73
20 Hoffman R W 1976 Stresses in thin films: the relevance of grain boundaries and impurities *Thin Solid Films* **34** 185–90
21 Ennos A E 1966 Stresses developed in optical film coatings *Appl. Opt.* **5** 51–61
22 Roll K and Hoffman H 1976 Michelson interferometer for deformation measurements in an UHV system at elevated temperatures *Rev. Sci. Instrum.* **47** 1183–5
23 Ledger A M and Bastien R C 1977 Intrinsic and thermal stress modeling for thin-film multilayers *Tech. Rep.* Contract DAA25-76-C-0410 (DARPA) (Norwalk, Conn: The Perkin Elmer Corporation)
24 Glass A J and Guenther A H 1978 Laser induced damage in optical materials: ninth ASTM symposium *Appl. Opt.* **17** 2386–2411. See p 2398 for summary of Ledger and Bastien results
25 Pulker H K 1982 Stress, adherence, hardness and density of optical thin films in *Optical Thin Films* ed R I Seddon (*Proc. SPIE* **325** 84–92)
26 Jacobsson R and Kruse B 1973 Measurement of adhesion of thin evaporated films on glass substrates by means of the direct pull method *Thin Solid Films* **15** 71–7
27 Heavens O S 1950 Some features influencing the adhesion of films produced by vacuum evaporation *J Phys. Rad.* **131** 355–60
28 Benjamin P and Weaver C 1960 Measurement of adhesion of thin films *Proc. R. Soc.* A **254** 163–76
29 Benjamin P and Weaver C 1960 Adhesion of metal films to glass *Proc. R. Soc.* A **254** 177–83
30 Benjamin P and Weaver C 1963 The adhesion of metal to crystal faces *Proc. R. Soc.* A **274** 267–73
31 Laugier M 1981 The development of the scratch test technique for the determination of the adhesion of coatings *Thin Solid Films* **76** 289–94
32 Laugier M 1981 Unusual adhesion-aging behaviour in ZnS thin films *Thin Solid Films* **75** L19–L20
33 Goldstein I S and DeLong R 1982 Evaluation of microhardness and scratch testing for optical coatings *J. Vac. Sci. Technol.* **20** 327–30
34 Heitman W and Ritter E 1968 Production and properties of vacuum evaporated films of thorium fluoride *Appl. Opt.* **7** 307–9
35 Jacobs M B 1967 *The analytical toxicology of industrial inorganic poisons* (New York: Wiley)
36 Browning E 1969 *Toxicosity of industrial metals* (London: Butterworth)
37 Bangert H and Pfefferkorn H 1980 Condensation and stability of ZnS thin films on glass substrates *Appl. Opt.* **19** 3878–9
38 Hass G, Ramsay J B and Thun R 1958 Optical properties and structure of cerium dioxide films *J. Opt. Soc. Am.* **48** 324–7

39 Hass G 1952 Preparation, properties and optical applications of thin films of titanium dioxide *Vacuum* **2** 331–45

40 1957 Improvements in or relating to the manufacture of thin light transmitting layers *UK Patent Specification* 775 002

41 1957 Method of producing titanium dioxide coatings *US Patent Specification* 2 784 115

42 1960 Process for the manufacture of thin films *US Patent Specification* 2 920 002

43 Pulker H K, Paesold G and Ritter E 1976 Refractive indices of TiO_2 films produced by reactive evaporation of various titanium-oxide phases *Appl. Opt.* **15** 2986–91

44 Apfel J H 1980 The preparation of optical coatings for fusion lasers *Proc. Int. Conf. Metall. Coatings* San Diego

45 Ritter E 1962 Zur Kentnis des SiO und Si_2O_3-Phase in dünnen Schichten *Opt. Acta* **9** 197–202

46 Bradford A P, Hass G, McFarland M and Ritter E 1965 Effect of ultraviolet irradiation on the optical properties of silicon oxide films *Appl. Opt.* **4** 971–6

47 Bradford A P and Hass G 1963 Increasing the far-ultra-violet reflectance of silicon oxide protected aluminium mirrors by ultraviolet irradiation *J. Opt. Soc. Am.* **53** 1096–1100

48 Heitmann W 1971 Reactive evaporation in ionized gases *Appl. Opt.* **10** 2414–18

49 Ebert J 1982 Activated reactive evaporation in *Optical Thin Films* ed R I Seddon (*Proc. SPIE* **325** 29–38)

50 Smith D and Baumeister P W 1979 Refractive index of some oxide and fluoride coating materials *Appl. Opt.* **18** 111–15

51 Stetter F, Esselborn R, Harder N, Friz M and Tolles P 1976 New materials for optical thin films *Appl. Opt.* **15** 2315–17

52 Baumeister P W and Arnon O 1977 Use of hafnium dioxide in mutilayer dielectric reflectors for the near uv *Appl. Opt.* **16** 439–44

53 Lubezky I, Ceren E and Klein Z 1980 Silver mirrors protected with Yttria for the 0.5 to 14 μm region *Appl. Opt.* **19** 1895

54 Cox J T and Hass G 1978 Protected Al mirrors with high reflectance in the 8–12 μm region from normal to high angles of incidence *Appl. Opt.* **17** 2125–6

55 Datta U 1979 Private communication

56 Moss T S 1952 Optical properties of tellurium in the infra-red *Proc. Phys. Soc.* **65** 62–6

57 Greenler R G 1955 Interferometry in the infrared *J. Opt. Soc. Am.* **45** 788–91

58 Smith S D and Seeley J S 1968 Multilayer filters for the region 0.8 to 100 microns *Final Report on Contract AF61(052)-833*, Air Force Cambridge Research Laboratories

59 Evans C S and Seeley J S 1968 Properties of thick evaporated layers of PbTe *Proc. Colloq: IV–VI compounds*, Paris

60 Seeley J S, Evans C S, Hunneman R and Whatley A 1976 Filters for the v2 band of CO_2: monitoring and control of layer deposition *Appl. Opt.* **15** 2736–45

61 Seeley J S, Hunneman R and Whatley A 1981 Far infrared filters for the Galileo–Jupiter and other missions *Appl. Opt.* **20** 31–9

62 Evans C S, Hunneman R and Seeley J S 1976 Optical thickness changes in freshly deposited layers of lead telluride *J. Phys. D: Appl. Phys.* **9** 321–8

63 Evans C S, Hunneman R and Seeley J S 1976 Increments at the interface between layers during infra-red filter manufacture *Opt. Acta* **23** 297–303
64 Hass G and Salzberg C D 1954 Optical properties of silicon monoxide in the wavelength region from 0.24 to 14.0 microns *J. Opt. Soc. Am.* **44** 181–7
65 Jacobsson R and Martensson J O 1966 Evaporated inhomogeneous thin films *Appl. Opt.* **5** 29–34
66 Fujiwara S 1963 Refractive indices of evaporated cerium dioxide–cerium fluoride films *J. Opt. Soc. Am.* **53** 880
67 Fujiwara S 1963 Refractive indices of evaporated cerium fluoride–zinc sulphide films *J. Opt. Soc. Am.* **53** 1317–18
68 1965 Surface coated optical elements *UK Patent Specification* 1 010 038
69 Yadava V N, Sharma S K and Chopra K L 1974 Optical dispersion of homogeneously mixed ZnS–MgF_2 films *Thin Solid Films* **22** 57–66
70 Yadava V N, Sharma S K and Chopra K L 1973 Variable refractive index optical coatings *Thin Solid Films* **17** 243–52
71 1947 Method of coating with quartz by thermal evaporation *UK Patent Specification* 632 442
72 1962 Improvements in and relating to the oxidation and/or transparency of thin partly oxidic layers *UK Patent Specification* 895 879
73 1962 Use of a rare earth metal in vaporizing metals and metal oxides *US Patent Specification* 3 034 924
74 Butterfield A W 1974 The optical properties of Ge_xSe_{1-x} thin films *Thin Solid Films* **23** 191–4
75 Jacobsson R 1975 Inhomogeneous and coevaporated homogeneous films for optical applications in *Physics of Thin Films* **8** 51–98 ed G Hass, M H Francombe and R W Hoffman
76 Heitman W 1968 The influence of various parameters on the refractive index of evaporated dielectric films *Appl. Opt.* **7** 1541–3
77 Pearson J M 1970 Electron microscopy of multilayer thin films *Thin Solid Films* **6** 349–58
78 Lissberger P H and Pearson J M 1976 The performance and structural properties of multilayer optical filters *Thin Solid Films* **34** 349–55
79 Pulker H K and Jung E 1971 Correlation between film structure and sorption behaviour of vapour deposited ZnS, cryolite and MgF_2 films *Thin Solid Films* **9** 57–66
80 Guenther K H and Pulker H K 1976 Electron microscopical investigations of cross sections of optical thin films *Appl. Opt.* **15** 2992–7
81 Movchan B A and Demchishin A V 1969 *Fiz. Met. Metalloved.* **28** 653–60
82 Thornton J A 1974 Influence of apparatus geometry and deposition conditions on the structure and topography of thick sputtered coatings *J. Vac. Sci. Technol.* **11** 666–70
83 Kinosita K and Nishibori M 1969 Porosity of MgF_2 films—evaluation based on changes in refractive index due to adsorption of vapors *J. Vac. Sci. Technol.* **6** 730–3
84 Harris M, Macleod H A, Ogura S, Pelletier E and Vidal B 1979 The relationship between optical inhomogeneity and film structure *Thin Solid Films* **57** 173–8
85 Ogura S 1975 Some features of the behaviour of optical thin films *PhD Thesis* Newcastle-upon-Tyne Polytechnic

86 Ogura S and Macleod H A 1976 Water sorption phenomena in optical thin films *Thin Solid Films* **34** 371–5

87 Schildt J, Steudel A and Walther H 1967 The variation of the transmission wavelength of interference filters by the influence of water vapour *J. Physique* **28** suppl C2 C2-276–C2-279

88 Koch H 1965 Optische Untersuchungen zur Wasserdampfsorption in Aufdampfschichten (inbesondere in MgF_2 Schichten) *Phys. Stat. Sol.* **12** 533–43

89 Koch H 1965 Uber Sorptionsvorgange beim Beluften von MgF_2 Schichten *Proc. Colloq. on Thin Films* Budapest

90 Macleod H A and Richmond D 1976 Moisture penetration patterns in thin films *Thin Solid Films* **37** 163–9

91 Richmond D 1976 Thin film narrow band optical filters *PhD Thesis* Newcastle-upon-Tyne Polytechnic

92 Lee C C 1983 Moisture adsorption and optical instability in thin film coatings *PhD Dissertation* University of Arizona

93 Roche P, Bertrand L and Pelletier E 1976 Influence of temperature on the optical properties of narrow-band interference filters *Opt. Acta* **23** 433–44

94 Elson J M 1976 Light scattering from surfaces with a single dielectric overlayer *J. Opt. Soc. Am.* **66** 682–94

95 Elson J M 1977 Infrared light scattering from surfaces covered with multiple dielectric overlayers *Appl. Opt.* **16** 2872–82

96 Elson J M 1979 Diffraction and diffuse scattering from dielectric multilayers *J. Opt. Soc. Am.* **69** 48–54

97 Elson J M and Bennett J M 1979 Relation between the angular dependence of scattering and the statistical properties of optical surfaces *J. Opt. Soc. Am.* **69** 31–47

98 Bousquet P, Flory F and Roche P 1981 Scattering from multilayer thin films: theory and experiment *J. Opt. Soc. Am.* **71** 1115–23

99 Gourley S J and Lissberger P H 1979 Optical scattering in multilayer thin films *Opt. Acta* **26** 117–43

100 Lowdermilk W H 1982 Coatings for laser fusion in *Optical Thin Films* ed R I Seddon (*Proc. SPIE* **325** 2–11)

101 Bennett H E, Guenther A H, Milam D and Newnam B E 1983 *Laser Induced Damage in Optical Materials: 1981. NBS Spec. Publ.* **638** (Washington, DC: US Gov. Print. Off.)

102 Maissel L I and Glang R 1970 *Handbook of thin film technology* (New York: McGraw-Hill)

103 Ritter E 1975 Dielectric film materials for optical applications in *Physics of Thin Films* **8** 1–49 ed G Hass, M H Francombe and R W Hoffman

104 Ritter E 1976 Optical film materials and their applications *Appl. Opt.* **15** 2318–27

105 Holland L 1956 *Vacuum Deposition of Thin Films* (London: Chapman and Hall)

106 Pulker H K 1984 *Coatings on Glass* (Oxford: Elsevier)

10 Layer uniformity and thickness monitoring

In the previous chapter we considered what is probably the most difficult aspect of thin-film coating and filter production, that of materials. As we saw, these are not always satisfactory, and there are still problems associated with their stability. Once the materials have been chosen, and their properties are known, the thin-film designer, using the methods discussed in chapters 3–7, can usually produce a design to meet a given specification. Given suitable materials and an acceptable design, however, there are still further difficulties to be overcome in the construction of a practical filter. The two most important remaining factors are, first, controlling the uniformity of layer thickness over the area of the substrate, and second, controlling the overall thickness of each layer. Lack of uniformity causes a shift of characteristic wavelength over the surface of the filter, without necessarily affecting the performance in other ways, while thickness errors usually cause a reduction in performance. The magnitude of the errors which can be tolerated will vary from one design to another and the estimation of this is dealt with briefly. The bulk of this chapter is concerned with the general problem of minimising these two sources of error. One other important topic is substrate preparation, and that is considered on pages 420–3.

UNIFORMITY

In the evaporation process, it is usual to maintain the pressure within the chamber sufficiently low to ensure that the molecules in the stream of evaporant will travel in straight lines until they collide with a surface. In order to calculate the thickness distribution in a plant, the assumption is usually made that every molecule of evaporant sticks where it lands. This assumption is not strictly correct, but it does allow uniformity calculations that are sufficiently accurate for most purposes. The distribution of thickness is then calculated in exactly the same way as intensity of illumination in an optical calculation. All that is required to enable the thickness to be estimated is a knowledge of the distribution of evaporant from the source.

Holland and Steckelmacher[1] published an early and detailed account of techniques for the prediction of layer thickness uniformity. Their expressions were extended later by Behrndt[2]. Holland and Steckelmacher divided sources into two broad types: those which have even emission in all directions and can be likened to a point source, and those which have a distribution similar to that from a flat surface, the intensity falling off as the cosine of the angle between the direction concerned and the normal to the surface. The expressions for the distribution of material emitted from the two types of source are:

$$\mathrm{d}M = [m/(4\pi)]\,\mathrm{d}\omega \qquad \text{for the point source}$$

and

$$\mathrm{d}M = [m/\pi]\cos\phi\,\mathrm{d}\omega \qquad \text{for the directed surface source}$$

where m is the total mass of material emitted from the source in all directions, and $\mathrm{d}M$ is the amount passing through solid angle $\mathrm{d}\omega$ (at angle ϕ to the normal to the surface in the case of the second type of source).

If the material is being deposited on a surface element $\mathrm{d}S$ of the substrate, which has its normal at an angle θ to the direction of the source from the element, then the amount which will condense on the surface will be given by

$$\mathrm{d}M = \frac{m}{4\pi}\frac{\cos\theta}{r^2}\,\mathrm{d}S \qquad \text{for the point source}$$

and

$$\mathrm{d}M = \frac{m}{\pi}\frac{\cos\phi\cos\theta}{r^2}\,\mathrm{d}S \qquad \text{for the directed surface source.}$$

In order to estimate the thickness of the deposit we need to know the density of the film. If this is denoted by μ then the thickness will be

$$t = \frac{m}{4\pi\mu}\frac{\cos\theta}{r^2} \qquad \text{for the point source}$$

and

$$t = \frac{m}{\pi\mu}\frac{\cos\phi\cos\theta}{r^2} \qquad \text{for the directed surface source.}$$

These are the basic equations used by Holland and Steckelmacher for estimating the thickness in uniformity calculations.

Flat plate

The simplest case is that of a flat plate held directly above and parallel to the source. Here the angle ϕ is equal to the angle θ and the thickness is

$$t = \frac{m}{4\pi\mu}\frac{\cos\theta}{r^2} = \frac{mh}{4\pi\mu(h^2+\rho^2)^{3/2}} \qquad \text{for the point source}$$

and

$$t = \frac{m}{\pi\mu}\frac{\cos^2\theta}{r^2} = \frac{mh^2}{\pi\mu(h^2+\rho^2)^2} \qquad \text{for the directed surface source}$$

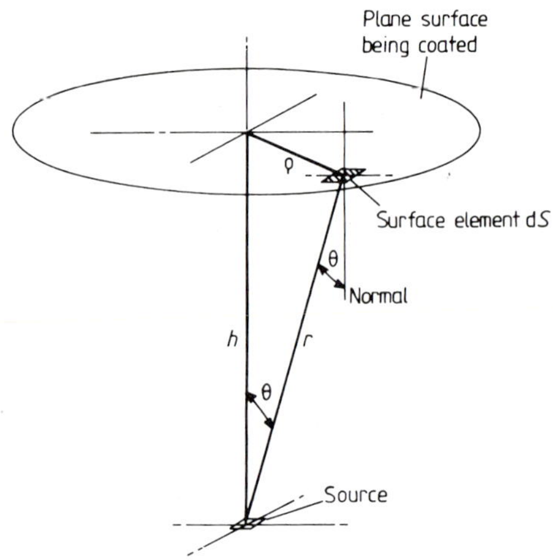

Figure 10.1 Diagram showing the goemetry of the evaporation from a central source on to a parallel plane surface.

with notation as in figure 10.1. These expressions simplify to

$$t/t_0 = [1 + (\rho/h)^2]^{-3/2} \qquad \text{for the point source}$$

and

$$t/t_0 = [1 + (\rho/h)^2]^{-2} \qquad \text{for the directed surface source}$$

and are plotted in figure 10.2. t_0 is the thickness immediately above the source where $\rho = 0$. In neither case is the uniformity at all good. Clearly the geometry is not suitable for any very accurate work unless the substrate is extremely small and in the centre of the plant.

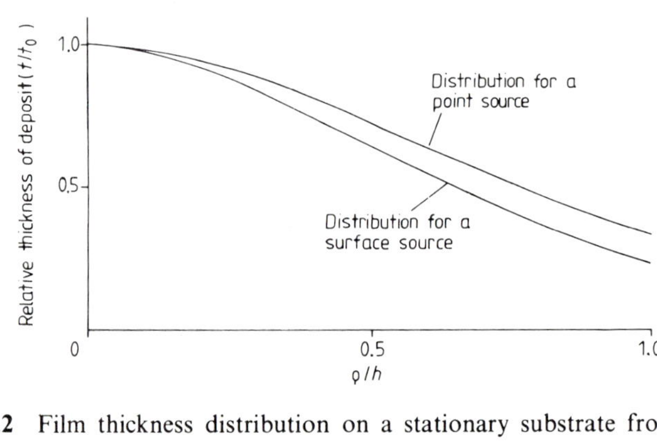

Figure 10.2 Film thickness distribution on a stationary substrate from a central source.

Spherical surface

A slightly better arrangement that can sometimes be used is a spherical geometry where the substrates lie on the surface of a sphere. A point source will give uniform thickness of deposit on the inside surface of a sphere when the source is situated at the centre. It can be shown that the directed surface source will give uniform distribution similarly when it is itself made part of the surface. In fact, it was the evenness of the coating within a sphere which led Knudsen[3] first to propose the cosine law for thin-film deposition. The method is often used in plants for simple blooming of components such as lenses where the uniformity need not be better than, say, 10% of the layer thickness at the centre of the component. However, for precise work, this uniformity is still not adequate.

A higher degree of uniformity involves rotation of the substrate carrier, which we shall now consider.

Rotating substrates

The situation here is as if, in figure 10.1, the surface for coating were rotated about a normal at distance R away from the source. As the surface rotates, the thickness deposited at any point will be equal to the average of the thickness which would be deposited on a stationary substrate around a ring centred on the axis of rotation, provided always that the number of revolutions during the deposition is sufficiently great to make the amount deposited in an incomplete revolution a very small proportion of the total thickness. By choosing the correct distance between source and axis of rotation, the uniformity can be made vastly superior to that for stationary substrates.

We shall consider first the directed surface source. Figure 10.3 shows the situation. The calculation is basically similar to that for the flat plate with a central source. Here we stop the plate and calculate the mean thickness around the circle containing the point in question and centred on the axis of rotation. The radius of the circle is ρ and if we define any point P on the circle by the angle ψ, then the thickness at the point is given by

$$t = \frac{m}{\pi\mu}\frac{h^2}{(h^2+\rho^2+R^2-2\rho R\cos\psi)^2}$$

where r, the distance from the source to the point, is given by

$$r^2 = h^2+\rho^2+R^2-2\rho R\cos\psi.$$

Then, taking the mean of the thickness around the circle, we have for the thickness of the deposit in the rotating case

$$t = \frac{m}{\pi\mu}\frac{1}{2\pi}\int_0^{2\pi}\frac{h^2\,\mathrm{d}\psi}{(h^2+\rho^2+R^2-2\rho R\cos\psi)^2}.$$

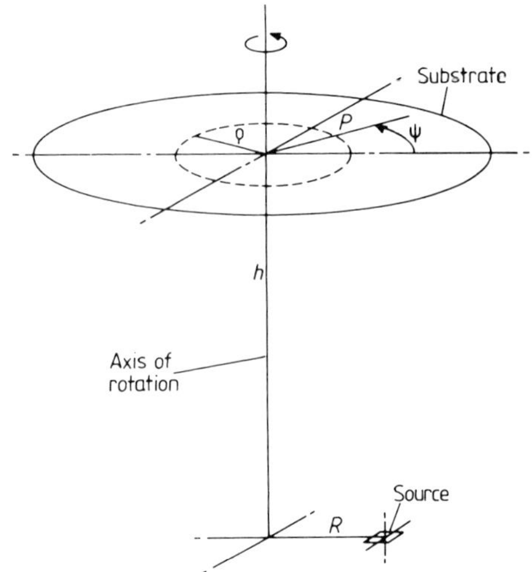

Figure 10.3 Diagram showing the geometry of the evaporation from a stationary offset source on to a rotating substrate.

Now the integral $\int_0^{2\pi} \mathrm{d}\psi/(1-a\cos\psi)^2$ can be evaluated by contour integration giving

$$\int_0^{2\pi} \frac{\mathrm{d}\psi}{(1-a\cos\psi)^2} = \frac{2\pi}{(1-a^2)^{3/2}}$$

so that the expression for thickness becomes

$$t = \frac{m}{\pi\mu} \frac{h^2}{(h^2+\rho^2+R^2)^2} \frac{1}{\{1-[2\rho R/(h^2+\rho^2+R^2)]^2\}^{3/2}}$$

$$\frac{t}{t_0} = \frac{(1+R^2/h^2)^2(1+\rho^2/h^2+R^2/h^2)}{[1+\rho^2/h^2+R^2/h^2-2(\rho/h)(R/h)]^{3/2}[1+\rho^2/h^2+R^2/h^2+2(\rho/h)(R/h)]^{3/2}}$$

where t/t_0 is, as before, the ratio of the thickness at the radius in question to that at the centre of the substrate holder.

Figure 10.4 shows this function plotted for several different dimensions which are typical of medium-sized coating plants. The distribution can immediately be seen to be vastly superior to that when the substrates are stationary. For one particular combination of dimensions, that corresponding to $R = 7$, the distribution is extremely even over the central part (radius 3.75) of the plant. This is the arrangement used in the production of narrowband filters where the uniformity must necessarily be very good. If the uniformity is not quite so important, where rather broader filters or perhaps antireflection coatings are concerned, then the sources can be moved outwards, allowing a larger area to be coated at the expense of a slight decline in uniformity.

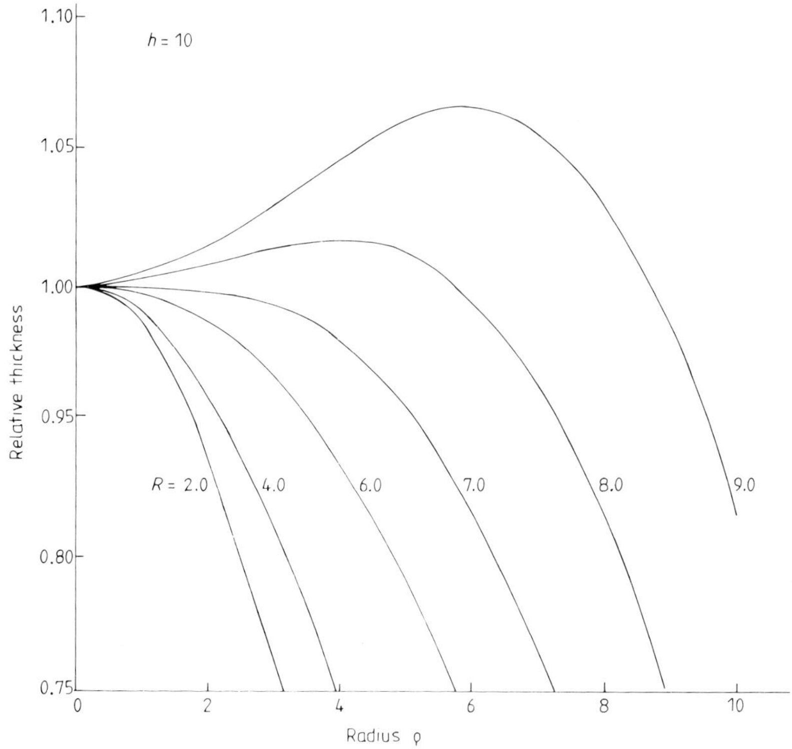

Figure 10.4 Theoretical film thickness distribution on substrates rotated about the centre of the plant for various source radii and substrate heights. The sources are assumed to be small directed surfaces parallel to the substrates.

A similar expression is found for a point source but this time involving elliptic integrals. The thickness at the point P, assuming that the substrate does not rotate, is given by

$$t = \frac{m}{4\pi\mu} \frac{h}{(h^2 + R^2 + \rho^2 - 2\rho R \cos\psi)^{3/2}}$$

and in the presence of rotation the thickness at any point around the ring of radius ρ will be the mean of this expression, i.e.

$$t = \frac{m}{4\pi\mu} \frac{1}{2\pi} \int_0^{2\pi} \frac{h\,\mathrm{d}\psi}{(h^2 + R^2 + \rho^2 - 2\rho R \cos\psi)^{3/2}}$$

$$= \frac{m}{4\pi^2\mu} \int_0^{\pi} \frac{h\,\mathrm{d}\psi}{(h^2 + R^2 + \rho^2 - 2\rho R \cos\psi)^{3/2}}.$$

Now let $(\pi-\psi)/2=\gamma$, then $\mathrm{d}\psi=-2\mathrm{d}\gamma$, and the expression for thickness becomes

$$t=\frac{m}{4\pi^2\mu}\int_{\pi/2}^{0}\frac{-h\,\mathrm{d}\gamma}{[h^2+(R+\rho)^2-4\rho R\sin^2\gamma]^{3/2}}$$

which can be written

$$t=\frac{m}{4\pi^2\mu}\frac{h}{[h^2+(R+\rho)^2]^{3/2}}\int_0^{\pi/2}\frac{\mathrm{d}\gamma}{\{1-[4\rho R/(h^2+(R+\rho)^2)]\sin^2\gamma\}^{3/2}}.$$

Now the integral in this expression is a standard form

$$\frac{1}{(1-k^2)}E(k,\alpha)=\int_0^{\alpha}\frac{\mathrm{d}\gamma}{(1-k^2\sin^2\gamma)^{3/2}}$$

where $E(k,\alpha)$ is an elliptic integral of the second kind, and is a tabulated function[4]. The expression for thickness then becomes:

$$T=\frac{hm}{4\pi^2\mu}\frac{E(k,\pi/2)}{[h^2+(R+\rho)^2]^{1/2}[h^2+(R-\rho)^2]}$$

where

$$k^2=4\rho R/[h^2+(R+\rho)^2].$$

Curves of this expression are given by Holland and Steckelmacher[1], and the shape is very similar to that for the directed surface source.

Almost all the sources used in the production of thin-film filters, especially the boat type, give distributions similar to the directed surface source. Holland and Steckelmacher also describe some experiments which they carried out to determine this point. Keay and Lissberger[5] have studied the distribution from a howitzer source loaded with zinc sulphide, and it appears that this is somewhere in between the point source and the directed surface source, probably due to scattering in the evaporant stream immediately above the heater where the pressure is high. The cloud of vapour which forms seems to act to some extent as a secondary point source. This behaviour of the howitzer probably depends to a considerable extent on the material which is being evaporated. Graper[6] has studied the distribution of evaporant from an electron gun and has found that this is somewhat more directional than the directed surface source. Its distribution can be described by a $\cos^x\theta$ law where x is somewhere between 1 and 3 and depends on the power input and on the amount of material in the hearth. Using zinc sulphide and cryolite, Richmond[7] found that the distribution from an electron gun source was best represented by a law of the form $\cos\theta$.

Normally, in calculating the distribution to be expected from a particular geometry, we assume that we are using directed surface sources, and then, when setting up a plant for the first time, the sources are placed at the theoretically best positions. The first few runs soon show whether or not any further adjustments are necessary, and if they are, they are usually very slight and can be made by trial and error. Once the best positions are found, it is important to ensure that the sources are always accurately set to reproduce them. Care should be taken to make sure that the angular alignment is correct. A source at the correct geometrical position but tilted away from the correct direction will give uniformity errors just as much as if it were laterally displaced. The frontispiece shows a plant that is being fitted with a flat plate work holder for the manufacture of narrowband filters.

Where uniformity must be good over as large an area as possible but where the ultimate is not required, it is possible to use a combination of a spherical surface and rotating plate. A domed work holder, or calotte, is rotated about its centre with the sources offset beneath it so that they are approximately on the surface of the sphere, with slight adjustments made during setting up. This gives very good results over a much larger area than would be possible with the simple rotating flat plate. Figure 10.5 shows the interior of a plant using this arrangement.

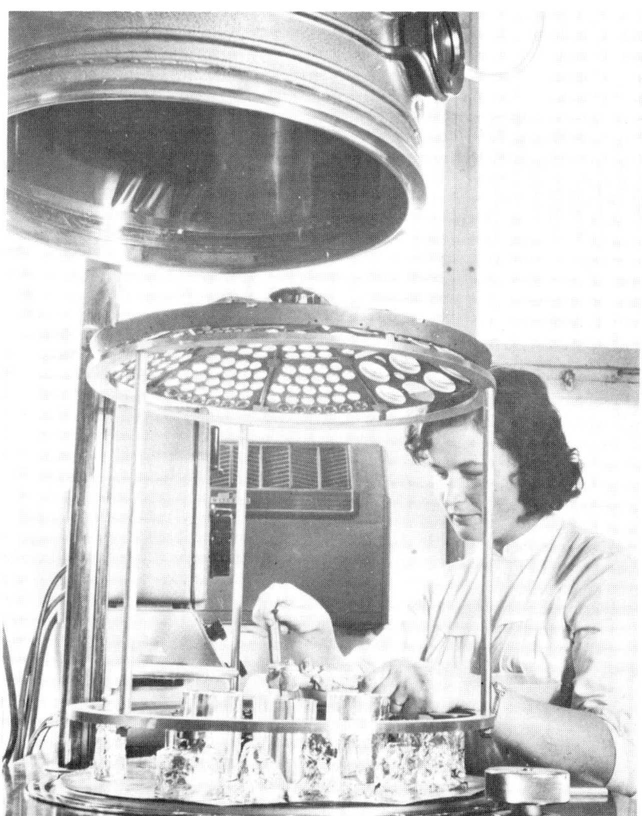

Figure 10.5 Photograph showing the interior of a plant with a domed calotte.

When still improved uniformity is required, it is possible to achieve it by what is known as a planetary jig. In this arrangement, the substrates not only rotate about the centre of the jig, but also about their own individual centres at much greater speed, so that they execute many revolutions for each single revolution of the jig as a whole. This carries a stage further the averaging process that occurs with the simple rotating jig.

Use of masks

It is possible to make corrections to distribution by careful use of masks. In their simplest form they are stationary and are placed just in front of the substrates that rotate on a single carrier about a single axis. The masks are cut so that they modify the radial distribution of thickness. Theoretical calculations give dimensions for masks of approximately the correct shape, which can then be trimmed according to experimental results to arrive at the final form. For a number of reasons, it is normal to leave the central monitor glass uncorrected. It is difficult to correct the central part of the plant where the mask width tends to zero, and, in any case, the monitor is usually stationary. Furthermore, in some monitoring arrangements, there is an advantage in having more material on the monitor than on the batch.

A further degree of freedom was introduced by Ramsay, Netterfield and Mugridge[8] in the form of a rotating mask. For a large flat substrate which is approaching the dimensions of the plant there is little other than simple rotation that can be done in terms of the carrier jig to improve uniformity. Planetary arrangements require much more room. Stationary masks are of some help but they are somewhat sensitive to the characteristic of the sources and are not therefore sufficiently stable for a very high degree of uniformity. A much more stable arrangement, that has been shown capable of uniformities of the order of 0.1 % over areas of around 200 mm diameter, involves rotating the mask about a vertical axis at a rotational speed considerably in excess of that of the substrate carrier. This effectively corrects the angular distribution of the source that can be positioned at the centre of the plant. The mask rotation axis is usually placed very near the source and positioned so that the line drawn from the source through the mask centre intersects the perimeter of the substrate carrier. In practice the axis of rotation and the rotating shutter are close to the source position and slight adjustment of the axis can be made for trimming purposes. It has been found to be an exceptionally stable arrangement.

SUBSTRATE PREPARATION

Before a substrate can be coated, it must be cleaned. The forces which hold films together and to the substrate are all short-range interatomic and intermolecular forces. These forces are extremely powerful, but their short range means that we can think of each atomic layer as being bound to the neighbouring layers only, and being little affected by material which is further removed from it. Thus, the adhesion of a thin film to the substrate depends critically on conditions at the substrate surface. Even a monomolecular layer of a contaminant on the surface can change the force of adhesion by orders of magnitude. Condensation of evaporant, too, is just as sensitive to surface conditions than can alter completely the characteristics of the subsequent layers. Substrate cleaning so that the condensing material attaches itself to the substrate and not an intervening layer of contaminant is therefore of paramount importance.

The typical symptoms of an inadequately cleaned substrate are a mottled, oily appearance of the coating, coupled usually with poor adhesion and optical performance. This can be caused also by such defects in the plant as backstreaming of oil from the pumps. When these symptoms appear it is usually advisable to extend any subsequent improvements in cleaning techniques to the plant as well.

A good account of various cleaning methods is given by Holland[9]. A more recent account is that of Mattox[10]. The best cleaning process will depend very much on the nature of the contamination that must be removed and, although it may seem self-evident, in all cleaning operations it is essential to avoid contaminating the surface rather than cleaning it. For laboratory work, when the substrates are reasonably clean to start with (microscope slide glass is usually in this condition), then for most purposes it will be found sufficient to wash the substrates thoroughly in detergent and warm water (not household detergent that sometimes has additives which cause smears to appear on the finished films), to rinse them thoroughly in running warm water (in areas where tap water is fairly pure, hot tap water will be found adequate), and then to dry them thoroughly and immediately with a clean towel or soft paper tissue, or, better still, to blow them dry with a jet of clean dry nitrogen. The substrates should never be allowed to dry themselves or stains will certainly occur which are usually impossible to remove. Substrates should be handled as little as possible after cleaning and, since they never remain clean for long, placed immediately in the coating plant and the coating operation started. Wax or grease will probably require treatment with an alcohol such as isopropyl, perhaps rubbing the surface with a clean fresh cotton swab soaked in the alcohol and then flooding the surface with the liquid. Care must be taken to ensure that the alcohol is really clean. A bottle of alcohol available to all in a laboratory seldom remains clean for long and a better arrangement is to keep it under lock and key and to allow the alcohol into the laboratory in wash bottles that emit the alcohol when squeezed.

This basic cleaning procedure can be modified and supplemented in various ways, especially if large numbers of substrates are to be handled automatically. Ultrasonic scrubbing in detergent solution or in alcohol is a very useful technique, although prolonged ultrasonic exposure is to be avoided since it can eventually cause surface damage. It is important that the substrates should be kept wet right through the cleaning procedure until they are dried as the final stage. Vapour cleaning is frequently used for this. The substrates are exposed to the vapour of alcohol or other degreasing agents so that initially it condenses and runs off, taking any residual contamination or the remains of the agent from the previous cleaning stage with it. The substrates gradually reach the temperature of the vapour and then no further condensation takes place, when the substrates can be withdrawn perfectly dry. Since the agent is condensing from the vapour phase, it is in an extremely pure form. An alternative end to the cleaning process is a rinse in deionised water followed by drying in a blast of dry, filtered nitrogen.

It is very difficult to see marks on the surface of the substrate with the naked eye. Dust can be picked up by oblique illumination, but wax and grease cannot. A common test for assessing the quality of a cleaning process is to breathe on one of the substrates so that moisture condenses on it in a thin layer. This tends to magnify the effects of any residue. The moisture acts in almost exactly the same way as a condensing film since the condensation pattern depends on the surface conditions. A surface examined in this way is said to exhibit a good or bad 'breath figure'. A contaminated surface gives a smeared pattern, while a clean surface is completely even. For those coatings that are to have minimum losses even this step cannot be guaranteed not to introduce slight residual contamination and is better omitted.

Once the substrates are in the plant, and they should always be loaded as soon as possible after cleaning, they can be given a final clean by a glow discharge. The equipment for this, which consists of a high-voltage supply, preferably DC, together with the necessary lead in electrodes, is fitted as standard in most plants. At a suitable pressure, which will vary with the particular geometry of the electrodes but which will usually be around 0.06 mb, a glow discharge is struck and, provided the geometry is correct, the surface of the substrates is bombarded with positive ions. This effectively removes any light residual contamination, although gross contamination will persist. It is not certain whether the cleaning action actually arises from a form of sputtering or whether the glow discharge is merely a convenient way of raising the temperature of the surfaces so that contaminants are baked off. Generally the glow discharge is limited in duration to 5 or perhaps 10 minutes. It has been suggested that, although glow discharge cleaning does remove grease, it does encourage dust particles; for coatings where minimum dust is required, such as high performance laser mirrors, glow discharge cleaning is frequently omitted. Lee[11] found that the omission of glow discharge cleaning led to a very great increase in the incidence of moisture penetration patches in his films and consequently to a fall in the performance of his filters.

The evaporation of the first layer should begin as soon as possible after the glow discharge has stopped. Cox and Hass[12] used a discharge current of 80 mA and a voltage of 5000 V for 5 minutes to clean substrates before coating them with zinc sulphide, and found that the time between finishing the discharge and starting the evaporation should be not greater than 3 minutes. If the time was allowed to exceed 5 minutes, then the quality of the films, especially their adhesion, deteriorated.

If, as sometimes happens, a filter is left for a period, say overnight, in an uncompleted state, it will often be found advisable to carry out a short period of glow discharge cleaning before starting to evaporate the remaining layers.

THICKNESS MONITORING

Given suitable materials, clean substrates, and plant with substrate-holder geometry to give the required distribution accuracy, the main problem which remains is that of controlling the deposition of the layers so that they have the characteristics required by the coating or filter design. Of course, many properties are required, but refractive index and optical thickness are the most important. There is no satisfactory way, at present, of measuring the refractive index of that portion of a film which is actually being deposited. Such measurements can be made later but for closed loop control, dynamic measurements are required. Normal practice, therefore, is simply to control, as far as possible, those deposition parameters that would affect refractive index so that the index produced for any given material is consistent. Measurements are made of the index and the value usually obtained is used in the coating design. This procedure, while it usually gives satisfactory results, is far from ideal and is used simply because, at the present time, there is no better way.

Film thickness can more readily be measured and, therefore, controlled. The simplest systems display a signal to a plant operator who is responsible for interpreting it and assessing the correct instant to terminate deposition. At the other end of the scale, there are completely automatic systems in which operator judgment plays no part and in which even operator intervention is rarely required.

There are many ways in which the thickness can be measured. All that is necessary is to find a parameter that varies in a suitable fashion with thickness and to devise a way of monitoring this parameter during deposition. Thus, parameters such as mass, electrical resistance, optical density, reflectance and transmittance have all been used. Of all the methods, those most frequently used involve either optical measurements of reflectance or transmittance or the measurement of total deposited mass by the quartz crystal microbalance.

The question of the best method for the monitoring of thin films is, of course, inseparable from that of how accurately the layers must be controlled. This second question is a surprisingly difficult one to answer. Indeed, it is impossible to separate the two questions: the tolerances which can be allowed and the method used for monitoring are closely related and one cannot be considered in depth independently of the other.

For convenience, however, we will consider some of the commoner arrangements for monitoring, including only the most rudimentary ideas of accuracy, and then, at a later stage, consider the question of tolerances along with some of the more advanced ideas of monitoring and its various classifications.

Optical monitoring techniques

Optical monitoring systems consists of some sort of light source illuminating a test substrate which may or may not be one of the filters in the batch, and a detector analysing the reflected or transmitted light. From the results of that analysis, the evaporation of the layer is stopped as far as possible at the correct point. Usually, so that the layer may be stopped as sharply as possible, the plant is fitted with a shutter which can be inserted in front of the evaporation sources. This is a much more satisfactory method than merely turning off the supply to the boats, which always take a finite time to stop emitting. Such a shutter can be seen in figure 9.3.

Almost all the early workers in the field used the eye as the detector, and the thicknesses of the films were determined by assessing their colour appearance in white light. In many cases they were concerned with simple single-layer coatings such as single-layer blooming, which are not at all susceptible to errors. When the blooming layer is of the correct thickness for visible light, the colour reflected from the surface in white light has a magenta tint, owing to the reduction of the reflectance in the green. The visual method is quite adequate for this purpose and is still being widely used. A very clear account of the method is given by Mary Banning[13], who compiled table 10.1.

In the production of other types of filter where the errors of the visual method would be too large, other methods must be used. An early paper by Polster[14] describes a photoelectric method which is basically the same as that used most often today. We saw in chapter 2 that if the film is without absorption, then its reflectance and transmittance measured at any one wavelength will vary with thickness in a cyclic manner, similar to a sine wave, although, for the higher indices, the waves will be flattened at their tops. The turning values correspond to those wavelengths for which the optical thickness of the film is an integral number of quarter wavelengths, the reflectance being equal to that of the substrate when the number is even and a maximum amount removed from the reflectance of the substrate when the number is odd. Figure 10.6 illustrates the behaviour of films of different values of refractive index. This affords the means for measurement. If the detector in the system is made highly selective, for example, by putting a narrow filter in front of it, then the measured reflectance or transmittance will vary in this cyclic way, and the film may be monitored to an integral number of quarter-waves by counting the number of turning points passed through in the course of the deposition. A typical arrangement to perform this operation is shown in figure 10.7. The filter

Table 10.1 (After Banning[13].)

Colour change for ZnS	Colour change for Na_3AlF_6	Optical thickness for green light
Bluish white	Yellow	
↓	↓	
White	Magenta	$\lambda/4$, first-order maximum
↓	↓	
Yellow	Blue	
↓	↓	
Magenta	White	$\lambda/2$, first-order minimum
↓	↓	
Blue	Yellow	
↓	↓	
Greenish white	Magenta	$3\lambda/4$, second-order maximum
↓	↓	
Yellow	Blue	
↓	↓	
Magenta	Greenish white	λ, second-order minimum
↓	↓	
Blue	Yellow	
↓	↓	
Green	Magenta	$5\lambda/4$, third-order maximum

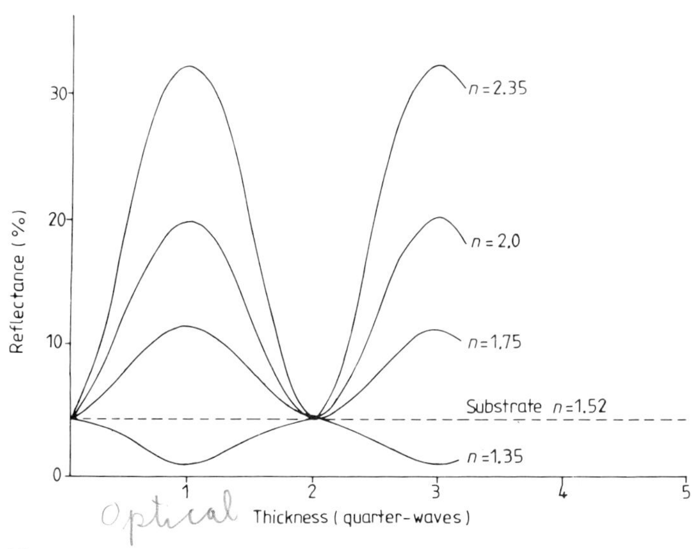

Figure 10.6 Curves showing the variation with thickness of the reflectance of several films with different refractive indices.

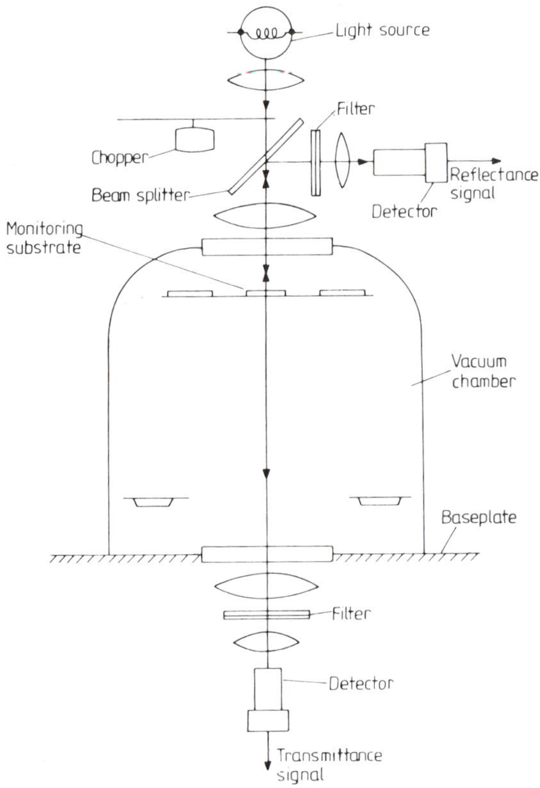

Figure 10.7 A possible arrangement of a monitoring system for reflectance and transmittance measurements.

may be an interference filter or, more flexible, an adjustable prism or grating monochromator.

Consider the deposition of a high-reflectance multilayer stack where all the layers are quarter-waves. Let the monitoring wavelength be the wavelength for which all the layers are one-quarter-wavelength thick. The reflectance of the test piece will vary as shown in figure 10.8. The example shown is typical of a stack for the visible region. The reflectance can be seen to increase during the deposition of the first layer, which is of high index, to a maximum where the deposition is terminated. During the second layer the reflectance falls to a minimum where the second layer is terminated. The third layer increases the reflectance once again and the fourth layer reduces it. This behaviour is superimposed on a trend towards a reflectance of unity so that the variable part of the signal becomes a gradually smaller part of the total. This puts a limit on the number of layers which can be monitored in reflectance in this way to around four, when a fresh monitoring substrate must be inserted. In transmission

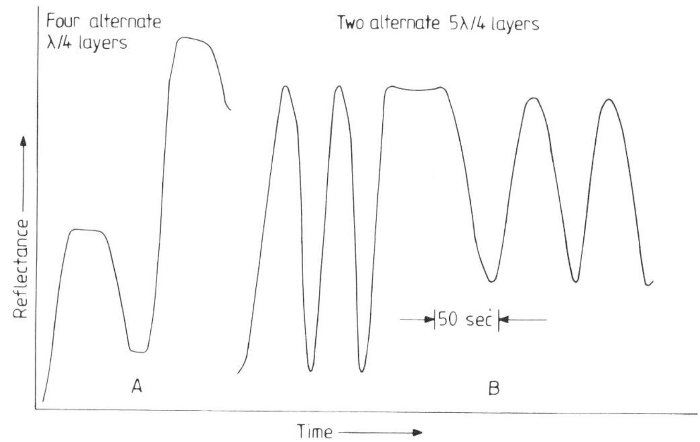

Figure 10.8 Record taken from a pen recorder of the reflectance of a monitor glass during film deposition. (After Perry[15].)

monitoring, this effect does not exist and the variable part of the signal remains a sufficiently large part of the whole. The only problem is that the overall trend of the signal is towards zero, so that eventually it will become too small in comparison with the noise in the system. With reasonable optics and a photomultiplier detector, the number of layers which may be dealt with in this way is around twenty-one. At this stage the noise usually becomes too great.

For the greatest accuracy from this method, the output of the detector should be displayed on a chart recorder; this makes it easier to determine the turning values. With this arrangement, a trained operator can usually terminate the layers to an accuracy on the monitoring substrate of around 5–10 %, depending on the index of the film. Of course, as we shall see, this does not necessarily mean that the actual thickness of the filters in the batch will be as accurate. Other sources of error operate to introduce differences between the monitor and the batch.

To improve the signal-to-noise ratio it is usual to chop the light before it enters the plant, partly because the evaporation process produces a great deal of light during the heating of the boats, but mainly because at the signal levels encountered, the electronic noise without some filtering would be impossibly great. The chopper should be placed immediately after the source of light but before the plant, and the filter should be inserted after the plant. This arrangement reduces the stray light to a greater extent than would placing either the filter before the plant or the chopper after it. It is, of course, always advisable to limit as far as possible the total light incident on the detector, partly because unchopped radiation can push the detector into a non-linear region and partly because it can cause damage to the device especially if it is a photomultiplier. If a filter rather than a monochromator is used, then great care should be taken to ensure that the sidebands are particularly well suppressed.

Photomultipliers have characteristics which vary considerably with wavelength, and if the monitoring wavelength lies in a rather insensitive region compared with the peak sensitivity, then small leaks in the more sensitive region, which might not be very noticeable in the characteristic curve of the filter, can cause considerable difficulties from stray light, even giving spurious signals of similar or greater magnitude than the true signal. Prism or grating monochromators are often safer for this work, besides being considerably more flexible.

We can investigate the errors likely to arise in this type of monitoring as follows: Suppose that in the monitoring of a single quarter-wave layer there is an error γ in the value of reflectance at the termination point. This will give rise to a corresponding error ϕ in the phase thickness of the layer δ where

$$\delta = (\pi/2) - \phi. \tag{10.1}$$

Because of the nature of the characteristic reflectance curve of the single layer, the error in phase thickness will be rather greater in proportion than the original error in reflectance. The admittance of the layer will be given by the characteristic matrix:

$$\begin{bmatrix} \cos\delta & (\mathrm{i}\sin\delta)/\eta \\ \mathrm{i}\eta\sin\delta & \cos\delta \end{bmatrix} \begin{bmatrix} 1 \\ N \end{bmatrix} \tag{10.2}$$

where

$$\cos\delta = \sin\phi \qquad \text{and} \qquad \sin\delta = \cos\phi.$$

This gives

$$Y = \frac{\sin\phi + \mathrm{i}(N\cos\phi)/\eta}{\mathrm{i}\eta\cos\phi + N\sin\phi}$$

where the symbols have their usual meaning. Introducing the approximations for $\sin\phi$ and $\cos\phi$ up to and including powers of the second order, we have

$$Y = \frac{\phi + \mathrm{i}(N/\eta)(1-\phi^2/2)}{\mathrm{i}\eta(1-\phi^2/2) + N\phi} \tag{10.3}$$

and the reflectance of the monitor *in vacuo* will be given by

$$R = \left| \frac{(N-1)\phi + \mathrm{i}(\eta - N/\eta)(1-\phi^2/2)}{(N+1)\phi + \mathrm{i}(\eta + N/\eta)(1-\phi^2/2)} \right|^2$$

which simplifies to

$$R = \frac{(\eta - N/\eta)^2}{(\eta + N/\eta)^2}\left(1 + \frac{(N^2 + 1 - \eta^2 - N^2/\eta^2)4N}{(\eta^2 - N^2/\eta^2)^2}\phi^2\right). \tag{10.4}$$

The values of γ and ϕ are related as follows

$$\gamma = \frac{(N^2 + 1 - \eta^2 - N^2/\eta^2)4N}{(\eta^2 - N^2/\eta^2)^2}\phi^2 = \sigma\phi^2 \tag{10.5}$$

since the first factor in equation (10.4) is just the reflectance when γ and ϕ are both zero.

Now, in most cases it will not be possible to determine the reflectance at the turning value to better than 1% of the true value. In many cases, especially where there is noise, it will not be possible even to do as well as this. However, assuming this value for γ, the expression for the error in the layer thickness becomes

$$\pm 0.01 = \sigma\phi^2$$

where the sign $\pm$ is taken to agree with $\sigma\phi^2$ and depends on whether or not the turning value is a maximum or a minimum. If the error is expressed in terms of a quarter-wave thickness which is equivalent to $\pi/2$ radians, the expression becomes

$$\text{Error} = \frac{\phi}{\pi/2} = \frac{0.1}{\pi/2\,|\sigma|^{1/2}}. \tag{10.6}$$

A typical case is the monitoring of a quarter wave of zinc sulphide on a glass substrate where $\eta = 2.35$ and $N = 1.52$. Substituting these values in expression (10.5) and using it in (10.6), the fractional error in the quarter wave becomes 0.08. This is a colossal error compared with the original error in reflectance, and illustrates the basic lack of accuracy inherent in this method.

In the infrared, it is often possible to use wavelengths for monitoring which are shorter than the wavelengths of the desired filter peaks by a factor of perhaps two or even four. This improves the basic accuracy by the same factor. For layers similar to that considered above, the errors would then be 0.04 or 0.02. These errors are on the limit of permissible errors, and it is clear that this simple system of monitoring is not really adequate for any but the simplest of designs.

What makes the method particularly difficult to apply is that it is only the portion of the signal before the turning point that is available to the operator, who has therefore to anticipate the turning value, and the fact that trained plant operators can achieve the theoretical figures for accuracy says much for their skill.

An alternative method, inherently more accurate, involves the termination of the layer at a point remote from a turning value where the signal changes much more rapidly. This consists of the prediction of the reflectance of the monitoring substrate when the layer is of the correct thickness and then the termination of the deposition at that point. One disadvantage is that the reflectance of the monitor, or the transmission, is not an easy quantity to measure absolutely, because of calibration drifts during the process due partly to such causes as the gradual coating of the plant windows—almost impossible to avoid. Another is that, whereas with turning value monitoring it is often possible to use just one single monitor, on which all the layers can be deposited, so that it becomes an exact replica of the other filters in the batch, in this alternative method the prediction of the reflectances used as termination values is very difficult if only one monitor is used, because small errors in early layers affect the shape of the curve for later layers.

Some of these difficulties may be avoided by using a separate monitor for each and every layer. To avoid the errors due to any shift in calibration which may occur in changing from one monitor to the next or in the coating of the plant windows, it is wise if at all possible to choose the parameters of the system so that the layer is thicker than a quarter-wave at the monitoring wavelength. This ensures that the termination point of the layer is beyond at least the first turning value, which can therefore be used as a calibration check. It will also be found necessary to set up the reflectance scale for each fresh monitoring substrate and the initial uncoated reflectance which will be known accurately can be used for this. Because a large number of monitor glasses is required, special monitor changers have been designed and are commercially available, which will accommodate stacks of 40 or so glasses.

The principal objection which most workers almost instinctively feel towards this system is that no longer is the monitor an exact replica of the batch of filters. This is to some extent a valid objection. The layer which is being deposited on an otherwise uncoated substrate is condensing on top of what may be quite a different structure from the partially finished filters of the batch. Behrndt and Doughty[16] have noticed a definite measurable difference between layers which are deposited on top of an already existing structure and those deposited on fresh substrates. They compared the deposition of zinc sulphide shown by a crystal monitor (this special type of monitor will be discussed shortly), which already had a number of layers on it, with the layer going down on a fresh glass substrate, and found that the layer began to grow on the crystal immediately the source was uncovered, but that the optical monitor took some time to register any deposition. The difference could amount to several tens of nanometres before the rates became equal. This, they decided, was due to the finite time for nuclei to form on the fresh glass surface and the rather small probability of sticking of the zinc sulphide until the nuclei were well and truly formed. Once the film started to grow, all the molecules reaching the surface would stick. On the crystal where a film already existed, not necessarily of zinc sulphide, nucleation sites were already there and the film started to grow immediately. The sticking coefficient of a material on a fresh monitor surface falls with rising vapour pressure, and zinc sulphide has a particularly large vapour pressure. Similar trouble was not experienced with thorium fluoride, which has a much lower vapour pressure. Behrndt and Doughty found that the problem could be solved by providing nucleation sites on the clean monitor slides by precoating them with thorium fluoride, which has a refractive index very close to that of glass. Some 20 nm or so of thorium fluoride was found to be sufficient and did not affect the monitoring of zinc sulphide deposited on top. This effect becomes greater the greater the surface temperature of the monitor. By changing the type of evaporation source to an electron beam unit, which produced less radiant heat for the same evaporation rate, it was found possible to operate at monitor temperatures low enough to cause the effect to disappear.

The authors also noticed an effect which is well known in thin-film optics. Thick substrates tend to have layers condensing on them which are thicker than those on thin substrates in the same or similar positions in the plant. In the case cited by the authors, the thin substrates were around 0.040 in while the thick ones were around half an inch thick. The difference in coating thickness was sufficient to shift the reflectance turning values by some 40–50 nm at 632.8 nm. This was shown, qualitatively, to be due to the difference in temperature between the two substrates. The thicker substrates took longer to heat up than the thin ones. The heating in this particular case was almost entirely due to radiation from the sources and, again when electron-beam sources were introduced, the effect was considerably reduced.

The accuracy of the monitoring process can be improved greatly if a system devised by Giacomo and Jacquinot[17], and known usually as the 'maximètre', is employed. This involves the measurement of the derivative of the reflectance versus wavelength curve of the monitor. At points where the reflectance is a turning value, the derivative of the reflectance with respect to wavelength is zero and is rapidly changing from a positive to a negative value in the case of a maximum and vice versa in the case of a minimum. The original apparatus consisted of a monochromator with a small vibrating mirror before the slits on the exit side so that a small spectral interval was scanned sinusoidally. The output signal from the detector consisted of a steady DC component, representing the mean reflectance, or transmittance, over the interval, a component of the same frequency as the scanning mirror representing the first derivative of the reflectance against wavelength, a component of twice the scanning frequency, representing the second derivative of the reflectance, and so on. A slight complication is the variation in sensitivity of the system with wavelength that appears as a change in the reflectance signal and hence the derivative, unless it is compensated. In their arrangement, Giacomo and Jacquinot produced an intermediate image of the spectrum within the monochromator, and a razor blade positioned along it made a linear correction to the intensity over a sufficiently wide region and was found to be accurate enough. A more usual technique today would be to make a correction electronically. The accuracy claimed for this system is a few tenths of a nanometre, typically 0.2–0.3 nm, and this is certainly achieved. A problem, as we have seen in chapter 9, is that the layers are frequently insufficiently stable themselves to retain optical thicknesses to this accuracy, especially when exposed to the atmosphere.

A method, similar in some respects, but with some definite advantages in interpretation, has been devised by Ring and Lissberger[18, 19]. It consists of measuring the reflectance or transmittance at two wavelengths and finding the difference. In the original system, a monochromator was used, containing a chopping system that switched the output of the monochromator from one wavelength to another and back again. The AC signal from the detector was a measure of the difference. Since the two wavelengths could be placed virtually anywhere within the region of sensitivity of the detector, the method had greater flexibility than the Giacomo and Jacquinot system. Greatest contrast in the two reflectance signals as a layer was being deposited could be obtained by placing the two wavelengths at the points of greatest opposite slope in the characteristic of the thin-film structure at the appropriate stage. When the signals at the two wavelengths were equal, the output of the system passed through a null, and, if displayed on a chart recorder, made detection of the terminal point of a particular layer, usually indicated by the null, particularly easy to detect.

More recently, the ideas inherent in these systems have been extended to broad spectral regions. Although the principles of these more modern methods are not new, it is the advances in detectors and in electronics and data analysis that have made them practical. Many of the systems have been developed in industry and frequently have not been published. In the cases of those that have been written up, detailed descriptions of the precise way in which they are used have often been lacking. Usually the technique involves a comparison between the spectral characteristic which is actually obtained at any instant, and that which is required at the instant of termination of the particular layer. In the earlier systems this was carried out visually by displaying both curves on a cathode ray tube. This works well when there is a close match between predicted and measured performance but frequently errors in earlier layers, and changes in the characteristics of layers from what is expected, cause the actual curves to differ to a greater or lesser extent from the predictions. In these circumstances, there can be great difficulty in assessing visually the correct moment to terminate a layer. The most recent systems, therefore, are usually linked to a computer which calculates a figure of merit which can either be displayed to a plant operator or, better still, used in the completely automatic termination of layers.

Details of scanning monochromator systems have been published by a number of authors. An early description of such a system is that of Hiraga *et al*[20], where the scanning was carried out by a rotating helical slit assembly. Pelletier and his colleagues in Marseilles[21,22] have developed two such systems. The first uses a stepping motor to rotate a grating and scan the system over a wide wavelength region, the second uses a holographic grating with a flat spectrum plane in which is situated a silicon photodiode array detector which can be scanned electronically.

The quartz crystal monitor

The normal modes of mechanical vibration of a quartz crystal have very high Q and can be transformed into electric signals by the piezoelectric properties of the quartz and vice versa. The crystal acts, therefore, as a very efficient tuned circuit that can be coupled into an electrical oscillator by adding appropriate electrodes. Any disturbance of its mechanical properties will cause a change in its resonant frequency. Such a disturbance might be an alteration of the temperature of the crystal or its mass. The principle of monitoring by the quartz crystal microbalance (as it is called) is to expose the crystal to the evaporant stream and to measure the change in frequency as the film deposits on its face and changes the total mass. In some arrangements the resonant frequency of the crystal is compared with that of a standard outside the plant and the difference frequency is measured, in others the number of vibrations in a given time interval is measured digitally. Usually the frequency shift will be converted internally into a measure of film thickness using film constants fed in by the operator. Since the signal from the quartz crystal monitor changes constantly in the same direction it can be more simply used in automatic systems than optical signals.

The mechanical vibrational modes of a slice of quartz crystal are very complicated. It has been found possible to limit the possible modes and the coupling between them by cutting the slice with respect to the axes of the crystal in a particular way, by proportioning the dimensions of the slice correctly and by supporting the crystal in its holder in the correct way. Quartz crystal vibrational modes also vary with temperature, some having positive temperature coefficient and some negative, and it has been found possible to cut the slice in such a way that modes which have opposite temperature dependence are intentionally coupled so that the combined effect is a resonant frequency independent of temperature over a limited temperature range. The usual cut of crystal which is used in thin-film monitors is the *AT* cut. This is cut from a slice which was oriented so that it contained the *x*-axis of the crystal and was at an angle of 35° 15′ to the *z*-axis. The mode of vibration is a high-frequency shear mode (figure 10.9) and the temperature coefficient is small over the range −40°C to +90°C, of the order of $\pm 10^{-6}\ ^{\circ}C^{-1}$ or slightly greater. The coefficient changes sign several times throughout the range so that the total fractional change in frequency over the complete range is only around[23] 5×10^{-5}. Usually the frequency chosen is around 5 MHz although the range could be anything from 0.5 MHz to 50 or 100 MHz.

As the thickness of the evaporant builds up, the frequency of the crystal falls and the reduction in frequency is proportional both to the square of the

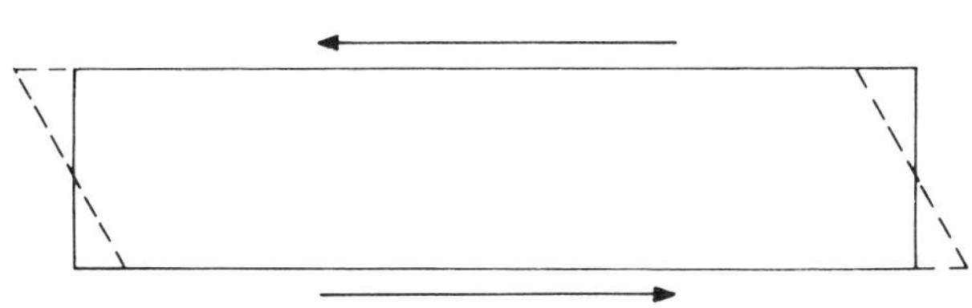

Figure 10.9 Quartz crystal operating in shear.

resonant frequency and to the mass of the film deposited. In a typical arrangement the measurement of mass thickness can be carried out to an accuracy of around 2%, which should be adequate for most optical filters. Unfortunately, the sensitivity of the crystal decreases with increasing build up of mass and the total amount of material which can be deposited before the crystal must be cleaned is limited. With existing crystals this makes them less useful for multilayer work, especially in the infrared, where in most cases a single crystal could not accommodate a complete filter. One way round this problem is to place a screen over the filter which cuts down the material reaching it to a fraction of that reaching the substrates in the batch. This, of course, reduces the accuracy of the system. Because the crystal measures mass and not optical thickness, it must be calibrated separately for each material used. One further difficulty, important only in some applications, is that the temperature of the crystal must be limited to below 120°C (otherwise the temperature coefficient becomes excessively large), so it may not always be possible to keep it at the same temperature as the other substrates in the plant.

Because of these factors the crystal monitor has not been much used for multilayer filter work except in the control of evaporation rate during deposition while the actual layer thickness control is carried out optically. However, for coatings involving a smaller number of layers, such as antireflection coatings, or for less critical multilayers, the quartz crystal can be and has been used quite satisfactorily, and possesses the two very real advantages of ease of installation—no windows required in the plant and no difficult lining-up problem—and of ease of interpretation of signal which just increases linearly with mass and does not oscillate like the optical signal. These advantages have made the quartz crystal popular for processes which are automatically controlled.

A useful set of instructions and tips on the quartz crystal monitor will be found in a paper by Riegert[23] which deals much more fully with the topics mentioned above.

TOLERANCES

The question of how accurately we must control the thickness of layers in the deposition of a given multilayer is surprisingly difficult to answer and has attracted a great deal of attention over the years.

One of the earliest approaches to the assessment of errors permissible in multilayers was devised by Heavens[24] who used an approximate method based on the alternative matrix formulation in equation (2.78). His method, useful mainly when calculations must be performed manually, consisted of a technique for recalculating fairly simply the performance of a multilayer with a small error in thickness in one of the layers. He showed that the final reflectance of a quarter-wave stack is scarcely affected by a 5% error in any one of the layers.

Lissberger[25] developed a method for calculating the performance of a multilayer involving the reflectances at the interfaces. In multilayers made up of quarter-waves, the expressions took on a fairly simple form which permitted the effects of small errors, in any or all of the layers, on the phase change caused in the light reflected by the multilayer to be estimated. Lissberger's results, applied to the all-dielectric Fabry–Perot filter, show that the most critical layer is the spacer. The layers on either side of the spacer layer are next most sensitive, and the remainder of the layers progressively less sensitive the further they are from the spacer.

We have already mentioned in chapter 7 the paper by Giacomo, Baumeister and Jenkins[26] where they examined the effects on the performance of narrowband filters of local variations in thickness, or 'roughness', of the films. This involved the study of the influence of thickness variations in any layer on the peak frequency of the complete filter. The treatment was similar in some respects to that of Lissberger. For the conventional Fabry–Perot filter, layers at the centre had the greatest effect. If all layers were assumed equally rough, the design least affected by roughness would have all the layers of equal sensitivity and attempts were made to find such a design. A phase-dispersion filter gave rather better results than the simple Fabry–Perot, but still fell short of ideal.

Baumeister[27] introduced the concept of sensitivity of filter performance to changes in the thickness of any particular layer. The method involved the plotting of sensitivity curves over the whole range of useful performance of a filter, curves which indicated the magnitude of performance changes due to errors in any one layer. His conclusions concerning a quarter-wave stack were that the central layer is the most sensitive and the outermost layers least sensitive. An interesting feature of these sensitivity curves for the quarter-wave stack is that the sensitivity is greatest nearest the edge wavelength. This is confirmed in practice with edge filters where errors usually produce more pronounced dips near the edge of the transmission zone than appear in the theoretical design.

Smiley and Stuart[28] adopted a different approach using an analogue computer. There were some difficulties involved in devising an analogue computer, but, once constructed, it possessed the advantage at the time that any of the parameters of the thin-film assembly could be easily varied. A particular filter which they examined was:

$$\text{Air} \,|\, 4H\,L\,4H \,|\, \text{Air}$$

with $n_H = 5.00$ and $n_L = 1.54$. Errors in one of the $4H$ layers and in the L layer were investigated separately. They found that errors greater than 1% in one $4H$ layer had a serious effect, errors of 5%, for example, caused a drop in peak transmittance to 70% and errors of 10% a drop to 50%, together with considerable degradation in the shape of the pass band. Errors of up to 10% in the L layer had virtually no effect on either the shape of the pass band or on the peak transmittance.

An investigation was performed by Heather Liddell as part of a study reported by Smith and Seeley[29] into some effects of errors in the monitoring of infrared Fabry–Perot filters of designs:

$$\text{Air}\,|\,HLHL\ HH\ LHLHL\,|\,\text{Substrate}$$

and

$$\text{Air}\,|\ HL\ HH\ LHL\ |\,\text{Substrate.}$$

A computer program to calculate the reflectance of a multilayer at any stage during deposition was used. Monitoring was assumed to be at or near a frequency of four times the peak frequency (i.e. a quarter of the desired peak wavelength) of the completed filter. It was shown that, if all layers were monitored on one single substrate, then, provided the form of the reflectance curve during deposition was predicted, and it was possible to terminate layers at reflectances other than turning values, there could be an advantage in choosing a monitoring frequency slightly removed from four times peak frequency. If no corrections were made for previous errors, then a distinct tendency for errors to accumulate in even-order monitoring (that is monitoring frequency an even integer times peak frequency) was noted.

The major problem in tolerancing is that real errors cannot be treated as small. That is to say that first order approximations are unrealistic. The error in one layer interacts non-linearly with the errors in other layers and it is not possible to treat them separately.

In recent years the most satisfactory approach for dealing with the effects of errors and the magnitude of permissible tolerances has been found to be the use of Monte Carlo techniques. In this method, the performance of the filter is calculated, first with no errors and then a number of times with errors introduced in all the layers. In the original form of the technique introduced by Ritchie[30], the errors are thickness errors and completely random and uncorrelated. They belong to the same infinite population, taken as normal with prescribed mean and standard deviation. The performance curves of the filter without errors and of the various runs with errors are calculated. Although statistical analyses of the results can be made, it is almost always sufficient simply to plot the various performance curves together, when visual assessment of the effects of errors of the appropriate magnitude can be made. The method really provides a set of traces which reproduce, as far as possible, what would actually be achieved in a succession of real production batches. The characteristics of the infinite normal population can be varied and the procedure repeated. It is sufficient to calculate some eight or perhaps ten curves for a set of error parameters. The level of error at which a satisfactory process yield would be achieved can then readily be determined. In the earliest version of the technique, the various errors were drawn manually from random number tables and converted into members of a normal population using a table of area under the error curve. (The procedure is described in textbooks of statistics—see Yule and Kendall[31], for example.) Later versions of the technique simply generate the random errors by computer. Although the errors are usually drawn from a normal population, the type of population has little effect on the order of the results. Normal distributions are convenient to program, and since there is no strong reason for not using them and because errors made up of a number of uncorrelated effects are well represented by normal distributions, most error analyses do make use of them.

Figure 10.10 shows some examples of plots where the errors are simple independent thickness errors of zero mean. From these and similar results we find that the errors which can be tolerated in a longwave-pass filter are normally of standard deviation 5 %, in a shortwave-pass filter around 2.5 %, and in an antireflection coating such as the quarter–half–quarter around 3 %. In a DHW filter of the type in figure 10.10, the permissible errors are not greater than 2 % while, for narrower filters or filters with greater number of cavities, the tolerances must be tighter. In fact, a rough guide is that the permissible standard deviation is not greater than the halfwidth of the filter. In a Fabry–Perot filter the main effect of random errors is a peak wavelength shift, the shape of the pass band being scarcely affected even by errors as large as 10 %. The standard deviation of the scatter in peak wavelength is slightly less than the standard deviation of the layer thickness errors so that some averaging process is operating, although the orders of magnitude are the same.

A system of monitoring in which the thickness errors in different layers are uncorrelated requires that each layer should be controlled independently of the others. In this type of monitoring, therefore, we cannot expect high precision in the centring of narrowband Fabry–Perot filters and we foresee great difficulties in being able to produce narrowband multiple-cavity filters at all.

This monitoring arrangement is what we have called indirect. Systems where each layer is controlled on a separate monitoring chip are of this type. There are difficulties with monitoring of low-index layers on a fresh glass substrate because of the small changes in transmittance or reflectance, and so the monitoring chips are usually changed after a low-index layer and before a high index, two or four layers per chip being normal. Sometimes these layers will be monitored to turning values. More frequently what is sometimes called level monitoring will be used. Here the layer reflectance or transmittance signal is terminated at a point removed from the turning value where the signal is still changing, leading to an inherently greater accuracy. This approach involves what is really an absolute measurement of reflectance or transmittance, and so the termination point is frequently chosen to be after a turning value rather than before, so that the extremum can be used as a calibration. This usually implies a shorter wavelength for monitoring or the introduction of a geometrical difference between batch and monitor, placing the monitor nearer the source or placing masks in front of the batch.

Narrowband filters are not normally monitored in this way. Instead, all the layers are monitored on the same substrate, usually the actual filter being

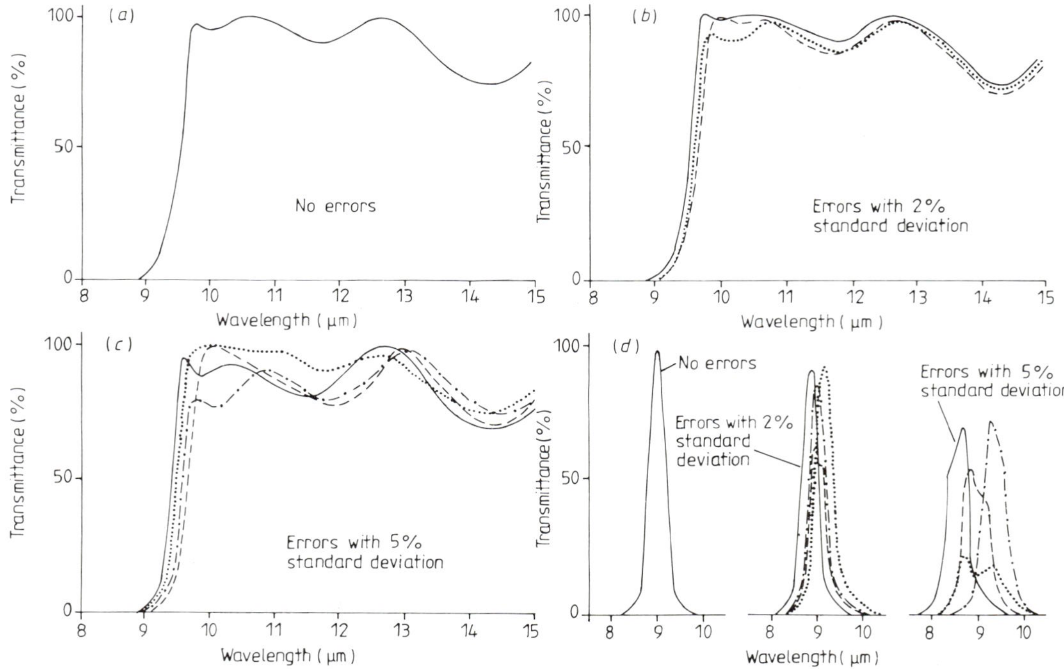

Figure 10.10 The effects of random errors in layer thickness on the performance of thin-film filters. (*a*), (*b*) and (*c*) A typical longwave-pass filter of design: Air$|$1.49 L (0.5 L H 0.5L)^7 1.49$H|$Ge where H = PbTe (n = 5.30) and L = ZnS (n = 2.35); (*d*) A DHW or two-cavity filter. Design: Air$|H$ LL $HHLHLH$ LL $HL|$Ge where L = ZnS, H = PbTe, $\lambda_0 = 9$ μm. (Some of the curves have been broken for clarity.) (Courtesy of F S Ritchie and Sir Howard Grubb, Parsons & Co Ltd.)

produced, a system known as direct monitoring. At the peak wavelength of the filter, the layers should all be quarter-waves or half-waves, and so we can expect a signal which reaches an extremum at each termination point. The accuracy cannot therefore be particularly high for any individual layer and, at first sight, it would appear that the achievable accuracy should be far short of what must be required. Since each layer is being deposited over all previous layers on the monitor substrate, then there is an interaction between the errors in any layer and those in the previous layers not included in the tolerancing calculation described above. We really require a technique which models the actual process as far as possible and this is a quite straightforward piece of computing. Each layer is simply considered to be deposited on a surface of optical admittance corresponding to that of the multilayer which precedes it, rather than on a completely fresh substrate. The results of such a simulation are shown in figure 10.11, taken from Macleod[32], which demonstrates the powerful error compensation mechanism that has been found to exist. The compensation has also been independently and simultaneously confirmed by Pelletier and his colleagues[33]. Its nature is perhaps best explained by the use of an admittance diagram.

Figure 10.12 shows such a diagram drawn for several quarter-waves. Since both the isoreflectance contours (see chapter 2) and the individual layer loci are

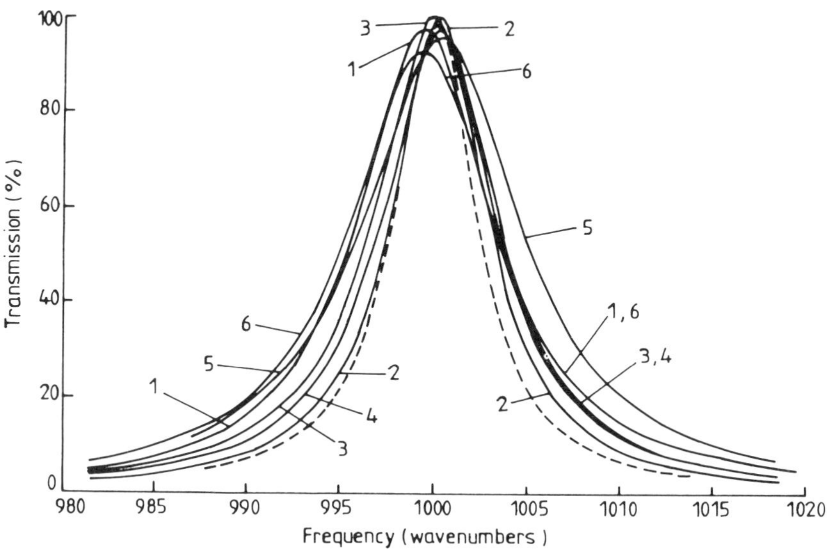

Figure 10.11 The effect of 1 % standard deviation reflectance error on the performance of the Fabry–Perot filter: Air |*HLHL HH LHLH*| Ge. The substrate is germanium ($n = 4.0$), *L* represents a quarter-wave of ZnS ($n = 2.3$) and *H* a quarter-wave of PbTe ($n = 5.4$). The monitoring is in first order. The dotted curve is the performance with no errors. (After Macleod[32].)

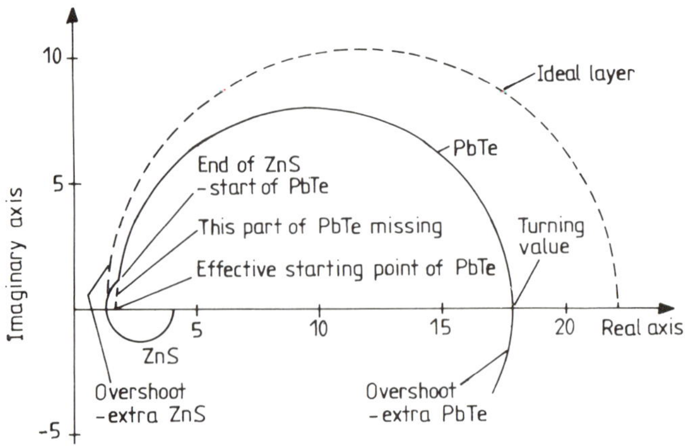

Figure 10.12 The admittance locus of the first two layers of the filter in figure 10.10 when there is an overshoot in the first layer of around 1/8th wave optical thickness. (After Macleod[32].)

circles centred on the real axis, the turning values must always occur at the intersections of the loci with the real axis, regardless of what has been deposited earlier. At the termination point of each layer there is the possibility of restoring the phase to zero or π. As far as any individual layer is concerned, it is principally the over- or undershoot of the previous layer that affects it. If the previous layer is too thick, the current one will tend to be thinner to compensate and vice versa. Of course it is impossible completely to cancel all effects of an error in a layer. The process is actually transforming the thickness errors into errors in reflectance at each stage since the loci will be slightly displaced from their theoretical position. This is not a serious error. As can be guessed from the shape of the diagram, the reflectance error is a second-order effect. Since the phase is self-corrected each time a layer is deposited, the peak wavelength of the filter will remain at the desired value, that of the monitoring wavelength. The remaining error, the residual one in reflectance, is then translated into changes in peak transmittance and halfwidth. Since the reflectance change is always a reduction, the bandwidth of an actual filter is invariably wider than theoretical. The peak transmittance falls to the extent that the reflectances on either side of the spacer layer are unbalanced. This is usually quite small and the reduction in peak transmittance is generally much less important than the increase in bandwidth.

In this monitoring arrangement, thickness errors in any individual layer are a combination of a compensation of the error in the previous layer together with the error committed in the layer itself. The magnitude of the thickness errors can be quite misleading in interpreting whether or not the filter can be made successfully. In figure 10.11, for example, thickness errors of the order of 50% occur in some layers and yet the filter characteristics are all useful ones.

The important characteristic is actually the error in reflectance or transmittance in determining the turning values, and it is possible to develop theoretical expressions which relate the reflectance or transmittance errors to the reduction in performance of the final filter[32]. This analysis includes an assessment of the sensitivity of each layer to errors which indicate those layers where the greatest care in monitoring should be exercised. These can be different from the thickness sensitivity of Lissberger[25] already mentioned. With high-index spacer layers, greatest sensitivity is found in the low-index layers following the spacer, while with low-index spacers, the spacer itself has the highest sensitivity. A feature of this analysis is that it demonstrates that for any particular error magnitude, there is a point where improved halfwidth does not result from an increase in the number of layers because the effect of errors is increasing more rapidly than the theoretical decrease in bandwidth. Then it is necessary to move to second- and higher-order spacers if decreased bandwidth is to result. This corresponds to what is found in practice. The error analysis also demonstrates that high-index spacers are to be preferred over low-index. We have already seen in chapter 7 that high-index spacers give decreased angular sensitivity and greater tuning range.

Formulae which permit the calculation of the errors in reflectance, in halfwidth and in peak transmittance as a function of the magnitude of the random errors in determining the turning values exist[32], but for most purposes a computer simulation will suffice. It should be noted that the compensation is effective only for the first order. Second-order monitoring, that is monitoring at the wavelength for which the layers are all half-waves, is not effective in preserving the peak wavelength. We can understand this because the admittance diagram is quite different and so the compensation is of a different nature. Likewise, third-order monitoring is not as effective as first-order, and, although the scatter in peak wavelength is less than that obtained with second-order monitoring, it is, nevertheless, quite large.

Multiple-cavity filters are similar in behaviour but there are some complications. The coupling layers in between the various Fabry–Perot sections of the filter turn out to be particularly sensitive to errors in a rather peculiar way. Preliminary examination of the admittance diagram for the various layers of a multiple-cavity filter and even the standard error analysis do not immediately reveal any marked difference in terms of error sensitivity between these layers and those of Fabry–Perot filters. Closer investigation shows that there is always one transition from one layer to the next occurring at or near to the central coupling layer where a thickness error is compensated by an error of the same rather than the opposite sense.[34] The condition is sketched in figure 10.13. An increase in thickness in the first layer results in an increase in thickness of the subsequent layer and vice versa. This condition must occur once between each pair of cavities. The net result is an increase or decrease in the relative spacing of the cavities causing the appearance of a multiple-peaked characteristic curve. The peaks become more pronounced, the greater the

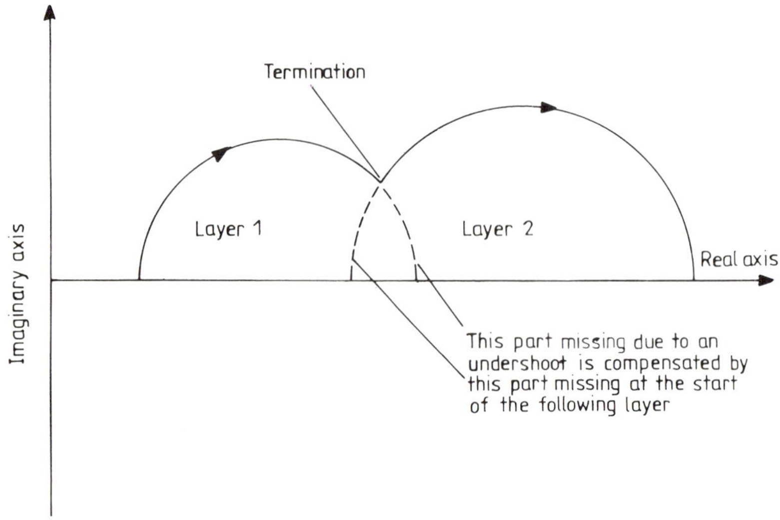

Figure 10.13 Error compensation when the admittance circles are on the same side of the real axis. (After Macleod and Richmond[34].)

relative error in spacing. One of the peaks always corresponds to the normal control wavelength and is close to the theoretical transmittance. The other peaks (one for a two-cavity, two for a three-cavity and so on) can appear on either side of the main peak depending on the nature of the particular errors. This false compensation can be destroyed if the second of the two layers concerned can be controlled independently of the others, either on a separate monitor plate or by a quartz crystal monitor, or even by simple timing. It is essential that it should also be deposited on the regular monitor as well, so that the compensation of the full filter should not be destroyed[34].

Pelletier and his colleagues[35] have studied theoretically the behaviour of the 'maximètre' types of monitoring systems in the production of narrowband filters. They conclude that, as we would expect, the accuracy of the system in the production of single layers is very much better than a single-wavelength system. In the monitoring of narrowband filters all on one substrate there is a compensation process operating like the turning value method but it is more complex in operation. For very small errors in most layers the system works adequately, but for large errors in most layers or small errors in certain critical layers, the errors accumulate in such a way as to cause a drastic broadening of the bandwidth of a Fabry–Perot filter or complete collapse of a multiple-cavity filter. Pelletier has introduced two concepts to describe this behaviour. Accuracy represents the error that will be committed in any particular layer without reference to the multilayer system as a whole. Stability represents the way in which the errors accumulate as the multilayer deposition proceeds. The accuracy of the 'maximètre' is excellent and greater than in the turning value method, but the stability in the control of narrowband filters is very poor and it can easily become completely unstable. Subsidiary measurements are therefore required to ensure stability if advantage is to be taken of the very great accuracy that is possible. Narrowband filters and their monitoring systems have been surveyed in reference 36.

The concepts of accuracy and stability and the discovery that the one does not ensure the other imply that different measurements may be necessary to ensure that both are simultaneously assured. This leads to the idea of broadband monitoring in which simultaneous measurements are made at a large number of wavelengths over a wide spectral region and a merit function representing the difference between actual and desired signals is computed. The merit function can then be used as a monitoring signal and layer deposition terminated when the merit function reaches a minimum. Although perfect deposition should ensure a minimum of zero in the figure of merit, inevitable errors in layer index and homogeneity will perturb the result. The accuracy and stability of such a broadband system in the monitoring of certain components such as beam splitters has been investigated by computer simulation[37] and evidence found for useful error compensation. Apart from the very qualitative justification discussed above no theory for such compensation yet exists and it may operate only in quite specific cases. Extensions of broadband monitoring to a system that would re-optimise those layers of a design yet to be deposited on the basis of errors measured in earlier layers appear possible and are under investigation in a number of laboratories. Even if successfully developed they are never likely to be able to reduce the need for stable reproducible materials.

Quartz crystal monitoring, in which the mass rather than optical thickness is measured, seems unlikely to possess powerful compensation. Yet simulation of a simple broadband system for antireflection coatings comparing optical monitoring with quartz crystal gave results which indicate that the quartz crystal is in no way inferior[38]. The relative merits of quartz crystal and optical monitoring form a subject of constant debate and published results for quartz crystal are impressive[39, 40]. It is clear that narrowband filters, if they are to be controlled in peak wavelength, do require direct optical monitoring, but quartz crystal monitoring is suitable for most other filter types. The general opinion, based to some extent on instinct, is that quartz crystal monitoring is most suitable for production of successive batches of identical components. For single runs of varying coating types, optical monitoring appears normally to be preferred. Optical monitoring is also preferred in applications such as filters for the far infrared, where very large thicknesses of materials are deposited in each coating run.

REFERENCES

1 Holland L and Steckelmacher W 1952 The distribution of thin films condensed on surfaces by the vacuum evaporation method *Vacuum* **2** 346–64

2 Behrndt K H 1963 Thickness uniformity on rotating substrates *Trans. 10th AVS Nat. Vac. Symp.* pp 379–84 (London: Macmillan)

3 Knudsen M 1915 Das Cosinusgesetz in der kinetischen Gastheorie *Annalen der Physik, IV. Folge* **48** 1113–21

4 Jancke E and Emde F 1952 *Tables of higher functions* 5th ed (Leipzig: Teubner). Note also the 4th edition (New York: Dover 1945). The conversion of the expression in the text to the required standard form appears in §6, p 57 of the fifth edition.

5 Keay D and Lissberger P H 1967 Application of the concept of effective refractive index to the measurement of thickness distributions of dielectric films *Appl. Opt.* **6** 727–30

6 Graper E B 1973 Distribution and apparent source geometry of electron-beam heated evaporation sources *J. Vac. Sci. Technol.* **10** 100–103

7 Richmond D 1976 Thin film narrow band optical filters *PhD Thesis* Newcastle upon Tyne Polytechnic

8 Ramsay J V, Netterfield R P and Mugridge E G V 1974 Large-area uniform evaporated thin films *Vacuum* **24** 337–40

9 Holland L 1956 *Vacuum Deposition of Thin Films* (London: Chapman and Hall)

10 Mattox D M 1978 Surface cleaning in thin film technology *Thin Solid Films* **53** 81–96

11 Lee C C 1983 Moisture adsorption and optical instability in thin film coatings *PhD Dissertation* University of Arizona

12 Cox J T and Hass G 1958 Antireflection coatings for germanium and silicon in the infrared *J. Opt. Soc. Am.* **48** 677–80

13 Banning M 1947 Practical methods of making and using multilayer filters *J. Opt. Soc. Am.* **37** 792–7

14 Polster H D 1952 A symmetrical all-dielectric interference filter *J. Opt. Soc. Am.* **42** 21–5

15 Perry D L 1965 Low loss multilayer dielectric mirrors *Appl. Opt.* **4** 987–91

16 Behrndt K H and Doughty D W 1966 Fabrication of multilayer dielectric films *J. Vac. Sci. Technol.* **3** 264–72

17 Giacomo P and Jacquinot P 1952 Localisation précise d'un maximum ou d'un minimum de transmission en fonction de la longeur d'onde. Application à la préparation des couches minces *J. Phys. et le Radium* **13** suppl to no 2; 59A–64A

18 Ring J 1957 *PhD Thesis* University of Manchester

19 Lissberger P H and Ring J 1955 Improved methods for producing interference filters *Opt. Acta* **2** 42–6

20 Hiraga R, Sugawara N, Ogura S and Amano S 1974 Measurement of spectral characteristics of optical thin film by rapid scanning spectrophotometer *Jpn. J. Appl. Phys.* suppl 2, pt 1, 689–92

21 Borgogno J P, Bousquet P, Flory F, Lazarides B, Pelletier E and Roche P 1981 Inhomogeneity in films: limitation of the accuracy of optical monitoring of thin films *Appl. Opt.* **20** 90–94

22 Flory F, Schmitt B, Pelletier E and Macleod H A 1983 Interpretation of wide band scans of growing optical thin films in terms of layer microstructure *Thin Film Technologies* ed J R Jacobsson (*Proc. SPIE* **401** 109–16)

23 Riegert R P 1968 Optimum usage of quartz crystal monitor based devices *Proc. IVth Int. Vac. Congr.* 527–30 (London: The Institute of Physics and the Physical

Society). Published in more detail by Sloan Instruments Corporation, Equipment Division, Alexandria, Virginia, USA
24 Heavens O S 1954 All-dielectric high-reflecting layers *J. Opt. Soc. Am.* **44** 371–3
25 Lissberger P H 1959 Properties of all-dielectric filters. I. A new method of calculation *J. Opt. Soc. Am.* **49** 121–5
26 Giacomo P, Baumeister P W and Jenkins F A 1959 On the limiting bandwidth of interference filters *Proc. Phys. Soc.* **73** 480–89
27 Baumeister P W 1962 Methods of altering the characteristics of a multilayer stack *J. Opt. Soc. Am.* **52** 1149–52
28 Smiley V N and Stuart F E 1963 Analysis of infrared interference filters by means of an analog computer *J. Opt. Soc. Am.* **53** 1078–83
29 Smith S D and Seeley J S 1968 Multilayer filters for the region 0.8 to 100 microns *Final Rep. Contr. AF61(052)-833* Air Force Cambridge Research Laboratories
30 Ritchie F S 1970 Multilayer filters for the infra-red region 10–100 microns *PhD Thesis* University of Reading
31 Yule G U and Kendall M G 1950 *Introduction to the theory of statistics* 14th ed (London: Griffen). See p 379.
32 Macleod H A 1972 Turning value monitoring of narrow-band all-dielectric thin-film optical filters *Opt. Acta* **19** 1–28
33 Bousquet P, Fornier A, Kowalczyk R, Pelletier E and Roche P 1972 Optical filters: monitoring process allowing the auto-correction of thickness errors *Thin Solid Films* **13** 285–90
34 Macleod H A and Richmond D 1974 The effect of errors in the optical monitoring of narrow-band all-dielectric thin film optical filters *Opt. Acta* **21** 429–43
35 Pelletier E, Kowalczyk R and Fornier A 1973 Influence du procédé de contrôle sur les tolérances de réalisation des filtres intérferentiels a bande étroite *Opt. Acta.* **20** 509–26
36 Macleod H A 1976 Thin film narrow band optical filters *Thin Solid Films* **34** 335–42
37 Vidal B, Fornier and Pelletier E 1979 Wideband optical monitoring of non-quarterwave multilayer filters *Appl. Opt.* **18** 3851–6
38 Macleod H A 1981 Monitoring of optical coatings *Appl. Opt.* **20** 82–9
39 Pulker H K 1978 Coating production: new ideas at a time of demand *Opt. Spectra* **12**(8) 43–6
40 van der Laan C J and Frankena H J 1977 Monitoring of optical thin films using a quartz crystal monitor *Vacuum* **27** 391–7

11 Specification of filters and environmental effects

Ideally, if a filter is to be manufactured for a customer for a given application, then the performance required by the customer, and the design, manufacturing and test methods, should all be defined, even if only implicitly. These details form different aspects of the specification of the filter.

There is no standard method for setting up the specification of an optical filter or coating, the problem being much the same as for any other device. There are three main aspects to be considered, the performance specification which lists the details of the performance required from the filter and is usually the customer's specification, the manufacturing specification which defines the design and details the steps involved in the manufacture of the filter, and the test specification laying down the tests which must be carried out on the filter to ensure that it meets the performance requirements, these latter aspects being mainly the concern of the manufacturer. In the following notes a few of the more important points are mentioned, but they do not form a complete guide to the writing of specifications, which is almost a subject in its own right.

Optical filter specifications can conveniently be divided into two sections, one concerned with optical properties and the other with physical or environmental properties. We shall first of all consider the optical properties.

OPTICAL PROPERTIES

Performance specification

The performance specification of a filter is really a statement of the capabilities of the filter in a language that can readily be interpreted by both system designer, or customer, and filter manufacturer alike. It can sometimes be prepared by a filter manufacturer from a knowledge of the performance which he knows he can achieve either for a customer or possibly without having a particular application in mind, as in the case of a standard product in a catalogue about which little need be said here. Probably more often, the performance specification will be written by the system designer and will state a level of performance which is required from a filter in order to achieve a desired level of performance from a system. In writing such a specification, an answer must first of all be given to the question: what is the filter for? The purpose of the filter must be set down as clearly and concisely as possible and this will form the basis for the work on the performance specification. There is really no systematic method for specifying the details of performance. Sometimes it happens that the performance of the system in which the filter is to be used must be of a certain definite level, otherwise there will be no point in proceeding further. The filter performance requirements can then be quite readily set down. Often, however, it will not be quite so simple. No absolute requirement for performance may exist, only that the performance should be as high as possible within allowable limits of complexity or perhaps price. In such a case, the performance of the system with different levels of filter performance must be balanced against cost and system complexity, and a decision made as to what is reasonable. The final specification will be a compromise between what is desirable and what is achievable. This will often need the input of much design and manufacturing information and close contact between customer and manufacturer. It should always be remembered in this that specifications which cannot be met in practice can be of only academic interest.

By way of an example let us briefly consider the case where a spectral line must be picked out against a continuum. Clearly a narrowband filter will be required, but what will be the required bandwidth and type of filter? The energy from the line which will be transmitted by the filter will depend on the peak transmission (assuming that the peak of the filter can always be tuned to the line in question), while the energy from the continuum will depend on the total area under the transmission curve, including the rejection region at wavelengths far removed from the peak. The narrower the pass band, the higher the contrast between the line and the continuum, especially as narrowing the pass band generally also improves the rejection. However, the narrower the pass band, because of the increased difficulty of manufacture, the higher the price, and, further, because of the increased sensitivity to lack of collimation, the larger the tolerable focal ratio. This latter point implies that, for the same field of view, a filter with a narrower bandwidth must be made larger to permit the same focal ratio to be retained, which in turn will increase still further the difficulties of manufacture and, possibly, the complexity of the entire system. Another way of improving the performance of the filter is by increasing the steepness of edge of the pass band while still retaining the same bandwidth. A rectangular pass band shape gives higher contrast than a simple Fabry–Perot of identical halfwidth and usually possesses the additional advantage that the rejection remote from the peak of the filter is also rather greater. This edge steepness can be specified by quoting the necessary tenth peak bandwidth or even the hundredth peak bandwidth. Again, inevitably, the steeper the edges, the more difficult the manufacture and the higher the price.

Because filters, as with any manufactured product, cannot be made exactly to a specification in absolute terms, some tolerances must always be stated. For a narrowband filter, the principal parameters which should be given tolerances are peak wavelength, peak transmission, and bandwidth. Since in almost all applications the higher the peak transmission the better, it is usually sufficient to state a lower limit for it. There are two aspects of peak wavelength tolerance. The first is uniformity of peak wavelength over the surface of the filter. There will always be some grading of the films, although perhaps small, and a limit must be put on this. The effect is similar to that of an incident cone of illumination (which has been discussed on pp 265–9) and it is usually best to limit the uniformity errors in the specification to not more than one third of the halfwidth. The second aspect is error in the mean peak wavelength measured over the whole area of the filter. The tolerance for this is usually made positive so that the filter can always be tuned to the correct wavelength by tilting. For a given bandwidth the amount of tilt which can be tolerated in any application will be determined to a great extent by the aperture and field of the system, since the total range of angles of incidence which can be accepted by a filter falls as the tilt angle is increased.

The bandwidth of the filter should also be specified and a tolerance put on it, but, because of the difficulty of controlling bandwidth very accurately, it is not usually desirable to tie it up too tightly and the tolerance should be kept as wide as possible, not normally less than 0.2 times the nominal figure unless there is a very good reason for it.

One other important parameter which will be involved in the optical performance specification is rejection in the stopping zones, which may be defined in a number of different ways. Either the average transmittance over a range, or absolute transmission at any wavelength in the range, can be given an upper limit. The first would usually apply where the interfering source is a continuum and the second where it is a line source, in which case the wavelengths involved should be stated, if known.

Yet another entirely different method of specifying filter performance is by drawing maximum and minimum envelopes of transmission against wavelength. The performance of the filter must not fall outside the region laid down by the envelopes. It is important that the acceptance angle of the filter be also stated. This type of specification is rather more definite than the first type mentioned above. A disadvantage, however, is that it may be rather too severe since everything is stated in absolute terms when average values can be just as good. A further point is that it is impossible to devise a test to determine whether or not a filter meets an absolute specification of this type. Finite bandwidth of the measuring apparatus will ultimately be involved. It is advisable, therefore, if specifying a filter in this way to include a note to the effect that the performance specified at each wavelength is the average over a certain definite interval.

There is little else that can be said in general terms about the optical performance specification. In any one application these factors will assume different relative importance and each case must to a very great extent be considered on its own merits. Clearly this is an area where it is of prime importance that the system designer works very closely with the filter designer.

Manufacturing specification

We shall now consider briefly the manufacturing specification which contains the filter design together with details of the manufacturing method. In most cases, this will be intended for the use of the plant operator.

First, the filter design, including the materials, will be given. Most filters contain not more than three different thin-film materials having relatively low, medium and high refractive index. Designs are usually written in terms of quarter-wave optical thicknesses at a reference wavelength, λ_0, using the symbols L, M, and H. Typical designs may be written:

$$L|\mathrm{Ge}|LHLHHLH \qquad L = \mathrm{ZnS} \quad H = \mathrm{Ge}$$

or

$$M|\mathrm{Si}|MHLHHLH \qquad L = \mathrm{CaF_2} \quad M = \mathrm{ZnS} \quad H = \mathrm{Ge}$$

the substrates being indicated by the symbols |Ge| and |Si|. Next the constructional details should be written down. These consist of the monitoring method to be used including the wavelengths, and the form of the signals together with other important details such as substrate temperature, special types of evaporation sources, and so on. It will be found useful to arrange the whole manufacturing specification in the form of a table which can be issued to the plant operators for use as a check list. Operators should always be encouraged to observe critically the operation of the plant so that faults or anomalies can be spotted at an early stage, and it is a help in this if they are expected to list comments in appropriate places on the form. It will also be found convenient to give each filter production batch a different reference number. Once the filters are produced, the completed specification form can then be filed by the plant operator to form the plant log book. Additional information such as pumping performance can also be recorded on the sheets, useful from the maintenance point of view.

Test specification

Probably the most important specification of all is the test specification. This lays down the complete set of tests which will be carried out on the filters to measure the performance. It should always be remembered that, although the filter will have been designed to meet a particular performance specification, it is only the performance laid down in the test specification that can actually be guaranteed, and, although it may seem obvious, the test specification must be written with the requirements of the performance specification always in mind. In fact it is possible simply to specify the performance of a filter as that which will pass the appropriate test specification. It will sometimes be found that the test specification, if it exists at all, is a rather loose document or that sometimes the customer's performance specification will serve both roles. If so, then someone somewhere along the line will be interpreting the performance specification to decide on the tests which have to be applied, and it is always better to have the tests and the method of interpretation in writing.

The first essential in any test specification is a definite statement of the performance or the make and type of the test equipment to be used. This ensures that results can be repeated if necessary, even if remote from the original testing site. Next, the various tests together with the appropriate acceptance levels can be set down.

It is in the measurement of such factors as uniformity where the tests and the method of interpretation are particularly important. Absolute uniformity is impossible to measure in the ordinary way. The peak wavelength would have to be measured at every point on the filter with an infinitesimally small measuring beam. A simpler and usually satisfactory method is to check the peak wavelength at the centre of the filter and at four approximately equally spaced areas around the circumference, using a specified area of measuring beam. The spread over the filter is taken to be the spread in the values of peak wavelength over the five separate measurements. The spectrometer used for the measurement will also have a finite bandwidth and features of the filter which are rather less than this will, in general, not be picked up. This applies particularly to the measurement of rejection. Rejection must be measured over a very wide region, and for the test to be completed in a reasonable time, a fast scanning speed must be used, which in turn requires a broad bandwidth. This averages the measurement over a finite region and is one of the reasons for stating the actual wavelengths of the lines if the energy which is to be rejected has a line rather than a continuous spectrum.

A technique for measuring the rejection of films using a Fourier transform spectrometer has been suggested by Bousquet and Richier[1]. While this is difficult to apply in the visible region, the availability of commercial Fourier transform spectrometers for the infrared makes it a feasible technique for infrared filters.

Of course, inevitably, the more extensive the testing which must be carried out on each individual filter, the more expensive that filter is going to be. Performance testing of low-price standard filters is, in the main, carried out on a batch basis, with only a few details being checked on each individual filter. This is a point which should be borne in mind by a prospective customer buying a standard filter from a catalogue, that a superlative level of performance cannot be absolutely guaranteed from a filter which, by its price, cannot have had more than the basic testing carried out on it. Only when very large numbers of batches of nominally similar filters are involved (solar-cell cover slips for example, which are described in chapter 12) can guaranteed superlative quality accompany low price.

So far we have dealt with the directly measurable optical performance of the filter, but there are additional properties which are of a subjective nature and rather more difficult to measure. These are connected with the quality and finish of the films and substrates. Substrates are specified as for any optically worked component, details such as flatness or curvature of surface, degree of polish and allowable blemishes, sleeks and the like, can all be stated. We shall not consider substrates further here. There is a specification, used particularly in the USA, MIL–0–13830 A, which gives a useful set of standards for optical components including substrates.

The quality of the coating can be measured by the presence or absence of defects such as pinholes, stains, spatter marks, and uncoated areas.

Pinholes are important for two reasons. First they are actually small uncoated, or partially uncoated, areas and as such will allow extra light to be transmitted in the rejection regions, reducing the overall performance of the filter. Second, and this is especially so for filters for the visible region, they are unsightly and detract from the appearance. In fact, they usually look worse to the eye than the effect they actually have on performance. Apart from the purely subjective appearance, the permissible level of pinholes can be defined on the basis of a given maximum number of a certain size per unit area, calculated to reduce the rejection in the stop bands by not more than a given amount. To calculate this figure, a minimum area of filter which will be used at any one time must be assumed. This will depend on the application, but in the absence of any definite information on this a suitable figure is 5 mm × 5 mm. Obviously the smaller this area, the lower the size of the largest pinhole. Of course, the actual counting of pinholes in any filter would involve a prohibitive amount of labour and in practice, with visible filters, the measurement is usually carried out visually, comparing the filter with limit samples. A simple fixture consisting of a light box with sets of filters laid out on it, some just inside, some on, and some just outside the limit, can be easily constructed. For infrared filters on transparent substrates this method can also be applied, but for filters on opaque substrates it is easier to measure actual rejection performance.

Spatter marks are caused by fragments of material ejected from the sources and, unless gigantic, do not affect the optical performance, the danger being that the fragments may be removed later, leaving pinholes. The incidence can be tied down just as with pinholes, but, as the optical performance is not affected unless the number of marks is enormous, the basis for deciding what is permissible is entirely subjective—although usually if the spatter is particularly bad it will be accompanied by pinholes. Often specifications will state that there must be no spatter marks visible to the naked eye, but this is vague, particularly when dealing with inspectors with no optical experience. Disagreements can arise between manufacturer and customer especially when, as can happen, the customer's inspectors use an eye glass to assist the naked eye. The best course is probably to relate the test to agreed limit samples when it can be carried out in exactly the same way as for pinholes, or else to omit it altogether.

Stains can be caused in a number of ways. The most common reason is a faulty substrate. One type of mark which is often seen, especially when antireflection coatings are involved, is due to a defect in the optical working. The polishing process consists partly of a smoothing out of irregularities in the surface by a movement of material. If the grinding, which always precedes the polishing, has been too coarse, then the deeper pits during the polishing are filled in with material which is only loosely bonded to the surface although the polish appears satisfactory to the eye. In the heating and then coating of the surface, this poorly bonded material breaks away, leaving a patch of surface which is etched in appearance and often possesses well defined boundaries. The only remedy for this type of blemish is improved polishing techniques. Other stains which may appear can be caused by faulty substrate cleaning. If water or even alcohol is allowed to dry without wiping on a surface, water marks appear. Droplets should always be wiped off the surface with a clean tissue or cloth during the cleaning process. Water should never be allowed to dry on the surface by itself. Stains, unless particularly bad, do not usually affect the optical performance to anything like the extent which their appearance would suggest (except in the case of very high performance components such as Fabry–Perot plates or laser mirrors), and the basis for judging them is again subjective.

Finally, the filter must be held in a jig during coating so that at least some uncoated areas must exist. These usually take the form of a ring around the periphery of the filter, perhaps around 0.5 mm wide. There will be a slight taper in the coating at the very edge which must also be allowed for, the combined taper and uncoated area forming a strip perhaps 1.0 mm in width. The uncoated area actually serves a useful purpose because mechanical mounts can grip the component at this point without damaging the coating. Damage near the edge is dangerous because it is there that delamination is frequently initiated. Uncoated areas should not occur within the boundary of the filter proper; when they do it is a sign of adhesion failures which may recur. They may be due to substrate contamination or to moisture penetration with weakening of adhesion, as described in chapter 9, but they are always cause for rejection of the component. Blisters, too, which are a slightly different version of the same fault, are also cause for immediate rejection.

PHYSICAL PROPERTIES

As far as the physical properties of the filter are concerned, there are two primary aspects. The first is the dimensions of the filter which must meet the requirements laid down. This is purely a matter of mechanical tolerances which we need not go into any further here. The second is that the filter must be capable of withstanding, as far as possible, the handling it will receive in service and also of resisting any attack from the environment. The assessment of the robustness of the coating will now be considered in greater detail.

The approach almost invariably used in defining and testing the robustness of a coating is to combine the performance and test specifications. A series of tests, which reproduce typical conditions likely to be met in practice, are set up, and then performance is defined as being a measure of the ability to pass these particular tests. This avoids the difficulty which would be met in setting up a more general performance specification.

There is one basic difference between the tests of optical performance and those which we are about to discuss. Optical tests are all nondestructive in nature while tests of robustness are, in the main, destructive. The filters are tested deliberately to cause damage, and the extent of the damage if it can be measured used as a measure of the robustness of the filter. It is thus not possible to carry out the whole series of tests on the actual filter which is to be supplied to the customer and it is normal to use a system of batch testing. A number of filters are made in a batch and either one or perhaps two are chosen at random for testing. Provided these test filters are found acceptable then the complete batch is assumed satisfactory. This arrangement is, of course, not peculiar to thin-film devices. Another aspect of this batch testing is involved in what is known as a type test. Often if a large number of filters, all of the same type and characteristic, are involved, a series of very extensive and severe tests will be carried out on a sample of filters from a number of production batches. The test results will then be assumed to apply to the entire production of this type of filter. Once the filters have passed this type test, normal production testing is carried out on a reduced scale. It is imperative that once the type test has been successful there are no subsequent changes, even of a minor nature in the production process, otherwise the type test would be invalidated.

Abrasion resistance

Coatings on exposed surfaces, such as the antireflection coating on a lens, will probably require cleaning from time to time. Cleaning usually consists of some sort of rubbing action with a cloth or perhaps lens tissue. Often there may be dust or grit on the surface of the lens which may not be removed before rubbing. The result of such treatment is abrasion and it is important to have the abrasion resistance of exposed coatings as high as possible. An absolute measure of abrasion resistance is not at all easy to establish because of the difficulty of defining it in absolute terms, and the approach which has been adopted has been to reproduce, under controlled conditions, abrasion, similar to that likely to be met in practice only rather more severe. The degree to which the coating withstands the treatment is then a guide to its performance in actual use. In the United Kingdom the work on this test has been carried out by Sira Institute (formerly the British Scientific Instrument Research Association), and the method which they established has been generally adopted, either in its original form, or, as we shall see, modified slightly. The basis of the Sira method is the adoption of a standard pad made from rubber loaded with emery powder, which, with a precise load, is drawn across the surface under test a given number of times—typically 20 times with a loading of 5 lb in^{-2}. The work has been directed mainly towards the assessment of the performance of magnesium fluoride single-layer antireflection coatings for the visible. It has been established that coatings of this type which are sufficiently robust do not show signs of damage under the normal test conditions given above. Abrasion resistance, however, has been found to be not just a function of the film material but also of the thickness. Multilayer coatings are generally much more prone to damage than either of the component materials in single-layer form. It is therefore necessary to establish fresh standards for each and every type of coating.

Unfortunately this test does not produce an actual measure of the abrasion resistance, but merely decides whether or not a given coating is acceptable. Because of this, some investigations into a better arrangement were carried out by Holland and van Dam[2]. In order to measure the resistance, the films must suffer some damage, and the measure of the damage can then be taken as a measure of the abrasion resistance. The method which these two workers devised was to subject the films to abrasive action, which varied in intensity over the surface and which was at its most intense point sufficiently severe to remove completely the coating. The point at which the coating just stopped being completely removed was then found. Of course this method is still relative in that a different standard must be set up for every thin-film combination, but it does permit comparison of the abrasion resistance of similar coatings, impossible with the previous method. The apparatus is shown in figure 11.1. It consists of a reciprocating arm carrying the abrasive pad of the Sira type, and is 0.25 inch in diameter, loaded with 5.5 lb. The table carrying the sample under test rotates approximately once for every three strokes of the

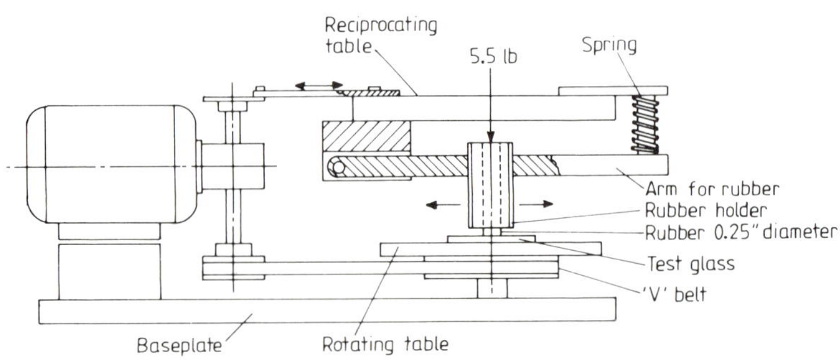

Figure 11.1 Schematic arrangement of an abrasion machine. The reciprocating table is supported by two horizontal bars not shown in the diagram. (After Holland and van Dam[2].)

pad. The pad traces out a series of spirals on the surface of the sample and the geometry is arranged so that the diameter of the area which is abraded is approximately 1.25 inches. The abrasion takes the form of a gradual fall off in intensity towards the outside of the circle, and the test is arranged to carry on for such a time that the central area of the coating is completely removed while the outside not at all. Holland and van Dam found that some 200 strokes were sufficient to do this with single layers of magnesium fluoride. They then defined the abrasion resistance measure of the coating by the formula

$$w = (d^2/D^2) \times 100\,\%$$

where d is the diameter of the circle where the coating has been completely removed and D is the diameter of the area which has been subjected to abrasion. Holland and van Dam studied particularly the case, as had Sira, of the single-layer magnesium fluoride antireflection coating for the visible region and they quote a wide range of most interesting results.

They investigated many different conditions of evaporation including angle of incidence and substrate temperature. A common value for the abrasion resistance of a typical magnesium fluoride layer of thickness to give antireflection in the green is between 2 and 5, depending on the exact conditions of deposition. Best results were obtained when the substrate temperature during evaporation was 300°C and the glow-discharge cleaning before coating lasted for 10 minutes. There was a significant reduction in abrasion resistance if either the temperature were allowed to drop to 260°C or if there were only 5 minutes of glow-discharge cleaning. They also found that the abrasion resistance of the film is increased considerably by burnishing with a Selvyt cloth or by baking further at 400°C in air after deposition. Another significant result obtained concerns the occurrence of a critical angle of vapour incidence during film deposition, beyond which the abrasion resistance falls off extremely rapidly. This critical angle varies slightly with film thickness but is approximately 40° for thicknesses in excess of 300 nm and rises as the thickness decreases.

So far, this test has not received general recognition in specifications. It should be extremely useful as a quality-control test in manufacture, especially as a reduction in quality can be detected long before it drops below the level of the normal abrasion test and remedial action can be taken before any coatings are even rejected.

Adhesion

Adhesion has already been discussed in chapter 9. In the simplest type of adhesion test, a piece of adhesive tape is stuck down on the surface of the coating and pulled off. Whether or not this removes the film is taken as an indication of whether the adhesion of the film to the substrate is less than or greater than that of the tape to the film. The test is again of the go–no-go type.

It is important if consistent results are to be obtained that some precautions are taken in carrying out the test. The first is that the tape should have a consistent peel adhesion rating, which should be stated in the specification. Peel adhesion is measured by sticking a freshly cut piece of tape on a clean surface, usually metal, and then steadily pulling it off, normal to the surface. The tension per unit tape width, usually expressed in grams per inch, is the measure of the peel adhesion rating of the tape. The rating obtained in this way is usually virtually the same as the rating obtained when the tape is removed from a thin-film coating. Some precautions in applying the test are necessary. Fresh tape should always be used. The tape should be stuck firmly to the coating, exerting a little pressure and smoothing it down. It should be removed steadily, pulling it at right angles to the surface, and never snatched off, which would put an uncontrolled impulsive load on the film and would certainly lead to inconsistent results. The same thickness of tape should be used for all testing. With thicker tape of the same peel adhesion rating, the test would be slightly less severe. The width of the tape, however, does not matter. A rating which is often used is 1200 gm/inch width. If necessary, the adhesion rating of any tape can easily be checked using a spring balance. For obvious reasons the test is often called the 'Scotch Tape test'.

Attempts have been made to devise quantitative techniques for adhesion measurement and a number of these have also been discussed in greater detail in chapter 9. The simplest and most straightforward is the direct-pull test, involving the attachment of the flat end of a cylindrical pin to the coating, followed by measurement of the force necessary to pull it off. Provided the coating is detached with the pin, the force required divided by the area of the pin is then the measure of adhesion. An alternative test that has some advantages as well as disadvantages is the scratch test, in which a loaded stylus is drawn across the coating with gradually increasing load. At each stroke the coating is examined under a microscope for signs of damage. The load at which the coating is completely removed is taken as the measure of adhesion. The Goldstein and Delong technique involving the use of a microhardness tester as a scratch tester has also been mentioned in chapter 9.

Environmental resistance

One further aspect of thin-film performance is also of very great importance. This is the resistance which the film assembly offers to environmental attack. Probably the most important aspect of the environmental performance of the filter is the resistance to the effects of humidity but the resistance to other agents such as temperature, vibration, shock, and corrosive fluids such as salt water, may all be important.

There are two possible approaches. Either the filter may be expected to operate satisfactorily while actually undergoing the test or it may only be expected to withstand the test conditions without suffering any permanent damage, although the performance need not be adequate during the actual application of the test. The latter is usual as far as interference filters are concerned, and in such a case the specification is known as a 'derangement specification' because it is sufficient that the performance is not permanently deranged by the application of the test conditions. In what follows we shall assume that the type of specification is the derangement type. Derangement specifications are easier to apply than the other type because the normal performance measuring equipment can be used remote from the environmental test chamber.

Of all the agents which are likely to cause damage, atmospheric moisture is probably the most dangerous. For most applications, particularly where severe environments are excluded, it will be found sufficient for the filter to be tested by exposing it for 24 hours to an atmosphere of relative humidity 98 % ± 2 % at a temperature of 50°C ± 2°C. It is often found that although the coatings are not removed by this test they are softened, and it is useful to carry out before the adhesion or abrasion-resistance tests, which can follow on immediately after.

A great deal of work has been carried out by Government bodies on the environmental testing of equipment and components for the Services. This has resulted in specifications that are equivalent to the most severe conditions ever likely to be met in both tropical and polar climates. These specifications include in the United Kingdom DEF 133 and DTD 1085 for aircraft equipment. Relevant specifications in the United States include MIL-C-675, MIL-C-14806, MIL-C-48497 and MIL-M-13508. The tests vary from one specification to another but can include exposure to the effects of high humidity and temperature cycling over periods of 28 days, exposure conditions equivalent to dust storms, exposure to fungus attack, vibration and shock, exposure to salt fog and rain and immersion in salt water. It is not always possible for coatings to meet all tests in these specifications and concessions are often given if the coatings are to be enclosed within an instrument. Humidity and exposure to salt fog and water are particularly severe tests. Fungus does not normally represent as severe a problem to the coatings as it does to the substrates. Certain types of glass can be damaged by fungus and in such cases coatings, even if they themselves are not attacked, will suffer along with the substrates. Most instruments likely to be exposed to sand or dust are adequately sealed since their performance is likely to suffer if dust or sand is permitted to enter. Thus dust storms are usually a danger only to those elements with surfaces on the outside of an instrument.

REFERENCES

1 Bousquet P and Richier R 1972 Etude du flux parasite transmis par un filtre optique à partir de la détermination de sa fonction de transfert *Opt. Commun.* **5** 27–30

2 Holland L and van Dam E W 1956 Wear resistance of magnesium fluoride films on glass *J. Opt. Soc. Am.* **46** 773–7

12 System considerations. Applications of filters and coatings

It is only rarely that thin-film filters or coatings are used by themselves. They usually form part of an optical system and it is in integrating coatings into such systems where many problems appear. There is an unfortunate tendency to leave coatings until late in the design process and some of the most severe problems occur during the attempted integration of coatings once the remainder of the design has been frozen. Such problems could frequently have been avoided had the incorporation of coatings been studied at a time when there was still some design flexibility.

Coatings cannot automatically be deposited with equal ease on any surface. Furthermore some tolerances must be permitted on coating performance. Then there is the shift in coating characteristics with angle of incidence, with temperature and with atmospheric humidity. Coatings often possess considerable intrinsic strain and the resulting stress can cause distortion that is significant in substrates of interferometric quality if they are not sufficiently thick. Lack of uniformity in coatings can also cause problems. Some of these difficulties arise from coating characteristics that show rapid change of phase with wavelength, characteristics frequently possessed by broadband reflectors. A lack of uniformity in the coating, if it is dielectric, is equivalent to a wavelength variation over the surface and if the phase dispersion is high then the resulting phase errors can be out of all proportion to the errors in thickness. The net result is an apparent loss of figure of the coated component that may show surprisingly large variations with wavelength. Extended-zone reflectors frequently exhibit rapid phase dispersion and so should be used with caution in applications where interferometric quality is required. All of these points have been discussed elsewhere in this book and the intention of repeating them here is simply to reinforce the point that coatings are like any other component and must be designed into the system as an integral part and not simply added at a later stage.

Coatings rarely stretch right to the edge of a substrate. Substrates must be held in jigs during coating and it is normal to do this by a lip that obscures the rim of the substrate leaving an uncoated ring. This is not entirely a disadvantage. Delamination is always most likely to start at the edge of a coating and the uncoated rim around the coating gives it a much more regular edge and reduces the risk of delamination. Further, the mount for the component need not make contact with the coating where it could damage it and increase the chance of spontaneous delamination. The uncoated ring can, however, be a disadvantage if the component is a filter that rejects certain wavelength regions because stray light can leak through the uncoated part unless precautions to baffle it are taken. The uncoated area can be considerably reduced by the use of wire clips to hold the substrates by the edges during deposition, a technique frequently used with components such as sunglasses, but problems with stray light leakage can sometimes lead to the requirement that there should be no uncoated area whatsoever. The normal method for achieving this is to cut the component after coating. This should be carried out only if absolutely necessary. It increases the cost considerably because of the risk of failure involved in the cutting operation and it inevitably leaves a coating edge that is uneven on a microscopic scale and more likely to include stress concentrators that can initiate delamination.

It is always more difficult to coat a curved surface than a plane one and the difficulties increase with the curvature. Difficult coatings with tight tolerances should wherever possible be deposited on plane surfaces. Narrowband filters can be tuned to shorter wavelengths by tilting. If small tilts can be permitted (by the use of wedged holders for example) then the tolerances on peak wavelength can be relaxed.

Standard size components are always to be preferred. The manufacturer already has the necessary jigs and fixtures and the substrates are available in quantity. Fewer test runs are required and there are fewer unexpected difficulties. When something goes wrong with the process an entire batch of components is usually lost. Such failures are more likely with components of unusual shape or size, and so a greater number of uncoated components must be produced to ensure the correct number of final coated components. All of this means that the cost of non-standard components is considerably greater than standard.

Most filters will consist of a series of components some of which are designed to reject radiation in regions outside the pass bands. Surprisingly disappointing performance can be achieved in cases where the rejected light is reflected rather than absorbed. We can illustrate this by considering two surfaces having reflectances and transmittances of R_1, T_1, R_2 and T_2. Light can be considered as being reflected backwards and forwards between the surfaces and being combined incoherently. The net transmittance is then given by equation (2.93) suitably adjusted:

$$T = \frac{T_1 T_2}{1 - R_1 R_2}.$$

If R_1 and R_2 are zero, that is, what is not transmitted is absorbed, then we have the expected result

$$T = T_1 T_2.$$

However, if $R_1 = 1 - T_1$ and $R_2 = 1 - T_2$ then the result becomes similar to (2.95):

$$T = \frac{1}{(1/T_1) + (1/T_2) - 1}.$$

Consider the case where $T_1 = T_2 = 0.01$. The first expression gives $T = (0.01)^2 = 0.0001$, a very satisfactory figure, while the second expression gives

$$T = \frac{1}{100 + 100 - 1} = \frac{1}{199} = 0.005$$

very disappointing from the point of view of rejection. The solution is somehow to reduce the effect of R_1 and R_2 either by ensuring that the reflected beams rapidly walk out of the system aperture, by for example tilting the components relative to each other, or by placing absorbing components in between the two surfaces so that the beams are rapidly attenuated.

Sometimes reflecting and absorbing components will be combined in a system. Examples of this might be a heat-reflecting filter coating consisting of an interference shortwave-pass filter deposited on a heat-absorbing glass or a narrowband filter consisting of an all-dielectric interference section, a metal–dielectric coating and an absorption glass. It is usually best in such cases to assemble the components such that the low-loss interference section faces the source. This ensures that the maximum amount of energy is rejected by reflection and minimises the temperature rise and possible resulting long-term damage. In the case of the narrowband filter assembly, the overall rejection performance of the filter is assisted by placing the absorbing glass component in between the two interference sections for the reasons discussed above.

Polarisation effects can sometimes be the cause of unexpected performance variation. We can illustrate this with the somewhat extreme case of a simple single-layer dielectric beam splitter shown in figure 12.1. The performance of such a coating, assuming a quarter-wave (monitored at normal incidence) of zinc sulphide ($n = 2.35$) immersed in glass ($n = 1.52$) at an angle of incidence of 45°, is given by

$$R_s = 33.15\% \qquad R_p = 4.03\% \qquad R_{MEAN} = 18.59\%$$
$$T_s = 66.85\% \qquad T_p = 95.97\% \qquad T_{MEAN} = 81.41\%.$$

Let us assume that the reflecting surface has a reflectance of 100 % and calculate the intensity of the output beam as a fraction of the input intensity. A simple calculation involves the unpolarised figures for T and R and yields $TR = (18.59\% \times 81.41\%) = 15.13\%$. However, this calculation has taken no account of the polarising effect of the beam splitter itself. The true figure for unpolarised incident light should be $0.5(R_s T_s + R_p T_p) = 13.01\%$ (a difference greater than 10 % of the previous figure). Polarisation of the input beam alters

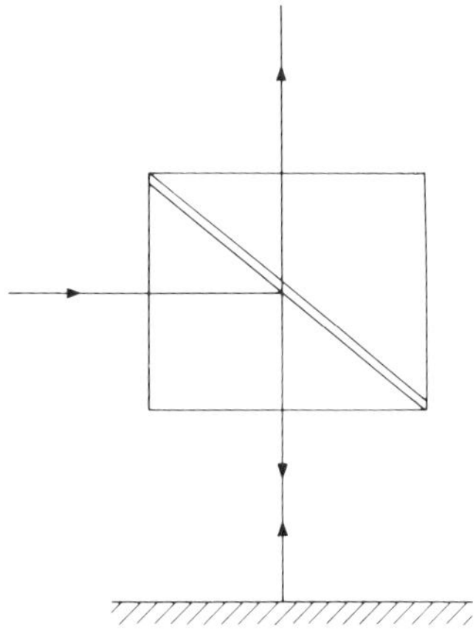

Figure 12.1 Arrangement of a single-layer dielectric beamsplitter used for the calculation of efficiency discussed in the text.

the results still further. With s-polarised input light the figure would be $R_sT_s = 22.16\%$ while with p-polarised light it would be as low as $R_pT_p = 3.87\%$. Thus with varying degrees of polarisation of the input light the efficiency of the system can vary from 3.87% to 22.16%. To avoid performance fluctuations resulting from such effects, a quarter-wave plate with axis at 45° to the plane of incidence is often inserted in the input side of a system to convert both s- and p-polarised light to circularly polarised, which makes the overall performance of the system equivalent at all times to unpolarised light. (It is unlikely that the input light should be already circularly polarised, but of course in that case the quarter-wave plate could make the situation worse.) Metal layers suffer less from polarisation effects, but they, too, do still have significant polarisation sensitive behaviour.

That was an example of an immersed coating. Note that immersed coatings always have very high effective angles of incidence since the important quantity for Snell's law is $n_0 \sin \theta_0$ rather than θ_0. Thus, in immersed coatings, angle-of-incidence effects are invariably enhanced. Polarisation effects are particularly pronounced but so also are the simple wavelength shifts associated with a change in angle of incidence.

Even in coatings that are not immersed, the changes in angle of incidence associated with a highly divergent or convergent beam can cause problems, especially if the component is tilted with respect to the axis. Sometimes the problems can be eased by deliberately introducing a variation in coating thickness over the surface of the component. This can be particularly effective when a point source is used close to a component when the small source dimensions ensure that only a small range of angles of incidence correspond to each point on the component surface.

A point to watch concerns polarisation effects associated with skew rays. p- and s-polarisation performance is calculated with respect to the plane of incidence. A skew ray possesses a plane of incidence that is usually rotated with respect to the principal plane of incidence containing the axial ray of the system. This can cause problems in large aperture polarisers, for example, where, although the s-transmittance for the skew rays can be very low, the corresponding plane of polarisation is actually rotated and can lead to an appreciably large leakage of light which is s-polarised with reference to the plane of incidence of the axial ray. As a rough example we can consider a cone of 1° half-angle incident at 45° on a polarising beam splitter. The plane of incidence of the marginal azimuthal rays will be rotated at an angle of approximately $1°/\sin 45°$, or 1.4°, with respect to the plane of incidence of the axial ray. Let us assume that both axial ray and marginal ray have zero transmittance for s-polarised light and unity for p-polarised light. Because of the rotation of the plane of incidence the effective transmittance of the marginal ray in the s-plane of the axial ray will then be $\sin^2(1.4°)$ or 0.06%.

A very useful account of problems associated with the integration of thin-film coatings into optical systems has been written by Matteucci and Baumeister[1].

POTENTIAL ENERGY GRASP OF INTERFERENCE FILTERS

It is worth while considering why interference filters are used in preference to other types of wavelength selecting devices such as prism and grating monochromators. Of course the size and mechanical stability of the thin-film filter are in themselves powerful arguments in favour of its use, and, especially in cases where space and weight are at a premium, in satellite-borne instruments for example, they are probably sufficient. However, there is an even more compelling reason for adopting thin-film filters and this is the greatly increased potential grasp of energy over dispersive systems.

Compared with a grating monochromator, for instance, the thin-film filter with the same bandwidth is capable, provided the rest of the system is correctly designed round it, of collecting several hundred, and in some cases thousand, times the amount of energy collected by the monochromator. This section, therefore, is devoted to a comparison of the interference filter with the diffraction grating, particularly from the point of view of the potential total energy grasp.

In order to compare the energy-gathering properties of various components, we have to assume that each is used in an ideal system designed to make maximum use of its energy-gathering powers, that the bandwidths of the various sytems are equal, and that any dispersive components are used well within their limiting resolutions so that their response functions are not complicated by large diffraction effects. We shall also assume that the source of illuminations is of equal brightness in all cases and that the collecting condensing optics are such that the entrance apertures of all systems are completely filled. The energy grasp under these conditions is then computed in each case as a function of the appropriate area of the component, and the comparison made on the basis of these figures.

In fact this analysis has been carried out by Jacquinot[2] for a diffraction grating, a prism and a Fabry–Perot interferometer. He has shown first that there is always a clear advantage in using a diffraction grating rather than a prism, the advantage varying from around 3 to perhaps 100 with the dispersion of the prism materials. Because of this, the comparison that primarily concerns us is between the interference filter and the diffraction grating. Jacquinot has also compared the Fabry–Perot interferometer having an air spacer with the diffraction grating, and showed that there is a clear gain of 30–400 times in the energy grasp of an interferometer over a grating of the same area. The case of an interference filter is similar but the spacer layer has an index appreciably greater than unity, especially in the infrared, which increases its grasp still further. In the analysis below, we shall follow the main lines of Jacquinot's argument, but shall extend the analysis to include a spacer of index other than unity.

Jacquinot considers a spectrometer consisting of an input slit, a collector and collimator of some description, a dispersive element which here is a grating, and an output element imaging the entrance slit on the exit slit, the final element in the system. It is assumed that the resolution is limited by the width of the slits and that the grating is capable of higher resolution if required. This means that we can define the resolution purely in terms of slit width and dispersion. In this condition the maximum luminosity for a given resolution will be achieved when the entrance and exit slit widths, expressed in terms of spectral interval, are equal, when a triangular response function will be obtained from the instrument. It is assumed that the source, which is an extended one, is monochromatic and of uniform brightness.

There will be some sort of imaging system which will produce an image of the source on the entrance slit. The brightness of the source image will be equal to that of the actual source, except for the transmission of the imaging system, which we can take to be unity without affecting the final result, since all systems to be compared will have a similar arrangement before the entrance aperture. Given that the brightness of the image is identical to that of the source, it only remains for the aperture of the imaging system to be made large enough for the aperture of the collector and collimator before the grating to be completely filled. Again we can assume that this has been carried out in all arrangements without any loss in generality. The situation is sketched in figure 12.2. The notation used here is, as far as possible, exactly that used by Jacquinot in his original paper to make the comparison easier. Let the brightness of the source image be denoted by B. Let the monochromator be adjusted so that the image of the entrance slit falls directly on the exit slit and let both slits have the same

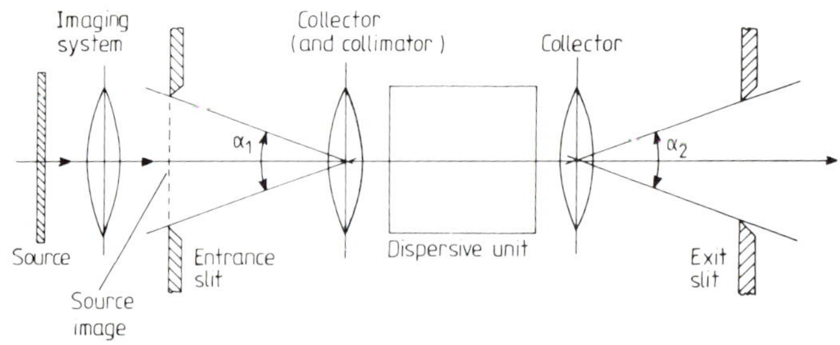

Figure 12.2 An idealised dispersive monochromator. (After Jacquinot[2].)

width and length. This corresponds to the apex of the triangle. The energy transmitted by the system will be given by

$$E = BS\omega T \tag{12.1}$$

where ω is the solid angle subtended by either slit at the appropriate collector element and S the area of the beam at the collector. $S\omega$ will be the same for both the entrance and the exit slit since we have arranged for the image of one to coincide with the other. T is the transmittance of the monochromator. If the width of the exit slit is α_2 and the length β_2, then the expression becomes

$$E = BST\beta_2\alpha_2. \tag{12.2}$$

If we denote the resolving power of the system by R, then we have that $\alpha_2 = \lambda D_2/R$ where D_2 is the angular dispersion of the system referred to the output slit, i.e.

$$E = BST\beta_2(\lambda D_2/R). \tag{12.3}$$

For the grating monochromator the angular dispersion is derived from the equation

$$\sigma(\sin i_1 + \sin i_2) = m\lambda \tag{12.4}$$

where σ is the grating constant, i.e. the interval between grooves, m is the order number, and i_1 and i_2 are the angles of incidence and diffraction respectively at the grating.

$$D_2 = \frac{\mathrm{d}i_2}{\mathrm{d}\lambda} = \frac{m}{\sigma\cos i_2} = \frac{\sin i_1 + \sin i_2}{\lambda\cos i_2}$$

i.e.

$$\lambda D_2 = \frac{\sin i_1 + \sin i_2}{\cos i_2}. \tag{12.5}$$

Now

$$S = A\cos i_2$$

where A is the area of the grating and we assume that it is completely illuminated and that no light is lost, so that

$$S\lambda D_2 = A(\sin i_1 + \sin i_2). \tag{12.6}$$

Jacquinot shows that $S\lambda D_2$ is a maximum for the Littrow mounting (where i_1 and i_2 are as nearly equal as possible) used on the blaze angle which we denote by ϕ. For that mounting

$$S\lambda D_2 = 2A\sin\phi \tag{12.7}$$

and

$$E = (BT\beta_2/R)\,2A\sin\phi. \tag{12.8}$$

ϕ we can take as 30°, say, when $\sin\phi = \frac{1}{2}$ and

$$E = BT\beta_2 A/R. \tag{12.9}$$

We shall now consider the interference filter and compare it with the diffraction grating. The case considered by Jacquinot is that of the conventional Fabry–Perot interferometer made up of a pair of plates in an etalon with a spacer of unity refractive index. Here we are more concerned with the interference filter where the spacer layer has an index greater than unity. As on p 261, we introduce the concept of an effective index of refraction which governs the angular behaviour of the filter. We shall use a similar analysis to that of Jacquinot, but recast it in the form of the results of chapter 7.

Jacquinot suggests that the filters be used with an acceptance angle such as to make the effective bandwidth of the filter $\sqrt{2}\times$ the value at normal incidence. Equation (7.40) gives

$$W_\Theta^2 = W_0^2 + (\Delta\nu')^2 \tag{12.10}$$

where W_0 and W_Θ are the halfwidths corresponding to collimated light at normal incidence and to a cone of semiangle Θ. If Θ is measured in air then

$$\Delta\nu' = \nu_0\Theta^2/2n^{*2}. \tag{12.11}$$

For $W_\Theta = \sqrt{2}W_0$ we must have $W_0 = \Delta\nu'$, i.e.

$$W_0 = \nu_0\Theta^2/2n^{*2} \tag{12.12}$$

and, from equation (7.41),

$$\hat{T}_\Theta = (W_0/\Delta\nu')\tan^{-1}(\Delta\nu'/W_0) = \tan^{-1}(1) = \pi/4 = 0.78 \tag{12.13}$$

where $\hat{T}_\Theta$ is the effective peak transmittance of the filter for a cone of incident light of semiangle Θ referred to the incident medium, which we are assuming is air.

If R_0 is the resolving power for perfectly collimated light at normal incidence and R_Θ that for a cone of semiangle Θ, then

$$R_0 = \nu_0/W_0$$

and, since $\Delta\nu'$ is small compared with ν_0,

$$R_\Theta = \nu_0/W_\Theta = R_0/\sqrt{2}. \tag{12.14}$$

But $W_0 = \Delta\nu'$ so that

$$R_0 = \nu_0/\Delta\nu' = 2n^{*2}/\Theta^2$$

and so

$$\Theta^2 = \sqrt{2}n^{*2}/R_\Theta. \tag{12.15}$$

If B is again the brightness of the source and A is that area of the filter that is fully illuminated, then the energy collected will be

$$E = BAT(\pi/4)\omega \tag{12.16}$$

where ω is the solid angle subtended by the aperture and T is the normal incidence transmittance. The factor $(\pi/4)$ is included from (12.13).

From figure 12.3,

$$\omega = 2\pi(1-\cos\Theta) = \pi\Theta^2. \tag{12.17}$$

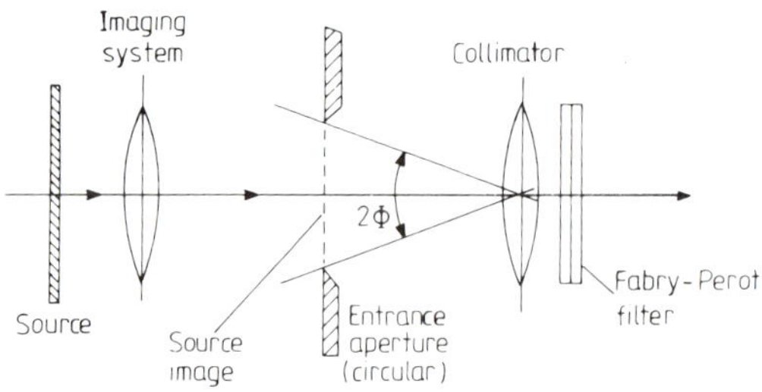

Figure 12.3 An arrangement of a monochromator using an interference filter.

Then, from equations (12.15), (12.16) and (12.17)

$$E = BAT\frac{\pi^2}{2}\frac{n^{*2}}{\sqrt{2}R_\Theta}. \tag{12.18}$$

This is similar to the form given by Jacquinot except for the factor n^{*2} which is missing in his expression.

We are now in a position to compare efficiencies. The relative energy grasp of the two systems is

$$\frac{E_{\text{filter}}}{E_{\text{grating}}} = \frac{BAT(\pi^2/2)n^{*2}/(R\sqrt{2})}{BT\beta_2 A/R}. \tag{12.19}$$

We can assume for this comparison that the resolution and areas and transmittances of the two systems are equal (that is transmittance at normal incidence in collimated light for the interference filter). Equation (12.19) then simplifies to

$$\frac{E_{\text{filter}}}{E_{\text{grating}}} = \frac{\pi^2}{2\sqrt{2}}\frac{n^{*2}}{\beta_2} = 3.4\frac{n^{*2}}{\beta}. \tag{12.20}$$

Jacquinot estimates the usual value of β to be 0.01 radian. With extreme care in design, values of 0.1 have been achieved, although this represents the very limit. For an n^* of unity, then, the value of the energy ratio varies between 34 and 340. However, n^* in the visible region is usually in excess of 1.5, which alters the range to 76–760. For the infrared the advantage of the filter is even greater, for n^* is usually of the order of 3.0, so that the range becomes 306–3060, a massive advantage. This means that we can happily make the interference filter much smaller than the grating and still have a very significant increase in energy grasp over it.

This analysis dealt with the single Fabry–Perot type of filter. The advantage with the DHW type of filter is slightly greater still, since the effective transmittance in a cone of illumination is higher than that of the Fabry–Perot.

NARROWBAND FILTERS IN ASTRONOMY

The problem of detecting faint astronomical objects is rendered even more difficult than it would otherwise be by the light of the night sky. This light consists mainly of starlight scattered by dust both in the atmosphere and in interstellar space (including light from our own Sun) together with emission from the upper atmosphere, and may be considered to be mainly of a continuous spectral nature although there are a number of emission lines as well. The sky light causes an overall fogging of the photographic plates, which are the most common detectors used in this work (although in recent years increasing use has been made of image tubes). Maximum contrast between the photographic image of a star or other object and the sky background is obtained when the sky fog is just apparent on the plate. The exposure time is chosen to give just this amount of fogging. The efficiency of the photographic detector falls off rapidly on either side of this optimum. The limit of detection of a faint object will be reached when the image is just discernable against the background.

The way in which the limit of detection varies with the parameters of the system has been studied particularly by Baum.[3] A simplified account of the analysis is given by Bowen[4] and it is this latter form that we follow here. The notation used by Bowen, which we also use here, differs slightly from that used by Baum.

The signal which is received from the object will consist of discrete photons arriving at a constant mean rate but randomly spaced. Provided the mean rate is sufficiently small (satisfied for the signals we are considering) we can consider the photons as forming a Poisson distribution (the distribution which deals with sequences of events where the probability of an occurrence in any particular time interval is vanishingly small, but where the total observing time is sufficiently long to ensure a finite number of events). For the Poisson distribution the standard deviation of successive measures of the number of photons N arriving in a certain constant time is simply $\sqrt{N}$.

Let D be the telescope aperture diameter, f the focal length of the telescope, t the observation time, β the diameter of the image of the object, n the number of photons from the object received per unit area of telescope aperture per second, s the number of background photons received per unit area of telescope aperture per unit solid angle of sky per second, p the limit of linear resolution of the emulsion, q the quantum efficiency of the entire system which includes the photographic emulsion and the transmission of the optical system, and m the number of photons recorded per unit area of photographic plate which will produce the correct level of background fog.

In his paper, Bowen defines the faintness of a star or object as $1/n$. We shall now examine the way in which the limiting detectable faintness varies with the parameters of the system. The fractional error in a measurement is denoted by B and is defined as the standard deviation associated with the measurement divided by the measurement itself. Thus in a measurement of a number of photons N, the fractional error would be $B = (\sqrt{N})/N = 1/\sqrt{N}$.

The number of photons recorded from the object and from an equal area of sky in time t is given by

$$D^2ntq + \beta^2 sD^2tq \tag{12.21}$$

where we are omitting factors of $\pi/4$. The standard deviation in successive measurements will be

$$(D^2ntq + \beta^2 D^2 stq)^{1/2} \tag{12.22}$$

and the fractional error in the measurement will be

$$\begin{aligned} \beta &= \frac{(D^2ntq + \beta^2 D^2 stq)^{1/2}}{D^2ntq} \\ &= \frac{(n + \beta^2 s)^{1/2}}{Dnt^{1/2}q^{1/2}}. \end{aligned} \tag{12.23}$$

For very faint objects, $n \ll \beta^2 s$ so that

$$B = \frac{\beta s^{1/2}}{Dnt^{1/2}q^{1/2}} \tag{12.24}$$

and the limiting faintness is given by

$$\left(\frac{1}{n}\right)_1 = \frac{B_1 Dt^{1/2}q^{1/2}}{\beta s^{1/2}} \tag{12.25}$$

where B_1 is the highest possible value of B where the object is still just detectable. Bowen suggests that B_1 should be 0.2. This formula applies as it stands to photoelectric detectors and shows how the faintness which can be detected increases with increasing aperture. For the photographic detector, however, the position is not quite the same. Here the time of exposure t must be chosen to give the correct background fog. The efficiency of the plate drops so quickly if the density of the background is incorrect that any other exposure time is of very much less value. This correct exposure time t_0 is given by

$$D^2 t_0 sq = mf^2$$

i.e.

$$t_0 = \frac{mf^2}{D^2 sq} \tag{12.26}$$

and, substituting in equation (12.25),

$$\left(\frac{1}{n}\right)_1 = \frac{B_1 D q^{1/2}}{\beta s^{1/2}} \sqrt{\left(\frac{mf^2}{D^2 sq}\right)} = \frac{B_1 m^{1/2} f}{\beta s}. \tag{12.27}$$

In the equation we are assuming that β is larger than the resolution limit of the plate. If this is not the case, where f is small for example, then β must be replaced by p/f, giving

$$\left(\frac{1}{n}\right)_1 = \frac{B_1 m^{1/2} f^2}{ps}. \tag{12.28}$$

These results, obtained by Bowen, are not what we might have expected, because they seem to show that the all-important parameter for photographic detection of faint objects is the focal length of the telescope and not the aperture. The longer the focal length, the greater the faintness which may be observed, regardless of the diameter of the aperture of the system. So far, however, we have neglected to notice that observation time is limited to one night. Increasing the focal length without a corresponding increase in aperture increases the necessary exposure time, which varies as the square of the focal length. Let t_m be the longest allowable exposure time. Then, for any given system, the largest value of focal length f_m will be given by

$$f_m^2 = \frac{t_m D^2 sq}{m} \qquad \text{i.e.} \qquad f_m = \frac{t_m^{1/2} D s^{1/2} q^{1/2}}{m} \tag{12.29}$$

which when substituted in equation (12.27) and (12.28) gives for f large or β large

$$\left(\frac{1}{n}\right)_1 = \frac{B_1 t_m^{1/2} D q^{1/2}}{\beta s^{1/2}} \tag{12.30}$$

and for f small and β small

$$\left(\frac{1}{n}\right)_1 = \frac{B_1 m^{1/2} t_m D^2 sq}{psm} = \frac{B t_m D^2 q}{pm^{1/2}}. \tag{12.31}$$

These expressions† show that, indeed as might be expected, there is a gain in going to larger telescopes.

† Reciprocity failure, which effectively means that q is reduced slightly as t increases, has been neglected in the derivation.

Given the maximum possible value of D and f, how can the situation be improved by the use of filters? If there is a difference in spectral distribution of the radiation from the object and the sky background, then it is possible that a filter inserted in the system might modify the ratio of photons received from the object to those received from the sky. If this process results in a reduction in n by a factor x to xn, and a reduction in s to ys, then the ratio n/s becomes xn/ys, and if x/y is sufficiently large, then a positive gain in faintness may result. Substituting these values in the expression for the case where the resolution of the emulsion is not the limiting factor, equation (12.30) becomes

$$\left(\frac{1}{n}\right)_1 = \frac{xB_1 t_{\mathrm{m}}^{1/2} D q^{1/2}}{y^{1/2} \beta s^{1/2}} \tag{12.32}$$

and, assuming we adjust the focal length of the system as before to give the longest exposure time t_{m}, then the result is obtained that a gain in $1/n$ is achievable provided that $x > \sqrt{y}$.†

For the case where the emulsion resolution is a limiting factor, the expression (12.31) for $1/n$ shows that there is no possibility of altering the situation by filtering. The filtering will work only when the object is extended, or when the focal length of the telescope is large enough, or when the grain of the plates is fine enough, to ensure that the plate resolution is not a limiting factor.

The great bulk of the sky light is scattered light which has a more or less continuous spectrum. Only the emission from the upper atmosphere has a component consisting of discrete lines. Since, for a gain due to filtering, it is not sufficient to ensure that $x > y$ but that $x > \sqrt{y}$, in cases where n has a continuous spectral distribution and there is no great difference between the distributions of n and s, there is probably very little to be gained by filtering. In fact, slight enhancement of the ratio of detected photons accompanied by a drop in transmittance could lead to a loss in performance rather than a gain. However, there are classes of objects which are characterised by line spectra and in these cases it is possible by using filters centred on the lines to retain n only slightly reduced, but to have s greatly reduced. Such a class of objects is the hydrogen emission nebulae. It is now known that hydrogen is one of the elements of interstellar gas—probably the most abundant. Where hydrogen clouds are near bright stars, the atomic hydrogen is ionised by the x-ray and extreme ultraviolet radiation from the stars, and, when the electrons and protons recombine, the characteristic hydrogen spectra are produced. The principal line emitted in the wavelength range detectable at the surface of the Earth is the first line of the Balmer series, H_α at 656.3 nm, which, although not always the brightest line, is the one where contrast can be greatly improved.

The use of an interference filter centred on 656.3 nm greatly increases the contrast between the nebulae and the night sky, and gives a large increase in the faintness of nebulae which can be detected.

† This, at first sight, odd result follows from the assumption made in deriving equation (12.24) that the object is faint so that $n \ll \beta^2 s$.

Equation (12.29) shows that when the interference filter is installed the focal ratio of the telescope must be adjusted to give the correct level of background fog.

$$\frac{f}{D}=\frac{t_{\mathrm{m}}^{1/2}}{m}(ys)^{1/2}q^{1/2}. \tag{12.33}$$

Generally, with typical interference filters, the focal ratio should be near unity. Such a focal ratio incident directly on a narrowband interference filter would have a disastrous effect on both the bandwidth and peak transmission. However, the optical arrangement of the big telescopes permits an alternative arrangement. The primary mirror of a large telescope usually produces a pencil of focal ratio around $f/5$. As we have seen in chapter 7, a narrowband filter for the visible region with a bandwidth of around 1% of peak wavelength will accept such a pencil quite satisfactorily and it is usual to insert the interference filter at or very near the prime focus. Beyond the prime focus a camera is installed which reduces the focal ratio of the system to the desired value. The arrangement is shown in figure 12.4(*a*). With this layout the variation with field angle of the pass band of the filter (due to angle of incidence variation) is kept very small. If necessary it could be eliminated altogether by use of an extra lens, as in figure 12.4(*b*).

In figure 12.4, the filter acts as a field stop and may limit the field of view of

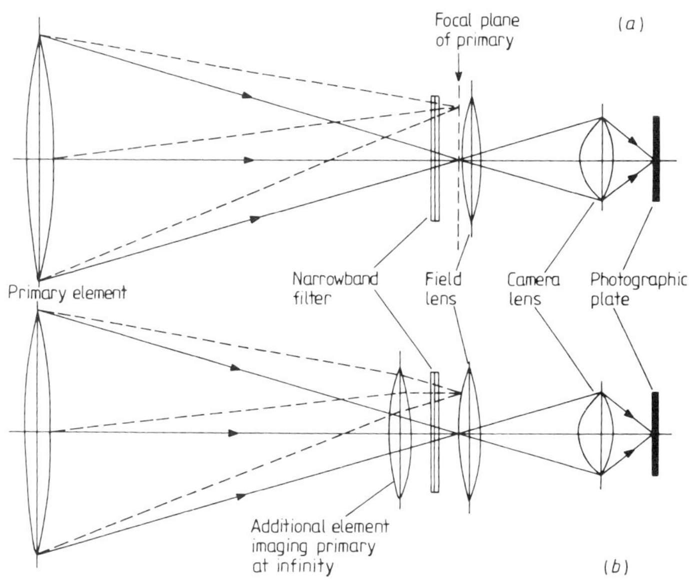

Figure 12.4 A narrowband filter in an astronomical telescope. The primary is shown here as a lens, but in the big telescopes would usually be a mirror. If necessary an additional element can be added as in (*b*) to alter the inclination of the off-axis pencils so that the effective peak wavelength of the filter is constant over the entire field.

the instrument. Filters up to 6 inches in diameter have been constructed, although 4 inches is probably a more usual figure. Filters with a diameter of 2 inches are readily available.

Some particularly fine examples of photographs taken with relatively broad combinations of coloured-glass filters and ones with interference filters of very much narrower bandwidths are given by Courtes[5]. Ring was the first successfully to use all-dielectric filters for this purpose, pioneering the development of these filters in the UK, and a paper by him[6] includes several photographs. A paper by Meaburn,[7] who took the excellent photographs in figure 12.5, illustrates extremely well the type of problem solved by interference filters and is well worth reading. Since this section appeared in the first edition, a particularly useful book by Meaburn[8] has been published and is worth consulting for further information.

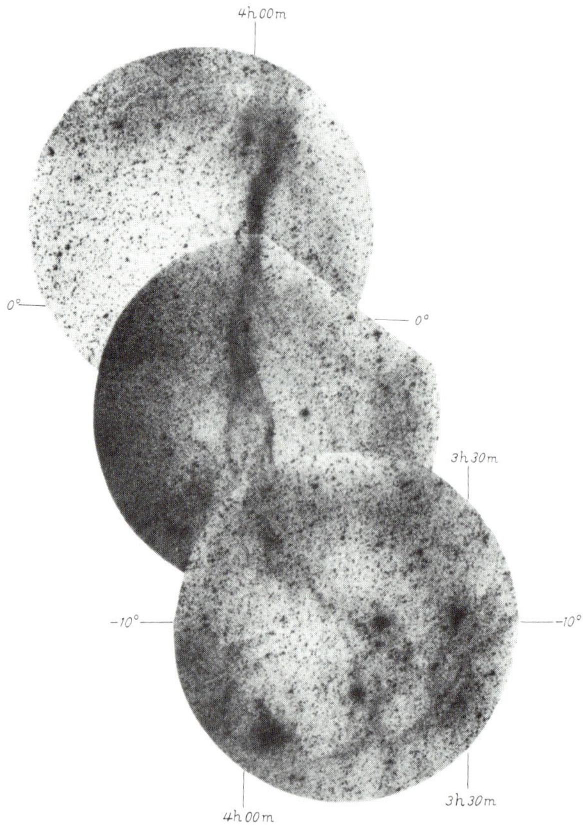

Figure 12.5(*a*) Nebulosities in the Cetus arc. H_α photographs of 1 hour exposure taken on a 6-inch $f/1$ Schmidt camera through a 4 nm bandwidth filter. (After Meaburn[7].)

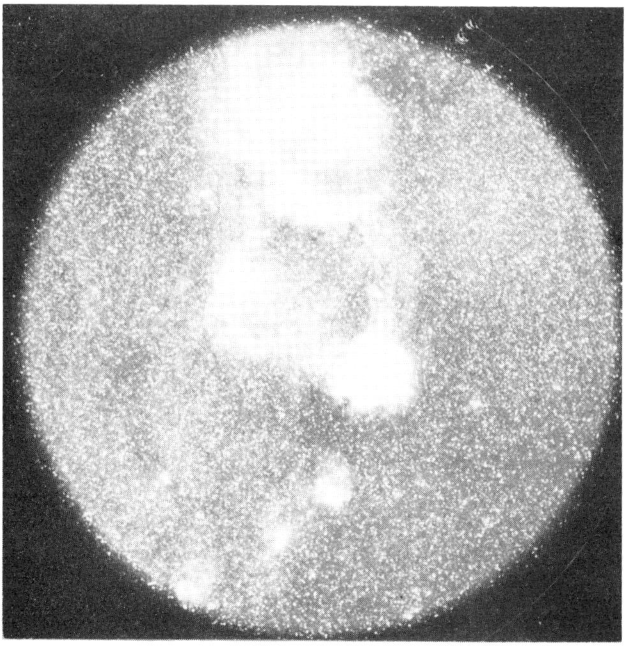

Figure 12.5(*b*) Nebulosities in the galactic anti-centre. Photograph taken through a 4 nm bandwidth filter centred on H_α (656.3 nm) with a 6-inch aperture Schmidt camera. The exposure was 1.75 hours. (Courtesy of Dr J Meaburn.)

ATMOSPHERIC TEMPERATURE SOUNDING

In the middle 1960s work began on a series of radiometers to be flown in satellites with the aim of measuring the distribution of temperature in the upper atmosphere. This programme was extremely successful. The first of these radiometers was designed by a joint team from the Universities of Oxford and Reading in the United Kingdom, the team at Reading moving to Heriot-Watt University at a late stage of the project. The radiometer was flown in the Nimbus IV spacecraft. The radiometer was known as the selective chopper radiometer (SCR) because of the basic principles of its design and it made extensive use of filters. It made measurements with a height resolution of 10 km of the temperature of that part of the atmosphere of height between 15 and 50 km, that is the troposphere and part of the stratosphere. The basic method used in the SCR and in other radiometers for temperature sounding is the detection and measurement of the radiation from atmospheric carbon dioxide.

Some ideas of the temperature structure of the atmosphere had already been formed, typical temperatures being of the order of 200 K at a height of 10 km rising to 240–280 K at heights of around 50 km. The peak of the black-body curve for a temperature of 200 K lies at a wavelength of 15 μm while that for 280 K is at 11 μm. The most favourable wavelength region for the measurement of the temperature of the atmosphere by detection of emitted radiation is therefore the band 11–15 μm. Of course the atmosphere will emit radiation only in the regions where it absorbs (the equivalence of thermal absorptance and emittance is a basic physical principle) and this, coupled with the fact that the radiation emitted from a given level must traverse the remainder of the atmosphere above that level to reach the detector in the spacecraft, allows an ingenious method to be used for the deduction of the temperature structure which was first suggested by Kaplan.[9]

Carbon dioxide is evenly distributed in the atmosphere and has extensive absorption bands around 15 μm so that it can be used as an indicator of the temperature of the atmosphere as a whole. Fortunately, over most of the important region, carbon dioxide is the only constituent of the atmosphere showing absorption (water vapour would interfere but is important only near the ground, and O_3 at 14 μm in the 25–40 km region can be avoided) which simplifies considerably the calculations. The absorption spectrum of CO_2 consists, at very low pressures, of a number of discrete lines which become gradually broader with increasing pressure. The detector in the spacecraft is arranged so that it responds to only a very narrow band of wavelengths in the CO_2 spectrum. If a waveband is chosen within which the absorption is high, then the radiation emitted at the bottom of the atmosphere will not reach space because the transmission of the atmosphere above it is low. At greater heights a much greater proportion of the energy emitted will reach the detector. However, also at greater heights, the energy emitted by the atmosphere will fall, because of decreasing density and pressure of CO_2, and, at a height which will depend on the absorption within the particular waveband chosen, the second process will overtake the first with the result that a major portion of the energy received by the detector will emanate from a narrow range of depths in the atmosphere. The mean depth can be changed by varying the centre wavelength of the band which is being detected, and so altering the variation of absorption with height. The experiment and apparatus are described in references 10–14. The following account is a much simplified version which follows directly work by J Houghton.

First we find the emittance of any layer by calculating the absorptance which is equivalent to the emittance. Consider a layer of the atmosphere situated at a depth z below the spacecraft. Let the transmittance of the atmosphere, at frequency ν, above this layer be T_z. In passing through a layer of thickness dz of the atmosphere the fractional intensity lost by unit intensity of radiation will be the absorptance of the layer. Next, consider radiation of initial intensity F at frequency ν at depth z. The fraction of this which appears at the detector in the spacecraft will be either FT_z or $(F-\mathrm{d}F)T_{(z-\mathrm{d}z)}$ and as these quantities will be equal we can write

$$(F-\mathrm{d}F)T_{(z-\mathrm{d}z)} = FT_z. \tag{12.34}$$

With some adjustment we find

$$A_{dz} = \frac{dF}{F} = -\frac{T_z - T_{(z-dz)}}{T_{(z-dz)}} = \frac{-(dT_z/dz)dz}{T_{(z-dz)}} \tag{12.35}$$

where A_{dz} is the absorptance and hence emittance of the layer. If $\bar{\mathscr{T}}$ is the mean temperature of the layer, then the black-body emission per unit frequency interval associated with it will be given by $B_\nu(\bar{\mathscr{T}})$ at frequency ν. The energy actually given out by the layer will be given by this expression multiplied by the emittance, i.e.

$$dI_z = KT_{(z-dz)}A_{dz}B_\nu(\bar{\mathscr{T}})$$

where dI_z is the energy per unit frequency interval received by the radiometer which emanates from a layer of thickness dz at depth z and K is a constant. Then

$$dI_z = -K\frac{dT_z}{dz}B_\nu(\bar{\mathscr{T}})dz. \tag{12.36}$$

If the detector in the spacecraft has a bandwidth of $\Delta\nu$, then the expression for the energy over this band becomes

$$\int_{\Delta\nu} dI_z d\nu = \int_{\Delta\nu} -K\frac{dT_z}{dz}B_\nu(\bar{\mathscr{T}})dz\,d\nu$$

and if R_ν/K is the response of the radiometer at frequency ν then the output of the instrument will be given by

$$D_z/dz = \int_{\Delta\nu} -R_\nu\frac{dT_z}{dz}B_\nu(\bar{\mathscr{T}})dz\,d\nu. \tag{12.37}$$

We can choose the frequency interval $\Delta\nu$ small enough for $B_\nu(\bar{\mathscr{T}})$ to be a constant over the interval. $B_\nu(\bar{\mathscr{T}})dz$ can then be moved outside the integral sign. What is left is the function

$$W_z = \int_{\Delta\nu} -R_\nu\frac{dT_z}{dz}d\nu \tag{12.38}$$

which is known as the weighting function, and represents the response of the system to radiation from depth z. We shall now look a little closer at the form of the weighting function, assuming that a single isolated absorption line is involved.

The absorption coefficient k_ν for radiation of frequency ν is defined by the relationship

$$dI_\nu = -k_\nu I_\nu du$$

where dI_ν is the change in intensity I_ν after traversing pathlength du of the absorbing gas. u is measured in terms of the quantity of gas traversed rather than physical distance and has such units as g cm^{-2} or atmo-cm (the equivalent pathlength in the gas at normal atmospheric pressure and temperature). The strength of the line S is defined as the absorption coefficient integrated over the whole width of the line.

For radiation of wavenumber ν near the centre of a single gaseous absorption line, k_ν is given by the Lorentz formula for pressure broadening:†

$$k_\nu = \frac{S}{\pi} \frac{\gamma}{(\nu - \nu_0)^2 + \gamma^2}. \tag{12.39}$$

γ is the halfwidth of the line which is proportional to pressure and can be written $\gamma = \gamma_0 (p/p_0)$ (γ is also inversely proportional to the square root of the absolute temperature, but, as this exhibits much less variation than pressure through the part of the atmosphere which we are considering, we can omit temperature from the calculation).

If the frequency ν is such that $\gamma^2 \ll (\nu - \nu_0)^2$ then we can write

$$k_\nu = \frac{S}{\pi} \frac{\gamma_0 p}{p_0 (\nu - \nu_0)^2} = \beta p. \tag{12.40}$$

Now CO_2 is uniformly mixed through the atmosphere so that the mass of CO_2 per unit area between the top of the atmosphere and depth z will be proportional to the atmospheric pressure at that depth, i.e.

$$u = cp \tag{12.41}$$

where c is a constant. The transmittance of the atmosphere above depth z, at which the pressure is p, will therefore be

$$\begin{aligned} T_z &= \exp\left(- \int_{p=0}^{p} k_\nu \mathrm{d}u \right) \\ &= \exp\left(- \int_{p=0}^{p} c k_\nu \mathrm{d}p \right) \\ &= \exp\left(-\tfrac{1}{2} c \beta p^2 \right). \end{aligned} \tag{12.42}$$

To simplify the analysis we can assume that p varies linearly with z, i.e.

$$p = fz \tag{12.43}$$

(or alternatively we could use p as the measure of the depth z since it is a single-valued function of z which increases continuously with z). The weighting function for a single monochromatic line of frequency ν, assuming that $R = 1$, is then

$$W_z = -\frac{\mathrm{d}T_z}{\mathrm{d}z} = \beta c f^2 z \exp\left(-\tfrac{1}{2} \beta f^2 c z^2 \right). \tag{12.44}$$

The form of this function is shown in figure 12.6. For the purposes of

† See for example p 47 of Houghton J and Smith S D 1966 *Infra-Red Physics* (Oxford: Oxford University Press).

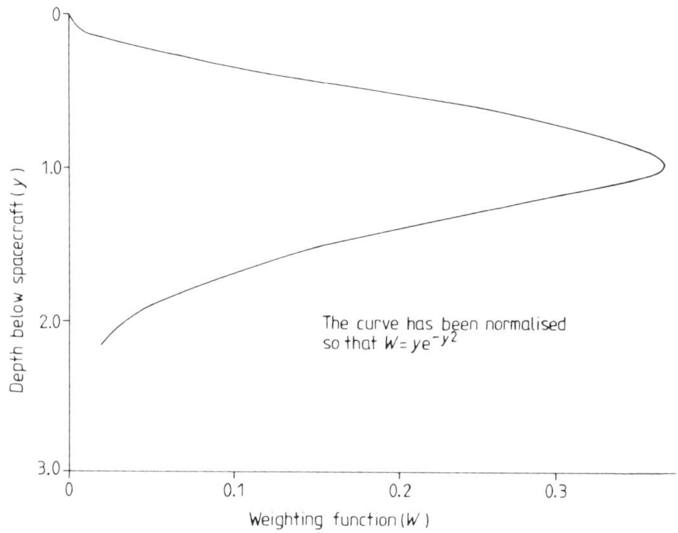

Figure 12.6 The form of the radiometer weighting function.

drawing this a new variable $y = (\frac{1}{2}\beta f^2 c)^{1/2} z$ has been introduced so that

$$-\frac{dT_z}{dz} = (2\beta c f^2)^{1/2} y e^{-y^2} \tag{12.45}$$

and the function which is actually plotted in figure 12.6 is ye^{-y^2}.

By choosing the appropriate wavelength, the form of the variation of the absorption coefficient can, to some extent, be controlled and the position of the maximum in terms of the height, or rather depth, varied. The absorption spectrum of CO_2 at 15 μm consists of a large number of separated lines. The teams at Oxford and Reading have made a special study of these, tabulating the positions and strengths, and have been able to choose a series of wavelengths to permit examination of the temperature structure of the atmosphere between 15 and 50 km with a resolution of 10 km.

One of the difficulties which exist is the finite bandwith of the radiometer. The bandwidths of practical filters cannot be made arbitrarily small and, because the CO_2 absorption coefficient varies with wavelength, the bandwidth of the radiometer will cause a reduction in the height resolution. For the channels designed to look deep into the atmosphere, the bandwidth does not affect the result too much and can be 10 cm^{-1}—well within the capabilities of an interference filter. The channels designed to look at the top of the atmosphere, however, must be positioned on the centres of the most intense lines, the Q-branch at 667 cm^{-1}, and the bandwidth must not effectively be greater than 1 cm^{-1}. This is beyond the current state of the art at 15 μm. The ingenious solution which has been adopted and which gives the radiometer its name is the use of a chopper filled with CO_2.

To explain the action of this selective chopper we shall first consider the operation of the simpler channels with the acceptable filters. In these channels, partly to ensure that the noise in the electronics is sufficiently low, and partly to ensure that the radiometer registers radiation from the atmosphere only and not from the components of the radiometer itself, which will all be emitting at 15 μm, a chopper is placed in the entrance aperture. Radiation emanating from the atmosphere will be chopped, while radiation from the radiometer itself will not and will escape detection. Of course the chopper will also radiate and so the usual method of alternately inserting and removing a blanking shutter in front of the radiometer entrance aperture would be quite useless, because the radiation from the shutter would also be chopped and detected along with the signal. The method which is used is extremely neat. The entrance aperture of each channel is divided into two equal parts so that one half of the aperture views, reflected in a fixed mirror, deep space, which can be assumed to be at a temperature of absolute zero and to represent a reference of zero radiation provided the reflectance of the mirror is sufficiently high, while the other half views the Earth's atmosphere reflected in a second mirror, which can be varied in position for calibration purposes. A chopper consisting of a vibrating black blade is arranged so that it obscures the fixed and variable mirrors alternately and, therefore, effectively chops the incoming signal. The radiation from the chopper blade is not detected because the blade remains within the aperture of the system all the time.

The selective chopper channel of the radiometer is similar to these other channels. However, a narrower filter is used, having a bandwidth of 3.2 cm^{-1} at 667 cm^{-1}, which is the narrowest yet obtainable at this frequency. In addition, a cell containing CO_2 is included in front of each section of the entrance aperture and the black blade of the chopper is exchanged for a mirror which looks at deep space. If the chopper mirror were completely removed, both parts of the entrance aperture would look at the atmosphere, reflected in a mirror, which again can be varied in position. With the chopper mirror in position and vibrating, one section of the aperture will look at deep space while the other section will look at the atmosphere through the appropriate CO_2 cell, and vice versa. The effect is just as if the input radiation were being chopped by alternate cells of CO_2. The simplest arrangement is to have one cell empty and one filled with CO_2, when, provided the CO_2 is at the correct pressure, the chopping will be effective only over the line centres. This, together with the narrowband filter, gives an effective bandwidth of around 1 cm^{-1}. Since the cells of CO_2 are within the aperture of the system all the time, the radiation from them will not be chopped and will not be detected. The radiation detected in this way originates from the very top of the atmosphere. The addition of a little CO_2 to the empty cell absorbs out the narrow-line centres, leaving an extremely narrow width on either side of centre and giving a still sharper weighting function which allows regions just below the top of the atmosphere to be examined. Various combinations of filter and chopper have been proposed and a set of weighting functions is shown in figure 12.7. Each satellite installation consists of six separate channels.

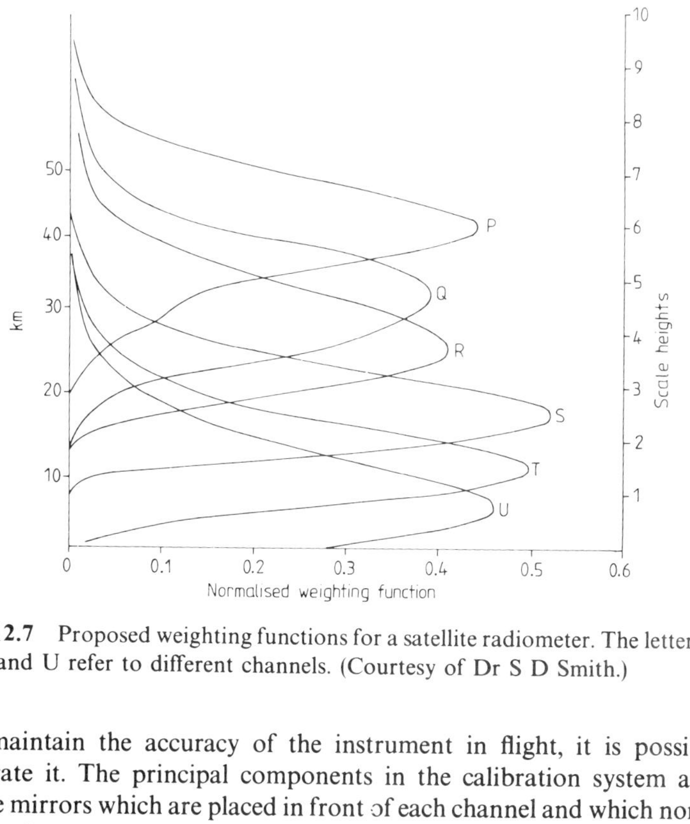

Figure 12.7 Proposed weighting functions for a satellite radiometer. The letters P, Q, R, S, T and U refer to different channels. (Courtesy of Dr S D Smith.)

To maintain the accuracy of the instrument in flight, it is possible to recalibrate it. The principal components in the calibration system are the variable mirrors which are placed in front of each channel and which normally reflect radiation from the atmosphere into the apertures. These mirrors are driven by small stepping motors and can be tilted to view the atmosphere, deep space, or a calibration black body giving a reference for both gain and zero in each channel. The proposed calibration sequence, which will repeat itself indefinitely in flight, is atmospheric radiation for 20 minutes, space for 2 minutes and calibration black body for 2 minutes. The channels having the extra CO_2 cells also have a balance calibration which ensures that the only difference between each half of the aperture is due to the CO_2 in the chopper cells. The narrowband filter which is used in the channel is replaced by a broadband filter at a wavelength outside the CO_2 absorption region which views the Earth's surface. Any signal detected under these circumstances is due to a difference in sensitivity between the two halves of the channel, which can be corrected if necessary.

Curves showing the measured transmittance of two of the basic filter elements are reproduced in figure 12.8. The sidebands are suppressed in the instrument by filters of the type shown in figure 6.20. The interference section of the blocking filter is deposited on one of the germanium lenses and an indium antimonide filter is fitted to the end of the light pipe over the detector. In addition, since it was found that the suppression in the wings of the Fabry–Perot filter was not quite high enough, a filter centred on the same wavelength but of the type.

$$L\,|\,\mathrm{Ge}\,|\,LHLHHLHLHLHHLH\,|\,\mathrm{Air}$$

which is a rather broader DHW type of around 20 cm^{-1} halfwidth, rather broader than that of figure 12.8(*b*), is placed in series with each Fabry–Perot. The composite filter possesses the narrow halfwidth of the Fabry–Perot together with the high sideband rejection typical of the DHW.

The construction of the radiometer is shown in figure 12.9. The optical

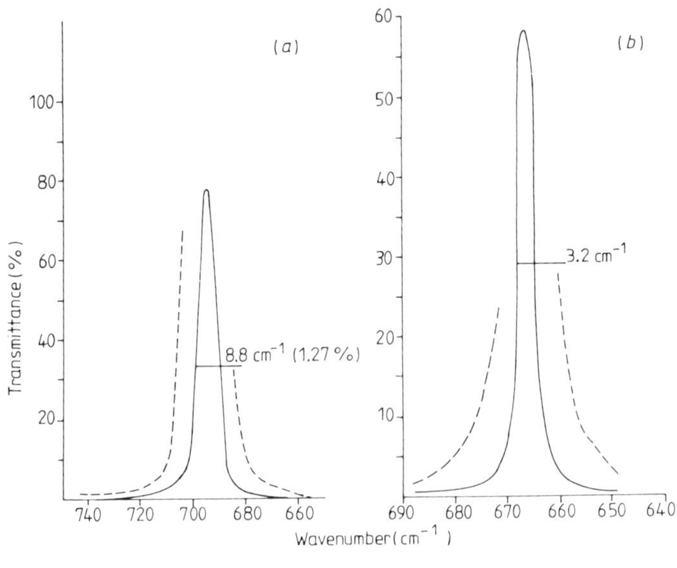

Figure 12.8 Measured transmittance of filters manufactured for the radiometer. The broken curves are merely the full line curves × 10.

(*a*) Air | *HLHHLH L HLHLHHLHL* | Ge substrate | *L* | Air.

Peak transmittance 78 % at 694.4 cm^{-1}.

(*b*) Air | *HLHLHHLHLHL* | Ge substrate | *L* | Air.

Peak transmittance 58 % at 666.4 cm^{-1}.

L = ZnS; *H* = PbTe. (Courtesy of Dr S D Smith and Sir Howard Grubb, Parsons & Co Ltd.)

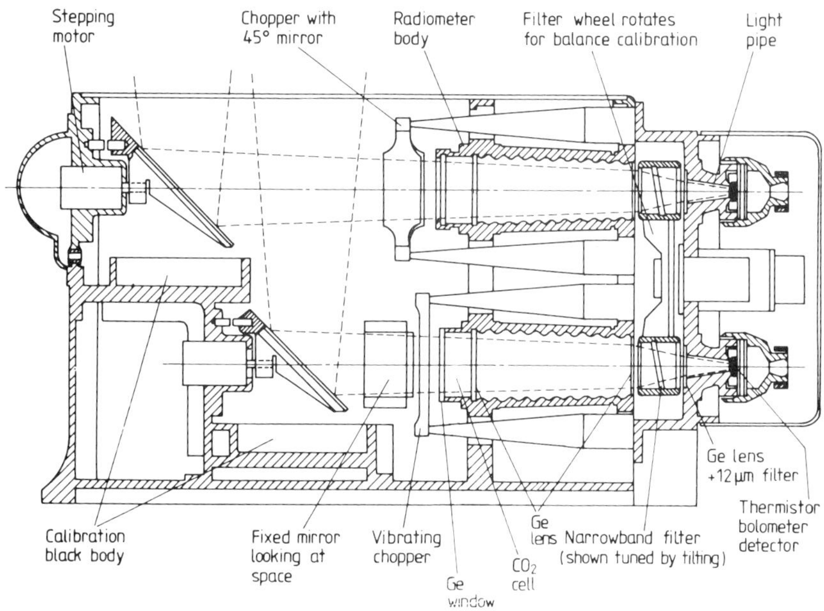

Figure 12.9 Schematic diagram of the selective chopper radiometer. (Courtesy of Dr S D Smith.)

system has been designed to use the full area of the narrowband filters together with the maximum range of angles which can be accommodated without destroying the spectral profile. It was this which prompted the work of Pidgeon and Smith on the angular dependence of filter characteristics discussed on pp 260–9.

The radiometer was successfully launched in April 1970 and made exceptionally useful temperature surveys of the upper atmosphere revealing much that was novel and unexpected. An early account of the instrument will be found in references 15 and 16.

ORDER SORTING FILTERS FOR GRATING SPECTROMETERS

There is a considerable advantage in using a diffraction grating rather than a prism for the selection of wavelengths in a monochromator or spectrometer, because the luminosity is so very much greater for the same resolution. A problem exists, however, with the diffraction grating which does not exist with the prism. This is the appearance of other orders in the spectrum which must be eliminated. The problem is particularly severe in the infrared, and the solution usually adopted has been the use of a low-resolution prism monochromator in series with the higher resolution grating monochromator. The lower resolution of the prism section, which is all that is necessary since order sorting is its sole function, means that the luminosity can be made as high as the grating section and the advantage associated with the grating thereby rctained. The grating and prism must be driven so that their respective wavelengths remain in step, a difficulty being that their angular dispersions vary in quite different ways. A simpler and attractive alternative is a longwave-pass thin-film filter. Recently several instruments have appeared on the market which use this system rather than the prism.

A paper by Alpert[17] gives an account of the various factors involved in the specification of such filters for infrared instruments. The most important feature is the rejection required in the stop regions. Before we can make an estimate of this rejection, we must first consider the way in which the energy varies in the various grating orders. Included in the assessment must be the characteristics of both the source and the detector.

The theory of the diffraction grating is considered in most textbooks on optics. For our present purpose it is sufficient to note two points. The first is that the angles of incidence and diffraction for any particular wavelength are given by the grating equation

$$\sin\theta + \sin\phi = \pm m\lambda/\sigma \tag{12.46}$$

where θ and ϕ are the angles of incidence and diffraction respectively, the sign convention being as shown in figure 12.10(*a*). σ is the grating constant, that is the spacing of the grooves, and m is the order number. From equation (12.46) we can see immediately the source of our present problem, that the angles

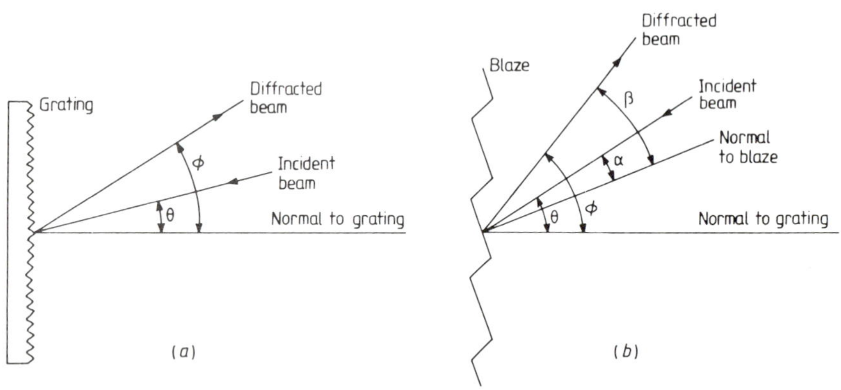

Figure 12.10(*a*) Sign convention for θ and ϕ. θ and ϕ have the same sign if they are on the same side of the grating normal. One side is chosen arbitrarily as positive. (*b*) Sign convention for α and β. α and β have the same sign if they are on the same side of the blaze normal. They are chosen positive when they have the same sense as the positive direction of θ and ϕ.

corresponding to any wavelength λ in the first order also exactly correspond to $\lambda/2$ in the second order, $\lambda/3$ in the third order, and so on. The second point is that the energy distribution in the various diffracted orders of any wavelength will be given by the pattern of lines in (12.46) modulated by the single-slit diffraction pattern of any one of the grooves at the appropriate wavelength. Modern diffraction gratings are invariably of the reflection type with the grooves 'blazed' or tilted, so that the single-slit diffraction pattern has its maximum at a particular wavelength in the first order, known as the blaze wavelength, rather than in the zero order, which increases considerably the efficiency of the grating over a range of wavelengths. In order to estimate the shape of the energy distribution we can assume the form of the grooves to be as in figure 12.10(*b*), although in practice the form may vary from that shown. α and β are the angles of incidence and diffraction referred to the normal to the groove, instead of the grating normal, but with the same sign convention applying. The intensity of the diffracted beam is given by an expression of the form

$$I = I_0 \frac{\sin^2[\pi\nu\sigma\cos\psi(\sin\alpha + \sin\beta)]}{[\pi\nu\sigma\cos\psi(\sin\alpha + \sin\beta)]^2} \tag{12.47}$$

where it is assumed that the grating will be sufficiently large to intercept the entire incident beam regardless of the angle of incidence. This expression is not strictly accurate over the entire range because at some angles the steps at the ends of the grooves may interfere slightly with the process, but it is good enough for our purpose. ψ is the angle between the grating and the blaze normals.

Most monochromators are of a type where the entrance and exit slits are fixed in position and the grating is rotated to scan the spectrum and where the angle of incidence is almost equal to the angle of diffraction. Little is lost by assuming that they are equal. With this assumption the curves shown in figure 12.11 have been derived for a typical grating and show how the

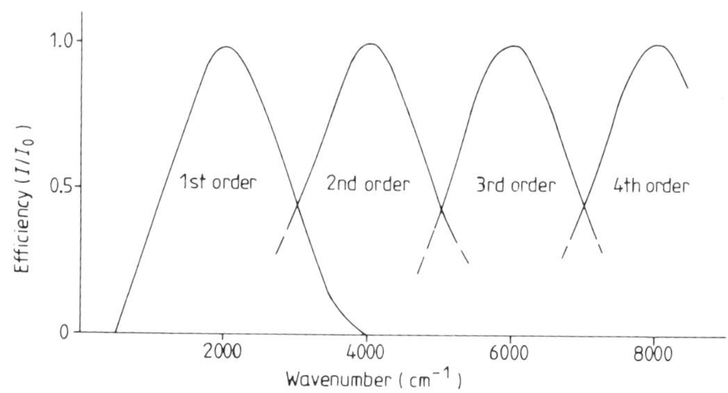

Figure 12.11 Intensity distribution in the various orders of a typical diffraction grating (theoretical) blazed for 2000 cm^{-1} (5 μm) in the first order.

intensities vary in the various orders. An important point is that, for the groove arrangement shown in figure 12.10(*b*), the dispersion, which is inversely proportional to the groove spacing, balances the alteration in the width of the diffraction pattern as the groove width varies, with the result that the variation of intensity with wavelength in any order depends solely on the blaze wavelength. A useful rule, which is generally used, is that the useful range of a grating which is blazed for a wavelength of λ_0 in the first order is from $2\lambda_0/3$ to $2\lambda_0$ in the first order, from $2\lambda_0/5$ to $2\lambda_0/3$ in the second order, and from $2\lambda_0/(2n+1)$ to $2\lambda_0/(2n-1)$ in the nth order. This is rather simpler in terms of wavenumber, the range being given by $\nu_0 \pm \frac{1}{2}\nu_0$ in the first order and $n\nu_0 \pm \frac{1}{2}\nu_0$ in the nth order. The bandwidth is more or less constant in terms of wavenumber. Measurements which have been made on gratings confirm the shape of the curves in figure 12.11. Some such measurements are reproduced by Alpert.

Now let us make the assumption that the diffraction grating is to be used in the first order and that the filter problem is the elimination of the second and higher orders. As far as the filter is concerned, the parameter which matters is the ratio of the detector signal in the first order to that in any of the other higher orders. The factors involved are, first of all, the variation of sensitivity of the detector; second, the variation in efficiency of the grating, already dealt with above; third, the dispersion of the grating in the various orders so that the energy in any order which is transmitted by the slits in the monochromator can be calculated; and last, the variation of output of the source. Of course, in some applications there may well be other factors which operate, such as the transmission of some optical components or the variation of reflectance of mirrors.

The detectors commonly used in this part of the infrared are thermal detectors which have reasonably flat response curves. In what follows we assume that they are perfectly flat. Any variation can be readily included in the analysis if required.

At any wavelength, the slits will pass a small band of wavelengths. If we assume that the slits are narrow enough so that energy variations over the range of wavelengths passed by the slit are negligible, then the energy transmitted in any order will be inversely proportional to the bandwidth of the slits in that order. From equation (12.46), the bandwidth is inversely proportional to the order number, which does help to reduce the requirements for filter performance.

In this part of the infrared, the sources which are generally used are either Nernst filaments or globars. For our present purpose we can assume, without too much error, that the source will be a black body probably peaking at around 2 μm, although this particular wavelength does not matter very much. The variation of energy with wavelength for a black-body source is given by Planck's equation:

$$E_\lambda = \frac{c_1}{\lambda^5[\exp(c_2/\lambda T) - 1]} \tag{12.48}$$

where E_λ is the spectral emissive power, and c_1 and c_2 are the first and second radiation constants with values 3.74×10^{-16} W m^2 and 1.4388×10^{-2} m K respectively.

For any wavelength λ, let the efficiency of the grating be denoted by ε_λ and the transmission of the order sorting filter by T_λ. Then the stray light due to the mth order, expressed as a fraction of the energy in the first order, will be given by

$$r_m = \frac{\varepsilon_{\lambda/m} E_{\lambda/m} T_{\lambda/m}}{m\varepsilon_\lambda E_\lambda T_\lambda} = \frac{\varepsilon_{\lambda/m} (\lambda/m) E_{\lambda/m} T_{\lambda/m}}{\varepsilon_\lambda \lambda E_\lambda T_\lambda}$$

where we have multiplied the numerator and denominator by λ. The permissible magnitude of r_m depends on the number of orders which are involved in producing significant interference. Let this number be N and let the total amount of permissible stray light be given by S, which is expressed as a fraction of the total first-order energy; then we can require that

$$r_m = S/N$$

and the maximum transmission which can be permitted at wavelength λ/m is given by

$$T_{\lambda/m} = T_\lambda \frac{S}{N} \frac{\lambda E_\lambda}{(\lambda/m) E_{\lambda/m}} \frac{\varepsilon_\lambda}{\varepsilon_{\lambda/m}}. \tag{12.49}$$

Now $\varepsilon_\lambda/\varepsilon_{\lambda/m}$ will be greater than unity except on the blaze wavelength. Without affecting the accuracy too greatly, we can make the assumption that each order m is effective only over the range $2\lambda_0/(2m+1)$ to $2\lambda/(2m-1)$ and that $\varepsilon_\lambda/\varepsilon_{\lambda/m}$ is unity over this range. Elsewhere we can assume that the mth order does not produce interference and omit it.

To complete the calculation, we need the value of $\lambda E_\lambda/(\lambda/m)E_{\lambda/m}$. The function λE_λ is plotted in figure 12.12. To make it possible to apply this figure generally, the variables have been normalised in the manner shown and the scales are logarithmic so that any particular set of conditions can be reproduced simply by sliding the scales along the axes.

The first step in drawing up the specification for a practical set of filters will be to decide on the required number of filters. Even one single diffraction grating has a useful wavelength range of 3:1, which is greater than the range which can be covered by just one filter.

Let the limits of the wavelength region over which the grating or set of gratings are to be used be λ_F and λ_S, where $\lambda_F > \lambda_S$. If we start with the longest wavelength, then the final filter in the series must block wavelengths $\lambda_F/2$ and shorter. An ideal longwave-pass filter would have a rectangular edge shape and it would be possible to use it over the whole of the range $\lambda_F/2$ to λ_F. Real filters have sloping edges and must be allowed some tolerance in edge position, otherwise manufacture becomes impossible. This means that the specification must show the start of the transmission region of the final filter as $(1+\alpha)\,\lambda_F/2$. Assuming that all the filters in the set are of more or less similar construction,

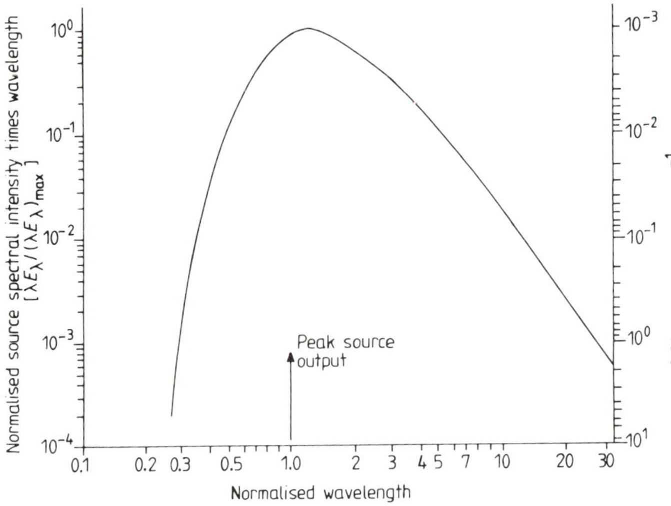

Figure 12.12 Curve showing the variation of λE_λ with wavelength for a black-body source.

then the same expression will also apply to the next filter in the set, which will have a transmission region specified to start at a wavelength given by $[(1+\alpha)^2/2^2]\lambda_F$ and to finish at $[(1+\alpha)/2]\lambda_F$. The regions for the other filters are found in exactly the same way. If there are n filters in the set, then the first filter must have the specified start wavelength at $[(1+\alpha)^n/2^n]\lambda_F$. We can equate this start wavelength to λ_S and solve for α:

$$\alpha = 2(\lambda_S/\lambda_F)^{1/n} - 1. \tag{12.50}$$

This expression can be evaluated in a practical case for several possible values of n and the set of filters giving the optimum arrangement of filters and the best degree of tightness of tolerance selected.

The advantage of using this type of specification is that any particular filter from any set of filters made to the specification is interchangeable with the corresponding filter in any other set. If this interchangeability is not required, it is possible to slacken the tolerances slightly, but this makes the problem of making up each individual set rather more of a puzzle.

To illustrate the method, let us consider the specification for a set of filters for use with a pair of gratings for the region 3–30 μm. The first grating can be the one already considered with blaze at 5 μm, while the second will be a similar one with blaze at 15 μm. The region 3–3.3 μm will not be covered with quite as great efficiency as the rest of the region, but the source will be rather more efficient here, which fact counterbalances the fall off in grating efficiency to some extent.

First we decide on the number of filters. By inspection we arrive at the conclusion that the minimum number of filters is four, but that this number leads to a specification which is rather tight, and it is better to use five filters. If we assume that the tolerances should be shared equally amongst them, then the limits of the pass regions and the edges of the rejection zones are as shown in table 12.1.

Table 12.1

Filter number	Pass region (μm)	Longwave edge of rejection zone (μm)
5	19–30	15
4	12–19	9.5
3	7.6–12	6
2	4.8–7.6	3.8
1	3–4.8	2.4

We then decide on the acceptable level of stray light in this case as, say, 1 % of the true first-order signal. We must also decide on the acceptable minimum transmission of the filters in the pass region, say 50 %. In practice the level will almost certainly be rather greater than this, but the use of a low figure in setting up the specification gives a pessimistic figure for the specified levels in the rejection region.

Next we compute the regions over which the various orders are effective in producing stray light. The results are shown in table 12.2. Both the actual wavelength of the interfering energy and the corresponding wavelengths in the first order with which it interferes are given. We can choose to use germanium as substrate material for the filters and therefore safely neglect all wavelengths shorter than 1.6 μm, because they will be effectively suppressed by the intrinsic absorption of the germanium.

The first filter we consider is filter number 4, which includes the blaze wavelength of the longer-wave grating in its transmission region. At the blaze wavelength the highest significant order is the ninth and N therefore is 8, i.e.

$$T_{\lambda_0}S/N = 0.5 \times 0.01/8 = 0.000\,625.$$

We therefore set the scale on the right-hand side of figure 12.12 to correspond to 0.000 625 at 15 μm and read off the allowable transmissions at the higher order wavelengths from the curve. This is shown in figure 12.13.

To simplify the task of setting up the specification, we assume that the transmission levels which are thus established apply to the complete range for each appropriate order, i.e. for the mth order, the transmission found in this way applies to the range $2\lambda_0/(2m+1)$ to $2\lambda_0/(2m-1)$, a slightly pessimistic result. The only exception which we make to this is that portion of the rejection zone immediately beside the edge of the transmission zone. Here it is important

Table 12.2

Order	Range (μm)	Corresponding range in the first order (μm)
15 μm grating		
1st	30 –10	30 –10
2nd	10 – 6	20 –12
3rd	6 – 4.29	18 –12.85
4th	4.29– 3.33	17.15–13.33
5th	3.33– 2.72	16.70–13.65
6th	2.72– 2.31	16.35–13.85
7th	2.31– 2.00	16.15–14.00
8th	2.00– 1.76	16.00–14.10
9th	1.76– 1.58	15.90–14.20
10th	10th and higher orders beyond germanium edge	
5 μm grating		
1st	10 – 3.33	10 – 3.33
2nd	3.33– 2.00	6.67– 4.00
3rd	2.00– 1.43	6.00– 4.28
4th	4th and higher orders beyond germanium edge	

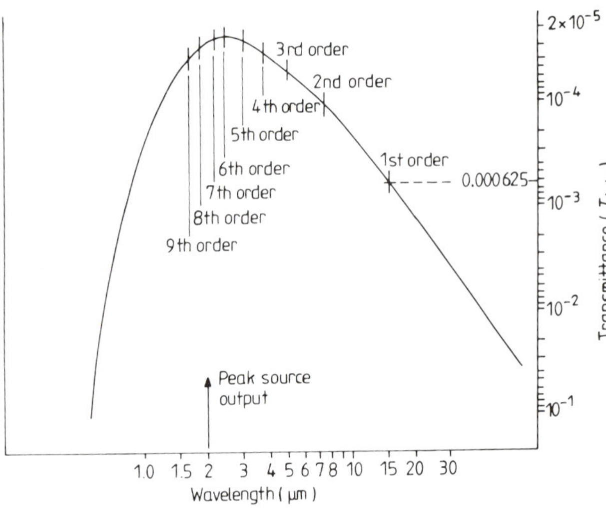

Figure 12.13 How the maximum transmittance levels are established for filter 4.

that the specification should not be tighter than is strictly necessary. The end of the transmission region is 19 μm. From the range of the higher order interference we see that only one order, the second, is effective at that wavelength. $T_\lambda S/N$ is therefore $0.5 \times 0.01/1.0 = 0.005$. Setting this value on the right-hand scale of figure 12.13 against the point on the curve corresponding to 19 μm, we read off 0.0009 against 9.5 μm, which is therefore the maximum allowable transmission at that wavelength. At 18 μm, the second and third orders are involved and the value of $T_\lambda S/N$ becomes 0.0025. Setting this against the point on the curve corresponding to 18 μm, we read off 0.0004 against 9 μm, which is the maximum allowable transmission at that point. At 17.15 μm there are three orders involved so that the transmission at 8.6 μm should be not greater than 0.0003. This procedure is repeated at each wavelength where a further order becomes significant until the full number of orders is reached. Points corresponding to these are plotted on a diagram and a horizontal line through each is linked with a vertical line through the adjacent point on the shortwave side. The specification for the filter is then completed by adding a minimum transmission level of 0.50 from 12–19 μm. Figure 12.14 shows the complete arrangement.

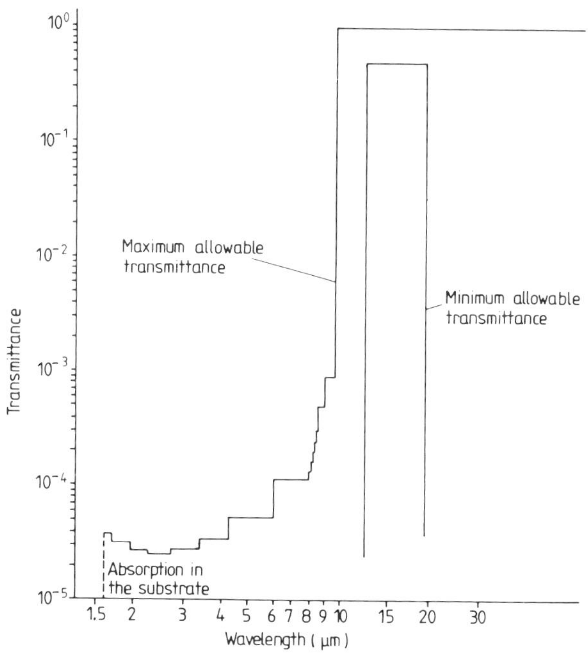

Figure 12.14 Specification of filter 4.

Next we consider the longest-wave filter, number 5. Here the conditions are not nearly so severe, because the filter is being used for a region which does not include the first-order blaze wavelength and there is therefore only slight higher-order interference. According to table 5.2, the second-order interference is falling off sharply beyond 20 μm and the third order is not effective anywhere within the pass region. The critical region is therefore 9.5–10 μm. $T_\lambda S/N$ is once again 0.005 and setting this value against the point corresponding to 20 μm in figure 12.12, we find the permissible transmission in the rejection region at 10 μm as 0.0009. Outside the 9.5–10 μm range the simple theory which predicts no interference at all is once again not sufficiently accurate. A convenient pessimistic assumption is that the transmission at the very edge of the rejection zone, i.e. at 15 μm, should be around 0.01 and then a straight line can be drawn from this point to that at 10 μm. On the shortwave side of 9.5 μm we can retain the transmission as 0.0009. The resulting transmission specification for the filter is given in figure 12.15.

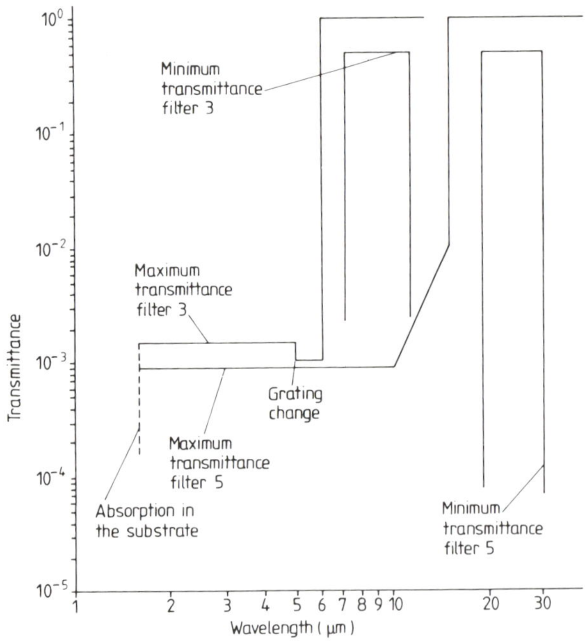

Figure 12.15 Specification of filters 3 and 5.

Filter number 3 covers the changeover from one grating to the next. Beyond 10 μm, the grating is blazed at 15 μm. The significant range for second-order interference is 12–20 μm so that, except just at 12 μm, second-order interference will be low. At 12 μm, $T_\lambda S/N$ is 0.005, and from figure 12.13 the permissible transmission at 6 μm is just over 0.001. We can specify this level of transmission as far as 5 μm, which corresponds to 10 μm in the first order, the grating changeover wavelength. On the short wavelength side of 10 μm, the 5 μm grating is used. Table 12.2 predicts that there will be no interference from the edge of the pass band at 7.6 μm right to 10 μm. However, to be safe, we assume that there will be second-order interference at 7.6 μm, and setting a value of 0.005 against 7.6 μm in figure 12.13, we establish a value for the transmittance at 3.8 μm, the second-order wavelength. This is shown on figure 12.15 and we further assume that it applies to the whole region between the germanium edge and 5 μm.

The specification for filter number 2 (figure 12.16) is set up in exactly the same way as for filter number 4 since it includes the blaze wavelength. However, the requirements are not nearly so severe, because both the peak of the source and the absorption edge of the germanium substrates are much closer to the pass band of the filter.

Filter 1 is similar to the others (figure 12.17). The short band from 1.6–2 μm, where the simple theory predicts no higher order interference (second order missing and third order corresponding to first-order wavelengths beyond 4.8 μm, the edge of the pass band), is filled in by a horizontal line at the same level as the allowable transmission at 2 μm.

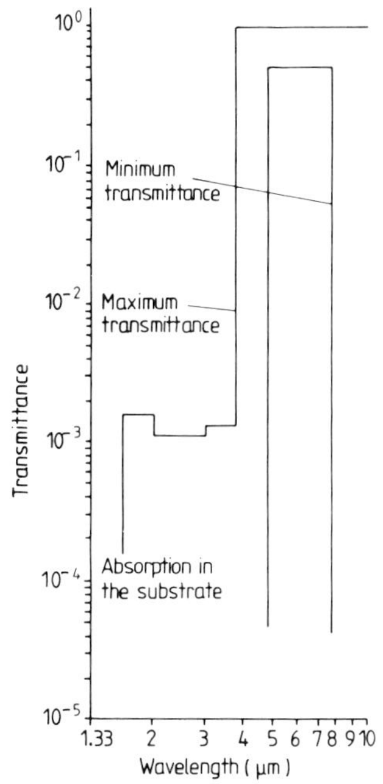

Figure 12.16 Specification of filter 2.

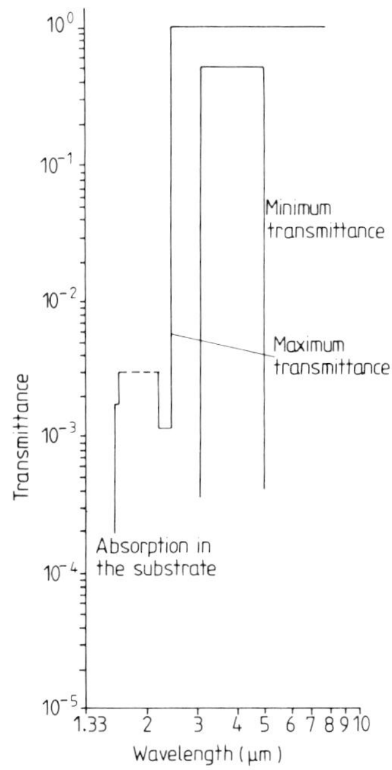

Figure 12.17 Specification of filter 1.

As far as the optical performance of the filters is concerned, there is only one further point to be specified, the bandwidth of the measuring spectrometer used for inspecting the filters. The requirement here is that the bandwidth should be not greater, nor appreciably less than the bandwidth of the final instrument in which the filters are to be used†. Any spikes of transmittance not resolved by this arrangement will not be resolved by the instrument itself. There is clearly no point in carrying out too strict a test, which would not only be an unnecessary waste of time and expense, but could also lead to a filter's being rejected when in fact it is perfectly satisfactory for the job.

Once the specification has been established, the design of the filters is just a straightforward application of the principles discussed in chapter 6. A study of the results suggests some general rules. The first is that the filters which include the first-order blaze wavelength in their pass regions are the most critical in their specifications, and to ease, as far as possible, their edge steepness the blaze wavelength should be arranged to be nearer the shortwave limit of the pass region than the longwave limit. The second point is that since the filters which do not include the first-order blaze have very much reduced rejection requirements, it is useful to make sure that the longest-wave filter, that will be the most difficult to fabricate, has a pass region clear of the blaze wavelength—even if in some applications it means an extra filter.

SOME COATINGS INVOLVING METAL LAYERS

Electrode films for Schottky-barrier photodiodes

A simple diode photodetector consists of a metal layer deposited over a semiconductor forming a Schottky barrier. High quantum efficiency can be achieved. The incident light passes through the metal layer into the depletion layer of the diode where it creates electron–hole pairs. The metal contact layer must transmit the incident light and since it has intrinsically high reflectance, it must be coated to reduce its reflection loss. We give here a very simple approach to the design of a combination of electrode and antireflection coating. A number of workers have made contributions in this area[18–20] with probably the most complete account of an analytical approach being that of Schneider[18].

The substrate for the thin films is the semiconducting part of the diode and it is fixed in its optical admittance. The metal layer goes directly over the semiconductor (in some arrangements there is a very thin insulating layer that has negligible optical interference effect) and so the potential transmittance is fixed entirely by the thickness of the metal. All that can be done to maximise actual transmittance is simply to reduce the reflectance to zero.

We take as an example a gold electrode layer deposited on silicon. We assume a wavelength of 700 nm and optical constants of 0.131–i3.842 for gold and 3.92–i0.05 for silicon[21]. The optical constants of silver and copper are quite similar to those of gold at this wavelength and the results apply almost equally

† i.e., the fractional bandwidth of the measuring instrument should be equal to the fractional bandwidth of the final instrument in the transmission region of the particular filter under test.

well to these two alternative metals. The admittance locus of a single gold film on silicon is shown in figure 12.18. An antireflection coating must bridge the gap between the appropriate point on the metal locus to the point (1, 0) corresponding to the admittance of air. We can assume that the maximum and minimum values of dielectric layer admittance available for antireflection coating are 2.35 and 1.35 respectively. Using these values, we can add to the admittance diagram two circles that pass through the point (1, 0) and correspond to admittance loci of dielectric materials of characteristic admittances 2.35 and 1.35 respectively. These loci define the limits of a region in the complex plane. Provided a metal locus ends within this region, then it will be possible to find a dielectric overcoat of admittance between 1.35 and 2.35 that, when the thickness is correctly chosen, will reduce the reflectance to zero. It is clear from the diagram that the thicker the metal film, the higher must be the admittance of the antireflection coating. Once the metal locus extends beyond this region, a single dielectric layer can no longer be used and a multilayer coating (or a single absorbing layer, although it would reduce transmittance and so would not be very useful in this particular application) becomes necessary. We have already considered multilayer coatings in the section on induced transmission filters. Here we limit ourselves to a single layer and take the highest available index of 2.35.

The remaining task in the design is then to find the thicknesses of metal and dielectric corresponding to the trajectories between the substrate and the point

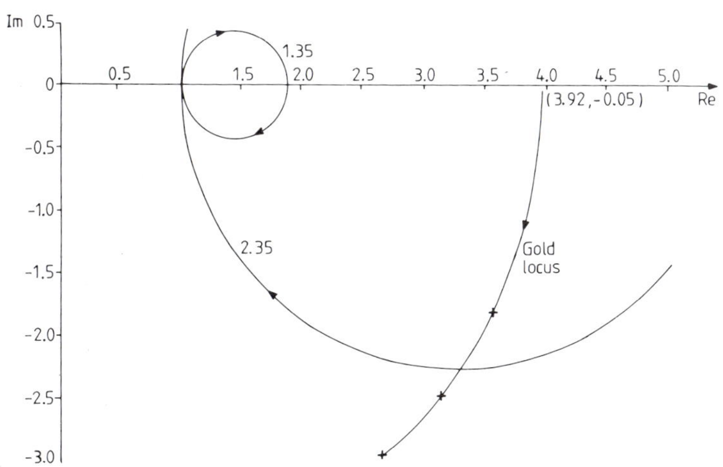

Figure 12.18 Admittance diagram showing some of the factors in the antireflection of a metal electrode layer in a photodiode. The optical constants of gold are assumed to be (0.131-i3.842) at a wavelength λ_0 of 700 nm. The gold is deposited on silicon with optical constants (3.92-i0.05). The crosses on the gold locus mark thickness increments of 0.005 λ_0, i.e. 3.5 nm. Also shown are loci corresponding to dielectric layers of indices 1.35 and 2.35 which terminate at the point 1.00.

of intersection and between the point of intersection and the point (1, 0) in figure 12.18. The points marked along the metal locus correspond to intervals of 0.005 λ_0 in geometrical thickness, that is to thickness intervals of 3.5 nm. Visual estimation suggests a value of 0.013 λ_0 for the thickness to the point of intersection. More accurate calculation gives 0.0133 λ_0, that is a thickness of 9.3 nm. The dielectric layer has an optical thickness of somewhere between an eighth- and a quarter-wave, and accurate calculation yields 0.186 λ_0.

The calculated performance of this coating is shown in figure 12.19. Of course, the thickness of the metal film is rather small and it is unlikely that the values of optical constants measured on thicker films would apply without correction, but the form of the curve and the basic principles of the coating are as discussed here.

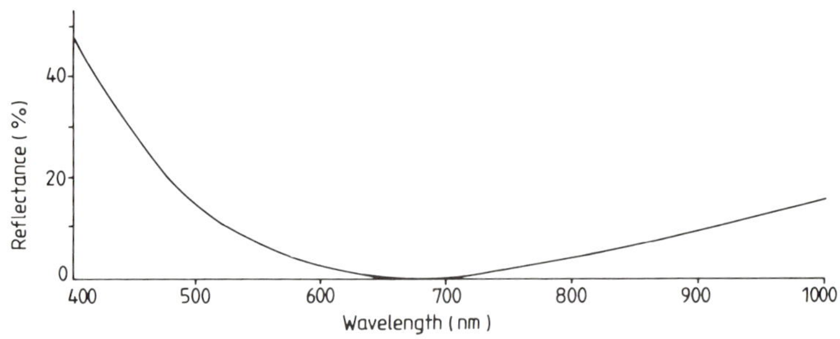

Figure 12.19 The calculated transmittance, including dispersion, of the gold electrode film and antireflection coating designed in the text.

Spectrally selective coatings for photothermal solar energy conversion

Coatings for application in the field of solar energy represent a complete subject in their own right. They have been discussed in detail by Hahn and Seraphin[22]. Here we consider simply a limited range of coatings based on antireflection coatings over metal layers which have much in common with the electrode film of the previous section.

Solar absorbers that operate at elevated temperatures can lose heat by radiation unless steps are taken to reduce their emittance in the infrared. Yet to operate efficiently they must have high solar absorptance in the visible and near infrared. Optimum results are obtained from an absorbing coating that exhibits a sharp transition from absorbing to reflecting at a wavelength in the near infrared that varies with the operating temperature of the absorber. One way of constructing such a coating is to start with a thick metal film or a metal substrate and apply an antireflection coating that is efficient over the visible but which becomes ineffective in the infrared, so that at longer wavelengths the reflectance is high and the thermal emittance, as a result, low. Fortunately, we are interested simply in a reduction of reflectance. Transmittance is unimportant. The energy that is not absorbed in the coating is absorbed in the substrate. Thus the antireflection coating can include absorbing layers.

A useful approach to the design is the use of a semiconducting layer over a metal. The semiconductor becomes transparent in the infrared beyond the intrinsic edge and so in that region the reflectance of the underlying metal predominates. In the visible and near infrared the absorption in the semiconductor is sufficient to suppress the metallic reflectance and to complete the design it is sufficient to add an antireflection coating to reduce the reflectance of the front face of the semiconductor. Since the metal is to dominate the infrared performance either the semiconductor layer must be relatively thin in the infrared or the metal must have sufficiently high k/n to be only slightly affected by the high index of the semiconductor in its transparent region. From the point of view of optical constants, silver is therefore the most favourable metal but it suffers from a lack of stability at elevated temperatures which cause it to agglomerate. Its optical constants are shifted and its reflectance reduced. Seraphin and his colleagues (see reference 22 for a readily available summary and more detailed references) have developed coatings in which the silver is stabilised by layers of chromium oxide (Cr_2O_3) which act as diffusion inhibitors. The silicon films are produced by chemical vapour deposition in which the silicon–hydrogen bonds in silane gas flowing over the substrate are broken by elevated substrate temperature and, as a result, silicon deposits. Adding oxygen or nitrogen to the gas stream gives an antireflection coating of silicon oxide or nitride which can be graded in composition by continuous variation of gas-stream composition. Such coatings can withstand temperatures in excess of 600°C without degradation.

The design of such coatings is straightforward. First of all, the thickness of silicon must be such that the visible absorption is sufficiently high to mask the underlying silver but not so thick that interference effects reduce reflectance and increase emittance in the infrared. In the visible region, the light that enters the silicon layer and is reflected from the silver at the rear surface should be sufficiently attenuated that only a very small proportion ever re-emerges. We can assume that the attenuation of this light depends on a law of the form $\exp(-4\pi kd/\lambda)$ and for the entire round trip from front surface to rear of film and back again to the front surface we should have a value roughly in the range 0.01–0.05. Let us choose a design wavelength of 500 nm in the first instance at which silicon in thin-film form has optical constants of 4.3–i0.74[21]. Then for $\exp(-4\pi kd/\lambda)$ to be 0.05, the value of d must be 160 nm. Since this is for the entire round trip, the film thickness should be half this value or 80 nm. An antireflection coating must then be added to reduce the visible reflectance of the front surface of the silicon layer. Since we have reduced the interference effects to a low level, the front surface will be similar to bulk silicon with optical constants characteristic of the film. Seraphin and his colleagues used a graded-index film of silicon nitride and silicon dioxide, but for simplicity we assume here a homogeneous film of roughly 2.0 admittance and a quarter-wave thick at 500 nm. We can take zirconium dioxide with its characteristic admittance of 2.07 as an example. The performance of the complete coating is shown in figure 12.20. The extra dip at 600 nm is a result of the thickness of the silicon. The silicon admittance locus spirals around, converging on the optical constants. At 600 nm, the spiral is somewhat shorter but the end point is passing through a region where the zirconium oxide layer can act as a reasonably efficient antireflection coating once again and so the dip appears. The silver begins to assert itself at around 700 nm in this design. We can shift the reflectance trough to a longer wavelength, say 750 nm, by carrying out a completely similar procedure but this time using 4.17–i0.37 for the optical constants. Now a double-pass reduction of 0.05 leads to a round-trip thickness of 480 nm, representing a film thickness of 240 nm. The performance is also shown in figure 12.20. In both traces the optical constants of silicon and silver were derived from reference 21.

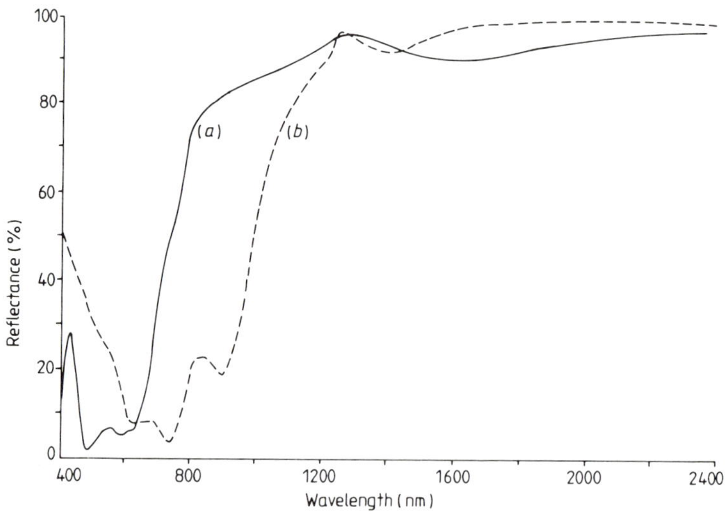

Figure 12.20 The calculated performance including dispersion of solar absorber coatings consisting of antireflected silicon over silver. Designs:

(*a*) Air | ZrO_2 | Si | Ag
1.0 | 0.25 λ_0 | 80 nm |

$\lambda_0 = 500$ nm.

(*b*) Air | ZrO_2 | Si | Ag
1.0 | 0.25 λ_0 | 240 nm |

$\lambda_0 = 750$ nm.

Further details are given in the text.

An alternative arrangement makes use of metal layers as part of an antireflection coating for silver. The great problem in designing an antireflection coating for a high-efficiency metal using entirely dielectric layers is that the admittance where the locus of the first dielectric layer, that is the layer next to the metal, first cuts the real axis is far from the point (1, 0) where we want to terminate the coating, and with each pair of subsequent quarter-waves we can modify that admittance by only $(n_H/n_L)^2$. Many quarter-waves are needed, as we have seen with the induced transmission filters. A metal layer, on the other hand, follows a different trajectory from a dielectric layer, cutting across dielectric loci, and can be used to bridge the gap between the large radius circle of the dielectric next to the metal and a dielectric locus that terminates at (1, 0). The metal locus itself can be arranged to pass through (1, 0) but the extra dielectric layer is capable of giving a slightly broader characteristic and also some protection to the metal layer. Silver could be used as the matching metal but its high k/n ratio leads to rather narrow spike-like characteristics even with the terminating dielectric layer and a metal with rather greater losses is better. We use chromium here as an illustration with aluminium oxide as dielectric. These materials have figured in published coatings (see reference 22 for further details). We choose a wavelength of 500 nm for the design and the optical constants we assume for our materials are silver: 0.05–i2.87; aluminium oxide: 1.67; and chromium; 2.86–i4.11. Again the optical constants of the metals were obtained from reference 21 with interpolation if necessary. An admittance diagram of a coating of design:

$$\text{Air}\left|\begin{array}{c}\text{Al}_2\text{O}_3\\0.184\,\lambda_0\end{array}\right|\begin{array}{c}\text{Cr}\\7.5\text{ nm}\end{array}\left|\begin{array}{c}\text{Al}_2\text{O}_3\\0.184\,\lambda_0\end{array}\right|\quad\text{Ag}\quad(\lambda_0 = 500\text{ nm})$$

is shown in figure 12.21. The chromium locus bridges the gap between the two dielectric layers. Because of its rather lower k/n ratio than silver its trajectory is flatter and the entire characteristic less sensitive to wavelength changes. The arrangement helps to keep the final end point of the coating in the vicinity of (1, 0) as the loci increase or decrease in length with changing wavelength or g. No attempt was made to refine this design although clearly, because of the wide range of possible thickness combinations that would lead to zero reflectance at the design wavelength, there must be scope for performance improvement by refinement. The characteristic of the coating is shown in figure 12.22. The reflectance minimum can be shifted to longer wavelengths by repeating the design process with appropriate values of the optical constants. This gives the desired zero but then at shorter wavelengths, where the dielectric loci are departing further and further from ideal and the chromium layer is unable to bridge the gap between them, a peak of high reflectance is obtained. At still shorter wavelengths, there is a second-order minimum where the dielectric layers make a complete revolution and are once again in the vicinity of the correct position. For the ideal values we have used in these calculations the central peak of high reflectance is very high indeed. Practical coatings also

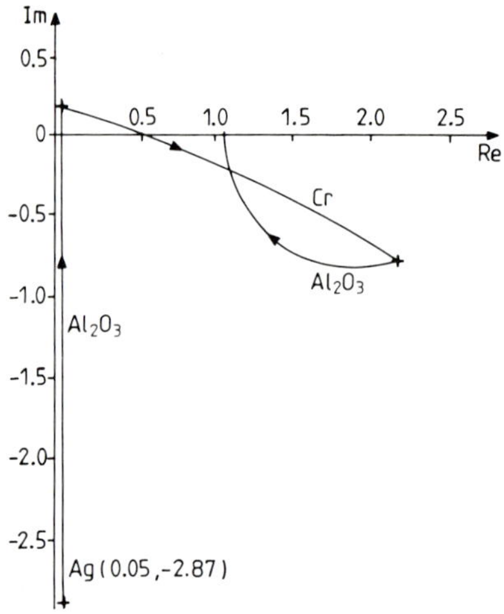

Figure 12.21 Admittance diagram at λ_0 of an absorber coating of design:

Air	Al_2O_3	Cr	Al_2O_3	Ag
	$0.184\lambda_0$	7.5 nm	$0.184\lambda_0$	

$\lambda_0 = 500$ nm.

See the text for an explanation.

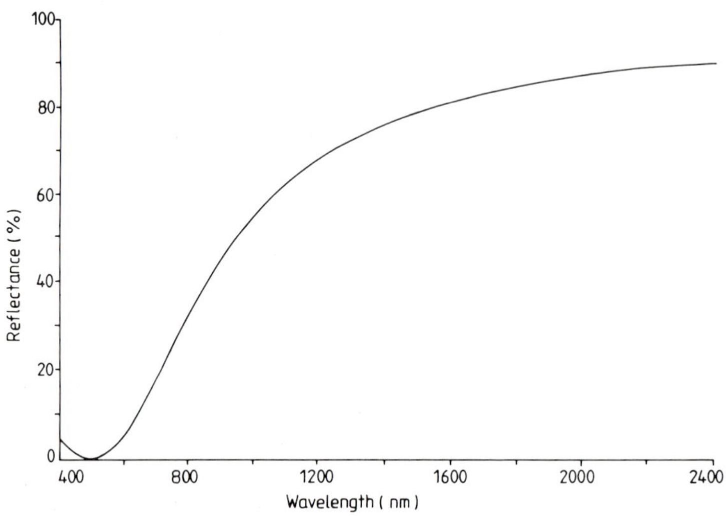

Figure 12.22 The calculated performance, including dispersion, of the absorber coating of figure 12.21.

show this double minimum (see Hahn and Seraphin[22]), but the central maximum is very much less prominent, the most likely explanation being that the layers in practice have much greater losses than we have assumed. In particular, the thin chromium layers are unlikely to have ideal optical constants. High losses would make the loci spiral in towards the centre of the diagram and reduce the wavelength sensitivity.

The major problems associated with such coatings are not their design but the necessary high-temperature stability. Spectrally selective solar absorbers are only economically viable when they are used to produce high temperatures and, indeed, it is only at high temperatures that they offer an advantage over the more conventional spectrally flat black absorbing surfaces that can be produced very much more cheaply. They are used under vacuum to eliminate gas conduction heat losses and so the major degradation mechanism is diffusion within the coatings. Silver is particularly prone to agglomerate at high temperatures and much development effort has resulted in the incorporation of thin diffusion barriers such as chromium oxide that inhibit diffusion and agglomeration of the components without affecting the optical properties. The achievements in terms of lifetime at high temperatures are impressive. Further details will be found in reference 22.

Heat reflecting metal–dielectric coatings

There are several applications where a cheap and simple heat-reflecting filter would be valuable. For example, a normal, spectrally flat solar absorber can be combined with such a filter so that the combination acts as a spectrally selective absorber. It is possible to construct a very simple band-pass filter that has the desired characteristics from a single metal layer surrounded by two dielectric matching layers[23–26]. The filter is similar in some respects to the induced transmission filter, although the maximum potential transmittance that is theoretically possible cannot usually be achieved. One design technique uses the admittance diagram and we can illustrate it with an example in which we consider a glass substrate and an incident medium of air or vacuum. Silver, with optical constants of 0.06–i3.75 at 600 nm can serve as metal and we assume a dielectric layer material of index 2.35. Zinc sulphide, which has such an index, has been used in this application, but the most durable and stable coatings are ones incorporating a refractory oxide. Figure 12.23 shows an admittance diagram in which one dielectric locus begins at the substrate and a second terminates at (1, 0) corresponding to the incident medium. If the complete coating is to have zero reflectance them the remaining layers must bridge the gap between these two loci. Once again, it is easy to see that a metal layer can do this and also that the particular optical constants of the metal are unimportant. They will simply alter somewhat the points of intersection with the two loci. The loci shown correspond approximately to the thickest silver film that will still give zero reflectance. To increase the silver thickness without sacrificing the

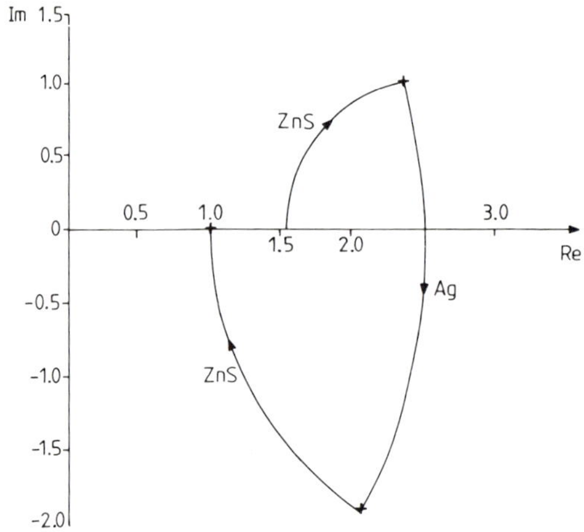

Figure 12.23 Admittance diagram of a metal–dielectric heat-reflecting filter. The diagram shows the locus at a wavelength of 600 nm of a ZnS | Ag | ZnS combination deposited on glass.

zero reflectance requires that the indices of the two dielectric layers be increased. A small increase in thickness of metal without a gross alteration in the design could be achieved by the insertion of a low-index quarter-wave layer next to the substrate to move the starting point of the next high-index dielectric layer, the upper one in the admittance diagram, further along the real axis towards the origin. The new locus would be outside the existing one demanding a thicker metal matching layer. In the absence of such a low-index layer, the final three-layer design is:

Air	ZnS	Ag	ZnS	Glass	
1.0	2.35	0.06–i3.75	2.35	1.52	($\lambda_0 = 600$ nm)
	$0.146\,\lambda_0$	15 nm	$0.141\,\lambda_0$		

with performance shown in figure 12.24. The steep fall towards the infrared is partly due to the drop in efficiency of the matching, but an inspection of the admittance diagram quickly reveals that the reduction in length of each locus accompanying an increase in wavelength should not by itself change the reflectance grossly. The dispersion of the silver, however, keeps the value of $(2\pi kd/\lambda)$ high and, hence, the locus long, and is primarily responsible for the increase in reflectance in the infrared. The coating could be based on virtually any metal with high infrared reflectance and high-index dielectric material. Gold and bismuth oxide have been successfully used[26].

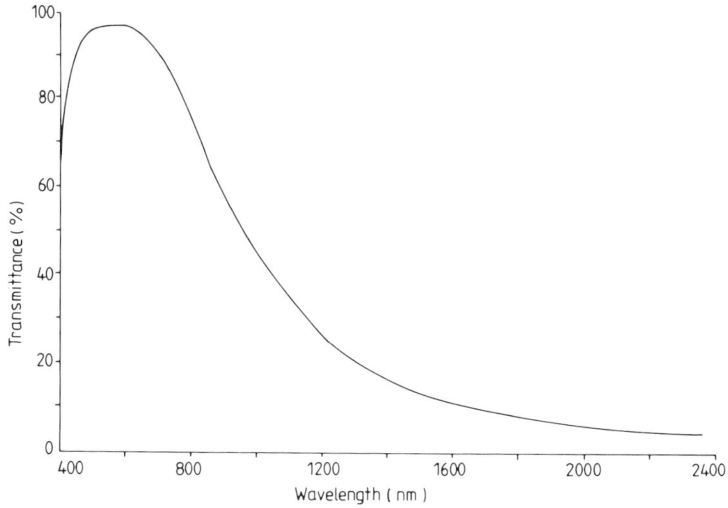

Figure 12.24 Transmittance, calculated with dispersion included, of the heat-reflecting coating of figure 12.23. Details of the design are given in the text.

REFERENCES

1 Matteucci J and Baumeister P W 1980 Integration of thin-film coatings into optical systems in *Int. Lens Design Conf.* ed R E Fischer (*Proc. SPIE* **237** 478–85)

2 Jacquinot P 1954 The luminosity of spectrometers with prisms, gratings or Fabry–Perot etalons *J. Opt. Soc. Am.* **44** 761–5

3 Baum W A 1962 The detection and measurement of faint astronomical sources: chapter 1 in *Astronomical Techniques* ed W A Hiltner, Vol 2 of *Stars and Stellar Systems* (Chicago: University of Chicago Press)

4 Bowen I S 1964 Telescopes *Astron. J.* **69** 816–25

5 Courtes G 1964 Interferometric studies of emission nebulosities *Astron. J.* **69** 325–33

6 Ring J 1956 The Fabry–Perot interferometer in astronomy in *Astronomical Optics and Related Subjects* ed Z Kopal pp 381–8 (Amsterdam: North Holland)

7 Meaburn J 1967 A search for nebulosity in the high galactic latitude radio spurs *Z. Astrophys.* **65** 93–104

8 Meaburn J 1976 *The detection and spectrometry of faint light* (Dordrecht: Reidel)

9 Kaplan L D 1959 Inference of atmospheric structure from remote radiation measurements *J. Opt. Soc. Am.* **49** 1004–7

10 Houghton J T 1961 The meteorological significance of remote measurements of infra-red emission from atmospheric carbon dioxide *Quart. J. R. Met. Soc.* **87** 102–4

11 Houghton J T and Shaw J H 1965 The deduction of stratospheric temperature from satellite observations of emission by the 15 micron CO_2 band *Mem. Soc. R. Sc. Liège., 5th ser.* **9** 350–6

12 Houghton J T 1963/4 Stratospheric temperature measurements from satellites *J. Br. Interplanet. soc.* **19** 381–5

13 Smith S D 1961 Design of interference filters for the observation of infra-red emission from atmospheric carbon dioxide by an earth satellite *Quart. J. R. Met. Soc.* **87** 431–4

14 Smith S D and Pidgeon C R 1965 Application of multiple beam interferometric methods to the study of CO_2 emission at 15 microns *Mem. Soc. R. Sc. Liège., 5th ser.* **9** 336–49

15 Houghton J T and Smith S D 1970 Remote sounding of atmospheric temperature from satellites. I. Introduction *Proc. R. Soc.* A **320** 23–33

16 Abel Z G, Ellis P J, Houghton J T, Peckham G, Rodgers C D, Smith S D and Williamson E J 1970 Remote sounding of atmospheric temperature from satellites. II. The selective chopper radiometer for Nimbus D *Proc. R. Soc.* A **320** 33–55

17 Alpert N L 1962 Infra-red filter grating spectrophotometers–design and properties *Appl. Opt.* **1** 437–42

18 Schneider M V 1966 Schottky barrier photodiode with antireflection coating *Bell System Tech. J.* **45** 1611–38

19 Yeh Y C M, Ernest F P and Stirn R J 1976 Practical antireflection coatings for metal–semiconductor solar cells *J. Appl. Phys.* **47** 4107–12

20 Hovel H J 1976 Transparency of thin metal films on semiconductor substrates *J. Appl. Phys.* **47** 4968–70

21 Hass G and Hadley L 1972 Optical constants of metals *Am. Inst. Phys. Handbook* ed D E Gray pp 6-124–56 (New York: McGraw-Hill)

22 Hahn R E and Seraphin B O 1978 Spectrally selective surfaces for photothermal solar energy conversion in *Physics of Thin films* **10** 1–69 (New York: Academic)

23 Fan J C C, Bachner F J, Foley G H and Zavracky P M 1974 Transparent heat-mirror films of TiO_2/Ag/TiO_2 for solar energy collection and radiation insulation *Appl. Phys. Lett.* **25** 693–5

24 Fan J C C and Bachner F J 1976 Transparent heat mirrors for solar energy applications *Appl. Opt.* **15** 1012–17

25 Bhargava B, Bhattacharyya R and Shah V V 1977 A broad band (visible) heat reflecting mirror *Thin Solid Films* **40** L9–L11

26 Holland L and Siddall G 1958 Heat-reflecting windows using gold and bismuth oxide films *Br. J. Appl. Phys.* **9** 359–61

Appendix Characteristics of thin-film materials

Characteristics of some thin-film materials

Materials	Evaporation technique	Refractive index	Region of transparency	Remarks	References
Aluminium oxide (Al_2O_3)	Electron bombardment	1.62 at 0.6 μm, 1.59 at 1.6 μm ($T_S = 300°C$) 1.59 at 0.6 μm, 1.56 at 1.6 μm ($T_S = 40°C$)		Can also be produced by anodic oxidation of Al in ammonium tartrate solution [1]	2
Antimony trioxide (Sb_2O_3)	Molybdenum boat	2.29 at 366 nm 2.04 at 546 nm	300 nm–1 μm	Important to avoid overheating otherwise decomposes	3
Antimony sulphide (Sb_2S_3)		3.0 at 589 nm	500 nm–10 μm	Brief note [4] p 189	5, 6
Beryllium oxide (BeO)	Tantalum boat. Reactive evaporation of Be metal in activated oxygen	1.82 at 193 nm 1.72 at 550 nm	190 nm–infrared	Highly toxic	7
Bismuth oxide (Bi_2O_3)	Reactive sputtering of bismuth in oxygen	2.45 at 550 nm			8
Bismuth trifluoride (BiF_3)	Graphite Knudsen cell	1.74 at 1 μm 1.65 at 10 μm	260 nm–20 μm		
Cadmium sulphide (CdS)	Quartz crucible with spiral filament in contact with charge	2.6 at 600 nm 2.27 at 7 μm	600 nm–7 μm	Avoid overheating. Filament temperature must be $\leqslant$ 1025°C	5, 9
Cadmium telluride (CdTe)	Molybdenum boat	3.05 in near IR			10 (brief)
Calcium fluoride (CaF_2)	Molybdenum or tantalum boat	1.23–1.26 at 546 nm	150 nm–12 μm		5, 10, 11

Ceric oxide (CeO_2)	Tungsten boat	2.2 at 550 nm [10] 2.18 at 550 nm. $T_S = 50°C$ 2.42 at 550 nm. $T_S = 350°C$ 2.2 in near IR [11]	400 nm–16 μm	Tends to form inhomogeneous layers. Suffers from moisture adsorption	12, 13, 14, 15
Cerous fluoride (CeF_3)	Tungsten boat	1.63 at 550 nm 1.59 at 2 μm	300 nm–> 5 μm	Hot substrate. Crazes on cold substrate [12]. High tensile stress	10, 12, 13, 16
Chiolite ($5NaF.3AlF_3$)	Howitzer or tantalum boat			Similar to cryolite	10
Cryolite (Na_3AlF_6)	Howitzer or tantalum boat	1.35 at 550 nm	< 200 nm–14 μm	Slightly hygroscopic. Soft, easily damaged	5, 10, 11, 12, 17, 18
Germanium (Ge)	Electron bombardment or graphite boat	4.25 in IR (usually slightly higher than bulk value)	1.7–100 μm	Absorption band centred at approx 25 mm	10, 12
Hafnium dioxide (HfO_2)	Electron beam	2.088 at 350 nm 2.00 at 500 nm	220 nm–12 μm		13, 19, 20
Lanthanum fluoride (LaF_3)	Tungsten boat	1.59 at 550 nm 1.57 at 2 μm	220 nm–> 2 μm	Sligty inhomogeneous Substrate heated	12, 13, 16, 21
Lanthanum oxide (La_2O_3)	Tungsten boat	1.95 at 550 nm 1.86 at 2 μm	350 nm–> 2 μm	Hot substrate (~ 300°C)	12, 13, 16
Lead chloride ($PbCl_2$)	Platinum boat or molybdenum boat [15]	2.3 at 550 nm 2.0 at 10 μm	300 nm–> 14 μm		10, 22
Lead fluoride (PbF_2)	Platinum boat	1.75 at 550 nm 1.70 at 1 μm	240 nm–> 20 μm		10, 12, 23, 24
Lead telluride (PbTe)	Tantalum boat	5.5 in IR	3.4–> 30 μm	Avoid overheating. Hot substrate (see text)	25, 26, 27

Materials	Evaporation technique	Refractive index	Region of transparency	Remarks	References
Lithium fluoride (LiF)	Tantalum boat	1.36 to 1.37 at 546 nm	110 nm–7 μm		5, 28
Magnesium fluoride (MgF_2)	Tantalum boat	1.38 at 550 nm 1.35 at 2 μm	210 nm–10 μm	Films on heated substrates much more rugged. High tensile stress.	5, 10, 11, 13, 19, 29 30, 31
Magnesium oxide (MgO)	Electron beam	1.7 at 550 nm. $T_S = 50°C$ 1.74 at 550 nm. $T_S = 300°C$	200 nm–8 μm		32
Neodymium fluoride (NdF_3)	Tungsten boat	1.60 at 550 nm 1.58 at 2 μm	220 nm–> 2 μm	Hot substrate 300°C	12, 13, 16
Neodymium oxide (Nd_2O_3)	Tungsten boat	2.0 at 550 nm 1.95 at 2 μm	400 nm–> 2 μm	Hot substrate 300°C Decomposes at high boat temperatures	12, 16
Praseodymium oxide (Pr_6O_{11})	Tungsten boat	1.92 at 500 nm 1.83 at 2 μm	400 nm–> 2 μm	Hot substrate 300°C	16
Scandium oxide (Sc_2O_3)	Electron beam	1.86 at 550 nm	350 nm–13 μm		33
Silicon (Si)	Electron bombardment with water-cooled hearth	3.5 in IR	1.1–14 μm		12
Silicon monoxide (SiO)	Tantalum boat or howitzer	2.0 at 550 nm 1.7 at 6 μm	500 nm–8 μm	Fast evaporation at low pressure	10 (brief) 2, 5, 12, 19 34, 35

Disilicon trioxide (Si_2O_3)	Tantalum boat or howitzer Reactive evaporation of SiO in oxygen [24] [30]. Evaporation of Si, $SiO_2 ZrO_2$ mixture [26]	1.52–1.55 at 550 nm	300 nm–8 μm		2, 12, 36, 37, 38, 39, 40, 42
Silicon dioxide (SiO_2)	Mixture in tungsten boat.[31]. Electron beam.	1.46 at 500 nm 1.445 at 1.6 μm	< 200 nm–8 μm (in thin films)		2, 12, 41, 43
Sodium fluoride (NaF)	Tantalum boat	1.34 in visible	< 250 nm–14 μm		5 (brief)
Tantalum pentoxide (Ta_2O_5)	Electron beam	2.16 at 550 nm	300 nm–10 μm		13
Tellurium (Te)	Tantalum boat	4.9 at 6 μm	3.4–20 μm		10, 12, 44, 45
Titanium dioxide (TiO_2)	Reactive evaporation of TiO or Ti_2O_3 in O_2. Electron beam reactive evaporation	2.2–2.7 at 550 nm depending on structure	350 nm–12 μm	Can also be produced by subsequent oxidation of Ti film	5, 12, 36, 41, 42, 46, 47, 48, 49, 50, 51
Thallous chloride (TlCl)	Tantalum boat	2.6 at 12 μm	Visible–> 20 μm		10, 52
Thorium oxide (ThO_2)	Electron beam	1.8 at 550 nm 1.75 at 2 μm	250 nm–> 15 μm	Radioactive	10, 12, 53, 54, 55
Thorium fluoride (ThF_4)	Tantalum boat	1.52 at 400 nm 1.51 at 750 nm	200 nm–> 15 μm	Radioactive (see [53]) Note: Thorium oxyfluoride ($ThOF_2$) actually forms ThF_4 when evaporated [53]	10, 12, 53, 54, 55, 56

Materials	Evaporation technique	Refractive index	Region of transparency	Remarks	References
Yttrium oxide (Y_2O_3)	Electron beam	1.82 at 550 nm	250 nm–> 2 μm		13, 19, 57
Zinc selenide (ZnSe)	Platinum boat or tantalum boat	2.58 at 633 nm	600 nm–> 15 μm		54
Zinc sulphide (ZnS)	Tantalum boat or howitzer	2.35 at 550 nm 2.2 at 2.0 μm	380 nm–~ 25 μm		5, 10, 12, 15 29, 55, 18 27
Zirconium dioxide (ZrO_2)	Electron bombardment	2.1 at 550 nm 2.0 at 2.0 μm	340 nm–12 μm		34, 13
Substance 1 ($ZrO_2 + ZrTiO_4$)	Tungsten boat Electron beam	2.1 at 500 nm	400 nm–7 μm		58

REFERENCES

1 Hass G 1949 On the preparation of hard oxide films with precisely controlled thickness on evaporated aluminium mirrors *J. Opt. Soc. Am.* **39** 532–40

2 Cox J T, Hass G and Ramsay J B 1964 Improved dielectric films for multilayer coatings and mirror protection *J. Phys.* **25** 250–4

3 Jenkins F A 1958 Extension du domaine spectral de pouvoir réflecteur élevé des couches multiples dielectriques *J. Phys. Rad.* **19** 301–6

4 Heavens O S, Ring J and Smith S D 1957 Interference filters for the infra-red *Spectrochim. Acta* **10** 179–94

5 Heavens O S 1960 Optical properties of thin films *Rep. Prog. Phys.* **23** 1–65 Includes brief reviews of the following materials: CaF_2 pp 42–3, Na_3AlF_6 pp 43–4, LiF p 44, ZnS pp 44–5, CdS p 45, CeO_2 pp 45–6, Sb_2S_3 p 46, SiO pp 46–7, TiO_2 p 47 and Al_2O_3 p 48

6 Billings S H and Hyman M Jr 1947 The infra-red refractive index and dispersion of evaporated stibnite films *J. Opt. Soc. Am.* **37** 119–21

7 Ebert J 1982 Activated reactive evaporation in *Optical Thin Films* ed R I Seddon (*Proc. SPIE* **325** 29–38)

8 Moravec T J, Skogman R A and Bernal G E 1979 Optical properties of bismuth trifluoride thin films *Appl. Opt.* **18** 105–10

9 Hall J F and Ferguson W F C 1955 Optical properties of cadmium sulphide and zinc sulphide from 0.6 micron to 14 micron *J. Opt. Soc. Am.* **45** 714–18

10 Ennos A E 1966 Stresses developed in optical film coatings *Appl. Opt.* **5** 51–61; Results on ZnS, MgF_2, $ThOF_2$, PbF_2, Na_3AlF_6, 5NaF.3AlF_3, CaF_2, CeF_3, SiO, $PbCl_2$, TlCl, TlI, Ge, Te, CdTe, Al, Cr, multilayers of ZnS–$ThOF_2$, ZnS–Na_3AlF_6, PbF_2–Na_3AlF_6

11 Heavens O S and Smith S D 1957 Dielectric thin films *J. Opt. Soc. Am.* **47** 469–72

12 Ritter E 1961 Gesichtspunkte bei der Stoffauswahl für dünne Schichten in der Optik *Z. angew. Math. Physik* **12** 275–6

13 Smith D and Baumeister P 1979 Refractive index of some oxide and fluoride coating materials *Appl. Opt.* **18** 111–15; Ta_2O_5, HfO_2, Y_2O_3, La_2O_3, ZrO_2, CeO_2, CeF_3, LaF_3, NdF_3, MgF_2

14 Hass G, Ramsay J B and Thun R 1958 Optical properties and structure of cerium dioxide films *J. Opt. Soc. Am.* **48** 324–7

15 Cox J T and Hass G 1958 Antireflection coatings for germanium and silicon in the infrared *J. Opt. Soc. Am.* **48** 677–80

16 Hass G, Ramsay J B and Thun R 1959 Optical properties of various evaporated rare earth oxides and fluorides *J. Opt. Soc. Am.* **49** 116–20

17 Pelletier E, Roche P and Vidal B 1976 Détermination automatique des constantes optiques et de l'épaisseur de couches minces: application aux couches diélectriques *Nouv. Rev. d'Optique* **7** 353–62

18 Netterfield R P 1976 Refractive indices of zinc sulphide and cryolite in multilayer stacks *Appl. Opt.* **15** 1969–73

19 Borgogno J P, Lazarides B and Pelletier E 1982 Automatic determination of the optical constants of inhomogeneous thin films *Appl. Opt.* **22** 4020–9; Results on Y_2O_3, TiO_2, MgF_2, HfO_2 and SiO_2

20 Baumeister P W and Arnon O 1977 Use of hafnium dioxide in multilayer dielectric reflectors for the near uv *Appl. Opt.* **16** 439–44

21 Bourg A, Barbaroux N and Bourg M 1965 Propriétés optiques et structure de couches minces de fluorure de lanthane *Optica Acta* **12** 151–60

22 Penselin S and Steudel A 1955 Fabry–Perot Interferometerverspiegelungen aus dielektrischem Vielfachschichten *Z. Phys.* **142** 21–41

23 *UK Patent Specification* 994 638 1965 Interference filters

24 Lès Z, Lès F and Gabla L 1963 Semitransparent metallic–dielectric mirrors with low absorption coefficient in the ultraviolet region of the spectrum, 3200–2400 Å *Acta Phys. Pol.* **23** 211–14

25 Smith S D and Seeley J S 1968 Multilayer filters for the region 0.8 to 100 microns *Final Report on Contract AF61(052)-833*, Air Force Cambridge Research Laboratories

26 Yi-Hsun Yen, Ling-Xin Zhu, Wen-De Zhang, Feng-Shan Zhang and Shou-Yin Wang 1984 Study of PbTe optical coatings *Appl. Opt.* **23** 3597–601

27 Ritchie F S 1970 Multilayer filters for the infra-red region 10–100 microns *PhD Thesis* University of Reading, England

28 Schulz L G 1949 The structure and growth of evaporation LiF and NaCl films on amorphous substrates *J. Chem. Phys.* **17** 1153–62

29 Hall J F and Ferguson W F C 1955 Dispersion of zinc sulphide and magnesium fluoride films in the visible spectrum *J. Opt. Soc. Am.* **45** 74–5

30 Wood O R, Craighead H G, Sweeney J E and Maloney P J 1984 Vacuum ultraviolet loss in magnesium fluoride films *Appl. Opt.* **23** 3644–9

31 Hall J F 1957 Optical properties of magnesium fluoride films in the ultra-violet *J. Opt. Soc. Am.* **47** 662–5

32 Pulker H K 1979 Characterization of optical thin films *Appl. Opt.* **18** 1969–77; Brief details of an enormous number of optical thin-film materials

33 Arndt D P *et al* 1984 Multiple determination of the optical constants of thin-film coating materials *Appl. Opt.* **23** 3571–96

34 Hass G and Salzberg C D 1954 Optical properties of silicon monoxide in the wavelength region from 0.24 to 14.0 microns *J. Opt. Soc. Am.* **44** 181–7

35 Novice M A 1964 Stresses in evaporated silicon monoxide films *Vacuum* **14** 385–92

36 *UK Patent Specification* 775 002 1957 Improvements in or relating to the manufacture of thin light transmitting layers

37 Ritter E 1962 Zur Kentnis des SiO und Si_2O_3-Phase in dünnen Schichten *Optica Acta* **9** 197–202

38 Okamoto E and Hishimuma Y 1965 Properties of evaporated thin films of Si_2O_3 *Trans. 3rd Int. Vac. Congress* **2**(2) 49–56; Production of Si_2O_3 films by evaporating a mixture of SiO_2, Si and ZrO_2

39 Bradford A P, Hass G, McFarland M and Ritter E 1965 Effect of ultraviolet irradiation on the optical properties of silicon oxide films *Appl. Opt.* **4** 971–6

40 Bradford A P and Hass G 1963 Increasing the far-ultra-violet reflectance of silicon oxide protected aluminium mirrors by ultraviolet irradiation *J. Opt. Soc. Am.* **53** 1096–1100

41 Reichelt W 1965 Fortschritte in der Herstellung von Oxydschichten für optische und elektrische Zwecke *Trans. 3rd Int. Vac. Congress* **2**(2) 25–9

42 *US Patent Specification* 2 920 002 1960 Process for the manufacture of thin films

43 *UK Patent Specification* 632 442 1947 Method of coating with quartz by thermal evaporation

44 Moss T S 1952 Optical properties of tellurium in the infra-red *Proc. Phys. Soc.* **65** 62–6

45 Greenler R G 1955 Interferometry in the infrared *J. Opt. Soc. Am.* **45** 788–91

46 Hass G 1952 Preparation, properties and optical applications of thin films of titanium dioxide *Vacuum* **2** 331–45

47 *US Patent Specification* 2 784 115 1957 Method of producing titanium dioxide coatings

48 *UK Patent Specification* 895 879 1962 Improvements in and relating to the oxidation and/or transparency of thin partly oxidic layers

49 *US Patent Specification* 3 034 924 1962 Use of a rare earth metal in vaporizing metals and metal oxides

50 Pulker H K, Paesold G and Ritter E 1976 Refractive indices of TiO_2 films produced by reactive evaporation of various titanium-oxide phases *Appl. Opt.* **15** 2986–91

51 Heitmann W 1971 Reactive evaporation in ionized gases *Appl. Opt.* **10** 2414–18

52 *UK Patent Specification* 970 071 1964 Infrared filters

53 Heitman W and Ritter E 1968 Production and properties of vacuum evaporated films of thorium fluoride *Appl. Opt.* **7** 307–9

54 Heitmann W 1966 Extrem hochreflektierende dielektrische Spiegelschichten mit Zincselenid *Z. angew. Phys.* **21** 503–8

55 Behrndt K H and Doughty D W 1966 Fabrication of multilayer dielectric films *J. Vac. Sci. Technol.* **3** 264–72

56 Ledger A M and Bastien R C 1977 Intrinsic and thermal stress modeling for thin-film multilayers *Technical Report, Contract DAA25-76-C-0410 (DARPA)* (Norwalk, Connecticut: The Perkin Elmer Corporation)

57 Lubezky I, Ceren E and Klein Z 1980 Silver mirrors protected with Yttria for the 0.5 to 14 μm region *Appl. Opt.* **19** 1895

58 Stetter F, Esselborn R, Harder N, Friz M and Tolles P 1976 New materials for optical thin films *Appl. Opt.* **15** 2315–17

59 Ritter E 1975 Dielectric film materials for optical applications in *Physics of Thin films* **8** 1–49 ed G Hass, M H Francombe and R W Hoffman (New York: Academic Press)

60 Ritter E 1976 Optical film materials and their applications *Appl. Opt.* **15** 2318–27

Index